Yves Nievergelt

Foundations of Logic and Mathematics

Applications to Computer Science and Cryptography

Birkhäuser
Boston • Basel • Berlin

Yves Nievergelt
Department of Mathematics
Eastern Washington University
Cheney, WA 99004-2418
U.S.A.

Library of Congress Cataloging-in-Publication Data

A CIP catalogue record for this book is available from the Library of Congress, Washington D.C., USA.

AMS Subject Classifications: Primary: 03-01; Secondary: 05-01, 11-01, 68-01, 96A60, 03B05, 03B10, 03B20, 03B25, 03B50, 03B70, 03E02, 03E04, 03E10, 03E15, 03E25, 03E30, 05A05, 05A10, 05C05, 05C12, 05C35, 05C38, 05C40, 05C45, 05C90, 11A05, 11A07, 11A41, 11A51, 11T71, 11Z05, 68P25, 68P30

Printed on acid-free paper

Birkhäuser 

ISBN 0-8176-4249-8 SPIN 10841173
ISBN 3-7643-4249-8

Typeset by the author in LaTeX2ε
Printed and bound by Hamilton Printing Company, Rensselaer, NY
Printed in the United States of America

9 8 7 6 5 4 3 2 1

Contents

Preface

This modern introduction to the foundations of logic, mathematics, and computer science answers frequent questions that mysteriously remain mostly unanswered in other texts:

- Why is the truth table for the logical implication so unintuitive?
- Why are there no recipes to design proofs?
- Where do these numerous mathematical rules come from?
- What are the applications of formal logic and abstract mathematics?
- What issues in logic, mathematics, and computer science still remain unresolved?

Answers to such questions must necessarily present both theory and significant applications, which explains the length of the book. The text first shows how real life provides some guidance for the selection of axioms for the basis of a logical system, for instance, Boolean, classical, intuitionistic, or minimalistic logic. From such axioms, the text then derives detailed explanations of the elements of modern logic and mathematics: set theory, arithmetic, number theory, combinatorics, probability, and graph theory, with applications to computer science. The motivation for such detail, and for the organization of the material, lies in a continuous thread from logic and mathematics to their uses in everyday life. As examples, the book presents several applications that have a substantial impact on civilization:

- philosophy,
- switching circuits,
- financial loans and mortgages,
- bar codes (Universal Product Codes),
- public-key cryptography (Rivest–Shamir–Adleman codes),
- the German design and the Allies' breaking of the `ENIGMA` cipher,
- shortest paths and maximal flows in transportation networks, and
- the determination of the shape of saturated hydrocarbon molecules.

To reveal the relations between such applications and the theory, all the intermediate results are proved, occasionally in several different ways to accommodate different learning styles. The text also shows that the foundations keep evolving. Indeed, by the end of the eighteenth century A.D., computer science did not exist yet, and construction of the foundations of logic and mathematics seemed forever finished:

> What is further remarkable about logic is that until now it has also been unable to take a single step forward, and therefore seems to all appearance to be finished and complete (Immanuel Kant [38, p. 106]).

Yet appearances proved deceiving, for nearly two centuries later, logic took a step forward with Kurt Gödel's Incompleteness Theorem [34]. To reflect the current state of the foundations and yet remain at an introductory level, the text shows how the differences among classical, intuitionistic, minimal, and implicational logics reflect common and scientific patterns of reasoning. The text also shows how the availability or the lack of a completeness theorem in different logical systems reveals certain requirements or difficulties in the automation of decision processes in computer science. Similarly, after the Zermelo–Fraenkel set theory, an outline of the theory of well-formed sets reveals a more restrictive but still practical set theory, where the "axiom of regularity" follows from the other axioms and hence avoids certain paradoxes. Such modern developments also demonstrate that the foundations of logic, mathematics, and computer science have not yet been definitely settled but are still being investigated, especially with respect to their relations to real life [74], which is the ultimate "acid" test:

> Until now it appears that where our logic is the most abstract it always gives correct results — it agrees with experiment. [...] The theory then that we are about to describe agrees with experiment wherever it has been tested [...] (Richard Phillips Feynman, [25, p. 6-2]).

Such unresolved issues leave many opportunities for further philosophical considerations and scientific investigations. However, experience indicates that investigations lead consistently to future successes only if the beginner has at least access to sufficient detail and precision. Moreover, after the course, former students often have ready access to only one reference: their former textbook. Therefore, this book is both a text and a reference, which also explains its length.

The intended audience includes university students of the mathematical sciences, computer science, or philosophy, as well as readers interested in logic, mathematics, and computer science, for instance, scientists, students, and teachers wanting to verify the foundations of specific topics. The text does not rely on any formal prerequisite, but it requires some ability to work with abstract concepts and formulae, for instance, from any previous training in any university level mathematics courses.

In various forms in the past fifteen years, this material has been taught at Eastern Washington University in undergraduate courses for students majoring in mathematics, computer science, or computer information systems, including students majoring in philosophy or mathematics education. I thank them for their patience in reading through several drafts, and in working through many of the more than 1000 exercises.

I also thank my colleague Dr. John E. Douglas, Retired Professor of Chemistry, for his many constructive criticisms, especially about the applications to the sciences.

I gratefully acknowledge the use of printed material (reference [89] in the bibliography) and electronic graphics from various United States Government agencies, as well as from the Uniform Code Council, Inc., as indicated on each figure. I also acknowledges the use of Donald Knuth's LaTeX mathematical typesetting language, the American Mathematical Society's fonts, Blue Sky Research's TeXtures, the MathWorks's MATLAB, and Wolfram Research's *Mathematica* software packages, all with an Apple Power Macintosh G3.

Eastern Washington University, Cheney, WA

YVES NIEVERGELT

June 2001

Outline

Part A (Theory) contains the core material on logic, set theory, and arithmetic. Readers willing to accept this theory temporarily can proceed to Part B at any time.

Chapter 0 (Boolean Algebraic Logic) explains the concepts of formulae and "truth" in Boolean logic, and then demonstrates the proof of tautologies with truth tables, and the synthesis of logical formulae with Karnaugh tables. Applications to computer science include the design of simple logical circuits. This material also provides a straightforward framework for the beginner to become acquainted with logic.

Chapter 1 (Logic and Deductive Reasoning) presents the propositional calculus, and the predicate calculus with universal and existential quantifiers. This chapter also explains the nature of proofs, and introduces several ways to design proofs. Such material lays the foundations for set theory and formal or computer languages based on rules. The goal of this introduction is not to make every student an expert in formal logic. Rather, the goal consists in revealing the bases of mathematics and computer science, and in showing what constitutes a complete proof. To this end, it does not seem necessary to ask beginners to design many new proofs. Rather, it suffices to make many such proofs available, for instance, as presented in this chapter, and then to ask students to design different proofs, for instance, shorter proofs, or proofs from different axioms, or different sequences of propositions leading to the same results.

Chapter 2 (Set Theory) shows how to apply logic to develop set theory. This development allows the beginner ample practice with straightforward proofs through the predicate calculus, with the same goals as those stated for the preceding chapter. The text also includes informal proofs to help the beginner make the transition from complete proofs in logic to outlines of proofs in mathematics and computer science.

Chapter 3 (Induction and Arithmetic) uses set theory to establish the concept of mathematical induction. The material then demonstrates how to derive the rules of integer and rational arithmetic, and how they apply to counting the elements of finite sets. The objective of this chapter is not to belabor the derivation of familiar results. Rather the objective consists in revealing where all the rules of arithmetic come from, tracing them back to logic and set theory. In particular, the introduction of the integers and rational numbers provides further examples of uses of equivalence relations from set theory. This subject also affords many opportunities for practice with proofs by induction, with pedagogical goals similar to those stated for the preceding two chapters. The chapter ends with applications to the arithmetic of financial loans and mortgages.

Chapter 4 (Decidability and Completeness) answers many frequent questions about differences between scientific patterns of reasoning and classical logic, including logical independence, consistency, completeness, and decidability. The text shows that there exist other logics — minimal and intuitionistic logics — that may correspond equally well to common and scientific arguments. These logics differ from one another mainly by allowing or forbidding any one of the laws of excluded middle, double negation, or converse contraposition. The same methods show that the minimal logic cannot be represented by finite truth tables. Then the Completeness Theorem provides a solution for the Decidability Problem in the form of an algorithm to design proofs for the full propositional calculus. Finally, transfinite induction proves the dependence of the "axiom of regularity" on the other axioms in the theory of well-formed sets, which also proves that a strict adherence to the other axioms avoids certain paradoxes.

Part B (Applications) contains a selection of real applications. These applications rely on Part A for their theoretical justifications, but they may be perused at any time.

Chapter 5 (Number Theory and Codes) proves the Euclidean algorithm, and establishes elements of number theory that are often taken for granted. The material then shows how modular arithmetic and number theory apply to the design of bar codes, book codes, and the Rivest–Shamir–Adleman codes in public key cryptography.

Chapter 6 (Ciphers, Combinatorics, and Probabilities) derives the basic formulae of combinatorics, and applies them to lay out the foundations of probability, ending with an analysis of the `ENIGMA` ciphering machine.

Chapter 7 (Graph Theory) shows how graph theory relies on the fundamental concepts of sets and relations, with applications to algorithms to find shortest routes or maximal flows in transportation networks, and spanning trees in chemistry.

The material has been used in the following ways.

At a more elementary level, some of part A (most of chapters 0, 1, 2, and section 3.1) and some applications from part B (parts of chapters 5, 6, 7) formed the contents of a one-trimester course — on the mathematical foundations of computer science — intended for sophomores and juniors majoring in computer information systems.

At an intermediate level, most of part A (with only brief outlines of sections 3.4 and 3.5) and some of chapter 4 or an application from part B, formed the contents of a one-trimester course — on the foundations of mathematics — intended for sophomores in mathematics or computer science.

Citations appear in the text to give credit where credit is due, and to document the terminology where there may be several terms for the same concept; however, consultations of the references should *not* be necessary to continue reading the text.

Part A

Theory

Chapter 0

Boolean Algebraic Logic

0.0 INTRODUCTION

Figure 1: Logic suggested the Earth's shape. Space travel confirmed it.

Uses of mathematics and logical reasoning can be traced back through several millennia to ancient civilizations in Babylonia, China, and India. Documents attributed to them show methods to calculate such items as taxes, the dimensions of altars, and the dates of future solstices or eclipses. More complicated problems arose, for example, the determination of the shapes and sizes of the Earth and the Moon, or the distances from Earth to the Moon and the Sun. (For a survey of these ancient records, consult, for instance, the texts by Dreyer [18] and van der Waerden [100].) The solutions to such problems required methods more sophisticated than mere calculation, and hence arose the need for a study of logic itself, which can be traced to the Greece of a couple of millennia ago. This study of logic continues. Indeed, moving beyond mere theoretical possibilities, demands for computable results have led to alternative logics (see Church's text [11]). For the purpose of an introduction, the following examples demonstrate how logic can help in resolving practical issues, and also how questions can arise about the validity of logical methods to reach conclusions. Yet an understanding of these examples will not be necessary for any of the subsequent material.

Example 1 (The status of Pluto) Textbooks have classified the celestial object Pluto as a planet since its discovery in 1930 [57, p. 213], but the question remains whether Pluto might rather be an asteroid or a comet. Various answers rely on various definitions and on logic.

For instance, one definition states that planets are bigger than moons. This definition can be stated in terms of a **hypothesis** (abbreviated by H), a **conclusion** (abbreviated by C), and a logical **implication** of the form "if H then C":

> If a celestial object P is a planet (H) ,
> then P must be bigger than every moon (C) .

Thus the logical implication "if H then C" is true by the foregoing definition of planets. With a logical implication ("if H then C") there are two other useful statements: its converse ("if C then H") and its contraposition ("if *not* C then *not* H").

In this example, the **converse** statement "if C then H" is false, because the Sun is bigger than every moon (C is true), but the Sun is not classified as a planet (H is false).

Still, the hypothesis H can be tested by the **contraposition** "if *not* C then *not* H":

> If a celestial object P is *not* bigger than every moon (*not* C) ,
> then P is *not* a planet (*not* H) .

Thus in 1978 measurements revealed that Pluto was smaller than the Moon [57, p. 213]; consequently, Pluto would no longer be a planet, by the foregoing definition.

The definition "if H then C" can also be tested in practice. For instance, textbooks classify Mercury as a planet, but they also classify Ganymede as a moon (of Jupiter), even though Mercury is smaller than Ganymede [57, p. 182 & 203]. Thus the statement "if Mercury is a planet, then Mercury is bigger than every moon" is false. Therefore the foregoing definition "if H then C" is false, and Pluto can remain a planet. Logic has thus resolved the issue by revealing that the question pertains not to the status of Pluto but to the definition of planets.

In practice the hypothesis H, the conclusion C, and the implication "if H then C" can all be tested. Nevertheless, the preceding discussion contains one logical principle that has been so successful that it remains widely accepted in theory and in practice: the **law of contraposition**, which states that if an implication "if H then C" holds, then its contraposition "if *not* C then *not* H" also holds.

The **converse law of contraposition** — which states that if the contraposition "if *not* C then *not* H" holds, then the implication "if H then C" also holds — was not used in the preceding discussion, and it is not accepted in some logical systems.

Example 2 (The shape of the Earth) In still air, the smooth surface of a small lake can look flat. Such an observation may have led to the impression that the Earth's surface is flat, like a flat disc (Thales of Miletus, about 640–562 B.C. [18, p. 11]) or a flat plane (Xenophanes of Kolophon, about 570–475 B.C. [18, p. 18]).

On a very large lake, if the surface were flat, then outbound ships would appear smaller and smaller but they would always remain in an observer's line of sight.

On the ocean, however, outbound ships seem to disappear from the bottom up, first the hull and then the masts, as though they were sinking "into" or "behind" the water.

Thus the Earth's surface seems to be spherical rather than flat. For such reasons, in Greece and India from the fifth century B.C., the type of surface fitted to such observations was a sphere [18, p. 39 & p. 242]. The reasoning just outlined reveals a logical implication "if H then C":

> If the Earth is flat (H),
> then outbound ships stay in the line of sight (C).

Yet observations lead to the negation ("*not* C") of the conclusion: outbound ships do *not* stay in the line of sight. Therefore the Earth is *not* flat, thanks to the contraposition:

> If outbound ships do *not* stay in the line of sight (*not* C) ,
> then the Earth is *not* flat (*not* H) .

As in example 1, the pattern of the foregoing reasoning begins with the logical **law of contraposition**, which states that if an implication "if H then C" holds, then its contraposition "if *not* C then *not* H" also holds. Then a practical experiment shows that the conclusion (C) fails, in other words, its negation (*not* C) holds, and therefore the negation of the hypothesis (*not* H) also holds. Also as in example 1, the converse law of contraposition has not been used.

Remark 3 (The need or the absence of need for logic) The solution to such problems as the determination of the shape of the Earth initially required logic because people could not observe the entire Earth. Therefore they had to rely on other observations, for instance, of outbound ships, and then they had to fit to the Earth a shape consistent with such observations. The first people on record to see the entire Earth were the crew of Apollo 8 , who took the photograph in figure 1 on 22 December 1968, between Earth and their first orbit around the Moon.

More generally, the need for logic, mathematics, and computing arises only if the solution of a problem or the answer to a question would otherwise remain unknown or impractical to obtain by other means. In other words (by contraposition), if you already know — or do not care to know — all the answers, then you might not need any logic.

Thus if you take any part of logic for granted, such as the first few chapters of the present text, then you may proceed to any other topic or part of the text.

Remark 4 (Difficulties with real examples) Some examples of uses of logic may enhance the effectiveness of the exposition. Such examples might consist of English sentences, for example, "The Earth is spherical." However, difficulties arise in determining whether and why such a practical sentence is true or false. Indeed, a statement as simple as "the Earth is spherical" can immediately be challenged to no end. Therefore, to focus on logic, mathematics, and computing, instead of debatable issues, the following discussion will also use "toy" examples with truth or falsity decided in advance.

Nevertheless, the material on the synthesis of Truth tables applies immediately to the design of electronic circuits [61, p. 108–111], [62, p. 460–463], as demonstrated by Claude Elwood Shannon [86], and as presented further in this chapter.

0.1 LOGICAL FORMULAE

0.1.1 The Empty Set

For purposes of demonstration, the forthcoming discussions will use an example from the start of the foundations of mathematics: the empty set $\varnothing$, also denoted by $\{\ \ \}$. In general, any set X might be an "element" (or a "member") of another set Y, a situation denoted by $X \in Y$ (read "X is an element of Y"). As one exception, the empty set $\varnothing$ has no element. Thus, one "toy" example of the use of logic is the formula $\varnothing \in \varnothing$, which is decreed "false" because it reads "the empty set is an element of the empty set." For the present purposes, its *only* relevant feature is that $\varnothing \in \varnothing$ is a "false" logical formula. There is nothing else to "understand" in this formula.

Also with sets, variations on the symbol $=$ can represent several relations. For instance, some sets Y and Z might have exactly the same elements: every element of Y is an element of Z, and, conversely, every element of Z is an element of Y, a relation denoted by $Y = Z$ (read "Y and Z have exactly the same elements"). For example, $\varnothing = \varnothing$, which is decreed to be "true" because both sides of the equation represent the same set. For the present purposes, its *only* relevant feature is that $\varnothing = \varnothing$ is a "true" logical formula. There is nothing else to "understand" in this formula. (There are also other variants of the concept of equality of sets [3, p. 6 & 53].)

0.1.2 An Alphabet for Logic

Table 5 Alphabet for logic and set theory.

TYPES	SYMBOLS (READ)
	Symbols for mathematical set theory
Constant:	$\varnothing$ (the empty set).
Relations:	$\in$ (in, belongs to, is an element of), $=$ (equals).
	Symbols for first order logic
Quantifiers:	$\forall$ (for each), $\exists$ (there exists).
	Symbols for Boolean algebraic logic
Values:	T (True), F (False).
Connectives:	$\neg$ (not), $\wedge$ (and), $\vee$ (OR, inclusive or), $\Rightarrow$ (implies, has the necessary consequence), $\Leftrightarrow$ (is equivalent to), $\downarrow$ (NOR, neither nor), $\mid$ (NAND, Sheffer's stroke), $\oplus$ (XOR, exclusive or).
Parentheses of various shapes:	((left parenthesis)) (right parenthesis), [[left bracket]] [right bracket], {{left brace}} {right brace}.
Variables:	any symbols different from all of the above, for instance, Roman, 𝔊𝔢𝔯𝔪𝔞𝔫, Γρεεκ, *italic*, **bold**, ... letters, UPPER or lower case, etc., optionally with one or more superscripts or subscripts: A, $A^{\sharp}$, $A^{\sharp\sharp}$, $A_{\flat}$, $A_{\flat\flat}$, $A_{\flat\flat\flat}$, ...

Logic can be outlined in English and other natural languages. Yet the sentences involved can become cumbersome (as in example 2), and for other applications certain logical statements fail to lend themselves to English or other natural languages. For instance, the phrase "one or the other" can elicit clarifications by such expressions as "either one or the other" (*excluding* the possibility of both) and "one and/or the other" (*allowing* the possibility of both). Therefore, besides English and natural languages, other symbols appear necessary to specify unambiguously the intended logic.

Several different systems of logical symbols exist, for example, all or only parts of the "alphabet" for Boolean algebraic logic in table 5. As for the letters of the alphabet in English and other natural languages, the symbols and names of the logical alphabet have *no intrinsic meaning*. In particular, the "variables" are only symbols, or characters, which do not "vary" but which inherit their name of "variables" from historical usage. Optional repeated superscripts or subscripts, as in A, $A^{\sharp}$, $A^{\sharp\sharp}$, $A_{\flat}$, $A_{\flat\flat}$, and $A_{\flat\flat\flat}$, allow for a supply of variables not limited by the alphabet selected for the variables. Similarly, the relations $\in$ and $=$ as well as the constant $\varnothing$ are symbols that have no meaning here but that will designate specific concepts in set theory. They must already appear in table 5, however, to be declared forbidden from the list of possible variables.

0.1.3 Well-Formed Formulae

As the letters of the Roman alphabet may combine into English words, phrases, clauses, and sentences, logical symbols may combine into logical formulae. For instance, the simplest "phrases" in the logical language of set theory are called "atomic formulae" from the Greek word "atomos" for indivisible. One way of presenting set theory includes only one type of atomic formula, $X \in Y$, with any variables X and Y. Yet a more common presentation includes the following types of atomic formulae.

Definition 6 An **atomic formula** from set theory consists of a sequence of three symbols: a variable or a constant, followed by a relation, followed by a variable or a constant. Thus here are the **rules** to form all the types of atomic formulae in set theory:

$(A1)$	$\varnothing = \varnothing$	(read "the empty set equals the empty set"),
$(A2)$	$E = \varnothing$	(read "the set E equals the empty set"),
$(A3)$	$\varnothing = E$	(read "the empty set equals the set E"),
$(A4)$	$X = Y$	(read "the set X equals the set Y"),
$(A5)$	$\varnothing \in \varnothing$	(read "the empty set is an element of the empty set"),
$(A6)$	$W \in \varnothing$	(read "the set W is an element of the empty set"),
$(A7)$	$\varnothing \in Z$	(read "the empty set is an element of the set Z"),
$(A8)$	$A \in B$	(read "the set A is an element of the set B"),

except for replacements of a variable by another variable; for instance, $V = W$ is a formula, by substitutions of other variables in rule A4.

Definition 6 corresponds to common usage in placing connectives between variables. For example, the phrase "X equals Y" parallels the notation $X = Y$, which is called an **infix** notation. There exist alternative but equivalent systems of notation, which list all the variables and then the relation, or vice versa.

For instance, the **prefix** notation lists the variables from right to left and then the relation. Thus $X = Y$ becomes $= XY$ in prefix notation, or $= (X, Y)$ with an optional comma and parentheses. The prefix notation is common in mathematics, especially in contexts that focus on the relation rather than on individual variables.

Similarly, the **postfix** notation, also called **reverse Polish** notation, lists the variables from left to right and then the relation. Thus $X = Y$ becomes $XY =$ in postfix notation. The postfix notation is common in computer science, because it parallels the order in which machines perform operations [43], [44], [45], [46], [47].

Regardless of the notation, as simple clauses may combine into complex sentences, atomic formulae may combine into other formulae, which, in turn, may then combine into more complicated formulae called "well-formed" logical formulae.

Definition 7 Every atomic formula is a **well-formed logical formula.** Also, for all well-formed logical formulae P and Q, and for every variable X, the following **rules** form expressions that are also well-formed logical formulae:

$(F1)$	$\neg(P)$	(read "not P"),
$(F2)$	$(P) \Rightarrow (Q)$	(read "P implies Q" or "if P then Q"),
$(F3)$	$(P) \vee (Q)$	(read "P or Q"),
$(F4)$	$(P) \wedge (Q)$	(read "P and Q"),
$(F5)$	$(P) \Leftrightarrow (Q)$	(read "P and Q are equivalent"),
$(F6)$	$\forall X(P)$	(read "for each X, P"),
$(F7)$	$\exists X(P)$	(read "there exists an X such that P").

Brackets [left and right] and braces {left and right} can replace parentheses (left and right). Only strings of symbols built from atomic formulae and rules F1–F7 are well-formed logical formulae. Formulae F1–F7, which involve negations, connectives, or quantifiers, are called **compound formulae**, whereas P and Q are called the **components** of the compound formulae. Such components as P and Q are also called **propositional variables**; for contrast, such a variable as X is then called an **individual variable**, or, if it occurs after a quantifier, an **operator variable**, as in $\forall X$ or $\exists X$.

Moreover, for all variables P, Q, and X, the symbols P and Q are **propositional forms**, the strings of symbols F1–F5 are **propositional forms**, and only expressions built in such a fashion represent propositional forms. Finally, a **logical formula** is either a well-formed logical formula or a propositional form.

Example 8 The string of symbols $\neg(\varnothing \in \varnothing)$ is a well-formed logical formula, by rules A5 and F1. The equivalent prefix formula would be $\neg \in \varnothing\varnothing$ while the equivalent postfix formula would be $\varnothing\varnothing \in \neg$. In set theory, the formula $\neg(\varnothing \in \varnothing)$ states that the empty set ($\varnothing$) is *not* ($\neg$) an element of itself ($\varnothing$).

Example 9 The string $\forall\exists(P)$ is *not* a well-formed logical formula, because rules F1–F7 allow only a variable after $\forall$, and table 5 prevents $\exists$ from being a variable.

Equivalent definitions can also use the prefix or the postfix notation. Moreover, the prefix and postfix notations do not require any parentheses, because each connective pertains to the results immediately next to it.

Example 10 The string of symbols $\{\neg[(P) \vee (Q)]\} \Rightarrow \{[\neg(P)] \vee [\neg(Q)]\}$ is a *propositional form*. The following sequence of operations demonstrates how to construct the equivalent postfix formula in the left-hand column, with the resulting intermediate infix subformulae in the right-hand column.

Postfix notation	Infix notation
$PQ\vee$	$(P) \vee (Q)$
$PQ \vee \neg$	$\neg[(P) \vee (Q)]$
$P\neg$	$\neg(P)$
$Q\neg$	$\neg(Q)$
$P\neg Q\neg\vee$	$[\neg(P)] \vee [\neg(Q)]$
$PQ \vee \neg P\neg Q\neg\vee \Rightarrow$	$\{\neg[(P) \vee (Q)]\} \Rightarrow \{[\neg(P)] \vee [\neg(Q)]\}$

The equivalent prefix formula is $\Rightarrow \neg \vee PQ \vee \neg P\neg Q$.

Example 11 The string $\exists E\,\{\forall Y\,[\neg(Y \in E)]\}$ is a *well-formed logical formula*:

$$\begin{array}{rl} Y \in E & \text{is a logical formula, by rule A8,} \\ \neg(Y \in E) & \text{is a logical formula, by rule F1,} \\ \forall Y\,[\neg(Y \in E)] & \text{is a logical formula, by rule F6,} \\ \exists E\,\{\forall Y[\neg(Y \in E)]\} & \text{is a logical formula, by rule F7.} \end{array}$$

This formula contains two individual operator variables, E and Y, but no propositional variables. The equivalent prefix notation and postfix notation would be

$$\exists E\forall Y\neg \in YE,$$
$$YE \in \neg Y\forall E\exists.$$

In set theory, the formula $\exists E\,\{\forall Y\,[\neg(Y \in E)]\}$ states that there exists ($\exists$) a set E such that for each ($\forall$) set Y it is false ($\neg$) that Y is an element of E.

Remark 12 Some authors read the symbol $\forall$ as "for all" but the requirement that it be followed by exactly one variable makes such a reading inconsistent with English grammar, which after "for all" would require a plural. Moreover, the reading of $\forall$ as "for all" can lead to confusion, because the reading of P in such formulae as $\forall X(P)$ can differ for *each* X, as demonstrated in the following chapters.

0.1.4 Exercises

The following exercises involve formulae that will appear repeatedly in the sequel. For these exercises, determine whether the strings of symbols are well-formed logical formulae. For each well-formed logical formula, explain how to form it from rules F1 – F7 and atomic formulae, as done in examples 8 and 11. For strings that are not well-formed logical formulae, including propositional forms, explain why they are not well-formed logical formulae, as done in example 9.

Exercise 1 $\varnothing \Leftrightarrow \varnothing$

Exercise 2 $\neg(\varnothing \Leftrightarrow \varnothing)$

Exercise 3 $\neg(\varnothing = \varnothing)$

Exercise 4 $(\varnothing \in \varnothing) \Rightarrow (\varnothing = \varnothing)$

Exercise 5 $(X = Z) \Leftrightarrow (Z = X)$

Exercise 6 $(P) \Rightarrow (Q)$

Exercise 7 $(P) \vee (Q)$

Exercise 8 $(\varnothing = \varnothing) \wedge (R)$

Exercise 9 $[\neg(\varnothing \in \varnothing)] \wedge [\neg(P)]$

Exercise 10 $\forall X[\exists Y(P)]$

Exercise 11 $\forall X[\neg(\exists Y)]$

Exercise 12 $\forall X(\exists Y)$

Exercise 13 $\forall X[\neg(X \in X)]$

Exercise 14 $\forall X(X = X)$

Exercise 15 $\forall X(X \Leftrightarrow X)$

Exercise 16 $\forall X(X = \varnothing)$

Exercise 17 $\exists Y[\neg(\varnothing \in Y)]$

Exercise 18 $\exists X[\neg(X = X)]$

Exercise 19 $\forall X[\neg(X \in \varnothing)]$

Exercise 20 $\neg[\exists X(X \in \varnothing)]$

Exercise 21 $\exists Z\{\forall X[\neg(X = Z)]\}$

Exercise 22 $\exists Z\{\forall X[\neg(X \Rightarrow Z)]\}$

Exercise 23 $\exists\varnothing\{\forall X[\neg(X \in \varnothing)]\}$

Exercise 24 $\exists X\{\forall Y[\exists Z(A \in B)]\}$

Exercise 25 $\forall X(\forall Y\{[(X \in Y) \wedge (Y \in X)] \Rightarrow (X = Y)\})$

Exercise 26 $\forall X[(P) \Rightarrow (Q)]$

Exercise 27 $\exists A\{[\neg(P)] \Rightarrow [\neg(\varnothing \in A)]\}$

Exercise 28 $\forall X\{[\neg(P)] \Rightarrow (X = X)\}$

Exercise 29 $\forall X\big\{\forall Y\big[\exists Z\big(\forall W\{(W \in Z) \Leftrightarrow [(W = X) \vee (W = Y)]\}\big)\big]\big\}$

Exercise 30 $\forall X\big[\forall Y\big(\exists Z\{(W \in Z) \Leftrightarrow [(W = X) \vee (W = Y)]\}\big)\big]$

Exercise 31 $\forall X\big\{\forall Y\big[(X = Y) \Leftrightarrow \{\forall W[(W \in X) \Leftrightarrow (W \in Y)]\}\big]\big\}$

Exercise 32 $\forall X\big\{\forall Y\big[(X = Y) \Leftrightarrow \{\forall Z[(X \in Z) \Leftrightarrow (Y \in Z)]\}\big]\big\}$

Exercise 33 $\forall X\big(\exists Y\{\forall Z[(Z \in Y) \Leftrightarrow (Z = X)]\}\big)$

Exercise 34 $\forall X\big(\exists Y\big\{\forall Z\big[(Z \in Y) \Leftrightarrow \{\forall E[(E \in Z) \Rightarrow (E \in X)]\}\big]\big\}\big)$

Exercise 35 $\forall X\big(\forall Y\{[(X \in A) \wedge (Y \in A)] \Rightarrow [(X \in Y) \vee (Y \in X) \vee (X = Y)]\}\big)$

Exercise 36 $\forall Z\big(\exists U\big\{\forall X\big[(X \in U) \Leftrightarrow \{\exists Y[(X \in Y) \wedge (Y \in Z)]\}\big]\big\}\big)$

Exercise 37 $\forall X\big[\forall Y\big(\forall Z\{[(X \in Y) \wedge (Y \in Z)] \Rightarrow (X \in Z)\}\big)\big]$

Exercise 38 $\forall Z\big([\neg(Z = \varnothing)] \Rightarrow \{\exists A[(A \in Z) \wedge \{\forall X \neg[(X \in A) \wedge (X \in Z)]\}]\}\big)$

Exercise 39

$$\forall Z\,\{[\neg(Z = \varnothing)] \Rightarrow \big(\exists W\big\{\forall X\big[(X \in W) \Leftrightarrow \{\forall Y[(Y \in Z) \Rightarrow (X \in Y)]\}\big]\big\}\big)\}$$

Exercise 40 $\forall P\big\{\forall A\big[\exists S\big(\forall X\,\{(X \in S) \Leftrightarrow [(X \in A) \wedge (P)]\}\big)\big]\big\}$

Write the following strings of symbols in postfix notation, as done in example 10.

Exercise 41 $(P) \Rightarrow (P)$

Exercise 42 $(P) \Leftrightarrow (P)$

Exercise 43 $(P) \Rightarrow \{\neg[\neg(P)]\}$

Exercise 44 $\{\neg[\neg(P)]\} \Rightarrow (P)$

Exercise 45 $(\neg\{\neg[\neg(P)]\}) \Rightarrow [\neg(P)]$

Exercise 46 $[\neg(P)] \Rightarrow (\neg\{\neg[\neg(P)]\})$

Exercise 47 $\neg\big(\neg\{(P) \vee [\neg(P)]\}\big)$

Exercise 48 $(P) \Leftrightarrow \{(P) \wedge [(P) \Rightarrow (P)]\}$

Exercise 49 $(P) \Rightarrow [(Q) \Rightarrow (P)]$

Exercise 50 $[(P) \Rightarrow (Q)] \Rightarrow (P)$

Exercise 51 $(P) \Rightarrow [(P) \vee (Q)]$

Exercise 52 $(Q) \Rightarrow [(P) \vee (Q)]$

Exercise 53 $[(P) \wedge (Q)] \Rightarrow [(P) \Rightarrow (Q)]$

Exercise 54 $(P) \Rightarrow \{(Q) \Rightarrow [(P) \wedge (Q)]\}$

Exercise 55 $[(P) \Rightarrow (Q)] \vee [(Q) \Rightarrow (P)]$

Exercise 56 $\big([(P) \Rightarrow (Q)] \wedge \{(P) \Rightarrow [\neg(Q)]\}\big) \Rightarrow [\neg(P)]$

Exercise 57 $\{(P) \wedge [(Q) \vee (R)]\} \Leftrightarrow \{[(P) \wedge (Q)] \vee [(P) \wedge (R)]\}$

Exercise 58 $\{(P) \vee [(Q) \wedge (R)]\} \Leftrightarrow \{[(P) \vee (Q)] \wedge [(P) \vee (R)]\}$

Exercise 59 $[(P) \Rightarrow (Q)] \Rightarrow \{[\neg(Q)] \Rightarrow [\neg(P)]\}$

Exercise 60 $\{[\neg(Q)] \Rightarrow [\neg(P)]\} \Rightarrow [(P) \Rightarrow (Q)]$

Exercise 61 $[(P) \Rightarrow (Q)] \Leftrightarrow \{[\neg(Q)] \Rightarrow [\neg(P)]\}$

Exercise 62 $\{(P) \Rightarrow [\neg(Q)]\} \Rightarrow \{(Q) \Rightarrow [\neg(P)]\}$

Exercise 63 $\{[(P) \Rightarrow (Q)] \Rightarrow (P)\} \Rightarrow (P)$

Exercise 64 $\{(P) \Rightarrow [(P) \Rightarrow (Q)]\} \Rightarrow \{(P) \Rightarrow (Q)\}$

Exercise 65 $[(P) \Leftrightarrow (Q)] \Rightarrow [(P) \Rightarrow (Q)]$

Exercise 66 $[(P) \Leftrightarrow (Q)] \Rightarrow [(Q) \Rightarrow (P)]$.

Exercise 67 $[(Q) \Rightarrow (R)] \Rightarrow \{[(P) \Rightarrow (Q)] \Rightarrow [(P) \Rightarrow (R)]\}$

Exercise 68 $[(P) \Rightarrow (Q)] \Rightarrow \{[(Q) \Rightarrow (R)] \Rightarrow [(P) \Rightarrow (R)]\}$

Exercise 69 $\{[(P) \Rightarrow (Q)] \wedge [(P) \Rightarrow (R)]\} \Leftrightarrow \{(P) \Rightarrow [(Q) \wedge (R)]\}$

Exercise 70 $\{[(P) \Rightarrow (Q)] \wedge [(R) \Rightarrow (S)]\} \Rightarrow \{[(P) \wedge (R)] \Rightarrow [(Q) \wedge (S)]\}$

Exercise 71 $\{[(P) \wedge (Q)] \Rightarrow (R)\} \Rightarrow \{(P) \Rightarrow [(Q) \Rightarrow (R)]\}$

Exercise 72 $\{[(P) \wedge (Q)] \Rightarrow (R)\} \Leftrightarrow \big(\{(P) \wedge [\neg(R)]\} \Rightarrow [\neg(Q)]\big)$

Exercise 73 $[(P) \Rightarrow (Q)] \Rightarrow \big([(P) \Rightarrow (R)] \Rightarrow \{(P) \Rightarrow [(Q) \wedge (R)]\}\big)$

Exercise 74 $\{(P) \Rightarrow [(Q) \Rightarrow (R)]\} \Rightarrow \{(Q) \Rightarrow [(P) \Rightarrow (R)]\}$

Exercise 75 $\{(P) \Rightarrow [(Q) \Rightarrow (R)]\} \Rightarrow \{[(P) \wedge (Q)] \Rightarrow (R)\}$

Exercise 76 $\{[(P) \Rightarrow (Q)] \Rightarrow (R)\} \Rightarrow \{[(R) \Rightarrow (P)] \Rightarrow [(S) \Rightarrow (P)]\}$

Exercise 77

$$(P) \Rightarrow \big[(R) \Rightarrow \big(\{[(P) \wedge (R)] \Rightarrow [(Q) \wedge (S)]\} \Rightarrow \{[(P) \Rightarrow (Q)] \wedge [(R) \Rightarrow (S)]\}\big)\big]$$

Exercise 78

$$\begin{aligned}\big\{\big[\big(\{(P) \Rightarrow (Q)\} \Rightarrow \{[\neg(R)] \Rightarrow [\neg(S)]\}\big) \Rightarrow (R)\big] \Rightarrow (U)\big\}\\ \Rightarrow \{[(U) \Rightarrow (P)] \Rightarrow [(S) \Rightarrow (P)]\}\end{aligned}$$

Exercise 79

$$\begin{aligned}\big\{\{[(R) \Rightarrow (P)] \Rightarrow [(S) \Rightarrow (P)]\} \Rightarrow \big[\{[(P) \Rightarrow (Q)] \Rightarrow (R)\} \Rightarrow (H)\big]\big\}\\ \Rightarrow \big\{(K) \Rightarrow \big[\{[(P) \Rightarrow (Q)] \Rightarrow (R)\} \Rightarrow (H)\big]\big\}\end{aligned}$$

Exercise 80

$$\big(\{[(R) \Rightarrow (P)] \Rightarrow (P)\} \Rightarrow [(S) \Rightarrow (P)]\big) \Rightarrow \big(\{[(P) \Rightarrow (Q)] \Rightarrow (R)\} \Rightarrow [(S) \Rightarrow (P)]\big)$$

For the following exercises, write each string of symbol in infix notation.

Exercise 81 $PP\neg\vee$

Exercise 82 $PP\neg \wedge \neg$

Exercise 83 $PP \wedge P \Leftrightarrow$

Exercise 84 $PP \vee P \Leftrightarrow$

Exercise 85 $PP\neg \Rightarrow P\neg \Rightarrow$

Exercise 86 $P\neg P \Rightarrow P \Rightarrow$

Exercise 87 $PQ \wedge P \Rightarrow$

Exercise 88 $PQ \wedge Q \Rightarrow$

Exercise 89 $PPPQ \wedge \vee \Rightarrow$

Exercise 90 $PPPQ \vee \wedge \Rightarrow$

Exercise 91 $PQ \vee QP\vee \Leftrightarrow$

Exercise 92 $PQ \wedge QP\wedge \Leftrightarrow$

Exercise 93 $PQR \vee \vee PQ \vee R\vee \Leftrightarrow$

Exercise 94 $PQR \wedge \wedge PQ \wedge R\wedge \Leftrightarrow$

Exercise 95 $PP\neg Q \Rightarrow\Rightarrow$

Exercise 96 $P\neg PQ \Rightarrow\Rightarrow$

Exercise 97 $PPQ \Rightarrow \wedge Q \Rightarrow$

Exercise 98 $PPQ \Rightarrow Q \Rightarrow\Rightarrow$

Exercise 99 $Q\neg PQ \Rightarrow \wedge P\neg \Rightarrow$

Exercise 100 $QP\neg Q\neg \Rightarrow \wedge P \Rightarrow$

Exercise 101 $PQ \Rightarrow PQ\neg \Rightarrow P\neg \Rightarrow\Rightarrow$

Exercise 102 $PP\neg QQ\neg\wedge \Rightarrow\Leftrightarrow$

Exercise 103 $PR \Rightarrow Q\neg R\neg \Rightarrow \wedge PQ \Rightarrow\Rightarrow$

Exercise 104 $PQQ\neg\wedge \Rightarrow P\neg \Rightarrow$

Exercise 105 $PQR \Rightarrow\Rightarrow PQ \Rightarrow PR \Rightarrow\Rightarrow\Rightarrow$

Exercise 106 $PQ \Rightarrow QP \Rightarrow PQ \Leftrightarrow\Rightarrow\Rightarrow$

Exercise 107 $PQ \Leftrightarrow PQ \Rightarrow QP \Rightarrow \wedge \Leftrightarrow$

Exercise 108 $PQ \Rightarrow RQ \Rightarrow \wedge PR \vee Q \Rightarrow\Leftrightarrow$

Exercise 109 $PR \Leftrightarrow QS \Leftrightarrow \wedge PQ \Rightarrow RS \Rightarrow\Leftrightarrow\Rightarrow$

Exercise 110 $PR \Leftrightarrow QS \Leftrightarrow \wedge PQ \Leftrightarrow RS \Leftrightarrow\Leftrightarrow\Rightarrow$

Exercise 111 $PQR \Rightarrow\Rightarrow PQ \wedge R \Rightarrow\Rightarrow$

Exercise 112 $PQR \Rightarrow\Rightarrow PQ \Rightarrow \wedge PR \Rightarrow\Rightarrow$

Exercise 113 $RQ \Rightarrow RQ \Rightarrow P \Rightarrow SP \Rightarrow\Rightarrow\Rightarrow$

Exercise 114 $PQ \Rightarrow PQ \Rightarrow Q \Rightarrow Q \Rightarrow\Rightarrow$

Exercise 115 $RQ \Rightarrow SP \Rightarrow\Rightarrow RP \Rightarrow SP \Rightarrow\Rightarrow\Rightarrow$

Exercise 116 $HR \Rightarrow SP \Rightarrow\Rightarrow RQ \Rightarrow P \Rightarrow SP \Rightarrow\Rightarrow\Rightarrow$

Exercise 117

$$PQ\neg RS \Rightarrow\Rightarrow SQ \Rightarrow HRQ \Rightarrow\Rightarrow\Rightarrow\Rightarrow\Rightarrow K\neg KL \Rightarrow\Rightarrow M \Rightarrow\Rightarrow M \Rightarrow$$

Exercise 118

$$LPQ \Rightarrow S \Rightarrow K \Rightarrow\Rightarrow MSH \Rightarrow P\neg R\neg \Rightarrow\Rightarrow\Rightarrow\Rightarrow MSP \Rightarrow RP \Rightarrow\Rightarrow\Rightarrow\Rightarrow$$

Exercise 119 Design a method to convert formulae from postfix into infix notation.

Exercise 120 Design a method to convert formulae from prefix into infix notation.

0.2 LOGICAL TRUTH AND CONNECTIVES

0.2.1 Logical Truth

In mathematics and applications, logical symbols can represent concepts or objects. For instance, the symbol $\varnothing$ and all the variables can represent sets, while the symbols $\in$ and $=$ can represent relations between sets. With such correspondences between symbols and concepts, some logical formulae can be "true" while others can be "false" in the application under consideration. Yet the notions of "true" and "false" can remain ambiguous or subject to interpretation. To avoid ambiguities and to provide a means to keep track of two mutually exclusive alternatives, the logic developed here includes two concepts, also called logical values or "Truth" values, denoted by True and False. The initial capitals distinguish the common "true" and "false" from their abstract logical counterparts True and False. Any other symbols could serve this purpose, for instance, T or 1 for True, and F or 0 for False.

Example 13 In set theory, the formula $\varnothing = \varnothing$ is True (subsection 0.1.1).

Example 14 The formula $\varnothing \in \varnothing$ is False by definition (subsection 0.1.1).

Example 15 Computers include facilities to *simulate* the logical concepts of True and False [43, p. 91], [44, p. 253], [45, p. 201], [46, p. 490], [47, p. 29–10]. For instance, computers include ways to define and then to test the simulated Truth values of formulae (which can include formulae other than well-formed logical formulae). Thus in the binary arithmetic of computers, the following seven formulae are all "True":

$$\begin{array}{rclcrcl} 0+0 &=& 0, & & 0*0 &=& 0, \\ 0+1 &=& 1, & & 1*0 &=& 0, \\ 1+0 &=& 1; & & 0*1 &=& 0, \\ & & & & 1*1 &=& 1. \end{array}$$

In contrast, the following eight formulae are all "False":

$$\begin{array}{rclcrcl} 0+0 &=& 1, & & 0*0 &=& 1, \\ 0+1 &=& 0, & & 1*0 &=& 1, \\ 1+0 &=& 0, & & 0*1 &=& 1, \\ 1+1 &=& 1; & & 1*1 &=& 0. \end{array}$$

Example 16 In set theory and in binary arithmetic, the formula $X = Y$ is neither True nor False, because X and Y have not been assigned to specific objects or concepts.

Thus some logical formulae can fail to have either of the logical values True or False. Those logical formulae that have a logical value have a special name.

Definition 17 A logical **proposition** is a logical formula that has a Truth value.

Example 18 In set theory, the formula $\varnothing = \varnothing$ is a proposition, because it is True.

Example 19 In set theory, the formula $\varnothing \in \varnothing$ is a proposition, because it is False.

Example 20 The formula $X = Y$ is *not* a proposition, for it has no Truth value.

Demonstrating that the logical values True and False are abstract concepts, not necessarily identical to the common notions of true and false, there exist alternative presentations of the same logic without Truth values, and there exist other logical systems with more than two Truth values, as explained in the next chapters.

0.2.2 Logical Connectives

One of the purposes of logic lies in establishing the logical Truth of compound logical formulae on the basis of the Truth of their components. In particular, for all logical propositions P and Q, the five compound formulae $\neg(P)$, $(P) \wedge (Q)$, $(P) \vee (Q)$, $(P) \Rightarrow (Q)$, and $(P) \Leftrightarrow (Q)$ are also propositions. Their Truth values depend on the particular logical system under consideration, but for the Boolean algebraic logic considered here, their Truth values are specified as follows.

Negation The operation of **negation**, denoted by the logical connective $\neg$, swaps the Truth value of logical formulae as indicated in table 21: if a proposition P is True, then $\neg(P)$ is False; similarly, if P is False, then $\neg(P)$ is True.

Table 21 Negation.

P	$\neg(P)$
T	F
F	T

Example 22 Because $\varnothing = \varnothing$ is True, it follows that $\neg(\varnothing = \varnothing)$ is False.

Example 23 Because $\varnothing \in \varnothing$ is False, it follows that $\neg(\varnothing \in \varnothing)$ is True.

Example 24 $\neg(0 + 0 = 0)$ is False, because $0 + 0 = 0$ is True (example 15).

Example 25 $\neg(0 + 0 = 1)$ is True, because $0 + 0 = 1$ is False (example 15).

Other symbols exist to denote the logical negation $\neg$, for instance, $\sim$ and $\overline{}$. Yet $\sim$ also denotes various relations in algebra [4, p. 78], [53, p. 6 & 15], [107, p. 17], and $\overline{}$ also denotes other operations in complex analysis [1, p. 7], [12, p. 7], and in topology [19, p. 69], [59, p. 42]. To avoid future confusion with such different concepts, this exposition denotes the logical negation by $\neg$, as in [73, p. 6], [85, p. 6].

Conjunction The **conjunction** of two formulae P and Q, denoted by $(P) \wedge (Q)$, or by $(P)\text{AND}(Q)$, or by $(P)(Q)$, or also by $(P)\&(Q)$, is True if both P and Q are True. It is False otherwise (if at least one of P or Q is False), as shown in table 26.

Table 26 Conjunction.

P	Q	$(P) \wedge (Q)$
T	T	T
T	F	F
F	T	F
F	F	F

Example 27 The proposition $(\varnothing = \varnothing) \wedge [\neg(\varnothing \in \varnothing)]$ is True, because both $\varnothing = \varnothing$ and $\neg(\varnothing \in \varnothing)$ are True.

Example 28 The proposition $(\varnothing = \varnothing) \wedge (\varnothing \in \varnothing)$ is False, because $\varnothing \in \varnothing$ is False.

Example 29 The formula $(0+0=0) \wedge (1+0=1)$ is True, because both $0+0=0$ and $1+0=1$ are True (example 15).

Example 30 The formula $(0+0=0) \wedge (0+0=1)$ is False, because $0+0=1$ is False (example 15).

Disjunction The **disjunction** of two logical formulae P and Q, denoted by $(P) \vee (Q)$ or by $(P)\mathtt{OR}(Q)$, is False if both P and Q are False, and it is True otherwise (if at least one of P or Q is True), as shown in table 31.

Table 31 Disjunction.

P	Q	$(P) \vee (Q)$
T	T	T
T	F	T
F	T	T
F	F	F

Example 32 The proposition $(\varnothing = \varnothing) \vee (\varnothing \in \varnothing)$ is True, because $\varnothing = \varnothing$ is True.

Example 33 The proposition $(\varnothing \in \varnothing) \vee [\neg(\varnothing = \varnothing)]$ is False, because both $\varnothing \in \varnothing$ and $\neg(\varnothing = \varnothing)$ are False.

Example 34 The formula $(0+0=0) \vee (0+0=1)$ is True, because $0+0=0$ is True (example 15).

Example 35 The formula $(0+0=1) \vee (0+1=0)$ is False, because both $0+0=1$ and $0+1=0$ are False (example 15).

Equivalence The **equivalence** of two formulae P and Q — denoted by $(P) \Leftrightarrow (Q)$ and read "P is logically equivalent to Q" — is True if P and Q have the same Truth value, whereas it is False if P and Q have different Truth values, as shown in table 36.

Table 36 Equivalence.

P	Q	$(P) \Leftrightarrow (Q)$
T	T	T
T	F	F
F	T	F
F	F	T

Example 37 The proposition $(\varnothing = \varnothing) \Leftrightarrow (\varnothing = \varnothing)$ is True, because $(\varnothing = \varnothing)$ and $(\varnothing = \varnothing)$ are *both* True.

Example 38 The proposition $(\varnothing \in \varnothing) \Leftrightarrow (\varnothing \in \varnothing)$ is True, because $\varnothing \in \varnothing$ and $\varnothing \in \varnothing$ are *both* False.

Example 39 The proposition $(\varnothing = \varnothing) \Leftrightarrow (\varnothing \in \varnothing)$ is False, because $\varnothing = \varnothing$ is True but $\varnothing \in \varnothing$ is False.

Example 40 The formula $(0 + 0 = 0) \Leftrightarrow (1 + 0 = 1)$ is True, because both $0 + 0 = 0$ and $1 + 0 = 1$ are True (example 15).

Example 41 The formula $(0 + 0 = 1) \Leftrightarrow (0 + 1 = 0)$ is True, because both $0 + 0 = 1$ and $0 + 1 = 0$ are False (example 15).

Example 42 The formula $(0 + 0 = 0) \Leftrightarrow (0 + 1 = 0)$ is False, because $0 + 0 = 0$ is True but $0 + 1 = 0$ is False (example 15).

Implication The **implication** of a logical formula Q by a logical formula P — denoted by $(P) \Rightarrow (Q)$ and read "P implies Q" — is False only if P is True and Q is False. It is True in all the other cases (if P is False or Q is True), as shown in table 43.

Table 43 Implication.

P	Q	$(P) \Rightarrow (Q)$
T	T	T
T	F	F
F	T	T
F	F	T

In a logical implication $(P) \Rightarrow (Q)$, the formula P is the **hypothesis,** or **antecedent,** whereas Q is the **conclusion,** or **consequent.** In particular, table 43 shows that a logical implication is True in all cases where the conclusion is True; table 43 also shows that a logical implication is True in all cases where the hypothesis is False.

Example 44 The implication $(\varnothing = \varnothing) \Rightarrow (\varnothing = \varnothing)$ is True, because the hypothesis $\varnothing = \varnothing$ and the conclusion $\varnothing = \varnothing$ are both True.

Example 45 The implication $(\varnothing = \varnothing) \Rightarrow (\varnothing \in \varnothing)$ is False, because the hypothesis $\varnothing = \varnothing$ is True but the conclusion $\varnothing \in \varnothing$ is False.

Example 46 The formula $(0 + 0 = 0) \Rightarrow (1 + 0 = 1)$ is True, because the hypothesis $0 + 0 = 0$ and the conclusion $1 + 0 = 1$ are both True (example 15).

Example 47 The formula $(0 + 0 = 0) \Rightarrow (0 + 1 = 0)$ is False, because the hypothesis $0 + 0 = 0$ is True but the conclusion $0 + 1 = 0$ is False (example 15).

Remark 48 True logical implications with a False hypothesis, as in examples 49 and 50, are convenient in classical mathematics, but other versions of mathematics avoid them [23]. Moreover, logical implications with a False hypothesis rarely occur in practical reasoning:

> Actually, the rule that any conditional is true if its antecedent is known to be false has almost no parallel in natural logic. Examples [of this type], which keep cropping up in textbooks, are only capable of confusing the student, since no natural subsystem in our language has expressions with this semantics [73, p. 36].

Correspondingly, some computers and logical circuits do not include any facility to test the Truth value of logical implications. Therefore the following four examples serve solely to demonstrate the difference between Boolean algebraic logic and practical reasoning. Reasons for such Truth tables appear further in remark 83. There exist other logics, which are closer to practical reasoning, but they are also more complicated than Boolean logic; examples of such logics appear in chapter 4.

Example 49 The implication $(\varnothing \in \varnothing) \Rightarrow (\varnothing = \varnothing)$ is True, because the hypothesis $\varnothing \in \varnothing$ is False, and the conclusion $\varnothing = \varnothing$ is True.

Example 50 The implication $(\varnothing \in \varnothing) \Rightarrow (\varnothing \in \varnothing)$ is True, because the hypothesis $\varnothing \in \varnothing$ and the conclusion $\varnothing \in \varnothing$ are both False.

Example 51 The formula $(0 + 0 = 1) \Rightarrow (1 + 0 = 1)$ is True, because the hypothesis $0 + 0 = 1$ is False, and the conclusion $1 + 0 = 1$ is True (example 15).

Example 52 The formula $(0 + 0 = 1) \Rightarrow (0 + 1 = 0)$ is True, because the hypothesis $0 + 0 = 1$ and the conclusion $0 + 1 = 0$ are both False (example 15).

Remark 53 A True implication $(P) \Rightarrow (Q)$ does *not* mean that its conclusion Q is True: Q might still be False, as in example 50. The guarantee that a conclusion Q is True requires that the implication $(P) \Rightarrow (Q)$ *and* its hypothesis P be *both* True.

Other symbols exist to denote the logical implication $\Rightarrow$, for instance, $\rightarrow$ and $\supset$. However, $\rightarrow$ also denotes mathematical functions and $\supset$ also denotes the mathematical inclusion of sets, as defined in subsequent chapters. To avoid future confusions with such different concepts, this exposition denotes the logical implication by $\Rightarrow$, as in [73, p. 6], [92, p. 39]. The following terminology will occasionally prove convenient.

Definition 54 (Singulary, unary) A logical connective is called **unary**, or also **singulary** [11, p. 12, n. 29] if and only if it applies to exactly one component.

Example 55 The negation $\neg$ is unary: it applies to a single component, as in $\neg(P)$.

Definition 56 (Binary) A logical connective is **binary** if and only if it applies to exactly two components.

Example 57 Applying to two components, as in $(P) \Rightarrow (Q)$, the connective $\Rightarrow$ is binary.

There also exist **ternary** connectives, which apply to exactly three components [11, §24, p. 130], but they will not be needed here.

0.2.3 Truth Tables

In the Boolean algebraic logic presented here, the logical connectives are *defined* by their Truth tables. In contrast, the Truth values of propositional forms with more than one connective are *derived* step by step from the Truth tables of their connectives.

Example 58 Consider the formula $[\neg(K)] \Rightarrow [\neg(H)]$. To construct its Truth table (shown in table 59), determine its Truth values as follows.

- List all the Truth values of its components, H and K, in the first two columns.
- In a third and fourth columns, line up the Truth values of $[\neg(K)]$ and $[\neg(H)]$, determined from the Truth table for the negation (table 21).
- In a fifth column, line up the Truth values of $[\neg(K)] \Rightarrow [\neg(H)]$, using the Truth table for the logical implication (table 43), with Truth values for $[\neg(K)]$ matching those of P, and with Truth values for $[\neg(H)]$ matching those of Q.

Definition 60 The form $[\neg(K)] \Rightarrow [\neg(H)]$ is the **contraposition** of $(H) \Rightarrow (K)$.

Table 59 A Truth table for $[\neg(K)] \Rightarrow [\neg(H)]$.

H	K	$\neg(K)$	$\neg(H)$	$[\neg(K)] \Rightarrow [\neg(H)]$
T	T	F	F	T
T	F	T	F	F
F	T	F	T	T
F	F	T	T	T

Example 61 The contraposition of the logical formula $(\varnothing \in \varnothing) \Rightarrow (\varnothing = \varnothing)$ is the logical formula $[\neg(\varnothing = \varnothing)] \Rightarrow [\neg(\varnothing \in \varnothing)]$.

Every logical implication has the same Truth value as that of its contraposition, as verified in Truth table 62. In contrast, merely reversing the direction of a logical implication does not preserve its Truth value, as shown in the following examples.

Table 62 Implication and contraposition.

H	K	$(H) \Rightarrow (K)$	$[\neg(K)] \Rightarrow [\neg(H)]$
T	T	T	T
T	F	F	F
F	T	T	T
F	F	T	T

Example 63 Consider the formula $(K) \Rightarrow (H)$. To construct its Truth table (shown in table 64), determine its Truth values as follows.

- List all the Truth values of its components, H and K, in the first two columns.
- In a third and fourth columns, line up the Truth values of K and H.
- In a fifth column, line up the Truth values of $(K) \Rightarrow (H)$, using the Truth table for the implication (table 43), with Truth values for K matching Truth values for P, and with Truth values for H matching Truth values for Q.

Table 64 A Truth table for $(K) \Rightarrow (H)$.

H	K	K	H	$(K) \Rightarrow (H)$
T	T	T	T	T
T	F	F	T	T
F	T	T	F	F
F	F	F	F	T

Definition 65 The propositional form $(K) \Rightarrow (H)$ is the **converse** of $(H) \Rightarrow (K)$.

Example 66 The converse of $(\varnothing \in \varnothing) \Rightarrow (\varnothing = \varnothing)$ is $(\varnothing = \varnothing) \Rightarrow (\varnothing \in \varnothing)$.

Example 63 reveals that the implication $(H) \Rightarrow (K)$ and its converse $(K) \Rightarrow (H)$ have *different* Truth values, as verified in Truth table 67.

Table 67 Implication and converse.

H	K	$(H) \Rightarrow (K)$	$(K) \Rightarrow (H)$
T	T	T	T
T	F	**F**	T
F	T	T	**F**
F	F	T	T

0.2.4 Exercises

For the following exercises, proceed as in examples 27, 28, 32, and 33.

Exercise 121 Determine the Truth value of $[\neg(\varnothing = \varnothing)] \wedge (\varnothing \in \varnothing)$ (True or False).

Exercise 122 Determine the Truth value of $[\neg(\varnothing = \varnothing)] \wedge [\neg(\varnothing \in \varnothing)]$.

Exercise 123 Determine the Truth value of $(\varnothing = \varnothing) \vee [\neg(\varnothing \in \varnothing)]$.

Exercise 124 Determine the Truth value of $[\neg(\varnothing = \varnothing)] \wedge [\neg(\varnothing \in \varnothing)]$.

Exercise 125 Determine the Truth value of $[\neg(\varnothing = \varnothing)] \vee [\neg(\varnothing \in \varnothing)]$.

Exercise 126 Determine the Truth value of $[\neg(\varnothing \in \varnothing)] \vee [\neg(\varnothing \in \varnothing)]$.

Exercise 127 Determine the Truth value of $[\neg(\varnothing \in \varnothing)] \wedge [\neg(\varnothing \in \varnothing)]$.

Exercise 128 Determine the Truth value of $\neg[(\varnothing \in \varnothing) \Rightarrow (\varnothing \in \varnothing)]$.

Exercise 129 Determine the Truth value of $\neg[(\varnothing \in \varnothing) \Leftrightarrow (\varnothing = \varnothing)]$.

Exercise 130 Determine the Truth value of $[\neg(\varnothing \in \varnothing)] \Leftrightarrow [\neg(\varnothing = \varnothing)]$.

For the following exercises, proceed as in examples 61 and 66.

Exercise 131 Write the converse of $(\varnothing = \varnothing) \Rightarrow (\varnothing \in \varnothing)$.

Exercise 132 Write the converse of $(\varnothing = \varnothing) \Rightarrow (\varnothing \in \varnothing)$.

Exercise 133 Write the contraposition of $(X \in Y) \Rightarrow (Y \in Z)$.

Exercise 134 Write the contraposition of $(X \in Y) \Rightarrow (Y \in Z)$.

Exercise 135 Write the converse of $(A = B) \Rightarrow [\neg(A = B)]$.

Exercise 136 Write the converse of $(A = B) \Rightarrow [\neg(A = B)]$.

Exercise 137 Write the contraposition of $[\neg(U = V)] \Rightarrow [\neg(W \in U)]$.

Exercise 138 Write the contraposition of $[\neg(U = V)] \Rightarrow (W \in U)$.

Exercise 139 Write the converse of $[\neg(X \in Y)] \Rightarrow [\neg(W = U)]$.

Exercise 140 Write the converse of $[\neg(X \in Y)] \Rightarrow [\neg(W = U)]$.

Draw Truth tables for the following propositional forms, as in examples 58 and 63.

Exercise 141 $[\neg(H)] \Rightarrow [\neg(K)]$

Exercise 142 $(H) \Rightarrow [\neg(K)]$

Exercise 143 $(H) \Rightarrow [(H) \wedge (K)]$

Exercise 144 $[(H) \vee (K)] \Rightarrow (K)$

Exercise 145 $(H) \Rightarrow [(K) \Rightarrow (L)]$

Exercise 146 $[(H) \Rightarrow (K)] \Rightarrow (L)$

Exercise 147 $\{(H) \wedge [\neg(K)]\} \Leftrightarrow \{\neg[(K) \vee (L)]\}$

Exercise 148 $\{[(H) \Rightarrow (K)] \Rightarrow [(K) \Rightarrow (L)]\} \Rightarrow \{(H) \Rightarrow [(K) \Rightarrow (L)]\}$

Exercise 149 $\{[(K) \Rightarrow (L)] \Rightarrow [(H) \Rightarrow (L)]\} \Rightarrow [(H) \Rightarrow (K)]$

Exercise 150 $\{[(H) \Rightarrow (K)] \Rightarrow [(H) \Rightarrow (L)]\} \Rightarrow [(K) \Rightarrow (L)]$

The following exercises do not follow the pattern of any previous example.

Exercise 151 Write a logical formula corresponding to the phrase "P but not Q."

Exercise 152 Write a logical formula for the phrase "either P or Q but not both."

Exercise 153 Write a logical formula for the phrase "P is necessary for Q."

Exercise 154 Write a logical formula for the phrase "P suffices for Q."

Exercise 155 Write a logical formula for the phrase "not P, unless Q."

Exercise 156 Write a logical formula for the phrase "neither P nor Q."

Exercise 157 Write a logical formula for the phrase "P does not suffice for Q."

Exercise 158 Write a logical formula for the phrase "P only if Q."

Exercise 159 Write a logical formula for the phrase "P if and only if Q."

Exercise 160 Write a logical formula for the phrase "if not Q, then P."

0.3 TAUTOLOGIES AND CONTRADICTIONS

0.3.1 Examples of Tautologies

Propositional forms that are True regardless of the Truth value of their components prove useful in both theory and applications of logic, because they allow for transformations of logical formulae into logically equivalent but more useful formulae.

Definition 68 A formula is a **tautology** if and only if it is True regardless of whether its components are True or False (provided each component has some Truth value).

One method for verifying that a propositional form is a tautology consists of drawing its Truth table, and then in verifying that its Truth value is True in all cases. Thus, Truth tables provide a method to prove certain tautologies, as illustrated in the following examples, which will recur throughout the text.

Example 69 To verify that $(B) \vee [\neg(B)]$ is a tautology, consider table 70.

Table 70 Law of excluded middle.

B	$\neg(B)$	$(B) \vee [\neg(B)]$
T	F	T
F	T	T

Table 70 confirms that $(B) \vee [\neg(B)]$ is True, regardless of whether its only component, B, is True or False. Hence, $(B) \vee [\neg(B)]$ is a tautology. The tautology $(B) \vee [\neg(B)]$ is called the **law of excluded middle,** because it asserts that for each proposition B, there is no other ("middle") True possibility than either B or $\neg(B)$.

Example 71 To verify that $\{\neg[\neg(P)]\} \Leftrightarrow (P)$ is a tautology, consider table 72.

Table 72 Complete law of double negation.

P	$\neg(P)$	$\neg[\neg(P)]$	$\{\neg[\neg(P)]\} \Leftrightarrow (P)$
T	F	T	T
F	T	F	T

Table 72 confirms that $\{\neg[\neg(P)]\} \Leftrightarrow (P)$ is True, regardless of whether its component, P, is True or False. Hence, $\{\neg[\neg(P)]\} \Leftrightarrow (P)$ is a tautology; it is sometimes called the **complete law of double negation**. More specifically, the formula $\{\neg[\neg(P)]\} \Rightarrow (P)$ is called the **law of double negation**, whereas its converse $(P) \Rightarrow \{\neg[\neg(P)]\}$ is called the **converse law of double negation** [11, p. 73, n. 163].

By example 71, P and $\neg[\neg(P)]$ are "logically equivalent" in the following sense.

Definition 73 Two logical formulae M and N are **logically equivalent** if and only if they have the same Truth tables: if and only if $(M) \Leftrightarrow (N)$ is a tautology.

Whereas the preceding examples involved only one component, the following examples show tautologies with two components.

Example 74 Consider the formula $\{[\neg(Q)] \Rightarrow [\neg(P)]\} \Leftrightarrow [(P) \Rightarrow (Q)]$.

Table 75 Complete law of contraposition: $[(P) \Rightarrow (Q)] \Leftrightarrow \{[\neg(Q)] \Rightarrow [\neg(P)]\}$.

P	Q	$\neg(Q)$	$\neg(P)$	$[\neg(Q)]\Rightarrow[\neg(P)]$	$(P)\Rightarrow(Q)$	$[\neg(Q)]\Rightarrow[\neg(P)]$ $\Updownarrow$ $(P)\Rightarrow(Q)$
T	T	F	F	T	T	T
T	F	T	F	F	F	T
F	T	F	T	T	T	T
F	F	T	T	T	T	T

Table 75 confirms that $\{[\neg(Q)] \Rightarrow [\neg(P)]\} \Leftrightarrow [(P) \Rightarrow (Q)]$ is True regardless of P and Q. Therefore, $\{[\neg(Q)] \Rightarrow [\neg(P)]\} \Leftrightarrow [(P) \Rightarrow (Q)]$ is a tautology; it is sometimes called the **complete law of contraposition**. More precisely, the formula

$$[(P) \Rightarrow (Q)] \Rightarrow \{[\neg(Q)] \Rightarrow [\neg(P)]\}$$

is called the **law of contraposition** [11, §15, p. 102] , while its converse

$$\{[\neg(Q)] \Rightarrow [\neg(P)]\} \Rightarrow [(P) \Rightarrow (Q)]$$

is called the **converse law of contraposition** [11, §20, p. 119] .

By example 74, every implication is logically equivalent to its contraposition. In contrast, reversing a logical implication does *not* preserve its Truth value.

Example 76 Consider the formula $[(P) \Rightarrow (Q)] \Leftrightarrow [(Q) \Rightarrow (P)]$.

Table 77 An implication is *not* logically equivalent to its converse.

P	Q	$(P) \Rightarrow (Q)$	$(Q) \Rightarrow (P)$	$[(P) \Rightarrow (Q)] \Leftrightarrow [(Q) \Rightarrow (P)]$
T	T	T	T	T
T	F	F	T	**F**
F	T	T	F	**F**
F	F	T	T	T

Table 77 reveals that the implication $(P) \Rightarrow (Q)$ and its converse $(Q) \Rightarrow (P)$ have *different* Truth values; hence they are *not* logically equivalent.

Remark 78 There exists an alternative arrangement of Truth tables, attributed to Willard Van Orman Quine [11, p. 95]. For instance, Truth table 77 in example 76 becomes table 79 in Quine's form. To draw a Truth table in Quine's form, proceed as follows.

Table 79 Quine's arrangement of Truth table 77.

$[(P)$	$\Rightarrow$	$(Q)]$	$\Leftrightarrow$	$[(Q)$	$\Rightarrow$	$(P)]$
T	T	T	T	T	T	T
T	F	F	**F**	F	T	T
F	T	T	**F**	T	F	F
F	T	F	T	F	T	F

- Write the propositional form on one line (here $[(P) \Rightarrow (Q)] \Leftrightarrow [(Q) \Rightarrow (P)]$).
- List all the Truth values of its components (here P and Q) in the first two columns where these components occur (here the first and third columns).
- Repeat all the Truth values of all the components, in the same respective order, in all the columns where these components occur. (Here the fifth column repeats the third column for Q, and the seventh column repeats the first column for P.)
- Record the Truth values of the simplest compound formulae in the columns below their connectives. (Here record the Truth values of $(P) \Rightarrow (Q)$ in the second column, below the first connective $\Rightarrow$; similarly record the Truth values of $(Q) \Rightarrow (P)$ in the sixth column, below the second connective $\Rightarrow$.)
- Repeat the preceding step for the compound formulae at the next levels. (Here record the Truth values of $[(P) \Rightarrow (Q)] \Leftrightarrow [(Q) \Rightarrow (P)]$ in the fourth column, below $\Leftrightarrow$, using the Truth values of $(P) \Rightarrow (Q)$ and $(Q) \Rightarrow (P)$ from the preceding step.)

While the previous examples involved only one or two components, the following example illustrates a Truth table with several components.

Example 80 The following formula is a tautology, as verified in table 81:

$$\{(P) \Rightarrow [(Q) \Rightarrow (R)]\} \Rightarrow \{[(P) \Rightarrow (Q)] \Rightarrow [(P) \Rightarrow (R)]\}$$

Table 81 Transitivity of the logical implication.

P	Q	R	$(Q) \Downarrow (R)$	$(P) \Downarrow (Q){\Rightarrow}(R)$	$(P) \Downarrow (Q)$	$(P) \Downarrow (R)$	$(P){\Rightarrow}(Q) \Downarrow (P){\Rightarrow}(R)$	$(P){\Rightarrow}[(Q){\Rightarrow}(R)] \Downarrow [(P){\Rightarrow}(Q)]{\Rightarrow}[(P){\Rightarrow}(R)]$
T	T	T	T	T	T	T	T	T
T	T	F	F	F	T	F	F	T
T	F	T	T	T	F	T	T	T
T	F	F	T	T	F	F	T	T
F	T	T	T	T	T	T	T	T
F	T	F	F	T	T	T	T	T
F	F	T	T	T	T	T	T	T
F	F	F	T	T	T	T	T	T

This tautology is one of the forms of the **transitivity** of the logical implication. Table 81 becomes table 82 in Quine's form.

Table 82 Quine's form of Truth table 81.

$\{(P) \Rightarrow [(Q) \Rightarrow (R)]\} \Rightarrow \{[(P) \Rightarrow (Q)] \Rightarrow [(P) \Rightarrow (R)]\}$												
T	T	T	T	T	T	T	T	T	T	T	T	T
T	F	T	F	F	T	T	T	T	F	T	F	F
T	T	F	T	T	T	T	F	F	T	T	T	T
T	T	F	T	F	T	T	F	F	T	T	F	F
F	T	T	T	T	T	F	T	T	T	F	T	T
F	T	T	F	F	T	F	T	T	T	F	T	F
F	T	F	T	T	T	F	T	F	T	F	T	T
F	T	F	T	F	T	F	T	F	T	F	T	F

Remark 83 Remark 48 has merely indicated that the last two lines of the Truth table for the Boolean logical implication rarely occur in common patterns of reasoning. Continuing beyond this mere indication, the present remark shows that the Truth table for the Boolean logical implication is the only Truth table that satisfy certain requirements. Specifically, the present considerations confirm that the law of contraposition

$$[(P) \Rightarrow (Q)] \Rightarrow \{[\neg(Q)] \Rightarrow [\neg(P)]\}$$

(from example 58), *and* the non-equivalence of an implication $(P) \Rightarrow (Q)$ with its converse $(Q) \Rightarrow (P)$ (example 63) hold only with implications defined as in table 43. To this end, denote by $\rightrightarrows$ any candidate connective for a logical implication. To reflect common experience, any concept of logical implication may have to satisfy the following two requirements.

- Firstly, if a hypothesis P holds, and if the conclusion Q holds, then the implication $(P) \rightrightarrows (Q)$ also holds.
- Secondly, if a hypothesis P holds, but if the conclusion Q fails, then the implication $(P) \rightrightarrows (Q)$ also fails.

These two requirements dictate the first two lines of Truth table 84. There remain only four

Table 84 A partial Truth table for $(P) \rightrightarrows (Q)$.

P	Q	$(P) \rightrightarrows (Q)$
T	T	T
T	F	F

possibilities for $\rightrightarrows$ in the last two rows, where the hypothesis is False. For convenience, denote these four connectives by $\Rightarrow$, $\looparrowright$, $\leadsto$, and $\curvearrowright$ respectively (these last three symbols are used in this manner only in the present discussion). Table 85 shows their Truth values.

Table 85 The connectives $\looparrowright$, $\leadsto$, $\curvearrowright$, and $\Rightarrow$.

P	Q	$(P) \looparrowright (Q)$
T	T	T
T	F	F
F	T	T
F	F	F

P	Q	$(P) \leadsto (Q)$
T	T	T
T	F	F
F	T	F
F	F	F

P	Q	$(P) \curvearrowright (Q)$
T	T	T
T	F	F
F	T	F
F	F	T

P	Q	$(P) \Rightarrow (Q)$
T	T	T
T	F	F
F	T	T
F	F	T

Examples 58 and 63 have already confirmed that the logical implication $\Rightarrow$ has the same Truth values as its contraposition has, but not as its converse has.

In contrast, $(P) \looparrowright (Q)$ and its contraposition $[\neg(Q)] \looparrowright [\neg(P)]$ do *not* have the same Truth values, as in table 86. Moreover, for the connective $\looparrowright$, neither the law of contraposition nor its converse hold, as verified in exercises 251 and 252.

Table 86 The connective $\looparrowright$ and its contraposition do *not* have the same Truth values.

P	Q	$(P) \looparrowright (Q)$	$[\neg(Q)] \looparrowright [\neg(P)]$
T	T	T	**F**
T	F	F	F
F	T	T	T
F	F	F	**T**

Similarly, $(P) \leadsto (Q)$ and its contraposition $[\neg(Q)] \leadsto [\neg(P)]$ do *not* have the same Truth values, as in table 87. Moreover, for the connective $\leadsto$, neither the law of contraposition nor its converse hold, as verified in exercises 253 and 254.

Table 87 The connective $\rightsquigarrow$ and its contraposition do *not* have the same Truth values.

P	Q	$(P) \rightsquigarrow (Q)$	$[\neg(Q)] \rightsquigarrow [\neg(P)]$
T	T	T	**F**
T	F	F	F
F	T	F	F
F	F	F	**T**

Finally, $(P) \curvearrowright (Q)$ and its converse $(Q) \curvearrowright (P)$ have the same Truth values, as in table 88. However, for the connective $\curvearrowright$, both the law of contraposition and its converse hold, as verified in exercises 255 and 256.

Table 88 The connective $\rightsquigarrow$ and its converse have the same Truth values.

P	Q	$(P) \curvearrowright (Q)$	$(Q) \curvearrowright (P)$
T	T	T	T
T	F	F	F
F	T	F	F
F	F	T	T

Thus the Truth table specified for $(P) \Rightarrow (Q)$ is the only one that reflects experience. As demonstrated in chapter 4, there exist other concepts of logical implication, but they do not lend themselves to Truth tables [11, p. 146, #26.12]. These logics will reveal yet another explanation of the Truth table for the Boolean logical implication (in remark 654). Specifically, the Truth table for the Boolean logical implication will follow from commonly accepted logical principles and either the law of excluded middle (example 69) or the law of double negation (example 71).

0.3.2 Contradictions

Propositional forms that are False regardless of the Truth value of their components can also prove useful in some contexts.

Definition 89 A formula is a **contradiction** if and only if it is False regardless of the Truth values of its components (provided each component has some Truth value).

Example 90 To verify that $(B) \wedge [\neg(B)]$ is a contradiction, set up its Truth table.

Table 91 A contradiction.

B	$\neg(B)$	$(B) \wedge [\neg(B)]$
T	F	F
F	T	F

Table 91 confirms that $(B) \wedge [\neg(B)]$ is False, regardless of whether its only component, B, is True or False. Hence, $(B) \wedge [\neg(B)]$ is a contradiction.

Whereas the contradiction $(B) \wedge [\neg(B)]$ has only one component, other contradictions can have several components.

Table 93 A contradiction with two components.

P	Q	$\neg(P)$	$\neg(Q)$	$(P)\wedge(Q)$	$[\neg(P)]\wedge[\neg(Q)]$	$[(P)\wedge(Q)]\vee\{[\neg(P)]\wedge[\neg(Q)]\}$
T	T	F	F	T	F	F
T	F	F	T	F	F	F
F	T	T	F	F	F	F
F	F	T	T	F	T	F

Example 92 Consider the formula $[(P) \wedge (Q)] \vee \{[\neg(P)] \wedge [\neg(Q)]\}$ in table 93.

Table 93 shows that $[(P) \wedge (Q)] \vee \{[\neg(P)] \wedge [\neg(Q)]\}$ is False regardless of the Truth values of P and Q. Therefore, $[(P) \wedge (Q)] \vee \{[\neg(P)] \wedge [\neg(Q)]\}$ is a contradiction.

Because an implication is True in all cases where its hypothesis is False regardless of its conclusion, it follows that a contradiction implies every conclusion.

Example 94 Table 95 shows that $\{(B) \wedge [\neg(B)]\} \Rightarrow (Q)$ is True regardless of the Truth value of Q.

Table 95 A contradictory hypothesis implies every conclusion.

B	Q	$(B) \wedge [\neg(B)]$	$\{(B) \wedge [\neg(B)]\} \Rightarrow (Q)$
T	T	F	T
T	F	F	T
F	T	F	T
F	F	F	T

0.3.3 Exercises

The tautologies presented in the following exercises will repeatedly play a role in the sequel. Verify these tautologies, for instance, through Truth tables, or through previously proved tautologies. The symbol T stands for any tautology or True proposition, and the symbol F stands for any contradiction or False proposition.

Exercise 161 $(P) \Rightarrow (T)$

Exercise 162 $(F) \Rightarrow (P)$

Exercise 163 $(P) \Rightarrow (P)$ (Reflexive law of material implication.)

Exercise 164 $(P) \Leftrightarrow (P)$

Exercise 165 $\{\neg[\neg(P)]\} \Rightarrow (P)$ (Law of double negation.)

Exercise 166 $(P) \Rightarrow \{\neg[\neg(P)]\}$ (Converse law of double negation.)

Exercise 167 $(\neg\{\neg[\neg(P)]\}) \Rightarrow [\neg(P)]$ (Law of triple negation.)

Exercise 168 $[\neg(P)]) \Rightarrow (\neg\{\neg[\neg(P)]\}$ (Converse law of triple negation.)

Exercise 169 $\neg\{(P) \wedge [\neg(P)]\}$ (Law of contradiction.)

Exercise 170 $(P) \Leftrightarrow [(P) \vee (P)]$ (Idempotence of $\vee$.)

Exercise 171 $(P) \Leftrightarrow [(P) \wedge (P)]$ (Idempotence of $\wedge$.)

Exercise 172 $(P) \Leftrightarrow [(F) \vee (P)]$ (Identity from $(F)\vee$.)

Exercise 173 $(P) \Leftrightarrow [(T) \wedge (P)]$ (Identity from $(T)\wedge$.)

Exercise 174 $\neg(\neg\{(P) \vee [\neg(P)]\})$ (Weak law of excluded middle.)

Exercise 175 $(P) \Leftrightarrow \{(P) \vee [(P) \wedge (Q)]\}$ (First law of absorption.)

Exercise 176 $(P) \Leftrightarrow \{(P) \wedge [(P) \vee (Q)]\}$ (Second law of absorption.)

Exercise 177 $[(P) \Rightarrow (Q)] \Rightarrow \{[\neg(Q)] \Rightarrow [\neg(P)]\}$
(Law of contraposition [11, §15, p. 102].)

Exercise 178 $\{[\neg(Q)] \Rightarrow [\neg(P)]\} \Rightarrow [(P) \Rightarrow (Q)]$
(Converse law of contraposition [11, §20, p. 119].)

Exercise 179 $[(P) \vee (Q)] \Leftrightarrow [(Q) \vee (P)]$ ("OR" commutes.)

Exercise 180 $(P) \Rightarrow [(P) \vee (Q)]$

Exercise 181 $[(P) \wedge (Q)] \Leftrightarrow [(Q) \wedge (P)]$ ("AND" commutes.)

Exercise 182 $(Q) \Rightarrow [(P) \vee (Q)]$

Exercise 183 $\{(P) \vee [(Q) \vee (R)]\} \Leftrightarrow \{[(P) \vee (Q)] \vee (R)\}$ ("OR" is associative.)

Exercise 184 $[(P) \wedge (Q)] \Rightarrow (P)$

Exercise 185 $\{(P) \wedge [(Q) \wedge (R)]\} \Leftrightarrow \{[(P) \wedge (Q)] \wedge (R)\}$ ("AND" is associative.)

Exercise 186 $[(P) \wedge (Q)] \Rightarrow (Q)$

Exercise 187 $\{(P) \wedge [(Q) \vee (R)]\} \Leftrightarrow \{[(P) \wedge (Q)] \vee [(P) \wedge (R)]\}$
("AND" distributes over "OR".)

Exercise 188 $[\neg(P)] \Rightarrow [(P) \Rightarrow (Q)]$ (Law of denial of the antecedent.)

Exercise 189 $\{(P) \vee [(Q) \wedge (R)]\} \Leftrightarrow \{[(P) \vee (Q)] \wedge [(P) \vee (R)]\}$
("OR" distributes over "AND".)

Exercise 190 $(P) \Rightarrow \{[\neg(P)] \Rightarrow (Q)\}$

Exercise 191 $(P) \Rightarrow [(Q) \Rightarrow (P)]$ (Law of affirmation of the consequent.)

Exercise 192 $\{(P) \Rightarrow [(P) \Rightarrow (Q)]\} \Rightarrow \{(P) \Rightarrow (Q)\}$

Exercise 193 $\{(P) \wedge [(P) \Rightarrow (Q)]\} \Rightarrow (Q)$

Exercise 194 $(P) \Rightarrow \{[(P) \Rightarrow (Q)] \Rightarrow (Q)\}$ (Law of assertion.)

Exercise 195 $\{[\neg(Q)] \wedge [(P) \Rightarrow (Q)]\} \Rightarrow [\neg(P)]$

Exercise 196 $\{(P) \Rightarrow [\neg(P)]\} \Rightarrow [\neg(P)]$
(Special law of reductio ad absurdum.)

Exercise 197 $[(P) \Rightarrow (Q)] \Rightarrow \big(\{(P) \Rightarrow [\neg(Q)]\} \Rightarrow [\neg(P)]\big)$
(Law of reductio ad absurdum.)

Exercise 198 $\big[(P) \Rightarrow \{(Q) \wedge [\neg(Q)]\}\big] \Rightarrow [\neg(P)]$ (Law of syllogism.)

Exercise 199 $(R) \Leftrightarrow \big([\neg(R)] \Rightarrow \{(S) \wedge [\neg(S)]\}\big)$ (Proof by contradiction.)

Exercise 200 $[(P) \wedge (Q)] \Rightarrow [(P) \Rightarrow (Q)]$

Exercise 201 $[(P) \Rightarrow (Q)] \vee [(Q) \Rightarrow (P)]$

Exercise 202 $[(P) \Leftrightarrow (Q)] \Leftrightarrow [(Q) \Leftrightarrow (P)]$
(Complete commutative law of equivalence.)

Exercise 203 $\{[\neg(P)] \Rightarrow (Q)\} \Rightarrow \big[\{[\neg(P)] \Rightarrow [\neg(Q)]\} \Rightarrow (P)\big]$
(Law of indirect proof.)

Exercise 204 $(P) \Rightarrow \{(Q) \Rightarrow [(P) \wedge (Q)]\}$

Exercise 205 $[(P) \Leftrightarrow (Q)] \Leftrightarrow \{[(P) \Rightarrow (Q)] \wedge [(Q) \Rightarrow (P)]\}$

Exercise 206 $[(P) \Rightarrow (Q)] \Rightarrow \big[\{[(P) \Rightarrow (R)] \Rightarrow (Q)\} \Rightarrow (Q)\big]$

Exercise 207 $[(P) \Rightarrow (Q)] \Rightarrow \{[(Q) \Rightarrow (P)] \Rightarrow [(P) \Leftrightarrow (Q)]\}$
(One of Tarski's axioms.)

Exercise 208 $[(P) \Rightarrow (Q)] \Rightarrow \{[(Q) \Rightarrow (R)] \Rightarrow [(P) \Rightarrow (R)]\}$
(Transitivity of implication.)

Exercise 209 $[(Q) \Rightarrow (R)] \Rightarrow \{[(P) \Rightarrow (Q)] \Rightarrow [(P) \Rightarrow (R)]\}$
(Transitivity of implication.)

Exercise 210 $[(P) \Rightarrow (Q)] \Rightarrow \big([(P) \Rightarrow (R)] \Rightarrow \{(P) \Rightarrow [(Q) \wedge (R)]\}\big)$

Exercise 211 $\{[(P) \Rightarrow (Q)] \wedge [(R) \Rightarrow (Q)]\} \Leftrightarrow \{[(P) \vee (R)] \Rightarrow (Q)\}$

Exercise 212 $\{[(P) \Rightarrow (Q)] \wedge [(P) \Rightarrow (R)]\} \Leftrightarrow \{(P) \Rightarrow [(Q) \wedge (R)]\}$
(Law of composition.)

Exercise 213 $\{[(P) \wedge (Q)] \Rightarrow (R)\} \Leftrightarrow \big(\{(P) \wedge [\neg(R)]\} \Rightarrow [\neg(Q)]\big)$
(Extended contraposition.)

Exercise 214 $\{(P) \Rightarrow [(Q) \Rightarrow (R)]\} \Rightarrow \{[(P) \wedge (Q)] \Rightarrow (R)\}$ (Law of importation.)

Exercise 215 $\{[(P) \wedge (Q)] \Rightarrow (R)\} \Rightarrow \{(P) \Rightarrow [(Q) \Rightarrow (R)]\}$ (Law of exportation.)

Exercise 216 $\big([(P) \Rightarrow (S)] \wedge \{[\neg(Q)] \Rightarrow [\neg(S)]\}\big) \Rightarrow [(P) \Rightarrow (Q)]$
(Special proof by contradiction with $(P) \Rightarrow (Q)$ instead of R.)

Exercise 217 $\{(P) \Rightarrow [(Q) \Rightarrow (R)]\} \Rightarrow \{[(P) \Rightarrow (Q)] \Rightarrow [(P) \Rightarrow (R)]\}$
(One of Łukasiewicz's axioms.)

Exercise 218 $\{(P) \Rightarrow [(Q) \Rightarrow (R)]\} \Rightarrow \{(Q) \Rightarrow [(P) \Rightarrow (R)]\}$
(Law of commutation.)

Exercise 219 $[(Q) \Rightarrow (R)] \Rightarrow \{[(P) \vee (Q)] \Rightarrow [(P) \vee (R)]\}$

Exercise 220 $\{[(P) \Rightarrow (Q)] \wedge [(R) \Rightarrow (S)]\} \Rightarrow \{[(P) \wedge (R)] \Rightarrow [(Q) \wedge (S)]\}$

Exercise 221

$$(P) \Rightarrow \big[(R) \Rightarrow \big(\{[(P) \wedge (R)] \Rightarrow [(Q) \wedge (S)]\} \Rightarrow \{[(P) \Rightarrow (Q)] \wedge [(R) \Rightarrow (S)]\}\big)\big]$$

Exercise 222 $\big[\{[(P) \Rightarrow (R)] \Rightarrow (Q)\} \Rightarrow (Q)\big] \Rightarrow \{[(Q) \Rightarrow (R)] \Rightarrow [(P) \Rightarrow (R)]\}$

Exercise 223 $\{[(P) \Leftrightarrow (R)] \wedge [(Q) \Leftrightarrow (S)]\} \Rightarrow \{[(P) \Rightarrow (Q)] \Leftrightarrow [(R) \Rightarrow (S)]\}$

Exercise 224 $\{[(P) \Leftrightarrow (R)] \wedge [(Q) \Leftrightarrow (S)]\} \Rightarrow \{[(P) \Leftrightarrow (Q)] \Leftrightarrow [(R) \Leftrightarrow (S)]\}$

Exercise 225 (Meredith's axiom.)

$$\begin{aligned}\big\{\big[\big(\{(P) \Rightarrow (Q)\} \Rightarrow \{[\neg(R)] \Rightarrow [\neg(S)]\}\big) \Rightarrow (R)\big] \Rightarrow (U)\big\}\\ \Rightarrow \{[(U) \Rightarrow (P)] \Rightarrow [(S) \Rightarrow (P)]\}\end{aligned}$$

Exercise 226 $\{[(P) \Rightarrow (Q)] \Rightarrow (R)\} \Rightarrow \{[(R) \Rightarrow (P)] \Rightarrow [(S) \Rightarrow (P)]\}$

Exercise 227 $[(R) \Rightarrow (Q)] \Rightarrow \big(\{[(R) \Rightarrow (Q)] \Rightarrow (P)\} \Rightarrow [(S) \Rightarrow (P)]\big)$

Exercise 228

$$\{[(H) \Rightarrow (R)] \Rightarrow [(S) \Rightarrow (P)]\} \Rightarrow \big(\{[(R) \Rightarrow (Q)] \Rightarrow (P)\} \Rightarrow [(S) \Rightarrow (P)]\big)$$

Exercise 229

$$\{[(R) \Rightarrow (Q)] \Rightarrow [(S) \Rightarrow (P)]\} \Rightarrow \{[(R) \Rightarrow (P)] \Rightarrow [(S) \Rightarrow (P)]\}$$

Exercise 230

$$\begin{aligned}\big\{\{[(R) \Rightarrow (P)] \Rightarrow [(S) \Rightarrow (P)]\} \Rightarrow \big[\{[(P) \Rightarrow (Q)] \Rightarrow (R)\} \Rightarrow (H)\big]\big\}\\ \Rightarrow \big\{(K) \Rightarrow \big[\{[(P) \Rightarrow (Q)] \Rightarrow (R)\} \Rightarrow (H)\big]\big\}\end{aligned}$$

Exercise 231

$$\big(\{[(R) \Rightarrow (P)] \Rightarrow (P)\} \Rightarrow [(S) \Rightarrow (P)]\big) \Rightarrow \big(\{[(P) \Rightarrow (Q)] \Rightarrow (R)\} \Rightarrow [(S) \Rightarrow (P)]\big)$$

Exercise 232

$$\begin{aligned}\Big[\big\{(P) \Rightarrow \big[\{[\neg(Q)] \Rightarrow [(R) \Rightarrow (S)]\} \Rightarrow \big([(S) \Rightarrow (Q)] \Rightarrow \{(H) \Rightarrow [(R) \Rightarrow (Q)]\}\big)\big]\big\}\\ \Rightarrow \big(\{[\neg(K)] \Rightarrow [(K) \Rightarrow (L)]\} \Rightarrow (M)\big)\Big] \Rightarrow (M)\end{aligned}$$

Exercise 233

$$\{[(P) \Leftrightarrow (R)] \wedge [(Q) \Leftrightarrow (S)]\} \Rightarrow \big(\{[\neg(P)] \Leftrightarrow [\neg(Q)]\} \Rightarrow \{[\neg(R)] \Leftrightarrow [\neg(S)]\}\big)$$

Exercise 234

$$\begin{aligned}\big(\big\{(L) \Rightarrow \big[\{[(P) \Rightarrow (Q)] \Rightarrow (S)\} \Rightarrow (K)\big]\big\} \Rightarrow\\ \big[(M) \Rightarrow \big([(S) \Rightarrow (H)] \Rightarrow \{[\neg(P)] \Rightarrow [\neg(R)]\}\big)\big]\big)\\ \Rightarrow \big[(M) \Rightarrow \{[(S) \Rightarrow (P)] \Rightarrow [(R) \Rightarrow (P)]\}\big]\end{aligned}$$

For each of the following exercises, determine whether the given formula is a tautology.

Exercise 235
$\{[(U) \vee (V)] \Leftrightarrow [(U) \vee (W)]\} \Rightarrow [(V) \Leftrightarrow (W)]$

Exercise 236
$\{[(U) \wedge (V)] \Leftrightarrow [(U) \wedge (W)]\} \Rightarrow [(V) \Leftrightarrow (W)]$

Exercise 237
$[(P) \Rightarrow (R)] \Rightarrow \{[(P) \Rightarrow (Q)] \wedge [(Q) \Rightarrow (R)]\}$

Exercise 238
$[(P) \Leftrightarrow (R)] \Rightarrow \{[(P) \Leftrightarrow (Q)] \wedge [(Q) \Leftrightarrow (R)]\}$

Exercise 239
$\{[(P) \Rightarrow (Q)] \wedge [(P) \Rightarrow (R)]\} \Rightarrow [(Q) \Rightarrow (R)]$

Exercise 240
$\{[(P) \Rightarrow (Q)] \wedge [(P) \Rightarrow (R)]\} \Rightarrow \{(P) \Rightarrow [(Q) \Rightarrow (R)]\}$

For each of the following exercises, proceed as in examples 90 and 94, and determine whether the given formula is a contradiction.

Exercise 241 $\neg[(P) \Rightarrow (P)]$

Exercise 242 $(P) \Rightarrow [\neg(P)]$

Exercise 243 $(P) \Leftrightarrow [\neg(P)]$

Exercise 244 $(P) \Rightarrow \{(P) \Rightarrow [\neg(P)]\}$

Exercise 245 $(P) \wedge \big(\{[\neg(P)] \vee (Q)\} \wedge [\neg(Q)]\big)$

Exercise 246 $\{[\neg(P)] \vee (Q)\} \wedge \big[\{[\neg(P)] \vee [\neg(Q)]\} \wedge (P)\big]$

Exercise 247 $[(P) \vee (Q)] \wedge \{[\neg(P)] \vee [\neg(Q)]\}$

Exercise 248 $[(P) \wedge (Q)] \vee \{[\neg(P)] \wedge [\neg(Q)]\}$

Exercise 249 $\{[\neg(P)] \vee (Q)\} \wedge \big(\{[\neg(Q)] \vee (R)\} \wedge \{(P) \wedge [\neg(R)]\}\big)$

Exercise 250 $\big([\neg(P)] \vee \{[\neg(Q)] \vee (R)\}\big) \wedge \big(\{[\neg(P)] \vee (Q)\} \wedge \{(P) \wedge [\neg(R)]\}\big)$

The following exercises pertain to the connectives $\looparrowright$, $\rightsquigarrow$, $\curvearrowright$ defined by table 85. Determine whether each formula is a tautology.

Exercise 251 $[(P) \looparrowright (Q)] \looparrowright \{[\neg(Q)] \looparrowright [\neg(P)]\}$

Exercise 252 $\{[\neg(Q)] \looparrowright [\neg(P)]\} \looparrowright [(P) \looparrowright (Q)]$

Exercise 253 $[(P) \rightsquigarrow (Q)] \rightsquigarrow \{[\neg(Q)] \rightsquigarrow [\neg(P)]\}$

Exercise 254 $\{[\neg(Q)] \rightsquigarrow [\neg(P)]\} \rightsquigarrow [(P) \rightsquigarrow (Q)]$

Exercise 255 $[(P) \curvearrowright (Q)] \curvearrowright \{[\neg(Q)] \curvearrowright [\neg(P)]\}$

Exercise 256 $\{[\neg(Q)] \curvearrowright [\neg(P)]\} \curvearrowright [(P) \curvearrowright (Q)]$

Exercise 257 $[(P) \looparrowright (Q)] \looparrowright [(Q) \looparrowright (P)]$

Exercise 258 $[(P) \rightsquigarrow (Q)] \rightsquigarrow [(Q) \rightsquigarrow (P)]$

Exercise 259 $(P) \looparrowright (P)$

Exercise 260 $(P) \rightsquigarrow (P)$

0.4 OTHER METHODS OF PROOF

0.4.1 Proofs by Tautologies

The only method of logical proof presented so far consists of drawing Truth tables. There also exist other methods of logical proof, but because they form the subject of detailed discussion in the next chapter, the present section merely outlines them briefly. One alternative method of logical proof relies on tautologies based on the "transitivity" of the logical connectives $\Rightarrow$ and $\Leftrightarrow$. The statements and proofs of their transitivity, and of a few other tautologies, first proceeds through Truth tables. Then subsequent examples will demonstrate this alternative method with tautologies.

Example 96 Truth table 97 shows that two propositions V and W are equivalent if and only if their negations $\neg(V)$ and $\neg(W)$ are equivalent.

Table 97 Equivalent propositions have equivalent negations.

V	W	$\neg(V)$	$\neg(W)$	$(V)\Leftrightarrow(W)$	$[\neg(V)]\Leftrightarrow[\neg(W)]$	$(V)\Leftrightarrow(W)$ $\Updownarrow$ $[\neg(V)]\Leftrightarrow[\neg(W)]$
T	T	F	F	T	T	T
T	F	F	T	F	F	T
F	T	T	F	F	F	T
F	F	T	T	T	T	T

Thus $[(V) \Leftrightarrow (W)] \Leftrightarrow \{[\neg(V)] \Leftrightarrow [\neg(W)]\}$ is a tautology.

Example 98 Truth table 99 shows that if two propositions V and W are equivalent, then $(U) \vee (V)$ and $(U) \vee (W)$ are also equivalent.

Table 99 Equivalent propositions lead to equivalent disjunctions.

U	V	W	(U) $\vee$ (V)	(U) $\vee$ (W)	(V) $\Updownarrow$ (W)	$(U)\vee(V)$ $\Updownarrow$ $(U)\vee(W)$	$(V)\Leftrightarrow(W)$ $\Downarrow$ $[(U)\vee(V)]\Leftrightarrow[(U)\vee(W)]$
T	T	T	T	T	T	T	T
T	T	F	T	T	F	T	T
T	F	T	T	T	F	T	T
T	F	F	T	T	T	T	T
F	T	T	T	T	T	T	T
F	T	F	T	F	F	F	T
F	F	T	F	T	F	F	T
F	F	F	F	F	T	T	T

Thus $[(V) \Leftrightarrow (W)] \Rightarrow \{[(U) \vee (V)] \Leftrightarrow [(U) \vee (V)]\}$ is a tautology.

Example 100 Truth table 101 shows that if two propositions V and W are equivalent, then $(U) \wedge (V)$ and $(U) \wedge (W)$ are also equivalent. Thus the following propositional form is a tautology: $[(V) \Leftrightarrow (W)] \Rightarrow \{[(U) \wedge (V)] \Leftrightarrow [(U) \wedge (V)]\}$.

Table 101 Equivalent propositions lead to equivalent conjunctions.

U	V	W	$(U) \wedge (V)$	$(U) \wedge (W)$	$(V) \Updownarrow (W)$	$(U)\wedge(V) \Updownarrow (U)\wedge(W)$	$(V)\Leftrightarrow(W) \Downarrow [(U)\wedge(V)]\Leftrightarrow[(U)\wedge(W)]$
T	T	T	T	T	T	T	T
T	T	F	T	F	F	F	T
T	F	T	F	T	F	F	T
T	F	F	F	F	T	T	T
F	T	T	F	F	T	T	T
F	T	F	F	F	F	T	T
F	F	T	F	F	F	T	T
F	F	F	F	F	T	T	T

Example 102 The formula $\{[(P) \Rightarrow (Q)] \wedge [(Q) \Rightarrow (R)]\} \Rightarrow [(P) \Rightarrow (R)]$ is a tautology, as verified in table 103. It is one of the forms of the **transitivity** of $\Rightarrow$.

Table 103 Transitivity of the logical implication.

P	Q	R	$(P) \Downarrow (Q)$	$(Q) \Downarrow (R)$	$(P)\Rightarrow(Q) \wedge (Q)\Rightarrow(R)$	$(P) \Downarrow (R)$	$[(P)\Rightarrow(Q)]\wedge[(Q)\Rightarrow(R)] \Downarrow (P)\Rightarrow(R)$
T	T	T	T	T	T	T	T
T	T	F	T	F	F	F	T
T	F	T	F	T	F	T	T
T	F	F	F	T	F	F	T
F	T	T	T	T	T	T	T
F	T	F	T	F	F	T	T
F	F	T	T	T	T	T	T
F	F	F	T	T	T	T	T

Example 104 The formula $\{[(P) \Leftrightarrow (Q)] \wedge [(Q) \Leftrightarrow (R)]\} \Rightarrow [(P) \Leftrightarrow (R)]$ is a tautology, as shown in table 105. It is one of the forms of the **transitivity** of $\Leftrightarrow$.

Table 105 Transitivity of the logical equivalence.

P	Q	R	$(P) \Updownarrow (Q)$	$(Q) \Updownarrow (R)$	$(P)\Leftrightarrow(Q) \wedge (Q)\Leftrightarrow(R)$	$(P) \Updownarrow (R)$	$[(P)\Leftrightarrow(Q)]\wedge[(Q)\Leftrightarrow(R)] \Downarrow (P)\Leftrightarrow(R)$
T	T	T	T	T	T	T	T
T	T	F	T	F	F	F	T
T	F	T	F	F	F	T	T
T	F	F	F	T	F	F	T
F	T	T	F	T	F	F	T
F	T	F	F	F	F	T	T
F	F	T	T	F	F	F	T
F	F	F	T	T	T	T	T

Example 106 Table 107 shows that the logical formula $\neg[(P) \wedge (Q)]$ has the same Truth table as $[\neg(P)] \vee [\neg(Q)]$; hence, $\{\neg[(P) \wedge (Q)]\} \Leftrightarrow [\neg(P)] \vee [\neg(Q)]$ is a tautology, called **de Morgan's first law.**

Table 107 De Morgan's first law.

P	Q	$\neg(P)$	$\neg(Q)$	$(P)\wedge(Q)$	$\neg[(P)\wedge(Q)]$	$[\neg(P)]\vee[\neg(Q)]$	$\neg[(P)\wedge(Q)] \Updownarrow [\neg(P)]\vee[\neg(Q)]$
T	T	F	F	T	F	F	T
T	F	F	T	F	T	T	T
F	T	T	F	F	T	T	T
F	F	T	T	F	T	T	T

The following examples demonstrate how to utilize the preceding considerations to prove further tautologies without Truth tables.

Example 108 The formula $[(P) \wedge (Q)] \Leftrightarrow (\neg\{[\neg(P)] \vee [\neg(Q)]\})$ is a tautology. Indeed, consider the following sequence of two equivalences:

$$\begin{array}{rl} (P) \wedge (Q) & \\ \Updownarrow & \text{by double negation (example 71),} \\ \neg\{\neg\,[(P) \wedge (Q)]\,\} & \\ \Updownarrow & \text{by de Morgan's first law (example 106).} \\ \neg\{\,[\neg(P)] \vee [\neg(Q)]\,\} & \end{array}$$

Example 104 now shows that $[(P) \wedge (Q)] \Leftrightarrow (\neg\{[\neg(P)] \vee [\neg(Q)]\})$ is a tautology.

Example 109 This example demonstrates the use of tautologies to prove **de Morgan's second law**, $\{\neg[(P) \vee (Q)]\} \Leftrightarrow \{[\neg(P)] \wedge [\neg(Q)]\}$:

$$\begin{array}{rl} [\neg(P)] \wedge [\neg(Q)] & \\ \Updownarrow & \text{by example 108 applied to } \neg(P) \text{ and } \neg(Q), \\ \neg(\,\{\neg[\neg(P)]\} \;\vee\; \{\neg[\neg(Q)]\}\,) & \\ \Updownarrow & \text{by substitutions (examples 96 and 98).} \\ \{\neg[(P) \vee (Q)]\} & \end{array}$$

0.4.2 Proofs by Contradictions

Besides proofs by Truth tables and by tautologies, some logical systems — including Boolean algebraic logic — also admit methods of proofs involving contradictions.

Example 110 One method of proof by contradiction establishes a proposition by proving that the negation of the proposition would imply a contradiction. Specifically, for every (False) contradiction F, and for every proposition R, the formula

$$(R) \Leftrightarrow \{[\neg(R)] \Rightarrow (F)\}$$

is a tautology, by table 111. In particular, if $[\neg(R)] \Rightarrow (F)$ is True, then R is True.

Table 111 Principle of proof by contradiction.

R	$\neg(R)$	F	$[\neg(R)] \Rightarrow (F)$	$(R) \Leftrightarrow \{[\neg(R)] \Rightarrow (F)\}$
T	F	F	T	T
F	T	F	F	T

The method of proof by **contradiction** — also called **reductio ad absurdum** — consists in proving that if the result R were False, then a "contradiction" (also called an "absurdity") would result; by contraposition, the result R must then be True. Specifically, the method consists in selecting a suitable proposition S to form a contradiction $(S) \wedge [\neg(S)]$ and then in proving the two logical implications

$$[\neg(R)] \Rightarrow (S),$$

$$[\neg(R)] \Rightarrow [\neg(S)].$$

The choice of S depends on the result R. If such a proposition S exists, then

$$[\neg(R)] \Rightarrow \{(S) \wedge [\neg(S)]\}.$$

Because the implication $[\neg(R)] \Rightarrow \{(S) \wedge [\neg(S)]\}$ is True but the conclusion $(S) \wedge [\neg(S)]$ is False, it follows that the hypothesis $\neg(R)$ must be False, and hence that R must be True. Thus, $(S) \wedge [\neg(S)]$ plays the rôle of the contradiction F in example 110.

In the particular case where R has the form of a logical implication $(P) \Rightarrow (Q)$, which is equivalent to $\neg\{(P) \wedge [\neg(Q)]\}$, then $\neg(R)$ becomes $(P) \wedge [\neg(Q)]$. In such a case, a particular way to prove $[\neg(R)] \Rightarrow \{(S) \wedge [\neg(S)]\}$ consists in proving both

$$(P) \Rightarrow (S),$$

$$[\neg(Q)] \Rightarrow [\neg(S)].$$

From their conjunction $\{(P) \Rightarrow (S)\} \wedge \{[\neg(Q)] \Rightarrow [\neg(S)]\}$ and the tautology

$$(\{(P) \Rightarrow (S)\} \wedge \{[\neg(Q)] \Rightarrow [\neg(S)]\}) \Rightarrow (\{(P) \wedge [\neg(Q)]\} \Rightarrow \{(S) \wedge [\neg(S)]\})$$

follows the conclusion $\{(P) \wedge [\neg(Q)]\} \Rightarrow \{(S) \wedge [\neg(S)]\}$. Hence $\neg\{(P) \wedge [\neg(Q)]\}$, which is equivalent to $(P) \Rightarrow (Q)$, is True by *reductio ad absurdum*.

Example 112 The present example proves the weak law of excluded middle,

$$\neg(\neg\{(P) \vee [\neg(P)]\}),$$

by contradiction. To this end, start with its negation:

$$\begin{array}{rl} \neg[\neg(\neg\{(P) \vee [\neg(P)]\})] & \\ \Updownarrow & \text{double negation,} \\ \neg\{(P) \vee [\neg(P)]\} & \\ \Updownarrow & \text{de Morgan's second law,} \\ [\neg(P)] \wedge \{\neg[\neg(P)]\} & \\ \Downarrow & \text{double negation and example 100,} \\ [\neg(P)] \wedge (P) & \end{array}$$

which is a contradiction. Therefore, $\neg(\neg\{(P) \vee [\neg(P)]\})$ is a tautology.

0.4.3 Exercises

Write propositional forms corresponding to the patterns of logical reasoning described in the following exercises. (The terminology is borrowed from Church [11, §15.9].)

Exercise 261 (*Reductio ad absurdum.*) Write a logical formula for the following sentence. "If the hypothesis implies the conclusion, and if the hypothesis implies the negation of the conclusion, then the hypothesis fails."

Exercise 262 (*Affirmation of the consequent.*) Write a logical formula for the following sentence. "If the conclusion holds, then the hypothesis implies the conclusion."

Exercise 263 (*Modus ponens.*) Write a logical formula for the following sentence. "If the hypothesis holds, and if the hypothesis implies the conclusion, then the conclusion holds."

Exercise 264 (*Modus tollens.*) Write a formula for the following sentence. "If the conclusion fails, and if the hypothesis implies the conclusion, then the hypothesis fails."

Exercise 265 (*Modus tollendo ponens.*) Write a logical formula for the following sentence. "If either the first or the second proposition holds, and if the first proposition fails, then the second proposition holds."

Exercise 266 (*Modus pollendo tonens.*) Write a logical formula for the following sentence. "If either the first or the second proposition holds, but not both, and if the first proposition holds, then the second proposition fails."

Exercise 267 (*Simple constructive dilemma.*) Write a logical formula for the following sentence. "If P then R, and if Q then R, and if P or Q, then R."

Exercise 268 (*Simple destructive dilemma.*) Write a logical formula for the following sentence. "If P then Q, and if Q then R, and if not Q or not R, then not P."

Exercise 269 (*Complex constructive dilemma.*) Write a logical formula for the following sentence. "If P then Q, and if R then S, and if P or R, then Q or S."

Exercise 270 (*Complex destructive dilemma.*) Write a formula for the following sentence. "If P then Q, and if R then S, and if not Q or not S, then not P or not R."

0.5 SYNTHESIS OF LOGICAL FORMULAE

0.5.1 Design of Logical Formulae with Specified Truth Values

Whereas the previous discussion has explained how to draw and analyze the Truth tables of given logical formulae, this discussion demonstrates the reverse process: how to design logical formulae with given Truth tables. The following examples will provide a basis for a general method. Specifically, the following examples provide logical formulae whose Truth tables contain each exactly one occurence of the value True, and the value False in all the other cases. Assembling such examples with disjunctions ($\vee$) will then produce logical formulae with any specified Truth tables.

Definition 113 (Minterm) A logical formula whose Truth table contains the value True exactly once is called a **minterm** (an abbreviation for "minimal term"). Table 114 shows all the minterms involving exactly two components.

Table 114 The four minterms with two components.

P	Q	$(P) \wedge (Q)$
T	T	**T**
T	F	F
F	T	F
F	F	F

P	Q	$(P) \wedge [\neg(Q)]$
T	T	F
T	F	**T**
F	T	F
F	F	F

P	Q	$[\neg(P)] \wedge (Q)$
T	T	F
T	F	F
F	T	**T**
F	F	F

P	Q	$[\neg(P)] \wedge [\neg(Q)]$
T	T	F
T	F	F
F	T	F
F	F	**T**

Example 115 Consider the problem of designing a yet unknown logical formula S with Truth values specified by the column labeled S in table 116. To this end, first identify the rows with the value True, here the first and last rows. For each such row, select the conjunction with the value True in this row only. Finally, connect the selected conjunctions by disjunctions, as shown in table 116. Thus, S can be the logical formula

Table 116 Design of a formula S with specified Truth values.

P	Q	S	$(P) \wedge (Q)$	$[\neg(P)] \wedge [\neg(Q)]$	$(P) \wedge (Q)$ $\vee$ $[\neg(P)] \wedge [\neg(Q)]$
T	T	**T**	**T**	F	**T**
T	F	F	F	F	F
F	T	F	F	F	F
F	F	**T**	F	**T**	**T**

$$[(P) \wedge (Q)] \vee \{[\neg(P)] \wedge [\neg(Q)]\},$$

which has the Truth values specified in the initial Truth table (column S in table 116).

The type of logical formula obtained in the foregoing examples is called a **disjunctive normal form,** which consists of a sequence of conjunctions ($\vee$) of smaller formulae that are all disjunctions ($\wedge$). In some instances, the logical formula obtained by the method just described simplifies further. For example, the logical formula

$$[(P) \wedge (Q)] \vee \{[\neg(P)] \wedge [\neg(Q)]\}$$

simplifies to

$$(P) \Leftrightarrow (Q).$$

Which of the two equivalent formulae proves "simpler" depends upon the context. In applications to the design of computers, for instance, the longer formula

$$[(P) \wedge (Q)] \vee \{[\neg(P)] \wedge [\neg(Q)]\}$$

may lend itself better to an implementation with "`OR, AND, NOT`" connectives than would the shorter formula $(P) \Leftrightarrow (Q)$.

Minterms also apply to formulae with more than two components.

Example 117 This example shows the design of a logical formula C with Truth values specified in table 118, with 1 instead of T, and with 0 instead of F.

Table 118 Design of a formula C with specified Truth values.

P	Q	R	C	Minterms
1	1	1	1	$(P) \wedge (Q) \wedge (R)$
1	1	0	1	$(P) \wedge (Q) \wedge [\neg(R)]$
1	0	1	1	$(P) \wedge [\neg(Q)] \wedge (R)$
1	0	0	0	
0	1	1	1	$[\neg(P)] \wedge (Q) \wedge (R)$
0	1	0	0	
0	0	1	0	
0	0	0	0	

The disjunctive normal form yields the following logical formula for C:

$$\{[\neg(P)] \wedge (Q) \wedge (R)\} \vee \{(P) \wedge [\neg(Q)] \wedge (R)\} \vee \{(P) \wedge (Q) \wedge [\neg(R)]\} \vee \{(P) \wedge (Q) \wedge (R)\}.$$

(The formula C represents the "carried" digit in the binary addition $P + Q + R$.)

0.5.2 Simplification by Distributivity and Excluded Middle

Whereas the preceding section has demonstrated how to form a logical formula satisfying a given Truth table, some situations require a subsequent simplification of the resulting logical formula. For instance, in the design of logical circuits a reduction of the number of connectives in the formula corresponds to a reduction in the number of electronic connectives (called "gates"), which then reduces the cost of the circuit.

One of the methods to simplify the disjunctive normal form of logical formulae consists in identifying pairs of conjunctions that differ from each other in only one component, such as $[(\neg(P)) \wedge (W)] \vee [(P) \wedge (W)]$. A simplification then results from distributivity and the three tautologies

$$\begin{array}{rl} (B) \vee [\neg(B)] & \text{(excluded middle)}, \\ [(T) \wedge (W)] \Leftrightarrow (W) & \text{(identity)}, \\ [(X) \vee (X)] \Leftrightarrow (X) & \text{(idempotency of } \vee). \end{array}$$

Example 119 The logical formula $[\neg(P)] \wedge (W)\} \vee [(P) \wedge (W)]$ simplifies as follows.

$$\begin{array}{rl} [\neg(P)] \wedge (W)\} \vee [(P) \wedge (W)] & \\ \Updownarrow & \text{by distributivity of } \wedge \text{ over } \vee, \\ \{[\neg(P)] \vee (P)\} \wedge (W) & \\ \Updownarrow & \text{by the law of excluded middle,} \\ (T) \wedge (W) & \\ \Updownarrow & \text{by the identity.} \\ W & \end{array}$$

Thus $\{[\neg(P)] \wedge (W)\} \vee [(P) \wedge (W)]$ simplifies to W.

Example 120 In the formula $\{[\neg(P)] \wedge (W)\} \vee [(P) \wedge (W)] \vee \{(P) \wedge [\neg(W)]\}$, the middle conjunction $[(P) \wedge (W)]$ can simplify with either of the other two conjunctions, but the idempotency of the disjunction enables both simplifications simultaneously, with the particular case $[(P) \wedge (W)] \vee [(P) \wedge (W)] \Leftrightarrow [(P) \wedge (W)]$:

$$\{[\neg(P)] \wedge (W)\} \vee [(P) \wedge (W)] \vee \{(P) \wedge [\neg(W)]\}$$

$\Updownarrow$ by idempotency applied to $(P) \wedge (W)$ instead of X,

$$\{[\neg(P)] \wedge (W)\} \vee [(P) \wedge (W)] \vee [(P) \wedge (W)] \vee \{(P) \wedge [\neg(W)]\}$$

$\Updownarrow$ by distributivity of $\wedge$ over $\vee$ twice,

$$(\{[\neg(P)] \vee (P)\} \wedge (W)) \vee ((P) \wedge \{[\neg(W)] \vee (W)\})$$

$\Updownarrow$ by the law of excluded middle twice,

$$\{(T) \wedge (W)\} \vee \{(P) \wedge (T)\}$$

$\Updownarrow$ by identity twice.

$$(W) \vee (P)$$

Thus $\{[\neg(P)] \wedge (W)\} \vee [(P) \wedge (W)] \vee \{(P) \wedge [\neg(W)]\}$ simplifies to $(W) \vee (P)$.

Example 121 Continuing example 117, This example illustrates the synthesis and the simplification of a logical formula C with Truth values specified in table 122 (repeating table 118). The disjunctive normal form yields the following logical formula for C:

Table 122 Design of a formula C with specified Truth values.

P	Q	R	C	Minterms
1	1	1	1	$(P) \wedge (Q) \wedge (R)$
1	1	0	1	$(P) \wedge (Q) \wedge [\neg(R)]$
1	0	1	1	$(P) \wedge [\neg(Q)] \wedge (R)$
1	0	0	0	
0	1	1	1	$[\neg(P)] \wedge (Q) \wedge (R)$
0	1	0	0	
0	0	1	0	
0	0	0	0	

$$\{[\neg(P)]\wedge(Q)\wedge(R)\} \vee \{(P)\wedge[\neg(Q)]\wedge(R)\} \vee \{(P)\wedge(Q)\wedge[\neg(R)]\} \vee \{(P)\wedge(Q)\wedge(R)\}.$$

By example 120, with $(P) \wedge (Q)$ instead of W, the last two conjunctions simplify as

$$(\{(P) \wedge (Q) \wedge [\neg(R)]\} \vee [(P) \wedge (Q) \wedge (R)]) \Leftrightarrow [(P) \wedge (Q)].$$

Similarly, with $(P) \wedge (R)$ instead of W, the second and last conjunctions simplify as

$$(\{(P) \wedge [\neg(Q)] \wedge (R)\} \vee [(P) \wedge (Q) \wedge (R)]) \Leftrightarrow [(P) \wedge (R)].$$

Finally, with $(Q) \wedge (R)$ instead of W, the first and last conjunctions simplify as

$$(\{[\neg(P)] \wedge (Q) \wedge (R)\} \vee [(P) \wedge (Q) \wedge (R)]) \Leftrightarrow [(Q) \wedge (R)].$$

Consequently, the disjunctive normal form for C simplifies to

$$[(P) \wedge (Q)] \vee [(P) \wedge (R)] \vee [(Q) \wedge (R)].$$

Moreover, a further simplification results from distributivity:

$$[(P) \wedge (Q)] \vee [(P) \wedge (R)] \vee [(Q) \wedge (R)]$$

$\Updownarrow$ by idempotency of the disjunction,

$$[(P) \wedge (Q)] \vee [(P) \wedge (R)] \vee [(P) \wedge (R)] \vee [(Q) \wedge (R)]$$

$\Updownarrow$ by distributivity of $\wedge$ over $\vee$ twice,

$$\{(P) \wedge [(Q) \vee (R)]\} \vee \{[(P) \vee (Q)] \wedge (R)\}$$

Thus, C becomes $\{(P) \wedge [(Q) \vee (R)]\} \vee \{[(P) \vee (Q)] \wedge (R)\}$.

The foregoing examples of simplifications of logical formulae through the idempotency of the disjunction demonstrate the practical usefulness of such a formula as $(A) \vee (A)$. In contrast, the comment

> "we can construct useless combinations, for example, $A \wedge A \wedge A \ldots \wedge A$"

[22, p. 180] again shows that the foundations of logic are not yet completely settled.

There exist other methods of simplification. For general formulae, the algorithm of McCluskey yields a general method suited for automated simplification by computers [76]. For small logical formulae, the graphical tables of Karnaugh provide a fast visual method [56], as explained in section 0.7.

0.5.3 Exercises

Exercise 271 Design a logical formula S with Truth values specified in table 123.

Table 123 Specified Truth values for S.

P	Q	S
T	T	F
T	F	T
F	T	F
F	F	T

Exercise 272 Design a logical formula U with Truth values specified in table 124.

Exercise 273 Consider three propositional variables, I, J, K, each of which may take the logical value True or False. Design a "consensus" propositional form C, which takes the value True if and only if all three variables I, J, K, are True.

Exercise 274 Consider four propositional variables, I, J, K, L, each of which may take the logical value True or False. Design a "consensus" propositional form C, which takes the value True if and only if all four variables I, J, K, L are True.

Table 124 Specified Truth values for U.

P	Q	U
T	T	T
T	F	F
F	T	T
F	F	T

Exercise 275 Consider two propositional variables, I and J, each of which may take the logical value True or False. Design a "tie" propositional form P, which detects an equal number of True and False values. In other words, the formula P is True if and only if any one of the two variables is True and the other is False.

Exercise 276 Consider four propositional variables, I, J, K, L, each of which may take the logical value True or False. Design a "tie" propositional form P, which detects an equal number of True and False values. In other words, P has the logical value True if and only if exactly two among I, J, K, L have the value True and exactly two among I, J, K, L have the value False.

Exercise 277 Consider three propositional variables, I, J, K, each of which may take the logical value True or False. Design a "majority" propositional form M, which takes the value True if and only if at least two of the three variables I, J, K, are True. Thus, the formula M is False if and only if more than one of the three variables are False.

Exercise 278 Consider five propositional variables, I, J, K, L, M, each of which may take the value True or False. Design a "supermajority" logical formula S, which takes the value True if and only if at least four of the five variables I, J, K, L, M are True. Thus, S is False if and only if more than one of the five variables are False.

Exercise 279 Consider four propositional variables, I, J, K, L, each of which may take the logical value True or False. Design a "majority" propositional form M, which takes the value True if and only if at least three of the four variables I, J, K, L are True. Thus, M is False if and only if more than one of the four variables are False.

Exercise 280 Consider six propositional variables, I, J, K, L, M, N, each of which may take the value True or False. Design a "supermajority" logical formula S, which takes the value True if and only if at least four of the six variables I, J, K, L, M, N are True. Thus, S is False if and only if more than two of the six variables are False.

0.6 OTHER CONNECTIVES AND APPLICATIONS

0.6.1 Other Logical Connectives

Besides the logical connectives $\neg$, $\Rightarrow$, $\wedge$, $\vee$, and $\Leftrightarrow$, which are common in mathematics, other logical connectives prove useful in such contexts as logic and computer science [24], for example, the connective "NOR" (for "neither ... nor"), also denoted by $\downarrow$.

Definition 125 The connective NOR is defined so that $(P)\texttt{NOR}(Q)$ has the same Truth values as $[\neg(P)] \wedge [\neg(Q)]$, which corresponds to the phrase "neither P nor Q". Thus,

$$[(P)\texttt{NOR}(Q)] \Leftrightarrow \{[\neg(P)] \wedge [\neg(Q)]\}$$

is a tautology. Equivalently, by de Morgan's second law,

$$\{[\neg(P)] \wedge [\neg(Q)]\} \Leftrightarrow \{\neg[(P) \vee (Q)]\},$$

so that $[(P)\text{NOR}(Q)] \Leftrightarrow \{\neg[(P) \vee (Q)]\}$, corresponding to "not 'P or Q' ".

Example 126 Truth table 127 for $(P)\text{NOR}(Q)$ uses the equivalent $[\neg(P)] \wedge [\neg(Q)]$.

Table 127 The connective NOR.

P	Q	$\neg(P)$	$\neg(Q)$	$[\neg(P)] \wedge [\neg(Q)]$	$(P)\text{NOR}(Q)$
T	T	F	F	F	F
T	F	F	T	F	F
F	T	T	F	F	F
F	F	T	T	T	T

Some logical connectives — for instance, NOR — can combine so as to play the role of every logical connective.

Definition 128 A logical connective is called **primitive**, or also **universal**, if and only if it can replace all the other connectives.

Universal logical connectives are useful in the design of logical circuits, because an ample supply of only one type of universal "gates" suffices to wire any logical circuit. For instance, the connective NOR can replace all the other logical connectives, $\neg$, $\wedge$, $\vee$, $\Rightarrow$, and $\Leftrightarrow$, as demonstrated by the following examples.

Example 129 Substituting P for Q in the definition of NOR confirms that $\neg(P)$ and $(P)\text{NOR}(P)$ have the same Truth table, as shown in table 130.

Table 130 The equivalence of $\neg(P)$ and $(P)\text{NOR}(P)$.

P	P	$\neg(P)$	$\neg(P)$	$(P)\text{NOR}(P)$	$[\neg(P)] \Leftrightarrow [(P)\text{NOR}(P)]$
T	T	F	F	F	T
F	F	T	T	T	T

The verification of the equivalence $[\neg(P)] \Leftrightarrow [(P)\text{NOR}(P)]$ can also use tautologies:

$$\begin{array}{rl} \neg(P) & \\ \Updownarrow & \text{by substitution in the tautology } (R) \Leftrightarrow [(R) \wedge (R)], \\ [\neg(P)] \wedge [\neg(P)] & \\ \Updownarrow & \text{by definition of NOR.} \\ (P)\text{NOR}(P) & \end{array}$$

Example 131 This example shows that NOR can replace $\wedge$, in the sense that

$$[(P) \wedge (Q)] \Leftrightarrow \{\ [(P)\text{NOR}(P)]\ \text{NOR}\ [(Q)\text{NOR}(Q)]\ \}.$$

Indeed,

$$\begin{array}{rl} (P) \wedge (Q) & \\ \Updownarrow & \text{by substitutions in } (R) \Leftrightarrow \{\neg[\neg(R)]\}, \\ \{\neg[\neg(P)]\} \wedge \{\neg[\neg(Q)]\} & \\ \Updownarrow & \text{by substitutions in the definition of NOR,} \\ [\neg(P)]\text{NOR}[\neg(Q)] & \\ \Updownarrow & \text{by example 129.} \\ \{\ [(P)\text{NOR}(P)]\ \text{NOR}\ [(Q)\text{NOR}(Q)]\ \} & \end{array}$$

Example 132 This example shows that NOR can replace $\vee$, in the sense that

$$\{(P) \vee (Q)\} \Leftrightarrow \{[(P)\text{NOR}(Q)]\text{NOR}[(P)\text{NOR}(Q)]\}.$$

Indeed,

$$\begin{array}{rl} (P) \vee (Q) & \\ \Updownarrow & \text{by de Morgan's laws and double negation,} \\ \neg\{[\neg(P)] \wedge [\neg(Q)]\} & \\ \Updownarrow & \text{by definition of NOR,} \\ \neg[(P)\text{NOR}(Q)] & \\ \Updownarrow & \text{by substitution in example 129.} \\ \{\ [(P)\text{NOR}(Q)]\ \text{NOR}\ [(P)\text{NOR}(Q)]\ \} & \end{array}$$

Example 133 Table 134 shows that the logical connective NOR commutes.

Table 134 The logical connective NOR commutes.

P	Q	$(P)\text{NOR}(Q)$	$(Q)\text{NOR}(P)$	$[(P)\text{NOR}(Q)] \Leftrightarrow [(Q)\text{NOR}(P)]$
T	T	F	F	T
T	F	F	F	T
F	T	F	F	T
F	F	T	T	T

Example 135 In contrast to the logical connectives $\wedge$ and $\vee$, which *are* associative, the logical connective NOR is *not* associative. In other words, the formulae

$$[(P)\text{NOR}(Q)]\text{NOR}(R),$$

$$(P)\text{NOR}[(Q)\text{NOR}(R)],$$

have *different* Truth tables. Indeed, their Truth tables have *different* columns in the row where P and Q are both True but R is False, as shown in table 136.

Table 136 The logical connective NOR is *not* associative.

P	Q	R	$[(P)\text{NOR}(Q)]\text{NOR}(R)$	$(P)\text{NOR}[(Q)\text{NOR}(R)]$
⋮	⋮	⋮	⋮	⋮
T	T	F	**T**	**F**
⋮	⋮	⋮	⋮	⋮

Yet another connective, "NAND" (for "not 'and' "), also denoted by "Sheffer's stroke" | in some texts [11, p. 134, n. 207], has properties similar to those of NOR.

Definition 137 The logical connective NAND is defined so that $(P)\text{NAND}(Q)$ has the same Truth values as $\neg[(P) \wedge (Q)]$ has. In other words, the formula

$$[(P)\text{NAND}(Q)] \Leftrightarrow \{\neg[(P) \wedge (Q)]\}$$

is a tautology. Equivalently, by de Morgan's first law,

$$\{\neg[(P) \wedge (Q)]\} \Leftrightarrow \{[\neg(P)] \vee [\neg(Q)]\},$$

it follows that $[(P)\text{NAND}(Q)] \Leftrightarrow \{[\neg(P)] \vee [\neg(Q)]\}$ is also a tautology.

Definition 138 The logical connective "exclusive 'or' " is abbreviated by "XOR" or by $\oplus$ and defined so that

$$[(P) \oplus (Q)] \Leftrightarrow ([(P) \vee (Q)] \wedge \{\neg[(P) \wedge (Q)]\})$$

is a tautology. Thus $[(P)\text{XOR}(Q)] \Leftrightarrow [(P) \oplus (Q)]$ is also a tautology.

Properties of the logical connectives XOR and NAND form the object of exercises.

0.6.2 Logical Connectives in Binary Arithmetic

As demonstrated by Claude Elwood Shannon [86], logical connectives help in the design of electronic circuits that perform logical operations. In such circuits, the logical values True and False then correspond to two electrical potentials, for instance, 1 and 0 respectively. The notation with 1 and 0 eases the transition from logic to arithmetic.

Example 139 The logical connective $\wedge$ corresponds to the ordinary multiplication $*$ of two numbers each equal to 0 or 1, as shown in table 140.

Table 140 The connective $\wedge$ corresponds to multiplication.

P	Q	$(P) * (Q)$	$(P) \wedge (Q)$	$[(P) * (Q)] \Leftrightarrow [(P) \wedge (Q)]$
1	1	1	1	T
1	0	0	0	T
0	1	0	0	T
0	0	0	0	T

Example 141 Consider the addition $P + Q = VW$ of two binary numbers P and Q, each equal to 0 or 1. The sum VW includes one or two binary digits, V and W, each equal to 0 or 1. The first digit, V, corresponds to the logical connective $\wedge$, as shown in table 142. For

Table 142 Addition of two binary digits.

P	Q	$(P) + (Q)$ $=$ VW	V $\Updownarrow$ $[(P) \wedge (Q)]$	W $\Updownarrow$ $[(P)\text{XOR}(Q)]$
1	1	10	1	0
1	0	01	0	1
0	1	01	0	1
0	0	00	0	0

the second digit W, the method of minterms gives

$$(W) \Leftrightarrow (\{(P) \wedge [\neg(Q)]\} \vee \{[\neg(P)] \wedge (Q)\}).$$

The digit W also corresponds to the connective XOR, as shown in the exercises.

0.6.3 Exercises

The following exercise completes the proof of the universality of the connective NOR.

Exercise 281 Through tautologies, Truth tables, or otherwise, find a logical formula equivalent to $(P) \Rightarrow (Q)$ but involving only the symbols P and Q and the sole logical connective NOR, with repetitions allowed.

The following exercises establish the universality of NAND (see definition 137).

Exercise 282 Determine whether the logical connective NAND is associative.

Exercise 283 Write the Truth table for $(P)\text{NAND}(Q)$.

Exercise 284 Verify that $\neg(P)$ and $(P)\text{NAND}(P)$ have the same Truth table.

Exercise 285 Identify a sequence of tautologies justifying all the following steps:

$$[\neg(P)] \Leftrightarrow \{\neg[(P) \wedge (P)]\} \Leftrightarrow [(P)\text{NAND}(P)]$$

Exercise 286 Either through tautologies or with Truth tables, verify that

$$[(P) \wedge (Q)] \Leftrightarrow \{[(P)\text{NAND}(Q)]\text{NAND}[(P)\text{NAND}(Q)]\}$$

Exercise 287 Through tautologies, Truth tables, or otherwise, find a logical formula equivalent to $(P) \vee (Q)$ but involving only the symbols P and Q and the sole logical connective NAND, with repetitions allowed.

Exercise 288 Through tautologies, Truth tables, or otherwise, find a logical formula equivalent to $(P) \Rightarrow (Q)$ but involving only the symbols P and Q and the sole logical connective NAND, with repetitions allowed.

The following exercises investigate relations between various connectives.

Exercise 289 Determine whether NAND distributes over NOR.

Exercise 290 Determine whether NOR distributes over NAND.

Exercise 291 Express $(P) \oplus (Q)$ as a formula built from P, Q, and F1 – F7.

Exercise 292 Verify that $\oplus$ commutes.

Exercise 293 Determine which of $(P) \oplus (\neg(P))$ and $(P) \oplus (P)$ is a tautology.

Exercise 294 Determine whether $\oplus$ is associative.

Exercise 295 Determine whether NAND distributes over XOR.

Exercise 296 Determine whether XOR distributes over NAND.

Exercise 297 Determine whether NOR distributes over XOR.

Exercise 298 Determine whether XOR distributes over NOR.

The following exercises establish the universality of $\neg$ and $\Rightarrow$ together.

Exercise 299 Verify that $[(P) \Rightarrow (Q)] \Leftrightarrow \{[\neg(P)] \vee (Q)\}$ is a tautology.

Exercise 300 Find a logical formula equivalent to $(P) \vee (Q)$ but in which only the connectives $\neg$ and $\Rightarrow$ may appear (as many times as necessary).

Exercise 301 Find a logical formula equivalent to $(P) \wedge (Q)$ but in which only the connectives $\neg$ and $\Rightarrow$ may appear (as many times as necessary).

Exercise 302 Verify that $[(P) \Leftrightarrow (Q)] \Leftrightarrow \{[(P) \Rightarrow (Q)] \wedge [(Q) \Rightarrow (P)]\}$ is a tautology.

Exercise 303 Find a logical formula equivalent to $(P) \Leftrightarrow (Q)$ but in which only the connectives $\neg$ and $\Rightarrow$ may appear (as many times as necessary).

Exercise 304 Find a logical formula equivalent to $(P) \Leftarrow (Q)$ but in which only the connectives $\neg$ and $\Rightarrow$ may appear (as many times as necessary).

Exercise 305 Find a logical formula equivalent to the minterm $(P) \wedge [\neg(Q)]$ but in which only the connectives $\neg$ and $\Rightarrow$ may appear (as many times as necessary).

Exercise 306 Find a logical formula equivalent to the minterm $[\neg(P)] \wedge (Q)$ but in which only the connectives $\neg$ and $\Rightarrow$ may appear (as many times as necessary).

Exercise 307 Find a logical formula equivalent to $(P)\text{NAND}(Q)$ but in which only the connectives $\neg$ and $\Rightarrow$ may appear (as many times as necessary).

Exercise 308 Find a logical formula equivalent to $(P)\text{NOR}(Q)$ but in which only the connectives $\neg$ and $\Rightarrow$ may appear (as many times as necessary).

Exercise 309 Find a logical formula equivalent to $(P)\text{XOR}(Q)$ but in which only the connectives $\neg$ and $\Rightarrow$ may appear (as many times as necessary).

Exercise 310 Determine whether $\Rightarrow$ distributes over $\Rightarrow$.

0.7 SYNTHESIS BY KARNAUGH TABLES

0.7.1 Karnaugh Tables

Rearrangements of Truth tables can help in the simplification of logical formulae containing only a few logical variables, for instance, with Marquand tables [101, p. 139–140 & Ch. XIV, p. 360–379], or with Karnaugh tables [56]. Karnaugh tables offer the advantage of displaying in adjacent locations minterms that differ from each other by only one Truth value. For a formula S that contains only one logical component, P, the Karnaugh table contains two cells, marked P or $\neg(P)$ repectively, as in table 143.

Table 143 Template

P	
$\neg(P)$	

- The cell marked P will show the Truth value of the formula S if P is True.
- The cell marked $\neg(P)$ will show the Truth value of S if $\neg(P)$ is True.

With only one logical component, P, there are only four non-equivalent logical formulae: P, $\neg(P)$, $(P) \vee [\neg(P)]$ (or, equivalently, **1**), $(P) \wedge [\neg(P)]$ (or, equivalently, **0**). Only one of these four formulae will match the Truth values in the Karnaugh table.

Example 144 This example shows the design of a yet unknown logical formula S with specified Truth values, first with a Truth table, then with a Karnaugh table.

Table 145 Truth table and Karnaugh table for S.

Truth table for S.

P	S	Minterms
1	**0**	
0	**1**	$\neg(P)$

Karnaugh table for S.

P	**0**
$\neg(P)$	**1**

With the method of Truth tables, table 145 contains only one occurence of the value 1 for the yet unknown formula S, in the row marked $\neg(P)$. Therefore, $\neg(P)$ is the logical formula for this Truth table. Similarly, the Karnaugh table contains two cells, one cell where P is True, and one cell where $\neg(P)$ is True. In each cell the Karnaugh table shows the Truth values of the yet unknown formula. The Karnaugh table contains only one occurence of the value 1, in the cell marked $\neg(P)$. Therefore, the logical formula for this Truth table consists of the single term $\neg(P)$.

Karnaugh tables for two logical components include one cell for each minterm, as in table 146. A Truth value within a cell will indicate whether to include the cell's minterm in the yet unknown formula. In the Karnaugh table, the simplifying feature is that within each row,

Table 146 Template for Karnaugh tables.

	Q	$\neg(Q)$
P	$(P) \wedge (Q)$	$(P) \wedge [\neg(Q)]$
$\neg(P)$	$[\neg(P)] \wedge (Q)$	$[\neg(P)] \wedge [\neg(Q)]$

and within each column, each entry differs from each adjacent entry by only *one* change in the Truth values of the components. Thus two adjacent entries can change from P to $\neg(P)$, with all the other components unchanged, thus allowing for the elimination of P and $\neg(P)$ by distributivity of $\wedge$ over $\vee$.

Example 147 The Truth table and Karnaugh table shown in table 148 correspond both to the same yet unknown logical formula U.
With the method of Truth tables, minterms lead to the logical formula

$$\{(P) \wedge [\neg(Q)]\} \vee \{[\neg(P)] \wedge [\neg(Q)]\},$$

which the distributivity of $\wedge$ over $\vee$ simplifies to

$$\{(P) \vee [\neg(P)]\} \wedge [\neg(Q)].$$

Because $(P) \vee [\neg(P)]$ is a tautology, the formula simplifies further to $\neg(Q)$. Correspondingly, the Karnaugh table shows the value 1 only in the entire column where $\neg(Q)$ is True, regardless of P and $\neg(P)$. Thus, the formula is $\neg(Q)$.

Table 148 Truth table and Karnaugh table for U.

Truth table for U.

P	Q	U	Minterms
1	1	**0**	
1	0	**1**	$(P) \wedge [\neg(Q)]$
0	1	**0**	
0	0	**1**	$[\neg(P)] \wedge [\neg(Q)]$

Karnaugh table for U.

	Q	$\neg(Q)$
P	**0**	**1**
$\neg(P)$	**0**	**1**

Table 149 Template for Karnaugh tables with three components.

	Q	Q	$\neg(Q)$	$\neg(Q)$
P	$(P)\wedge(Q)\wedge(R)$	$(P)\wedge(Q)\wedge[\neg(R)]$	$(P)\wedge[\neg(Q)]\wedge[\neg(R)]$	$(P)\wedge[\neg(Q)]\wedge(R)$
$\neg(P)$	$[\neg(P)]\wedge(Q)\wedge(R)$	$[\neg(P)]\wedge(Q)\wedge[\neg(R)]$	$[\neg(P)]\wedge[\neg(Q)]\wedge[\neg(R)]$	$[\neg(P)]\wedge[\neg(Q)]\wedge(R)$
	R	$\neg(R)$	$\neg(R)$	R

Karnaugh tables for three components have a cell for each minterm, as in table 149.

Example 150 This example demonstrates the use of Karnaugh tables to *simplify* an existing logical formula *defined by minterms*, for instance, the following formula:

$$\begin{array}{ll} & \{[\neg(P)] \wedge (Q) \wedge (R)\} \\ \vee & \{(P) \wedge [\neg(Q)] \wedge (R)\} \\ \vee & \{(P) \wedge (Q) \wedge [\neg(R)]\} \\ \vee & [(P) \wedge (Q) \wedge (R)] \end{array}$$

For each minterm in the given formula, identify its location in the template in table 149, and insert the value **1** at every such location.

	Q	Q	$\neg(Q)$	$\neg(Q)$
P	**1**	**1**	**0**	**1**
$\neg(P)$	**1**	**0**	**0**	**0**
	R	$\neg(R)$	$\neg(R)$	R

The *first column* contains two adjacent occurrences of **1**, corresponding to $(P)\wedge(Q)\wedge(R)$ or $[\neg(P)]\wedge(Q)\wedge(R)$, which simplifies to $(Q) \wedge (R)$.

	Q	Q	$\neg(Q)$	$\neg(Q)$
P	**1**			
$\neg(P)$	**1**			
	R	$\neg(R)$	$\neg(R)$	R

The *leftmost* two entries in the *first row* contain two adjacent occurrences of **1**, corresponding to $(P)\wedge(Q)\wedge(R)$ or $(P)\wedge(Q)\wedge[\neg(R)]$, which simplifies to $(P)\wedge(Q)$.

	Q	Q	$\neg(Q)$	$\neg(Q)$
P	**1**	**1**		
$\neg(P)$				
	R	$\neg(R)$	$\neg(R)$	R

The *extreme left* and *extreme right* entries in the *first row* contain two occurrences of **1**; these extreme entries would be adjacent after a rearrangement of the Karnaugh table; they correspond to $(P) \wedge (Q) \wedge (R)$ or $(P) \wedge [\neg(Q)] \wedge (R)$, which simplifies to $(P) \wedge (R)$.

	Q	Q	$\neg(Q)$	$\neg(Q)$
P	**1**			**1**
$\neg(P)$				
	R	$\neg(R)$	$\neg(R)$	R

Thus, the initial formula simplifies to $[(Q) \wedge (R)] \vee [(P) \wedge (Q)] \vee [(P) \wedge (R)]$.

Example 151 This example shows how to *design* a logical formula with Truth values specified by a Karnaugh table.

In this example, all the occurrences of **1** fill the *entire first row,* including all the cells where P is True. Therefore, the corresponding formula is P.

	Q	Q	$\neg(Q)$	$\neg(Q)$
P	**1**	**1**	**1**	**1**
$\neg(P)$	**0**	**0**	**0**	**0**
	R	$\neg(R)$	$\neg(R)$	R

Example 152 This example also shows how to design a logical formula with Truth values specified by a Karnaugh table.

In this example, all the occurrences of **1** fill the *first and last columns,* including all the cells where R is True. (These cells would be adjacent after a rearrangement of the Karnaugh table.) Therefore, the corresponding formula is R.

	Q	Q	$\neg(Q)$	$\neg(Q)$
P	**1**	**0**	**0**	**1**
$\neg(P)$	**1**	**0**	**0**	**1**
	R	$\neg(R)$	$\neg(R)$	R

Example 153 This example shows how to design a logical formula specified by a Karnaugh table with *four* components.

This Karnaugh table shows the value 1 exactly in all the cells where all three variables P, Q, and R are True. Therefore, the corresponding formula is $(P) \wedge (Q) \wedge (R)$.

	Q	Q	$\neg(Q)$	$\neg(Q)$	
H	**1**	**0**	**0**	**0**	P
H	**0**	**0**	**0**	**0**	$\neg(P)$
$\neg(H)$	**0**	**0**	**0**	**0**	$\neg(P)$
$\neg(H)$	**1**	**0**	**0**	**0**	P
	R	$\neg(R)$	$\neg(R)$	R	

Example 154 This example also shows how to design *and simplify* a logical formula specified by a Karnaugh table with four components.

This Karnaugh table shows the value **1** exactly in all the cells for $\neg(P)$ and $\neg(R)$ both True. Therefore, the corresponding formula is $[\neg(P)] \wedge [\neg(R)]$, which, by de Morgan's law, is logically equivalent to $\neg[(P) \vee (R)]$.

	Q	Q	$\neg(Q)$	$\neg(Q)$	
H	**0**	**0**	**0**	**0**	P
H	**0**	**1**	**1**	**0**	$\neg(P)$
$\neg(H)$	**0**	**1**	**1**	**0**	$\neg(P)$
$\neg(H)$	**0**	**0**	**0**	**0**	P
	R	$\neg(R)$	$\neg(R)$	R	

Example 155 This example shows how to design and simplify a logical formula V with Truth values specified by a Truth table, for instance, table 156. The corresponding Karnaugh

Table 156 Specified Truth table for V, and corresponding Karnaugh table.

H	P	Q	R	V
1	1	1	1	**1**
1	1	1	0	**0**
1	1	0	1	**1**
1	1	0	0	**0**
1	0	1	1	**0**
1	0	1	0	**0**
1	0	0	1	**0**
1	0	0	0	**0**
0	1	1	1	**1**
0	1	1	0	**0**
0	1	0	1	**1**
0	1	0	0	**0**
0	0	1	1	**0**
0	0	1	0	**0**
0	0	0	1	**0**
0	0	0	0	**0**

	Q	Q	$\neg(Q)$	$\neg(Q)$	
H	**1**	**0**	**0**	**1**	P
H	**0**	**0**	**0**	**0**	$\neg(P)$
$\neg(H)$	**0**	**0**	**0**	**0**	$\neg(P)$
$\neg(H)$	**1**	**0**	**0**	**1**	P
	R	$\neg(R)$	$\neg(R)$	R	

table for V contains the value **1** only in its corners, which is where P and R are True, regardless of H, $\neg(H)$, Q, and $\neg(Q)$. Therefore, the formula is $(P) \wedge (R)$.

0.7.2 Exercises

Table 157 Values for A, B, C, D.

H	P	Q	R	A	B	C	D
1	1	1	1	1	0	0	1
1	1	1	0	0	1	1	0
1	1	0	1	0	0	1	1
1	1	0	0	0	0	0	0
1	0	1	1	0	1	1	0
1	0	1	0	0	1	0	0
1	0	0	1	0	0	1	0
1	0	0	0	0	0	0	0
0	1	1	1	0	0	1	1
0	1	1	0	0	0	1	0
0	1	0	1	0	0	0	1
0	1	0	0	0	0	0	0
0	0	1	1	0	0	0	0
0	0	1	0	0	0	0	0
0	0	0	1	0	0	0	0
0	0	0	0	0	0	0	0

Values for P, Q.

K	L	M	N	P	Q
1	1	1	1	0	1
1	1	1	0	0	1
1	1	0	1	1	1
1	1	0	0	0	0
1	0	1	1	0	0
1	0	1	0	0	1
1	0	0	1	1	0
1	0	0	0	0	0
0	1	1	1	0	0
0	1	1	0	0	0
0	1	0	1	0	1
0	1	0	0	0	0
0	0	1	1	0	0
0	0	1	0	0	0
0	0	0	1	0	0
0	0	0	0	0	0

Exercise 311 As specified in table 157, find logical formulae for each of A, B, C, D in terms of H, P, Q, R, with any of the logical connectives $\neg$, $\wedge$, $\vee$, $\Rightarrow$, or $\Leftrightarrow$.

Exercise 312 As specified in table 157, find logical formulae for each of P, Q in terms of K, L, M, N, with any of the logical connectives $\neg$, $\wedge$, $\vee$, $\Rightarrow$, or $\Leftrightarrow$.

For each of the following exercises, find a logical formula with the fewest connectives that has the corresponding Karnaugh table.

Exercise 313

P	**1**
$\neg(P)$	**1**

Exercise 314

P	**0**
$\neg(P)$	**0**

Exercise 315

	Q	$\neg(Q)$
P	**1**	**1**
$\neg(P)$	**0**	**0**

Exercise 316

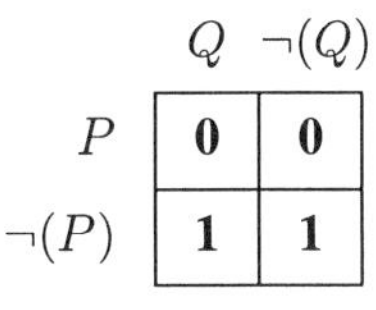

	Q	$\neg(Q)$
P	**0**	**0**
$\neg(P)$	**1**	**1**

Exercise 317

	Q	$\neg(Q)$
P	**0**	**1**
$\neg(P)$	**1**	**0**

Exercise 318

	Q	$\neg(Q)$
P	**1**	**0**
$\neg(P)$	**0**	**1**

Exercise 319

	Q	$\neg(Q)$
P	**1**	**1**
$\neg(P)$	**1**	**1**

Exercise 320

	Q	$\neg(Q)$
P	**0**	**0**
$\neg(P)$	**0**	**0**

Exercise 321

	Q	Q	$\neg(Q)$	$\neg(Q)$
P	**1**	**1**	**0**	**0**
$\neg(P)$	**0**	**0**	**1**	**1**
	R	$\neg(R)$	$\neg(R)$	R

Exercise 322

	Q	Q	$\neg(Q)$	$\neg(Q)$
P	**0**	**1**	**1**	**0**
$\neg(P)$	**0**	**1**	**1**	**0**
	R	$\neg(R)$	$\neg(R)$	R

Exercise 323

	Q	Q	$\neg(Q)$	$\neg(Q)$
P	**0**	**1**	**0**	**1**
$\neg(P)$	**1**	**0**	**0**	**1**
	R	$\neg(R)$	$\neg(R)$	R

Exercise 324

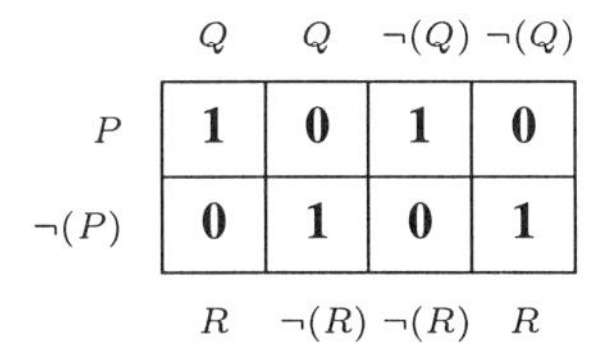

	Q	Q	$\neg(Q)$	$\neg(Q)$
P	**1**	**0**	**1**	**0**
$\neg(P)$	**0**	**1**	**0**	**1**
	R	$\neg(R)$	$\neg(R)$	R

Exercise 325

	Q	Q	$\neg(Q)$	$\neg(Q)$	
H	**0**	**0**	**0**	**0**	P
H	**0**	**0**	**0**	**0**	$\neg(P)$
$\neg(H)$	**1**	**1**	**1**	**1**	$\neg(P)$
$\neg(H)$	**1**	**1**	**1**	**1**	P
	R	$\neg(R)$	$\neg(R)$	R	

Exercise 326

	Q	Q	$\neg(Q)$	$\neg(Q)$	
H	**0**	**1**	**1**	**0**	P
H	**0**	**0**	**0**	**0**	$\neg(P)$
$\neg(H)$	**0**	**0**	**0**	**0**	$\neg(P)$
$\neg(H)$	**0**	**1**	**1**	**0**	P
	R	$\neg(R)$	$\neg(R)$	R	

Exercise 327

	Q	Q	$\neg(Q)$	$\neg(Q)$	
H	**1**	**1**	**0**	**0**	P
H	**0**	**0**	**1**	**1**	$\neg(P)$
$\neg(H)$	**1**	**1**	**0**	**0**	$\neg(P)$
$\neg(H)$	**0**	**0**	**1**	**1**	P
	R	$\neg(R)$	$\neg(R)$	R	

Exercise 328

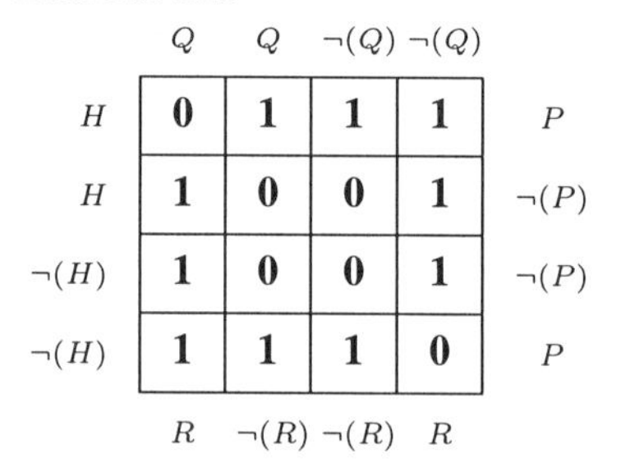

	Q	Q	$\neg(Q)$	$\neg(Q)$	
H	**0**	**1**	**1**	**1**	P
H	**1**	**0**	**0**	**1**	$\neg(P)$
$\neg(H)$	**1**	**0**	**0**	**1**	$\neg(P)$
$\neg(H)$	**1**	**1**	**1**	**0**	P
	R	$\neg(R)$	$\neg(R)$	R	

Exercise 329

	Q	Q	$\neg(Q)$	$\neg(Q)$	
H	**0**	**0**	**0**	**0**	P
H	**0**	**1**	**1**	**0**	$\neg(P)$
$\neg(H)$	**1**	**0**	**0**	**1**	$\neg(P)$
$\neg(H)$	**0**	**0**	**0**	**0**	P
	R	$\neg(R)$	$\neg(R)$	R	

Exercise 330

	Q	Q	$\neg(Q)$	$\neg(Q)$	
H	**1**	**0**	**0**	**0**	P
H	**0**	**1**	**0**	**0**	$\neg(P)$
$\neg(H)$	**0**	**0**	**1**	**0**	$\neg(P)$
$\neg(H)$	**0**	**0**	**0**	**1**	P
	R	$\neg(R)$	$\neg(R)$	R	

0.8 AN APPLICATION TO CIRCUITS

A compendium of several authors' contributions edited by Murray S. Klamkin ([61, p. 108–111], [62, p. 460–463]) includes the statement and several solutions for the following problem, posed by Raphael Miller (at the Hermes Electronic Corporation).

Applied problem 158 (Minimum Switching Circuit) The problem consists in designing a "minimum switching circuit" to open a valve if and only if the data flowing into the circuit lie within a specified range. The data consist of seven binary variables $HIJKLMN$, each of which can take only either of the values 0 and 1. Moreover, the circuit must open the valve only for the values of the data listed in table 159.

The first part of the problem amounts to designing a logical formula F with value 1 for the data in table 159, and 0 for all the other combinations of $HIJKLMN$. The second part consists in simplifying the formula F so that it contains only negations and the smallest number of conjunctions and disjunctions (to be implemented with diodes for AND and OR

Table 159 Design of a formula F with the value 1 only in the following cases.

H	I	J	K	L	M	N	F
0	0	0	1	0	0	0	1
1	0	0	1	0	0	0	1
0	1	0	1	0	0	0	1
1	1	0	1	0	0	0	1
0	0	1	1	0	0	0	1
1	0	1	1	0	0	0	1
0	1	1	1	0	0	0	1

gates). The last part consists in proving that there is no shorter logically equivalent formula. Raphael Miller's first and second solutions contained 66 and 15 diodes respectively. A. H. McMorris (University of Houston) reduced the number of diodes to 9. Layton E. Butts (System Laboratories) later supplied a solution with only 8 diodes. Finally, A. H. McMorris proved that 8 was the minimum number of diodes. For the circuit manufacturer, the reduction from 66 to 8 represents a saving of 88% in the number of electronic gates, and the proof of the minimality of 8 means that any additonal research into further reductions would be wasted.

0.9 PROJECTS

Project 1 Determine whether NAND and NOR are the *only* universal binary logical connectives. (See also [88, p. 14]).

Project 2 Design a logical formula F with the values specified in table 159. Then reduce it to an equivalent formula with six negations ($\neg$) and a total of eight disjunctions and conjunctions ($\vee, \wedge$). Finally, prove, or locate and study a proof, that there is no shorter logically equivalent formula. (See also [61, p. 108–111], [62, p. 460–463]).

Chapter 1

Logic and Deductive Reasoning

1.0 INTRODUCTION

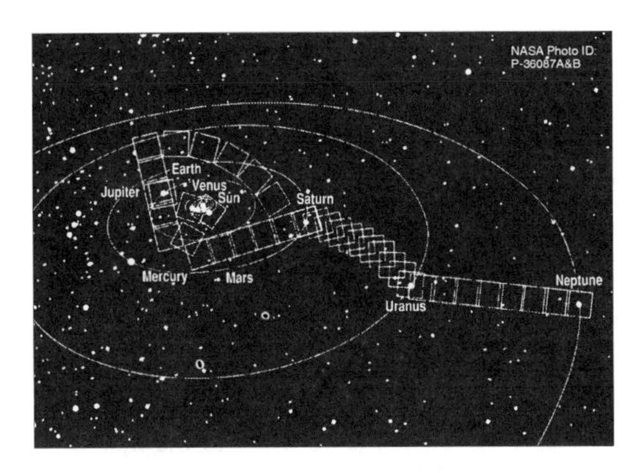

Figure 1.1: Kepler derived the shape of orbits. Space probes confirmed it.

Some natural phenomena include features that cannot be directly "seen" or otherwise perceived through the senses. Nevertheless, starting from other observations, deductive reasoning can lead to conclusions about such hidden features. For illustration purposes, the following example demonstrates how logic captures the methods used in the reasoning. The same example also documents the difficulties in using real applications, where any statement can be widely considered "true" for thousands of years, and suddenly become "false" in the light of new evidence. (Yet an understanding of this example will not be necessary for any of the subsequent material.)

Example 160 (The shape of orbits) The problem of determining the orbits of the planets can be traced back through several millennia. Several types of orbits have been fitted to various kinds of ideas. Most of these ideas eventually succumbed to the law of contraposition. For example, rectilinear motions fitted the poetical ideas of Xenophanes of Kolophon, about 570–475 B.C. [18, p. 18]; yet rectilinear motions fell into oblivion because they did not fit astronomical observations. A century later, Philolaus of Thebes proposed circular orbits for the Earth, the Moon, the planets, and the Sun, all around a "central fire" reflected by the Sun toward the Earth; such orbits fitted coarse observations of planetary motions [18, p. 40–49]. Nevertheless, the "central fire" hypothesis expired because it did not fit finer astronomical observations. In the third century B.C., Aristarchus of Samos outlined a heliocentric system with a circular orbit for the Earth around the Sun [18, p. 137]. Such a system fits the observation that some planets seem bigger and brighter when they appear opposite to the Sun, for instance, in the western sky at dawn or in the eastern sky at dusk (the planets are then "on the same side" of the Sun as the Earth is, and hence closer to the Earth); the same planets also seem smaller and dimmer when they appear close to the Sun (the planets are then about to pass "on the other side" of the Sun from the Earth, and hence farther from the Earth).

In the sixteenth century A.D., Nicolaus Copernicus (1473–1543) again proposed a heliocentric system, in which all the planets revolve along circles around the Sun. Galileo Galilei (1564–1642) then observed through a telescope that planets go through phases, as the Moon goes through decreasing and increasing phases. Viewed from the Earth, the phases of the planets could be explained by the direction in which one half of the planet faces the Sun, thus corroborating the heliocentric system. By then, Tycho Brahe (1546–1601) had instruments to measure the position of planets with sufficient accuracy, and based on such measurements, enough geometry and algebra was available to calculate and predict future positions of the same planets. The hypothesis of *circular* heliocentric orbits had thus been considered "true" for nearly two millenia. Still finer astronomical data would soon throw it off center.

Indeed, discrepancies between Tycho Brahe's observations of Mars and Copernicus's heliocentric model led Johann Kepler (1571–1630) to abandon circles for the orbits, and finally to substitute *ellipses* with a focus at the Sun (about 18 December 1604 A.D.) [18, p. 389–392]. Corroborating this conclusion, figure 1.1 combines photographs of the solar system from the space probe Galileo.

The forms of reasoning just outlined reveal the following patterns.

> If planets revolve along circles,
> then mathematics predicts their positions, and
> if mathematics predicts their positions,
> then Mars will be at its predicted position.

Hence the reasoning connects the hypothesis directly to the final conclusion, in a pattern called the **transitivity** of the logical implication:

> if planets revolve along circles,
> then Mars will be at its predicted position.

However, accurate observations later negated the conclusion, and hence the **law of contraposition** leads to the rejection of circular orbits:

> if Mars was *not* at its predicted position,
> then Mars did *not* revolve along a circle.

As in examples 1 and 2 (chapter 0), the converse law of contraposition was not used.

1.1 PROPOSITIONAL CALCULUS

1.1.1 Formulae, Axioms, Rules, and Proofs

A logic can admit of more than one description. For instance, Boolean algebraic logic can be described with Truth tables or through the "propositional calculus" — also called "sentential calculus" — presented here.

Whereas Truth tables may suffice to clarify many logical questions, Truth tables do not directly reflect common uses of logical reasoning, which usually consists of sentences rather than Truth tables. For example, the first axiom of Euclidean geometry consists not of Truth tables, but, instead, of a sentence (paraphrased from [49, p. 3]):

> For each pair of distinct points there exists exactly one line passing through both points.

Inserting such an axiom into a Truth table may, depending upon the context, require a separate Truth table for each "point" and thus infinitely many Truth tables. Moreover, Truth tables do not lend themselves to some of the logical questions involving predicates ("features") and quantifiers ("for each" and "there exists"), for instance, in geometry, in set theory, and in mathematics in general. Furthermore, Truth tables cannot accommodate certain logical systems, for instance, the intuitionistic logic of L. E. J. Bouwer and Arend Heyting [11, p. 146, # 26.12] (as also proved in chapter 4). Hence arises the need for a method of deduction different from Truth tables, for instance, the propositional calculus outlined here. Propositional calculus replaces Truth tables by certain logical formulae called **axioms**. Different selections of axioms can lead to different kinds of logic, but the present chapter focuses mainly on **classical logic**, which has been successful for several millennia.

Several choices for the initial axioms and formulae lead to the same classical logic. Because the principal concepts of logic consist of "negation" and "implication" several common choices of initial axioms and formulae involve only the connectives $\neg$ and $\Rightarrow$.

Definition 161 (Well-formed formulae) Every variable can be a well-formed formula — also called a **well-formed propositional form** — of the propositional calculus. Moreover, for all well-formed formulae P and Q, the following two strings of symbols are also well-formed formulae of the propositional calculus:

$$\begin{array}{lll} (W1) & \neg(P) & \text{(read "not } P\text{")}, \\ (W2) & (P) \Rightarrow (Q) & \text{(read "} P \text{ implies } Q\text{" or "if } P\text{, then } Q\text{")}. \end{array}$$

Furthermore, only strings of symbols built from variables through applications of the rules W1 and W2 can be well-formed formulae of the propositional calculus.

Several choices of well-formed propositional forms can serve as axioms.

Example 162 (Meredith's single axiom) A selection of only one axiom identifies concisely the basis for classical logic, for instance,

Axiom M1 (Meredith's axiom) [79, p. 132–133, #8.50]

$$\{[(\{(P) \Rightarrow (Q)\} \Rightarrow \{[\neg(R)] \Rightarrow [\neg(S)]\}) \Rightarrow (R)] \Rightarrow (U)\} \\ \Rightarrow \{[(U) \Rightarrow (P)] \Rightarrow [(S) \Rightarrow (P)]\}$$

Yet one axiom does not differentiate the roles of such separate connectives as $\neg$ and $\Rightarrow$.

Example 163 (Tarski's Axioms) In contrast, Tarski lists seven axioms [96, p. 147]:

Axiom I.	$(P) \Rightarrow [(Q) \Rightarrow (P)]$.
Axiom II.	$\{(P) \Rightarrow [(P) \Rightarrow (Q)]\} \Rightarrow [(P) \Rightarrow (Q)]$.
Axiom III.	$[(P) \Rightarrow (Q)] \Rightarrow \{[(Q) \Rightarrow (R)] \Rightarrow [(P) \Rightarrow (R)]\}$.
Axiom IV.	$[(P) \Leftrightarrow (Q)] \Rightarrow [(P) \Rightarrow (Q)]$.
Axiom V.	$[(P) \Leftrightarrow (Q)] \Rightarrow [(Q) \Rightarrow (P)]$.
Axiom VI.	$[(P) \Rightarrow (Q)] \Rightarrow \{[(Q) \Rightarrow (P)] \Rightarrow [(P) \Leftrightarrow (Q)]\}$.
Axiom VII.	$\{[\neg(Q)] \Rightarrow [\neg(P)]\} \Rightarrow [(P) \Rightarrow (Q)]$.

Tarski's seven axioms lead more directly to the same classical logic, but it may not appear evident that logical reasoning must depend upon so many first principles.

A system that remains concise and differentiates the roles of separate connectives consists of the following three axioms [11, §20, p. 119], [73, p. 31], [75, p. 165].

Definition 164 (Łukasiewicz's axioms) A logical formula is an **axiom** of the classical propositional calculus if and only if it is one of the following three formulae, attributed to Łukasiewicz [50, p. 29], where P, Q, and R may be any logical propositions:

Axiom P1 $(P) \Rightarrow [(Q) \Rightarrow (P)]$.

Axiom P2 $\{(P) \Rightarrow [(Q) \Rightarrow (R)]\} \Rightarrow \{[(P) \Rightarrow (Q)] \Rightarrow [(P) \Rightarrow (R)]\}$.

Axiom P3 $\{[\neg(Q)] \Rightarrow [\neg(P)]\} \Rightarrow [(P) \Rightarrow (Q)]$.

The three axioms just stated reflect common notions about logic.

- Axiom P1, $(P) \Rightarrow [(Q) \Rightarrow (P)]$, called the **law of affirmation of the consequent**, states that once P has been verified, then *any* proposition Q implies P.
- Axiom P2, $\{(P) \Rightarrow [(Q) \Rightarrow (R)]\} \Rightarrow \{[(P) \Rightarrow (Q)] \Rightarrow [(P) \Rightarrow (R)]\}$, the **law of self-distributivity of implication**, states the transitivity of implication.
- Axiom P3, $\{[\neg(Q)] \Rightarrow [\neg(P)]\} \Rightarrow [(P) \Rightarrow (Q)]$, called the **converse law of contraposition** [11, §20, p. 119], states that a contraposition, $[\neg(Q)] \Rightarrow [\neg(P)]$ suffices to establish a classical logical implication $(P) \Rightarrow (Q)$.

The converse law of contraposition distinguishes classical logic from several other systems of logic, which will also be described farther in the present exposition. Still, all these logical systems include Łukasiewicz's first two axioms, P1 and P2.

With Truth tables replaced by axioms, Truth values can be replaced by the following rules of inference, and tautologies can be replaced by theorems. Thus, the logical system of the propositional calculus need not include any concept of True or False; instead, the propositional calculus includes the following concepts of theorem and proof.

Definition 165 A well-formed propositional form is a **theorem** of the propositional calculus if and only if it is obtained by the following **rules of inference**:

Rule 166 (Axiom)

Every axiom is a theorem.

Rule 167 (Substitution)

For each component K of a theorem R (which is a propositional form),
and for each well-formed propositional form L,
the propositional form obtained by replacing in R every occurrence of K by L
is again a theorem.

Rule 168 ("Modus Ponens" (abbreviated by M. P.), or "Detachment")

For all propositional forms P and Q,
if P is a theorem and
if $(P) \Rightarrow (Q)$ is a theorem,
then Q is a theorem.

With *Modus Ponens*, P is the **minor premise** while $(P) \Rightarrow (Q)$ is the **major premise**.

A **proof** of a theorem R is a finite sequence of logical formulae $P, Q, \ldots, R$, in which each formula is either a substitution in an axiom or in a previously proven formula, or results from the rule of *Modus Ponens*.

Example 169 (Substitution) The formula $(P) \Rightarrow [(Q) \Rightarrow (P)]$ is an instance of axiom P1; hence it is a theorem. Substituting $\neg(L)$ for P in $(P) \Rightarrow [(Q) \Rightarrow (P)]$ yields $[\neg(L)] \Rightarrow \{(Q) \Rightarrow [\neg(L)]\}$, which is another instance of axiom P1, and hence also a theorem. Because such substitutions in an axiom yield other axioms, each axiom is also called an **axiom schema**. Thus both $(P) \Rightarrow [(Q) \Rightarrow (P)]$ and $[\neg(L)] \Rightarrow \{(Q) \Rightarrow [\neg(L)]\}$ result from the axiom schema P1.

Definition 170 For every logical formula R, the notation

$$\vdash R$$

means that there exists a proof of R. An alternative notation, $P1, P2, P3, \vdash R$, also specifies the list of axioms, here $P1, P2, P3$, from which R is a theorem.

More generally, for all logical formulae P and R, the notation $P \vdash R$ means that with P added to the list of axioms, there exists a proof of R. The corresponding alternative notation, $P1, P2, P3, P \vdash R$, again specifies the list of axioms. In other words, R is a **theorem** for the logic with axioms $P1, P2, P3, P$. With a different terminology, $P \vdash R$ means that R is **derivable** from P and the axioms.

Yet more generally, for all logical formulae $P, Q, \ldots, R$, either notation $P, Q, \ldots \vdash R$ or $P1, P2, P3, P, Q, \ldots \vdash R$, means that with $P, Q, \ldots$ added to the list of axioms, there exists a proof of R. In other words, R is a **theorem** for the logic with axioms $P1, P2, P3, P, Q, \ldots$ The formula R is then **derivable** from $P, Q, \ldots$, if and only if $P, Q, \ldots \vdash R$. In the notation of Smullyan [88, p. 17] and Stolyar [92, p. 63], $P, Q, \ldots \vdash R$ is also denoted by

$$\frac{P, Q, \ldots}{R}.$$

Verifying a proof reduces to checking that each step conforms to the foregoing definition of proof. In contrast, *constructing* a proof may require some creativity, which may involve trying some rules and some axioms in various combinations, some of which may fail whereas others may succeed. For the propositional calculus presented here, there is an algorithm (a recipe) to design proofs, but its justification first requires most of the proofs presented here [92, p. 193–197]. Moreover, the algorithm is cumbersome and would generate proofs longer than the ones explained here. Nevertheless, the collection of all the proofs shown here will demonstrate the steps that the algorithm would involve. With such an understanding of the algorithm, a user might then automate the algorithm with a computer. Finally, the following proofs also provide some practice in creating proofs without using an algorithm, a practice that corresponds more closely to the situation in mathematics, and for which there can be no algorithm [34].

1.1.2 Examples of Proofs with Axioms P1 and P2

The following three theorems provide examples of proofs by propositional calculus. The first theorem is called a "derived rule" because it involves a hypothesis, T, which can be any formula that has already been proved. For example, T can be any axiom.

Theorem 171 (Derived rule) *For each logical formula S and for each theorem T, the implication $(S) \Rightarrow (T)$ is a theorem. Thus, $T \vdash [(S) \Rightarrow (T)]$.*

Proof. Apply axiom P1 and *Modus Ponens* as follows:

$\vdash T$	hypothesis (minor premise),
$\vdash (T) \Rightarrow [(S) \Rightarrow (T)]$	substitution in axiom P1 (major premise),
$\vdash (S) \Rightarrow (T)$	*Modus Ponens* and preceding two formulae.

□

The second theorem also involves a hypothesis, $(H) \Rightarrow [(K) \Rightarrow (L)]$, which can be any formula that has already been proved.

Theorem 172 (Derived rule) *For all logical propositions H, K, L, if*

$$(H) \Rightarrow [(K) \Rightarrow (L)]$$

is a theorem, then

$$[(H) \Rightarrow (K)] \Rightarrow [(H) \Rightarrow (L)]$$

is also a theorem. Thus, $\{(H) \Rightarrow [(K) \Rightarrow (L)]\} \vdash \{[(H) \Rightarrow (K)] \Rightarrow [(H) \Rightarrow (L)]\}$.

Proof. Apply axiom P2 and *Modus Ponens*:

$\vdash (H) \Rightarrow [(K) \Rightarrow (L)]$	hypothesis,
$\vdash \{(H) \Rightarrow [(K) \Rightarrow (L)]\} \Rightarrow \{[(H) \Rightarrow (K)] \Rightarrow [(H) \Rightarrow (L)]\}$	axiom P2,
$\vdash [(H) \Rightarrow (K)] \Rightarrow [(H) \Rightarrow (L)]$	*Modus Ponens.*

□

The following theorem does not involve any hypothesis other than the axioms of the classical propositional calculus. Thus it represents a universal pattern of reasoning.

Theorem 173 (Reflexive law of implication) *The formula $(P) \Rightarrow (P)$ is a theorem.*

Proof. Apply axioms P1, theorem 172, and the rule of *Modus Ponens*, as follows:

$\vdash (\underbrace{P}_{H}) \Rightarrow \{\overbrace{\underbrace{[(P) \Rightarrow (P)]}_{K}}^{Q} \Rightarrow (\underbrace{P}_{L})\}$	substitution in axiom P1,
$\vdash \{(\overbrace{P}^{H}) \Rightarrow \overbrace{[(P) \Rightarrow (P)]}^{K}\} \Rightarrow [(\overbrace{P}^{H}) \Rightarrow (\overbrace{P}^{L})]$	theorem 172,
$\vdash (P) \Rightarrow [(P) \Rightarrow (P)]$	substitution in axiom P1,
$\vdash (P) \Rightarrow (P)$	*Modus Ponens.*

□

1.1.3 Exercises

The following two exercises pertain to P. T. Johnstone's three axioms J1, J2, J3 [54, p. 12]:

Axiom J1 $(P) \Rightarrow [(Q) \Rightarrow (P)]$.

Axiom J2 $\{(P) \Rightarrow [(Q) \Rightarrow (R)]\} \Rightarrow \{[(P) \Rightarrow (Q)] \Rightarrow [(P) \Rightarrow (R)]\}$.

Axiom J3 (Law of double negation) $\{\neg[\neg(P)]\} \Rightarrow (P)$.

Exercise 331 Using *only* Johnstone's three axioms (J1, J2, J3) and the rules of substitution and *Modus Ponens,* prove that $(P) \Rightarrow (P)$ is a theorem.

Exercise 332 Using *only* Johnstone's three axioms (J1, J2, J3) and the rules of substitution and *Modus Ponens,* prove that if a logical proposition T is a theorem, then $(Q) \Rightarrow (T)$ is a theorem for every logical proposition Q.

The following two exercises pertain to Łukasiewicz's alternative axioms Ł1, Ł2, Ł3 [79, p. 132, #8.49]:

Axiom Ł1 $[(A) \Rightarrow (B)] \Rightarrow \{[(B) \Rightarrow (C)] \Rightarrow [(A) \Rightarrow (C)]\}$.

Axiom Ł2 $\{[\neg(P)] \Rightarrow (P)\} \Rightarrow (P)$.

Axiom Ł3 $(P) \Rightarrow \{[\neg(P)] \Rightarrow (Q)\}$.

Exercise 333 Using *only* Łukasiewicz's alternative axioms (Ł1, Ł2, Ł3) and the rules of substitution and *Modus Ponens,* prove that $(P) \Rightarrow (P)$ is a theorem.

Exercise 334 Using *only* Łukasiewicz's alternative axioms (Ł1, Ł2, Ł3) and the rules of substitution and *Modus Ponens,* prove that if a logical proposition T is a theorem, then $[\neg(T)] \Rightarrow (Q)$ is a theorem for every logical proposition Q.

For the following two exercises, consider Tarski's seven axioms:

Axiom I.	$(P) \Rightarrow [(Q) \Rightarrow (P)]$.
Axiom II.	$\{(P) \Rightarrow [(P) \Rightarrow (Q)]\} \Rightarrow [(P) \Rightarrow (Q)]$.
Axiom III.	$[(P) \Rightarrow (Q)] \Rightarrow \{[(Q) \Rightarrow (R)] \Rightarrow [(P) \Rightarrow (R)]\}$.
Axiom IV.	$[(P) \Leftrightarrow (Q)] \Rightarrow [(P) \Rightarrow (Q)]$.
Axiom V.	$[(P) \Leftrightarrow (Q)] \Rightarrow [(Q) \Rightarrow (P)]$.
Axiom VI.	$[(P) \Rightarrow (Q)] \Rightarrow \{[(Q) \Rightarrow (P)] \Rightarrow [(P) \Leftrightarrow (Q)]\}$.
Axiom VII.	$\{[\neg(Q)] \Rightarrow [\neg(P)]\} \Rightarrow [(P) \Rightarrow (Q)]$.

Exercise 335 Prove that $(P) \Rightarrow (P)$ is a theorem, using only the two rules of inference (substitution and *Modus Ponens*) and Tarski's seven axioms.

Exercise 336 Using only the two rules of inference and Tarski's seven axioms, prove that if T and $[\neg(R)] \Rightarrow [\neg(T)]$ are theorems, then R is a theorem.

The following exercises refer to various axioms and to the following statements:

if the hypothesis holds (H: planets revolve along circles),
then a first conclusion holds (K: mathematics predicts their positions);
if the first conclusion holds (L: mathematics predicts their positions),
then a second conclusion holds (M: Mars will be at a certain position).

Exercise 337 From the preceding statements (H, K, L, M), determine how Łukasiewicz's alternative axioms (Ł1, Ł2, Ł3) yield the statements

if the hypothesis holds (H: planets revolve along circles),
then the final conclusion holds (M: Mars will be at a certain position).

Exercise 338 From the preceding statements (H, K, L, M), determine how Tarski's seven axioms (I–VII) yield the statements

if the hypothesis holds (H: planets revolve along circles),
then the final conclusion holds (M: Mars will be at a certain position).

Exercise 339 From the preceding statements (H, K, L, M), write a well-formed formula corresponding to the statement

Mars was *not* at the predicted position,
therefore Mars does *not* revolve along a circle.

Exercise 340 From the preceding statements (H, K, L, M), write a well-formed formula corresponding to the statements

> if planets revolve along circles,
> then Mars will be at a certain position;
> however, Mars was *not* at the predicted position,
> therefore Mars does *not* revolve along a circle.

1.2 CLASSICAL IMPLICATIONAL CALCULUS

1.2.1 Derived Rules: Implications Subject to Hypotheses

With the two rules of inference (*Modus Ponens* and substitution), the first two axioms of classical propositional calculus (P1 and P2) pertain only to logical implications ($\Rightarrow$); they form the **classical implicational calculus**, as presented here. In contrast, the concept of negation ($\neg$) does not belong to the classical implicational calculus.

Theorems of the classical implicational calculus are variants of the transitivity of implications: they shorten sequences of implications into one implication, from the initial hypothesis to the last conclusion. The same theorems also allow for the substitution of any proposition by any logically equivalent proposition. Moreover, their proofs constitute steps of an algorithm known as the Deduction Theorem.

For instance, the following **derived rules of inference** extend the rule of *Modus Ponens* to situations where K and $(K) \Rightarrow (L)$ might hold only under some hypothesis H: then the conclusion L also holds under the same hypothesis H.

Theorem 174 (Derived rule) *For all logical propositions H, K, L, if*

$$(H) \Rightarrow (K) \quad \text{and}$$
$$(H) \Rightarrow [(K) \Rightarrow (L)]$$

are theorems, then

$$(H) \Rightarrow (L)$$

is also a theorem. Thus, $[(H) \Rightarrow (K)], \{(H) \Rightarrow [(K) \Rightarrow (L)]\} \vdash [(H) \Rightarrow (L)]$.

Proof. Apply axiom P2 and *Modus Ponens*:

$\vdash (H) \Rightarrow [(K) \Rightarrow (L)]$	hypothesis,
$\vdash [(H) \Rightarrow (K)] \Rightarrow [(H) \Rightarrow (L)]$	substitution in theorem 172,
$\vdash (H) \Rightarrow (K)$	hypothesis,
$\vdash (H) \Rightarrow (L)$	*Modus Ponens.*

□

The following theorem simplifies the use of theorem 174 if one of the logical implications is already a theorem.

Theorem 175 (Derived rule) *For all logical propositions H, K, L, if*

$$(H) \Rightarrow (K) \quad \textit{and}$$
$$(K) \Rightarrow (L)$$

are theorems, then

$$(H) \Rightarrow (L)$$

is also a theorem. Thus, $[(H) \Rightarrow (K)], [(K) \Rightarrow (L)] \vdash [(H) \Rightarrow (L)]$.

Proof. Apply theorems 171 and 174:

$\vdash (K) \Rightarrow (L)$	hypothesis,
$\vdash (H) \Rightarrow [(K) \Rightarrow (L)]$	substitution in theorem 171,
$\vdash (H) \Rightarrow (K)$	hypothesis,
$\vdash (H) \Rightarrow (L)$	theorem 174.

□

Similarly, the following theorem simplifies the use of theorem 174 if one of the components is already a theorem.

Theorem 176 (Derived rule) *For all logical propositions H, K, L, if*

$$(K) \quad \textit{and}$$
$$(H) \Rightarrow [(K) \Rightarrow (L)]$$

are theorems, then

$$(H) \Rightarrow (L)$$

is also a theorem. Thus, $(K), \{(H) \Rightarrow [(K) \Rightarrow (L)]\} \vdash [(H) \Rightarrow (L)]$.

Proof. Apply theorems 171 and 174:

$\vdash K$	hypothesis,
$\vdash (H) \Rightarrow (K)$	theorem 171,
$\vdash (H) \Rightarrow [(K) \Rightarrow (L)]$	hypothesis,
$\vdash (H) \Rightarrow (L)$	theorem 174.

□

The following theorem demonstrates how the rule of *Modus Ponens* extends to a sequence of several logical implications.

Theorem 177 (Derived rule) *For all logical propositions P, Q, R, S, if*

$$(P) \Rightarrow (Q),$$
$$(P) \Rightarrow [(Q) \Rightarrow (R)], \quad \textit{and}$$
$$(P) \Rightarrow [(R) \Rightarrow (S)]$$

are theorems, then

$$(P) \Rightarrow (S)$$

is also a theorem. Thus,

$$[(P) \Rightarrow (Q)], \; \{(P) \Rightarrow [(Q) \Rightarrow (R)]\}, \; \{(P) \Rightarrow [(R) \Rightarrow (S)]\} \vdash [(P) \Rightarrow (S)].$$

Proof. Apply theorem 174 twice:

$\vdash (P) \Rightarrow (Q)$	hypothesis,
$\vdash (P) \Rightarrow [(Q) \Rightarrow (R)]$	hypothesis,
$\vdash (P) \Rightarrow (R)$	theorem 174,
$\vdash (P) \Rightarrow [(R) \Rightarrow (S)]$	hypothesis,
$\vdash (P) \Rightarrow (S)$	theorem 174.

□

Similar theorems hold for a sequence of more than three consecutive implications, but their need will not arise here. The following theorem simplifies the use of theorem 177 if two of the logical implications are already theorems.

Theorem 178 (Derived rule) *For all logical propositions P, Q, R, S, if*

$$\begin{array}{ll} (P) \Rightarrow (Q), & \\ (Q) \Rightarrow (R), & \textit{and} \\ (R) \Rightarrow (S) & \end{array}$$

are theorems, then

$$(P) \Rightarrow (S)$$

is also a theorem. Thus,

$$[(P) \Rightarrow (Q)],\ [(Q) \Rightarrow (R)],\ [(R) \Rightarrow (S)] \vdash [(P) \Rightarrow (S)].$$

Proof. Apply theorem 175 twice:

$$\begin{array}{ll} \vdash (P) \Rightarrow (Q) & \text{hypothesis,} \\ \vdash (Q) \Rightarrow (R) & \text{hypothesis,} \\ \vdash (P) \Rightarrow (R) & \text{theorem 175,} \\ \vdash (R) \Rightarrow (S) & \text{hypothesis,} \\ \vdash (P) \Rightarrow (S) & \text{theorem 175.} \end{array}$$

□

The following theorems demonstrate the transitivity of implications subject to a common hypothesis.

Theorem 179 (Derived rule) *For all logical propositions H, K, L, M, if*

$$\begin{array}{ll} (H) \Rightarrow [(K) \Rightarrow (L)] & \textit{and} \\ (H) \Rightarrow \{(K) \Rightarrow [(L) \Rightarrow (M)]\} & \end{array}$$

are theorems, then

$$(H) \Rightarrow [(K) \Rightarrow (M)]$$

is also a theorem. Thus,

$$\{(H) \Rightarrow [(K) \Rightarrow (L)]\},\ [(H) \Rightarrow \{(K) \Rightarrow [(L) \Rightarrow (M)]\}] \vdash \{(H) \Rightarrow [(K) \Rightarrow (M)]\}.$$

Proof. Use axiom P2 and theorems 171 and 174:

$$\begin{array}{ll} \vdash (H) \Rightarrow \{(K) \Rightarrow [(L) \Rightarrow (M)]\} & \text{hypothesis,} \\ \vdash \{(K) \Rightarrow [(L) \Rightarrow (M)]\} \Rightarrow \{[(K) \Rightarrow (L)] \Rightarrow [(K) \Rightarrow (M)]\} & \text{axiom P2,} \\ \vdash (H) \Rightarrow \{[(K) \Rightarrow (L)] \Rightarrow [(K) \Rightarrow (M)]\} & \text{theorem 175.} \\ \vdash (H) \Rightarrow [(K) \Rightarrow (L)] & \text{hypothesis,} \\ \vdash (H) \Rightarrow [(K) \Rightarrow (M)] & \text{theorem 174.} \end{array}$$

□

The proof of the following theorem shows the use of the foregoing derived rules.

Theorem 180 *The formula $\{(P) \Rightarrow [(P) \Rightarrow (Q)]\} \Rightarrow [(P) \Rightarrow (Q)]$ is a theorem.*

Proof. Apply theorems 171, 173, and 174:

$$\vdash \underbrace{(P) \Rightarrow (P)}_{K} \qquad \text{theorem 173,}$$

$$\vdash \{\overbrace{(P) \Rightarrow [(P) \Rightarrow (Q)]}^{H}\} \Rightarrow \{[\overbrace{(P) \Rightarrow (P)}^{K}] \Rightarrow [\overbrace{(P) \Rightarrow (Q)}^{L}]\} \qquad \text{axiom P2,}$$

$$\vdash \{\underbrace{(P) \Rightarrow [(P) \Rightarrow (Q)]}_{H}\} \Rightarrow [\underbrace{(P) \Rightarrow (Q)}_{L}] \qquad \text{theorem 176.}$$

□

1.2.2 Examples of Proofs of Implicational Theorems

The preceding derived rules of inference involved a hypothesis and a conclusion, and stated that *if* the hypothesis holds, *then* the conclusion holds. In contrast, the following theorems are theorems of the implicational calculus, without requiring any hypotheses other than axioms P1 and P2. In other words, the following theorems correspond to universal patterns of deductive reasoning in classical logic.

For instance, in patterns of deductive reasoning involving two premises, the following theorems confirm that the order of the premises does not matter.

Theorem 181 (Transitive law of implication) *The following formula is a theorem:*

$$[(Q) \Rightarrow (R)] \Rightarrow \{[(P) \Rightarrow (Q)] \Rightarrow [(P) \Rightarrow (R)]\}.$$

Proof. The variables H, K, L refer to theorem 174:

$$\vdash \underbrace{[(Q) \Rightarrow (R)]}_{H} \Rightarrow \underbrace{\{(P) \Rightarrow [(Q) \Rightarrow (R)]\}}_{K} \qquad \text{axiom P1,}$$

$$\vdash \underbrace{\{(P) \Rightarrow [(Q) \Rightarrow (R)]\}}_{K} \Rightarrow \underbrace{\{[(P) \Rightarrow (Q)] \Rightarrow [(P) \Rightarrow (R)]\}}_{L} \qquad \text{axiom P2,}$$

$$\vdash \underbrace{[(Q) \Rightarrow (R)]}_{H} \Rightarrow \underbrace{\{[(P) \Rightarrow (Q)] \Rightarrow [(P) \Rightarrow (R)]\}}_{L} \qquad \text{theorem 174.}$$

□

Swapping the premises $(P) \Rightarrow (Q)$ and $(Q) \Rightarrow (R)$ also yields $(P) \Rightarrow (R)$.

Theorem 182 (Transitive law of implication) *The following formula is a theorem:*

$$[(P) \Rightarrow (Q)] \Rightarrow \{[(Q) \Rightarrow (R)] \Rightarrow [(P) \Rightarrow (R)]\}.$$

Proof. Apply theorems 173, 181, and 179:

$$\vdash [(P) \Rightarrow (Q)] \Rightarrow [(P) \Rightarrow (Q)] \qquad \text{theorem 173,}$$

$$\underbrace{\{(P) \Rightarrow (Q)\}}_{H} \Rightarrow \{\underbrace{[(Q) \Rightarrow (R)]}_{A} \Rightarrow \underbrace{[(P) \Rightarrow (Q)]}_{H}\} \qquad \text{axiom P1,}$$

$$\vdash \{\underbrace{(Q) \Rightarrow (R)}_{A}\} \Rightarrow \{\underbrace{[(P) \Rightarrow (Q)]}_{B} \Rightarrow \underbrace{[(P) \Rightarrow (R)]}_{C}\} \qquad \text{theorem 181,}$$

$$\vdash \underbrace{[(P) \Rightarrow (Q)]}_{H} \Rightarrow \{\underbrace{[(Q) \Rightarrow (R)]}_{A} \Rightarrow \underbrace{[(P) \Rightarrow (R)]}_{L}\} \qquad \text{theorem 179.}$$

□

The following derived rules allow for substitutions within implications subject to hypotheses, for instance, a substitution within an intermediate hypothesis.

Theorem 183 (Derived rule) *For all logical propositions H, K, L, M, if*

$(H) \Rightarrow [(L) \Rightarrow (M)]$ *and*
$(K) \Rightarrow (L)$

are theorems, then

$(H) \Rightarrow [(K) \Rightarrow (M)]$

is also a theorem.

Proof. Apply theorems 182 and 175:

$$\begin{array}{ll} \vdash (K) \Rightarrow (L) & \text{hypothesis,} \\ \vdash [(K) \Rightarrow (L)] \Rightarrow \{[(L) \Rightarrow (M)] \Rightarrow [(K) \Rightarrow (M)]\} & \text{theorem 182,} \\ \vdash [(L) \Rightarrow (M)] \Rightarrow [(K) \Rightarrow (M)] & \textit{Modus Ponens.} \\ \vdash (H) \Rightarrow [(L) \Rightarrow (M)] & \text{hypothesis,} \\ \vdash (H) \Rightarrow [(K) \Rightarrow (M)] & \text{theorem 175.} \end{array}$$

□

The second theorem allows for a substitution in the conclusion.

Theorem 184 (Derived rule) *For all logical propositions H, L, M, N, if*

$$\begin{array}{l} (H) \Rightarrow [(L) \Rightarrow (M)] \quad \textit{and} \\ (M) \Rightarrow (N) \end{array}$$

are theorems, then

$$(H) \Rightarrow [(L) \Rightarrow (N)]$$

is also a theorem.

Proof. Apply theorems 182, 175, and 176:

$$\begin{array}{ll} \vdash (H) \Rightarrow \underbrace{[(L) \Rightarrow (M)]}_{P} & \text{hypothesis,} \\ \vdash \underbrace{[(L) \Rightarrow (M)]}_{P} \Rightarrow \underbrace{\{[(M) \Rightarrow (N)] \Rightarrow [(L) \Rightarrow (N)]\}}_{Q} & \text{theorem 182,} \\ \vdash (H) \Rightarrow \underbrace{\{[(M) \Rightarrow (N)] \Rightarrow [(L) \Rightarrow (N)]\}}_{Q} & \text{theorem 175,} \\ \vdash (M) \Rightarrow (N) & \text{hypothesis,} \\ \vdash (H) \Rightarrow [(L) \Rightarrow (N)] & \text{theorem 176.} \end{array}$$

□

The following theorem allows for yet another change in the order of hypotheses.

Theorem 185 (Law of commutation) *The following formula is a theorem:*

$$\{(P) \Rightarrow [(Q) \Rightarrow (R)]\} \Rightarrow \{(Q) \Rightarrow [(P) \Rightarrow (R)]\}.$$

Proof. Apply theorem 183:

$$\begin{array}{ll} \vdash \underbrace{\{(P) \Rightarrow [(Q) \Rightarrow (R)]\}}_{H} \Rightarrow \{\underbrace{[(P) \Rightarrow (Q)]}_{L} \Rightarrow \underbrace{[(P) \Rightarrow (R)]}_{M}\} & \text{axiom P2,} \\ \vdash (\underbrace{Q}_{K}) \Rightarrow \underbrace{[(P) \Rightarrow (Q)]}_{L} & \text{axiom P1,} \\ \vdash \underbrace{\{(P) \Rightarrow [(Q) \Rightarrow (R)]\}}_{H} \Rightarrow \{(\underbrace{Q}_{K}) \Rightarrow \underbrace{[(P) \Rightarrow (R)]}_{M}\} & \text{theorem 183.} \end{array}$$

□

The following theorem combines the rule of *Modus Ponens* into a single formula.

Theorem 186 (Law of assertion) *The formula $(A) \Rightarrow \{[(A) \Rightarrow (B)] \Rightarrow (B)\}$ is a theorem.*

Proof. Apply theorem 173 and the law of commutation (theorem 185):

$$\vdash \underbrace{[(A) \Rightarrow (B)]}_{P} \Rightarrow [(\underbrace{A}_{Q}) \Rightarrow (\underbrace{B}_{R})] \quad \text{theorem 173,}$$

$$\vdash (\underbrace{A}_{Q}) \Rightarrow \{\underbrace{[(A) \Rightarrow (B)]}_{P} \Rightarrow (\underbrace{B}_{R})\} \quad \text{theorem 185 and } \textit{Modus Ponens}.$$

□

The foregoing theorems involve only implications but no negation, and their proofs do not involve any negation either. Nevertheless, there exist other theorems involving only $\Rightarrow$ but not $\neg$ for which there does not exist any proof involving only implications. Examples of such theorems are hidden in the exercises, to be revealed later.

1.2.3 Exercises

Investigate whether the formulae in the following exercises are theorems, using any of the axioms P1 and P2, any rules of inference, and any of the theorems just proved.

Exercise 341 $[(H) \Rightarrow (L)] \Rightarrow \{(H) \Rightarrow [(K) \Rightarrow (L)]\}$

Exercise 342 $[(K) \Rightarrow (L)] \Rightarrow \{(H) \Rightarrow [(K) \Rightarrow (L)]\}$

Exercise 343 $[(A) \Rightarrow (B)] \Rightarrow \big[\{[(A) \Rightarrow (B)] \Rightarrow (B)\} \Rightarrow (B)\big]$

Exercise 344 $[(A) \Rightarrow (B)] \Rightarrow \big[\{[(A) \Rightarrow (B)] \Rightarrow (A)\} \Rightarrow (A)\big]$

Exercise 345 $\{[(P) \Rightarrow (P)] \Rightarrow (P)\} \Rightarrow (P)$

Exercise 346 $(P) \Rightarrow \{[(P) \Rightarrow (P)] \Rightarrow (P)\}$

Exercise 347 $\{[(P) \Rightarrow (Q)] \Rightarrow (P)\} \Rightarrow (P)$ (Peirce's law.)

Exercise 348 $[(P) \Rightarrow (R)] \Rightarrow \big[\{[(P) \Rightarrow (Q)] \Rightarrow (R)\} \Rightarrow (R)\big]$

Exercise 349 $\{[(R) \Rightarrow (Q)] \Rightarrow (P)\} \Rightarrow \{[(R) \Rightarrow (Q)] \Rightarrow [(S) \Rightarrow (P)]\}$

Exercise 350 $[(R) \Rightarrow (Q)] \Rightarrow \big(\{[(R) \Rightarrow (Q)] \Rightarrow (P)\} \Rightarrow [(S) \Rightarrow (P)]\big)$

1.3 PROOFS BY CONTRAPOSITION

1.3.1 Examples of Proofs with Axiom P3

The classical implicational calculus belongs to several logical systems, which differ from one another by their different axioms about negation. For instance, classical logic defines its concept of negation by the converse law of contraposition (axiom P3):

$$\{[\neg(Q)] \Rightarrow [\neg(P)]\} \Rightarrow [(P) \Rightarrow (Q)].$$

The following proofs demonstrate the use of the converse law of contraposition.

Theorem 187 (Law of denial of the antecedent) *The formula* $[\neg(P)] \Rightarrow [(P) \Rightarrow (Q)]$ *is a theorem.*

Proof. Apply axioms P1 and P3 with the transitivity of implication (theorem 175):

$\vdash [\neg(P)] \Rightarrow \{[\neg(Q) \Rightarrow [\neg(P)]\}$	substitution in axiom P1,
$\vdash \{[\neg(Q) \Rightarrow [\neg(P)]\} \Rightarrow [(P) \Rightarrow (Q)]$	axiom P3,
$\vdash [\neg(P)] \Rightarrow [(P) \Rightarrow (Q)]$	theorem 175.

□

Theorem 188 *The formula* $(P) \Rightarrow \{[\neg(P)] \Rightarrow (Q)\}$ *is a theorem.*

Proof. Apply theorem 187) and the law of commutation (theorem 185):

$\vdash [\neg(P)] \Rightarrow [(P) \Rightarrow (Q)]$	theorem 187,
$\vdash \{[\neg(P)] \Rightarrow [(P) \Rightarrow (Q)]\} \Rightarrow \big[(P) \Rightarrow \{[\neg(P)] \Rightarrow (Q)\}\big]$	theorem 185,
$\vdash (P) \Rightarrow \{[\neg(P)] \Rightarrow (Q)\}$	*Modus Ponens.*

□

The following two theorems establish the complete law of double negation.

Theorem 189 (Law of double negation) *The formula* $[\neg\neg(P)] \Rightarrow (P)$ *is a theorem.*

Proof. Apply the transitivity of implication (theorem 178) and theorem 180:

$\vdash \{\neg[\neg(P)]\} \Rightarrow \{[\neg\neg\neg\neg(P)] \Rightarrow [\neg\neg(P)]\}$	axiom P1,
$\vdash (\{\neg[\neg\neg\neg(P)]\} \Rightarrow \{\neg[\neg(P)]\}) \Rightarrow \{[\neg(P)] \Rightarrow [\neg\neg\neg(P)]\}$	axiom P3,
$\vdash ([\neg(P)] \Rightarrow \{\neg[\neg\neg(P)]\}) \Rightarrow \{[\neg\neg(P)] \Rightarrow (P)\}$	axiom P3,
$\vdash [\neg\neg(P)] \Rightarrow \{[\neg\neg(P)] \Rightarrow (P)\}$	theorem 178,
$\vdash ([\neg\neg(P)] \Rightarrow \{[\neg\neg(P)] \Rightarrow (P)\}) \Rightarrow \{[\neg\neg(P)] \Rightarrow (P)\}$	theorem 180,
$\vdash [\neg\neg(P)] \Rightarrow (P)$	*Modus Ponens.*

□

Theorem 190 (Converse law of double negation) *The formula* $(P) \Rightarrow [\neg\neg(P)]$ *is a theorem.*

Proof. Apply the law of double negation (theorem 189) and contraposition (P3):

$\vdash \big(\neg\{\neg[\neg(P)]\}\big) \Rightarrow [\neg(P)]$	theorem 189,
$\vdash \big\{\big(\neg\{\neg[\neg(P)]\}\big) \Rightarrow [\neg(P)]\big\} \Rightarrow \big[(P) \Rightarrow \{\neg[\neg(P)]\}\big]$	axiom P3,
$\vdash (P) \Rightarrow \{\neg[\neg(P)]\}$	*Modus Ponens.*

□

The following three derived rules of inference will simplify subsequent proofs.

Theorem 191 (Derived rule) *If* $(K) \Rightarrow (L)$ *is a theorem, then the following formula is also a theorem:* $[(H) \Rightarrow (K)] \Rightarrow [(H) \Rightarrow (L)]$.

Proof. Apply the transitivity of implication in the form of theorem 182:

$\vdash (K) \Rightarrow (L)$	hypothesis,
$\vdash [(K) \Rightarrow (L)] \Rightarrow \{[(H) \Rightarrow (K)] \Rightarrow [(H) \Rightarrow (L)]\}$	theorem 182,
$\vdash [(H) \Rightarrow (K)] \Rightarrow [(H) \Rightarrow (L)]$	*Modus Ponens.*

□

Theorem 192 (Derived rule) *If* $(I) \Rightarrow (H)$ *is a theorem, then the following formula is also a theorem:* $[(H) \Rightarrow (K)] \Rightarrow [(I) \Rightarrow (K)]$.

Proof. Apply the transitivity of implication in the form of theorem 181:

$$\begin{array}{ll} \vdash (I) \Rightarrow (H) & \text{hypothesis,} \\ \vdash [(I) \Rightarrow (H)] \Rightarrow \{[(H) \Rightarrow (K)] \Rightarrow [(I) \Rightarrow (K)]\} & \text{theorem 181,} \\ \vdash [(H) \Rightarrow (K)] \Rightarrow [(I) \Rightarrow (K)] & \textit{Modus Ponens.} \end{array}$$

□

Theorem 193 (Derived rule) *If* $(P) \Rightarrow (Q)$ *and* $(R) \Rightarrow (S)$ *are theorems, then the following formula is also a theorem:* $[(Q) \Rightarrow (R)] \Rightarrow [(P) \Rightarrow (S)]$.

Proof. Apply theorems 192, 191, and the transitivity of implication (theorem 175):

$$\begin{array}{ll} \vdash (P) \Rightarrow (Q) & \text{hypothesis,} \\ \vdash \underbrace{[(Q) \Rightarrow (R)]}_{U} \Rightarrow \underbrace{[(P) \Rightarrow (R)]}_{V} & \text{theorem 192,} \\ \vdash (R) \Rightarrow (S) & \text{hypothesis,} \\ \vdash \underbrace{[(P) \Rightarrow (R)]}_{V} \Rightarrow \underbrace{[(P) \Rightarrow (S)]}_{W} & \text{theorem 191,} \\ \vdash \underbrace{[(Q) \Rightarrow (R)]}_{U} \Rightarrow \underbrace{[(P) \Rightarrow (S)]}_{W} & \text{theorem 175.} \end{array}$$

□

With axiom P3, the following theorem gives the complete law of contraposition.

Theorem 194 (Law of contraposition, principle of transposition) *The following formula is a theorem:* $[(P) \Rightarrow (Q)] \Rightarrow \{[\neg(Q)] \Rightarrow [\neg(P)]\}$.

Proof. Apply a derived rule of inference (theorem 193) and transitivity (theorem 175):

$$\begin{array}{ll} \vdash \{\neg[\neg(P)]\} \Rightarrow (P) & \text{theorem 189,} \\ \vdash (Q) \Rightarrow \{\neg[\neg(Q)]\} & \text{theorem 190,} \\ \vdash [(P) \Rightarrow (Q)] \Rightarrow \{[\neg\neg(P)] \Rightarrow [\neg\neg(Q)]\} & \text{theorem 193,} \\ \vdash \{[\neg\neg(P)] \Rightarrow [\neg\neg(Q)]\} \Rightarrow \{[\neg(Q)] \Rightarrow [\neg(P)]\} & \text{axiom P3,} \\ \vdash [(P) \Rightarrow (Q)] \Rightarrow \{[\neg(Q)] \Rightarrow [\neg(P)]\} & \text{theorem 175.} \end{array}$$

□

1.3.2 Proofs by Reductio ad Absurdum

Within classical logic, a proposition and its negation together form an "absurdity" that cannot hold. In particular, if a hypothesis implies a conclusion and its negation — an absurdity — then the hypothesis may be rejected. The following theorems establish the validity of such a pattern of reasoning, called **reduction to the absurd**.

Theorem 195 (Special law of reductio ad absurdum) *The following formula is a theorem:* $\{(P) \Rightarrow [\neg(P)]\} \Rightarrow [\neg(P)]$.

Proof. Start with theorem 194 and the denial of the antecedent (theorem 187):

$$\begin{array}{ll}
\vdash \{(P) \Rightarrow [\neg(P)]\} \Rightarrow (\{\neg[\neg(P)]\} \Rightarrow [\neg(P)]) & (194), \\
\vdash \{\neg[\neg(P)]\} \Rightarrow ([\neg(P)] \Rightarrow \{\neg[(P) \Rightarrow (P)]\}) & (187), \\
\vdash [\{\neg[\neg(P)]\} \Rightarrow ([\neg(P)] \Rightarrow \{\neg[(P) \Rightarrow (P)]\})] & \\
\quad \Rightarrow [(\{\neg[\neg(P)]\} \Rightarrow [\neg(P)]) \Rightarrow (\{\neg[\neg(P)]\} \Rightarrow \{\neg[(P) \Rightarrow (P)]\})] & (\text{P2}), \\
\vdash (\{\neg[\neg(P)]\} \Rightarrow [\neg(P)]) \Rightarrow (\{\neg[\neg(P)]\} \Rightarrow \{\neg[(P) \Rightarrow (P)]\}) & (\textit{M.P.}), \\
\vdash (\{\neg[\neg(P)]\} \Rightarrow \{\neg[(P) \Rightarrow (P)]\}) \Rightarrow \{[(P) \Rightarrow (P)] \Rightarrow [\neg(P)]\} & (\text{P3}), \\
\vdash \{(P) \Rightarrow [\neg(P)]\} \Rightarrow \{[(P) \Rightarrow (P)] \Rightarrow [\neg(P)]\} & (178), \\
\vdash (P) \Rightarrow (P) & (173), \\
\vdash \{(P) \Rightarrow [\neg(P)]\} \Rightarrow [\neg(P)] & (176).
\end{array}$$

□

Theorem 196 (Law of reductio ad absurdum) *The following formula is a theorem:*

$$[(P) \Rightarrow (Q)] \Rightarrow (\{(P) \Rightarrow [\neg(Q)]\} \Rightarrow [\neg(P)]).$$

Proof. Apply theorems 194, 171, 185 (with *Modus Ponens*), 175, 195, 184:

$$\begin{array}{ll}
\vdash [(P) \Rightarrow (Q)] \Rightarrow \{[\neg(Q)] \Rightarrow [\neg(P)]\} & (194), \\
\vdash (P) \Rightarrow ([(P) \Rightarrow (Q)] \Rightarrow \{[\neg(Q)] \Rightarrow [\neg(P)]\}) & (171), \\
\vdash [(P) \Rightarrow (Q)] \Rightarrow [(P) \Rightarrow \{[\neg(Q)] \Rightarrow [\neg(P)]\}] & (185), \\
\vdash [(P) \Rightarrow \{[\neg(Q)] \Rightarrow [\neg(P)]\}] \Rightarrow (\{(P) \Rightarrow [\neg(Q)]\} \Rightarrow \{(P) \Rightarrow [\neg(P)]\}) & (\text{P2}), \\
\vdash [(P) \Rightarrow (Q)] \Rightarrow (\{(P) \Rightarrow [\neg(Q)]\} \Rightarrow \{(P) \Rightarrow [\neg(P)]\}) & (175), \\
\vdash \{(P) \Rightarrow [\neg(P)]\} \Rightarrow [\neg(P)] & (195), \\
\vdash [(P) \Rightarrow (Q)] \Rightarrow (\{(P) \Rightarrow [\neg(Q)]\} \Rightarrow [\neg(P)]) & (184).
\end{array}$$

□

1.3.3 Exercises

For the following four exercises, use the rules of inference and only the following six axioms, attributed to Frege [29] in [50, p. 29].

Axiom F1 $(P) \Rightarrow [(Q) \Rightarrow (P)]$.

Axiom F2 $\{(P) \Rightarrow [(Q) \Rightarrow (R)]\} \Rightarrow \{[(P) \Rightarrow (Q)] \Rightarrow [(P) \Rightarrow (R)]\}$.

Axiom F3 $\{(P) \Rightarrow [(Q) \Rightarrow (R)]\} \Rightarrow \{(Q) \Rightarrow [(P) \Rightarrow (R)]\}$.

Axiom F4 $[(P) \Rightarrow (Q)] \Rightarrow \{[\neg(Q)] \Rightarrow [\neg(P)]\}$.

Axiom F5 $\{\neg[\neg(P)]\} \Rightarrow (P)$.

Axiom F6 $(P) \Rightarrow \{\neg[\neg(P)]\}$.

Exercise 351 Prove that $F1$–$F6 \vdash (P) \Rightarrow \{[\neg(P)] \Rightarrow [\neg(Q)]\}$.

Exercise 352 Prove that $F1$–$F6 \vdash [\neg(P)] \Rightarrow \{(P) \Rightarrow [\neg(P)]\}$.

Exercise 353 Prove that $F1$–$F6 \vdash \{[\neg(Q)] \Rightarrow [\neg(P)]\} \Rightarrow [(P) \Rightarrow (Q)]$.

Exercise 354 Explain why Frege's six axioms F1–F6 are logically equivalent to Łukasiewicz's three axioms P1, P2, P3.

The following exercises outline a proof that the axioms P1, P2, P3 of classical logic are logically equivalent to the following three axioms C1, C2, C3, due to Church [11, §10, p. 72]. Because the two logical systems have the same first two axioms, they also have the same implicational calculus. The two logical systems differ from each other only by their third axiom, where F stands for False, so that $\neg(F)$ is a theorem.

Axiom C1 $(P) \Rightarrow [(Q) \Rightarrow (P)]$.

Axiom C2 $\{(P) \Rightarrow [(Q) \Rightarrow (R)]\} \Rightarrow \{[(P) \Rightarrow (Q)] \Rightarrow [(P) \Rightarrow (R)]\}$.

Axiom C3 $\{[(P) \Rightarrow (F)] \Rightarrow (F)\} \Rightarrow (P)$.

Exercise 355 This exercise establishes the converse law of contraposition in Church's system. Prove the tautology $\{[(B) \Rightarrow (F)] \Rightarrow [(A) \Rightarrow (F)]\} \Rightarrow [(A) \Rightarrow (B)]$ within Church's system, using only results from the *implicational* calculus (axioms C1 and C2) and axiom C3 (*not* axiom P3). Hint: start from axiom C2, with $(B) \Rightarrow (F)$ for P, with A for Q, and with F for R. Then use the transitivity of implications.

To show the equivalence of Church's logical system and classical logic, the following exercise establishes the equivalence of $\neg(P)$ and Church's $(P) \Rightarrow (F)$.

Exercise 356 Prove that $[(P) \Rightarrow (F)] \Rightarrow [\neg(P)]$ and $[\neg(P)] \Rightarrow [(P) \Rightarrow (F)]$ are theorems, using the theorem $\neg(F)$, and the classical axioms P1, P2, P3.

Exercise 357 Define a "negation" $\sim (P)$ to be an abbreviation for $(P) \Rightarrow (F)$. Explain why every theorem of classical logic is a theorem of Church's logic.

Exercise 358 Using the theorem $\neg(F)$ and axioms P1, P2, P3, prove the theorems $\{[(P) \Rightarrow (F)] \Rightarrow (F)\} \Rightarrow \{\neg[\neg(P)]\}$ and $\{\neg[\neg(P)]\} \Rightarrow \{[(P) \Rightarrow (F)] \Rightarrow (F)\}$.

Exercise 359 Using axioms P1, P2, P3 and any of the classical theorems already proved, prove $[\neg(P)] \Rightarrow \big[(P) \Rightarrow \{\neg[(S) \Rightarrow (S)]\}\big]$ and $\big[(P) \Rightarrow \{\neg[(S) \Rightarrow (S)]\}\big] \Rightarrow [\neg(P)]$.

Exercise 360 In classical logic define a constant f to be an abbreviation for $\neg[(S) \Rightarrow (S)]$. Explain why every theorem of Church's logic is a theorem of classical logic .

1.4 OTHER CONNECTIVES

1.4.1 Definitions of Other Connectives

The logical connectives $\neg$ and $\Rightarrow$ suffice to define all the other logical connectives, for instance, the conjunction $\wedge$, the disjunction $\vee$, and the equivalence $\Leftrightarrow$, as outlined here.

Definition 197 (Conjunction, disjunction, and equivalence)

$(P) \wedge (Q)$ stands for $\neg\{(P) \Rightarrow [\neg(Q)]\}$;
$(P) \vee (Q)$ stands for $[\neg(P)] \Rightarrow (Q)$;
$(P) \Leftrightarrow (Q)$ stands for $\big(\neg\{[(P) \Rightarrow (Q)] \Rightarrow \neg[(Q) \Rightarrow (P)]\}\big)$,
which is equivalent to $[(P) \Rightarrow (Q)] \wedge [(Q) \Rightarrow (P)]$.

1.4.2 Examples of Proofs of Theorems with Conjunctions

The following theorems pertain to the connective $\wedge$ (conjunction). The first theorem shows that the conjunction $\wedge$ commutes.

Theorem 198 (Commutativity of $\wedge$) *The formula* $[(P) \wedge (Q)] \Rightarrow [(Q) \wedge (P)]$ *is a theorem.*

Proof. Apply contraposition, double negation, transitivity, and the definition:

$$\begin{array}{ll} \vdash \{(Q) \Rightarrow [\neg(P)]\} \Rightarrow (\{\neg[\neg(P)]\} \Rightarrow [\neg(Q)]) & \text{contraposition,} \\ \vdash (P) \Rightarrow \{\neg[\neg(P)]\} & \text{theorem 190,} \\ \vdash \{(Q) \Rightarrow [\neg(P)]\} \Rightarrow \{(P) \Rightarrow [\neg(Q)]\} & \text{theorem 183,} \\ \vdash [\neg(\{(P) \Rightarrow [\neg(Q)]\})] \Rightarrow [\neg(\{(Q) \Rightarrow [\neg(P)]\})] & \text{contraposition,} \\ \vdash [(P) \wedge (Q)] \Rightarrow [(Q) \wedge (P)] & \text{definition of } \wedge. \end{array}$$

Swapping the roles of P and Q yields the converse: $\vdash [(Q) \wedge (P)] \Rightarrow [(P) \wedge (Q)]$. □

The following theorems show that if $(P) \wedge (Q)$ holds, then P holds and Q holds.

Theorem 199 *The formula* $[(P) \wedge (Q)] \Rightarrow (Q)$ *is a theorem.*

Proof. Apply contraposition, double negation, transitivity, and the definition:

$$\begin{array}{ll} \vdash [\neg(Q)] \Rightarrow \{(P) \Rightarrow [\neg(Q)]\} & \text{axiom P1,} \\ \vdash (\neg\{(P) \Rightarrow [\neg(Q)]\}) \Rightarrow \{\neg[\neg(Q)]\} & \text{contraposition,} \\ \vdash [(P) \wedge (Q)] \Rightarrow \{\neg[\neg(Q)]\} & \text{definition of } \wedge, \\ \vdash \{\neg[\neg(Q)]\} \Rightarrow (Q) & \text{theorem 189,} \\ \vdash [(P) \wedge (Q)] \Rightarrow (Q) & \text{theorem 175.} \end{array}$$

□

Theorem 200 *The formula* $[(P) \wedge (Q)] \Rightarrow (P)$ *is a theorem.*

Proof. Apply theorems 199, 198, 175:

$$\begin{array}{ll} \vdash [(P) \wedge (Q)] \Rightarrow [(Q) \wedge (P)] & \text{theorem 198,} \\ \vdash [(Q) \wedge (P)] \Rightarrow (P) & \text{theorem 199,} \\ \vdash [(P) \wedge (Q)] \Rightarrow (P) & \text{theorem 175.} \end{array}$$

□

The next theorem demonstrates a "converse" to the preceding two theorems. so that if P and Q hold, then their conjunction $(P) \wedge (Q)$ also holds.

Theorem 201 *If P and Q are both theorems, then $(P) \wedge (Q)$ is a theorem; equivalently,* $\vdash (P) \Rightarrow \{(Q) \Rightarrow [(P) \wedge (Q)]\}$.

Proof. Apply a reductio ad absurdum, contraposition, and transitivity:

$$\begin{array}{ll} \vdash (P) \Rightarrow (\{(P) \Rightarrow [\neg(Q)]\} \Rightarrow [\neg(Q)]) & \text{theorem 196,} \\ & \\ \vdash (\{(P) \Rightarrow [\neg(Q)]\} \Rightarrow [\neg(Q)]) & \text{contraposition} \ldots \\ \quad \Rightarrow \{(Q) \Rightarrow (\neg\{(P) \Rightarrow [\neg(Q)]\})\} & \ldots \text{continued,} \\ & \\ \vdash (P) \Rightarrow \{(Q) \Rightarrow (\neg\{(P) \Rightarrow [\neg(Q)]\})\} & \text{theorem 175,} \\ \vdash (P) \Rightarrow \{(Q) \Rightarrow [(P) \wedge (Q)]\} & \text{definition of } \wedge. \end{array}$$

Hence, if P and Q are both theorems, then $(P) \wedge (Q)$ is a theorem:

$\vdash (P) \Rightarrow \{(Q) \Rightarrow [(P) \wedge (Q)]\}$	just proved,
$\vdash (P)$	hypothesis,
$\vdash (Q) \Rightarrow [(P) \wedge (Q)]$	*Modus Ponens*,
$\vdash (Q)$	hypothesis,
$\vdash (P) \wedge (Q)$	*Modus Ponens*.

□

Theorem 201 allows for the following derived rule of inference.

Theorem 202 *If* $(H) \Rightarrow (K)$ *and* $(H) \Rightarrow (L)$ *are theorems, then* $(H) \Rightarrow [(K) \wedge (L)]$ *is a theorem:*

$$\vdash [(H) \Rightarrow (K)] \Rightarrow \big([(H) \Rightarrow (L)] \Rightarrow \{(H) \Rightarrow [(K) \wedge (L)]\}\big);$$

conversely, if $(H) \Rightarrow [(K) \wedge (L)]$ *is a theorem, then* $(H) \Rightarrow (K)$ *and* $(H) \Rightarrow (L)$ *are theorems.*

Proof. Apply theorem 201 and transitivity (theorem 184) with M for $(K) \wedge (L)$:

$\vdash [(H) \Rightarrow \{(K) \Rightarrow [(L) \Rightarrow (M)]\}] \Rightarrow \big([(H) \Rightarrow (K)] \Rightarrow \{(H) \Rightarrow [(L) \Rightarrow (M)]\}\big)$	axiom P2,
$\vdash \{(H) \Rightarrow [(L) \Rightarrow (M)]\} \Rightarrow \{[(H) \Rightarrow (L)] \Rightarrow [(H) \Rightarrow (M)]\}$	axiom P2,
$\vdash [(H) \Rightarrow \{(K) \Rightarrow [(L) \Rightarrow (M)]\}] \Rightarrow \big([(H) \Rightarrow (K)] \Rightarrow \{[(H) \Rightarrow (L)] \Rightarrow [(H) \Rightarrow (M)]\}\big)$	theorem 184,
$\vdash (K) \Rightarrow [(L) \Rightarrow (M)]$	theorem 201,
$\vdash (H) \Rightarrow \{(K) \Rightarrow [(L) \Rightarrow (M)]\}$	theorem 171,
$\vdash [(H) \Rightarrow (K)] \Rightarrow \{[(H) \Rightarrow (L)] \Rightarrow [(H) \Rightarrow (M)]\}$	*Modus Ponens*,

with M for $(K) \wedge (L)$. The converse results from theorems 199 and 171:

$\vdash [(K) \wedge (L)] \Rightarrow (L)$	theorem 199,
$\vdash (H) \Rightarrow \{[(K) \wedge (L)] \Rightarrow (L)\}$	theorem 171,
$\vdash \{(H) \Rightarrow [(Q) \Rightarrow (R)]\} \Rightarrow \{[(H) \Rightarrow (Q)] \Rightarrow [(H) \Rightarrow (R)]\}$	axiom P2,
$\vdash \{(H) \Rightarrow [(K) \wedge (L)]\} \Rightarrow [(H) \Rightarrow (L)]$	*Modus Ponens*.

Replacing $[(K) \wedge (L)] \Rightarrow (L)$ (theorem 199) by $[(K) \wedge (L)] \Rightarrow (K)$ (theorem 200) in the foregoing proof yields a proof of $\{(H) \Rightarrow [(K) \wedge (L)]\} \Rightarrow [(H) \Rightarrow (K)]$. □

For instance, theorem 202 yields the following theorem.

Theorem 203 (Idempotency of $\wedge$) *The formula* $(P) \Rightarrow [(P) \wedge (P)]$ *is a theorem.*

Proof. Substitute P for each of H, K, L in theorem 202:

$\vdash [(H) \Rightarrow (K)] \Rightarrow \big([(H) \Rightarrow (L)] \Rightarrow \{(H) \Rightarrow [(K) \wedge (L)]\}\big)$	theorem 202,
$\vdash [(P) \Rightarrow (P)] \Rightarrow \big([(P) \Rightarrow (P)] \Rightarrow \{(P) \Rightarrow [(P) \wedge (P)]\}\big)$	substitutions,
$\vdash (P) \Rightarrow (P)$	theorem 173,
$\vdash [(P) \Rightarrow (P)] \Rightarrow \{(P) \Rightarrow [(P) \wedge (P)]\}$	*Modus Ponens*,
$\vdash (P) \Rightarrow [(P) \wedge (P)]$	*Modus Ponens*.

□

1.4.3 Examples of Proofs of Theorems with Equivalences

A particular instance of the foregoing theorem allows for any proof of any equivalence $(I) \Leftrightarrow (J)$ to be split into two separate proofs of $(I) \Rightarrow (J)$ and $(J) \Rightarrow (I)$.

Theorem 204 *If $(I) \Rightarrow (J)$ and $(J) \Rightarrow (I)$ are theorems, then so is $(I) \Leftrightarrow (J)$. Conversely, if $(I) \Leftrightarrow (J)$ is a theorem, then so are $(I) \Rightarrow (J)$ and $(J) \Rightarrow (I)$.*

Proof. Apply theorem 201 with $(I) \Rightarrow (J)$ for P, and with $(J) \Rightarrow (I)$ for Q, so that if $(I) \Rightarrow (J)$ and $(J) \Rightarrow (I)$ are theorems, then $[(I) \Rightarrow (J)] \wedge [(J) \Rightarrow (I)]$ is also a theorem, which is $(I) \Leftrightarrow (J)$ by definition, and conversely. □

Hence the following theorem establishes the symmetry of equivalence.

Theorem 205 (Symmetry of $\Leftrightarrow$) *If $(H) \Leftrightarrow (K)$ is a theorem, then so is $(K) \Leftrightarrow (H)$.*

Proof. Apply the definition of $\Leftrightarrow$ and the commutativity of $\wedge$ (theorem 198):

$\vdash (H) \Leftrightarrow (K)$	hypothesis,
$\vdash [(H) \Rightarrow (K)] \wedge [(K) \Rightarrow (H)]$	definition of $\Leftrightarrow$,
$\vdash [(K) \Rightarrow (H)] \wedge [(H) \Rightarrow (K)]$	commutativity of $\wedge$ and *Modus Ponens*,
$\vdash (K) \Leftrightarrow (H)$	definition of $\Leftrightarrow$.

□

Similarly, the following theorem establishes the transitivity of equivalence.

Theorem 206 (Transitivity of $\Leftrightarrow$) *If $(H) \Leftrightarrow (K)$ and $(K) \Leftrightarrow (L)$ are theorems, then $(H) \Leftrightarrow (L)$ is a theorem.*

Proof. Apply the definition of $\Leftrightarrow$ and theorems 204 and 175:

$\vdash (H) \Leftrightarrow (K)$	hypothesis,
$\vdash (H) \Rightarrow (K)]$	theorem 204,
$\vdash (K) \Leftrightarrow (L)$	hypothesis,
$\vdash (K) \Rightarrow (L)]$	theorem 204,
$\vdash (H) \Rightarrow (L)$	theorem 175,

By symmetry (theorem 205) $(K) \Leftrightarrow (H)$ and $(L) \Leftrightarrow (K)$ are also theorems, whence $(L) \Rightarrow (H)$ is a theorem. From $(H) \Rightarrow (L)$ and $(L) \Rightarrow (H)$ it then follows that $(H) \Leftrightarrow (L)$ is a theorem. □

The following theorem demonstrates the use of the transitivity of equivalence in the proof that the conjunction $\wedge$ is associative.

Theorem 207 (Associativity of $\wedge$) *The following formula is a theorem:*

$$\{(P) \wedge [(Q) \wedge (R)]\} \Rightarrow \{[(P) \wedge (Q)] \wedge (R)\}.$$

Proof. Apply the law of commutation (theorem 185) and theorem 202:

$$
\begin{array}{rl}
\vdash [(P) \Rightarrow \{(R) \Rightarrow [\neg(Q)]\}] & \text{twice theorem 185 \ldots} \\
\Leftrightarrow [(R) \Rightarrow \{(P) \Rightarrow [\neg(Q)]\}] & \text{\ldots and theorem 202,} \\
\Updownarrow & \text{contrapositions,} \\
\vdash \{\neg[(P) \Rightarrow \{(Q) \Rightarrow [\neg(R)]\}]\} & \\
\Leftrightarrow \{\neg[(R) \Rightarrow \{(P) \Rightarrow [\neg(Q)]\}]\} & \\
\Updownarrow & \text{definition of } \wedge, \\
\vdash \{\neg[(P) \Rightarrow \{\neg[(Q) \wedge (R)]\}]\} & \\
\Leftrightarrow \{\neg[(R) \Rightarrow \{\neg[(P) \wedge (Q)]\}]\} & \\
\Updownarrow & \text{definition of } \wedge, \\
\vdash \{(P) \wedge [(Q) \wedge (R)]\} & \\
\Leftrightarrow \{(R) \wedge [(P) \wedge (Q)]\} & \\
\Updownarrow & \text{commutativity of } \wedge. \\
\vdash \{(P) \wedge [(Q) \wedge (R)]\} & \\
\Leftrightarrow \{[(P) \wedge (Q)] \wedge (R)\} &
\end{array}
$$

□

1.4.4 Examples of Proofs of Theorems with Disjunctions

The first theorem establishes the law of excluded middle in classical logic.

Theorem 208 (Excluded middle) *The formula* $(B) \vee [\neg(B)]$ *is a theorem.*

Proof. Apply theorem 173 and the definition of the disjunction $(\vee)$:

$$
\begin{array}{ll}
\vdash [\neg(B)] \Rightarrow [\neg(B)] & \text{theorem 173,} \\
\vdash (B) \vee [\neg(B)] & \text{definition of } \vee \text{ with } \neg \text{ and } \Rightarrow.
\end{array}
$$

□

The next theorem shows that the disjunction $\vee$ commutes:

Theorem 209 (Commutativity of $\vee$) *The formula* $[(P) \vee (Q)] \Rightarrow [(Q) \vee (P)]$ *is a theorem.*

Proof. Apply contraposition and theorems 189 and 184:

$$
\begin{array}{ll}
\vdash \{[\neg(P)] \Rightarrow (Q)\} \Rightarrow ([\neg(Q)] \Rightarrow \{\neg[\neg(P)\}) & \text{contraposition,} \\
\vdash \{\neg[\neg(P)\} \Rightarrow (P) & \text{theorem 189,} \\
\vdash \{[\neg(P)] \Rightarrow (Q)\} \Rightarrow \{[\neg(Q)] \Rightarrow (P)\} & \text{theorem 184.}
\end{array}
$$

□

Theorem 210 *The formulae* $(P) \Rightarrow [(P) \vee (Q)]$ *and* $(Q) \Rightarrow [(P) \vee (Q)]$ *are theorems.*

Proof. Apply theorem 188 and the definitions:

$$
\begin{array}{ll}
\vdash (P) \Rightarrow \{[\neg(P)] \Rightarrow (Q)\} & \text{theorem 188,} \\
\vdash (P) \Rightarrow [(P) \vee (Q)] & \text{definition of } \vee \text{ with } \neg \text{ and } \Rightarrow.
\end{array}
$$

For the second formula, swap P and Q to get the theorem $(Q) \Rightarrow [(Q) \vee (P)]$. Then apply the commutivity of $\vee$, $[(Q) \vee (P)] \Rightarrow [(P) \vee (Q)]$, and finally the transitivity of $\Rightarrow$ to complete the proof of $(Q) \Rightarrow [(P) \vee (Q)]$. □

The following theorem shows that the disjunction $\vee$ is associative:

Theorem 211 (Associativity of $\vee$) *The following formula is a theorem:*

$$\{(P) \vee [(Q) \vee (R)]\} \Leftrightarrow \{[(P) \vee (Q)] \vee (R)\}.$$

Proof. Apply the law of commutation (theorem 185) and theorem 202:

$$
\begin{array}{rl}
\vdash \left([\neg(P)] \Rightarrow \{[\neg(R)] \Rightarrow (Q)\}\right) & \text{twice theorem 185 \ldots} \\
\Leftrightarrow \left([\neg(R)] \Rightarrow \{[\neg(P)] \Rightarrow (Q)\}\right) & \text{\ldots and theorem 202,} \\
\Updownarrow & \text{definition of } \vee, \\
\vdash \{(P) \vee [(R) \vee (Q)]\} \Leftrightarrow \{(R) \vee [(P) \vee (Q)]\} & \\
\Updownarrow & \text{commutativity of } \vee, \text{ etc.} \\
\vdash \{(P) \vee [(Q) \vee (R)]\} \Leftrightarrow \{[(P) \vee (Q)] \vee (R)\} &
\end{array}
$$

□

1.4.5 Examples of Proofs with Conjunctions and Disjunctions

The following theorems establish de Morgan's laws in classical logic.

Theorem 212 (De Morgan's first law) *The following formula is a theorem:*

$$\{\neg[(P) \wedge (Q)]\} \Leftrightarrow \{[\neg(P)] \vee [\neg(Q)\}.$$

Proof. Apply the definitions of $\wedge$ and $\vee$ and double negations:

$$
\begin{array}{rl}
\neg[(P) \wedge (Q)] & \\
\Updownarrow & \text{definition of } \wedge, \\
[\neg(\neg\{(P) \Rightarrow [\neg(Q)]\})] & \\
\Updownarrow & \text{double negation (theorems 189 and 190),} \\
(P) \Rightarrow [\neg(Q)] & \\
\Updownarrow & \text{double negation (theorems 189 and 190),} \\
\{\neg[\neg(P)]\} \Rightarrow [\neg(Q)] & \\
\Updownarrow & \text{definition of } \vee. \\
[\neg(P)] \vee [\neg(Q)]. &
\end{array}
$$

□

Theorem 213 (De Morgan's second law) *The following formula is a theorem:*

$$\{\neg[(P) \vee (Q)]\} \Leftrightarrow \{[\neg(P)] \wedge [\neg(Q)\}.$$

Proof. Apply the definitions of $\wedge$ and $\vee$ and double negations:

$$
\begin{array}{rl}
\neg[(P) \vee (Q)] & \\
\Updownarrow & \text{definition of } \vee, \\
\neg\{[\neg(P)] \Rightarrow (Q)\} & \\
\Updownarrow & \text{double negation (theorems 189 and 190),} \\
\neg([\neg(P)] \Rightarrow \{\neg[\neg(Q)]\}) & \\
\Updownarrow & \text{definition of } \wedge. \\
[\neg(P)] \wedge [\neg(Q)]. &
\end{array}
$$

□

The following theorem shows that disjunctions distribute over conjunctions.

Theorem 214 (Distributivity of $\vee$ over $\wedge$) *The following formula is a theorem:*

$$\{(P) \vee [(Q) \wedge (R)]\} \Leftrightarrow \{[(P) \vee (Q)] \wedge [(P) \vee (R)]\}.$$

Proof. Apply the definition of $\vee$ and theorem 202:

$$
\begin{array}{rl}
[(P) \vee (Q)] \wedge [(P) \vee (R)] & \\
\Updownarrow & \text{definition of } \vee, \\
\{[\neg(P)] \Rightarrow (Q)\} \wedge \{[\neg(P)] \Rightarrow (R)\} & \\
\Updownarrow & \text{theorem 202}, \\
[\neg(P)] \Rightarrow [(Q) \wedge (R)] & \\
\Updownarrow & \text{definition of } \vee. \\
(P) \vee [(Q) \wedge (R)] &
\end{array}
$$

□

The following theorem shows that conjunctions distribute over disjunctions.

Theorem 215 (Distributivity of $\wedge$ over $\vee$) *The following formula is a theorem:*

$$\{(P) \wedge [(Q) \vee (R)]\} \Leftrightarrow \{[(P) \wedge (Q)] \vee [(P) \wedge (R)]\}.$$

Proof. Apply the definition of $\wedge$ and theorem 202:

$$
\begin{array}{rl}
[(P) \wedge (Q)] \vee [(P) \wedge (R)] & \\
\Updownarrow & \text{definition of } \wedge, \\
(\neg\{(P) \Rightarrow [\neg(Q)]\}) \vee (\neg\{(P) \Rightarrow [\neg(R)]\}) & \\
\Updownarrow & \text{de Morgan's first law}, \\
\neg(\{(P) \Rightarrow [\neg(Q)]\} \wedge \{(P) \Rightarrow [\neg(R)]\}) & \\
\Updownarrow & \text{theorem 202}, \\
\neg[(P) \Rightarrow \{[\neg(Q)] \wedge [\neg(R)]\}] & \\
\Updownarrow & \text{de Morgan's second law}, \\
\neg[(P) \Rightarrow \{\neg[(Q) \vee (R)]\}] & \\
\Updownarrow & \text{definition of } \wedge. \\
(P) \wedge [(Q) \vee (R)] &
\end{array}
$$

□

1.4.6 Exercises

For the following exercises, prove that the stated formulae are theorems, using the classical propositional calculus and any of the results just proved.

Exercise 361 $\{[(P) \Rightarrow (Q)] \Rightarrow (P)\} \Rightarrow (P)$

Exercise 362 $[(P) \Rightarrow (R)] \Rightarrow [\{[(P) \Rightarrow (Q)] \Rightarrow (R)\} \Rightarrow (R)]$

Exercise 363 $\{[(P) \Rightarrow (Q)] \Rightarrow (R)\} \Rightarrow \{[(R) \Rightarrow (P)] \Rightarrow (P)\}$

Exercise 364 $\{[(P) \Rightarrow (Q)] \Rightarrow (R)\} \Rightarrow \{[(P) \Rightarrow (R)] \Rightarrow (R)\}$

Exercise 365 $(P) \Rightarrow ([\neg(Q)] \Rightarrow \{\neg[(P) \Rightarrow (Q)]\})$

Exercise 366 $[\neg(P)] \Rightarrow \{\neg[(P) \wedge (Q)]\}$

Exercise 367 $(P) \Leftrightarrow \{[(P) \Rightarrow (Q)] \Rightarrow (P)\}$

Exercise 368 $[\neg(P)] \Rightarrow ([\neg(Q)] \Rightarrow \{\neg[(P) \vee (Q)]\})$

Exercise 369 $[(P) \Rightarrow (Q)] \Leftrightarrow (\neg\{(P) \wedge [\neg(Q)]\})$

Exercise 370 $[(P) \Rightarrow (Q)] \Leftrightarrow \{[\neg(P)] \vee (Q)\}$

1.5 OTHER FORMS OF DEDUCTIVE REASONING

1.5.1 Conjunctions of Implications

The following theorems establish derived rules of inference based on conjunctions of implications. The first theorem shows that if either of two hypotheses leads to a conclusion, then the *disjunction* of both hypotheses also leads to the same conclusion.

Theorem 216 *If* $\vdash (U) \Rightarrow (W)$ *and* $\vdash (V) \Rightarrow (W)$, *then* $\vdash [(U) \vee (V)] \Rightarrow (W)$. *Hence*

$$\vdash \{[(U) \Rightarrow (W)] \wedge [(V) \Rightarrow (W)]\} \Rightarrow \{[(U) \vee (V)] \Rightarrow (W)\}.$$

Proof. The first part of the proof assumes the two hypotheses.

$\vdash (U) \Rightarrow (W)$ hypothesis,

$\vdash [\neg(W)] \Rightarrow [\neg(U)]$ contraposition and *Modus Ponens*,

$\vdash (V) \Rightarrow (W)$ hypothesis,

$\vdash [\neg(W)] \Rightarrow [\neg(V)]$ contraposition and *Modus Ponens*,

$\vdash [\neg(W)] \Rightarrow \{[\neg(U)] \wedge [\neg(V)]\}$ theorem 202,

$\vdash [\neg(W)] \Rightarrow \{\neg[(U) \vee (V)]\}$ de Morgan's second law,

$\vdash [(U) \vee (V)] \Rightarrow (W)$ axiom P3 and *Modus Ponens*.

The second part of the proof dispenses with the two hypotheses.

$\vdash \{[(U) \Rightarrow (W)] \wedge [(V) \Rightarrow (W)]\} \Rightarrow [(U) \Rightarrow (W)]$ theorem 200,

$\vdash [(U) \Rightarrow (W)] \Rightarrow \{[\neg(W)] \Rightarrow [\neg(U)]\}$ transposition,

$\vdash \{[(U) \Rightarrow (W)] \wedge [(V) \Rightarrow (W)]\} \Rightarrow \{[\neg(W)] \Rightarrow [\neg(U)]\}$ theorem 175;

$\vdash \{[(U) \Rightarrow (W)] \wedge [(V) \Rightarrow (W)]\} \Rightarrow [(V) \Rightarrow (W)]$ theorem 200,

$\vdash [(V) \Rightarrow (W)] \Rightarrow \{[\neg(W)] \Rightarrow [\neg(V)]\}$ transposition,

$\vdash \{[(U) \Rightarrow (W)] \wedge [(V) \Rightarrow (W)]\} \Rightarrow \{[\neg(W)] \Rightarrow [\neg(V)]\}$ theorem 175;

$\vdash \{[(U) \Rightarrow (W)] \wedge [(V) \Rightarrow (W)]\}$
$\quad \Rightarrow (\{[\neg(W)] \Rightarrow [\neg(U)]\} \wedge \{[\neg(W)] \Rightarrow [\neg(V)]\})$ theorem 202;

$\vdash (\{[\neg(W)] \Rightarrow [\neg(U)]\} \wedge \{[\neg(W)] \Rightarrow [\neg(V)]\})$
$\quad \Rightarrow ([\neg(W)] \Rightarrow \{[\neg(U)] \wedge [\neg(V)]\})$ theorem 202;

$\vdash ([\neg(W)] \Rightarrow \{[\neg(U)] \wedge [\neg(V)]\}) \Rightarrow [\{[(U) \vee (V)]\} \Rightarrow (W)]$ contraposition;

$\vdash \{[(U) \Rightarrow (W)] \wedge [(V) \Rightarrow (W)]\} \Rightarrow [\{[(U) \vee (V)]\} \Rightarrow (W)]$ theorem 202.

□

Similarly, the second theorem shows that if either of two hypotheses leads to a conclusion, then the *conjunction* of both hypotheses also leads to the same conclusion.

Theorem 217 *If* $\vdash (U) \Rightarrow (W)$ *and* $\vdash (V) \Rightarrow (W)$, *then* $\vdash [(U) \wedge (V)] \Rightarrow (W)$. *Hence*

$$\vdash \{[(U) \Rightarrow (W)] \wedge [(V) \Rightarrow (W)]\} \Rightarrow \{[(U) \wedge (V)] \Rightarrow (W)\}.$$

Proof. This proof relies on theorems 183, 200, 199, 210, 216:

$$\vdash \underbrace{\{[(U) \Rightarrow (W)] \wedge [(V) \Rightarrow (W)]\}}_{H} \Rightarrow \{\underbrace{[(U) \vee (V)]}_{L} \Rightarrow (\underbrace{W}_{M})\} \quad \text{theorem 216,}$$

$$\vdash \underbrace{[(U) \wedge (V)]}_{K} \Rightarrow \underbrace{[(U) \vee (V)]}_{L} \quad \text{200, 199, 210,}$$

$$\vdash \underbrace{\{[(U) \Rightarrow (W)] \wedge [(V) \Rightarrow (W)]\}}_{H} \Rightarrow \{\underbrace{[(U) \wedge (V)]}_{K} \Rightarrow (\underbrace{W}_{M})\} \quad \text{theorem 183.}$$

□

The third theorem shows that if one hypothesis leads to either of two conclusions, then the same hypothesis also leads to the disjunction of both conclusions.

Theorem 218 *If* $\vdash (P) \Rightarrow (Q)$ *or* $\vdash (P) \Rightarrow (S)$, *then* $\vdash (P) \Rightarrow [(Q) \vee (S)]$. *Hence*

$$\vdash \{[(P) \Rightarrow (Q)] \vee [(P) \Rightarrow (S)]\} \Rightarrow \{(P) \Rightarrow [(Q) \vee (S)]\},$$

and conversely.

Proof. This proof uses contraposition and previously established equivalences.

$$\begin{array}{rl}
\neg\{(P) \Rightarrow [(Q) \vee (S)]\} & \\
\Updownarrow & \text{definition of } \wedge \text{ and equivalences,} \\
(P) \wedge \{\neg[(Q) \vee (S)]\} & \\
\Updownarrow & \text{de Morgan's second law,} \\
(P) \wedge \{[\neg(Q)] \wedge [\neg(S)]\} & \\
\Updownarrow & \text{idempotence of } \wedge, \\
[(P) \wedge (P)] \wedge \{[\neg(Q)] \wedge [\neg(S)]\} & \\
\Updownarrow & \text{associativity and commutativity of } \wedge, \\
\{(P) \wedge [\neg(Q)]\} \wedge \{(P) \wedge [\neg(S)]\} & \\
\Updownarrow & \text{definition of } \wedge, \\
\{\neg[(P) \Rightarrow (Q)]\} \wedge \{\neg[(P) \Rightarrow (S)]\} & \\
\Updownarrow & \text{de Morgan's first law.} \\
\neg\{[(P) \Rightarrow (Q)] \vee [(P) \Rightarrow (S)]\} &
\end{array}$$

□

The fourth theorem shows that if two implications hold, then the conjunction of their hypotheses leads to the *conjunction* of their conclusions.

Theorem 219 *If* $\vdash (P) \Rightarrow (Q)$ *and* $\vdash (R) \Rightarrow (S)$, *then* $\vdash [(P) \wedge (R)] \Rightarrow [(Q) \wedge (S)]$. *Hence*

$$\vdash \{[(P) \Rightarrow (Q)] \wedge [(R) \Rightarrow (S)]\} \Rightarrow \{[(P) \wedge (R)] \Rightarrow [(Q) \wedge (S)]\}.$$

Proof.

$$\vdash [(P) \wedge (R)] \Rightarrow (P) \qquad \text{theorem 200,}$$
$$\vdash \{[(P) \wedge (R)] \Rightarrow (P)\} \Rightarrow ([(P) \Rightarrow (Q)] \Rightarrow \{[(P) \wedge (R)] \Rightarrow (Q)\}) \qquad \text{theorem 182,}$$
$$\vdash [(P) \Rightarrow (Q)] \Rightarrow \{[(P) \wedge (R)] \Rightarrow (Q)\} \qquad \textit{Modus Ponens};$$

$$\vdash \{[(P) \Rightarrow (Q)] \wedge [(R) \Rightarrow (S)]\} \Rightarrow [(P) \Rightarrow (Q)] \qquad \text{theorem 200,}$$
$$\vdash \underbrace{\{[(P) \Rightarrow (Q)] \wedge [(R) \Rightarrow (S)]\}}_{H} \Rightarrow \underbrace{\{[(P) \wedge (R)] \Rightarrow (Q)\}}_{K} \qquad \text{theorem 174;}$$

$$\vdash [(P) \wedge (R)] \Rightarrow (R) \qquad \text{theorem 200,}$$
$$\vdash \{[(P) \wedge (R)] \Rightarrow (R)\} \Rightarrow ([(R) \Rightarrow (S)] \Rightarrow \{[(P) \wedge (R)] \Rightarrow (S)\}) \qquad \text{theorem 182,}$$
$$\vdash [(R) \Rightarrow (S)] \Rightarrow \{[(P) \wedge (R)] \Rightarrow (S)\} \qquad \textit{Modus Ponens};$$

$$\vdash \{[(P) \Rightarrow (Q)] \wedge [(R) \Rightarrow (S)]\} \Rightarrow [(R) \Rightarrow (S)] \qquad \text{theorem 200,}$$
$$\vdash \underbrace{\{[(P) \Rightarrow (Q)] \wedge [(R) \Rightarrow (S)]\}}_{H} \Rightarrow \underbrace{\{[(P) \wedge (R)] \Rightarrow (S)\}}_{L} \qquad \text{theorem 174;}$$
$$\vdash \underbrace{\{[(P) \Rightarrow (Q)] \wedge [(R) \Rightarrow (S)]\}}_{H} \Rightarrow (\underbrace{\{[(P) \wedge (R)] \Rightarrow (Q)\}}_{K} \wedge \underbrace{\{[(P) \wedge (R)] \Rightarrow (S)\}}_{L}) \qquad \text{theorem 202;}$$

$$\vdash (\{\underbrace{[(P) \wedge (R)]}_{H} \Rightarrow (Q)\} \wedge \{\underbrace{[(P) \wedge (R)]}_{H} \Rightarrow (S)\}) \Rightarrow \{\underbrace{[(P) \wedge (R)]}_{H} \Rightarrow [(Q) \wedge (S)]\} \qquad \text{theorem 202;}$$

$$\vdash \{[(P) \Rightarrow (Q)] \wedge [(R) \Rightarrow (S)]\} \Rightarrow \{[(P) \wedge (R)] \Rightarrow [(Q) \wedge (S)]\} \qquad \text{theorem 174.}$$

□

Similarly, the fifth theorem shows that if two implications hold, then the conjunction of their hypotheses leads to the *disjunction* of their conclusions.

Theorem 220 *If* $\vdash (P) \Rightarrow (Q)$ *and* $\vdash (R) \Rightarrow (S)$, *then* $\vdash [(P) \wedge (R)] \Rightarrow [(Q) \vee (S)]$. *Hence*

$$\vdash \{[(P) \Rightarrow (Q)] \wedge [(R) \Rightarrow (S)]\} \Rightarrow \{[(P) \wedge (R)] \Rightarrow [(Q) \vee (S)]\}.$$

Proof. This proof relies on theorem 219:

$$\vdash \underbrace{\{[(P) \Rightarrow (Q)] \wedge [(R) \Rightarrow (S)]\}}_{H} \Rightarrow \{\underbrace{[(P) \wedge (R)]}_{L} \Rightarrow \underbrace{[(Q) \wedge (S)]}_{M}\} \qquad 219,$$
$$\vdash \underbrace{[(Q) \wedge (S)]}_{M} \Rightarrow \underbrace{[(Q) \vee (S)]}_{N} \qquad 200, 199, 210,$$
$$\vdash \underbrace{\{[(P) \Rightarrow (Q)] \wedge [(R) \Rightarrow (S)]\}}_{H} \Rightarrow \{\underbrace{[(P) \wedge (R)]}_{L} \Rightarrow \underbrace{[(Q) \vee (S)]}_{N}\} \qquad 184.$$

□

1.5.2 Proofs by Cases or by Contradiction

The following theorems establish further derived rules of inference. The first theorem forms a part of the basis for an algorithm — called the Completeness Theorem — to design proofs within the propositional calculus, as explained in chapter 4.

Theorem 221 (Proof by cases) *If* $(H) \Rightarrow (R)$ *and* $[\neg(H)] \Rightarrow (R)$ *are theorems, then* R *is a theorem. Hence the following formula is also a theorem:*

$$([(H) \Rightarrow (R)] \wedge \{[\neg(H)] \Rightarrow (R)\}) \Rightarrow (R).$$

Proof. Apply theorems 216 and 208:

$\vdash (H) \Rightarrow (R)$	hypothesis,
$\vdash [\neg(H)] \Rightarrow (R)$	hypothesis,
$\vdash \{(H) \vee [\neg(H)]\} \Rightarrow (R)$	theorem 216,
$\vdash (H) \vee [\neg(H)]$	theorem 208,
$\vdash R$	*Modus Ponens*;

$\vdash ([(H) \Rightarrow (R)] \wedge \{[\neg(H)] \Rightarrow (R)\}) \Rightarrow [\{(H) \vee [\neg(H)]\} \Rightarrow (R)]$	theorem 216,
$\vdash (H) \vee [\neg(H)]$	theorem 208,
$\vdash ([(H) \Rightarrow (R)] \wedge \{[\neg(H)] \Rightarrow (R)\}) \Rightarrow (R)$	theorem 176.

□

The second theorem establishes the classical principle of proof by contradiction.

Theorem 222 (Proof by contradiction) *If* $[\neg(R)] \Rightarrow (S)$ *and* $[\neg(R)] \Rightarrow [\neg(S)]$ *are theorems, then* R *is a theorem. Hence the following formula is also a theorem:*

$$(\{[\neg(R)] \Rightarrow (S)\} \wedge \{[\neg(R)] \Rightarrow [\neg(S)]\}) \Rightarrow (R).$$

Proof. Apply theorem 219, de Morgan's first law, and contraposition:

$\vdash [\neg(R)] \Rightarrow (S)$	hypothesis,
$\vdash [\neg(R)] \Rightarrow [\neg(S)]$	hypothesis,

$$\vdash (\{[\neg(R)] \Rightarrow (S)\} \wedge \{[\neg(R)] \Rightarrow [\neg(S)]\}) \Rightarrow (\{[\neg(R)] \wedge [\neg(R)]\} \Rightarrow \{(S) \wedge [\neg(S)]\}) \quad \text{theorem 219,}$$

$$\vdash [\neg(R)] \Rightarrow \{[\neg(R)] \wedge [\neg(R)]\} \quad \text{theorem 203,}$$

$$\vdash (\{[\neg(R)] \Rightarrow (S)\} \wedge \{[\neg(R)] \Rightarrow [\neg(S)]\}) \Rightarrow ([\neg(R)] \Rightarrow \{(S) \wedge [\neg(S)]\}) \quad \text{theorem 183,}$$

$\vdash ([\neg(R)] \Rightarrow \{(S) \wedge [\neg(S)]\}) \Rightarrow [(\neg\{(S) \wedge [\neg(S)]\}) \Rightarrow \{\neg[\neg(R)]\}]$	contraposition,
$\vdash ([\neg(R)] \Rightarrow \{(S) \wedge [\neg(S)]\}) \Rightarrow [\{[\neg(S)] \vee (S)\} \Rightarrow (R)]$	de Morgan,
$\vdash [\neg(S)] \vee (S)$	theorem 208 ,
$\vdash ([\neg(R)] \Rightarrow \{(S) \wedge [\neg(S)]\}) \Rightarrow (R)$	theorem 176 ,
$\vdash (\{[\neg(R)] \Rightarrow (S)\} \wedge \{[\neg(R)] \Rightarrow [\neg(S)]\}) \Rightarrow (R)$	theorem 175.

□

1.5.3 Exercises

Exercise 371 Determine whether the following derived rule holds. "If $(V) \Rightarrow (W)$ and $(R) \Rightarrow (S)$ are theorems, then $[(V) \vee (R)] \Rightarrow [(W) \vee (S)]$ is also a theorem":

$$\{[(V) \Rightarrow (W)] \wedge [(R) \Rightarrow (S)]\} \Rightarrow \{[(V) \vee (R)] \Rightarrow [(W) \vee (S)]\}.$$

Exercise 372 Determine whether the following derived rule holds. "If $(I) \Rightarrow (R)$ and $(I) \Rightarrow [\neg(S)]$ are both theorems, then $\neg[(R) \Rightarrow (S)]$ is also a theorem":

$$\big([(I) \Rightarrow (R)] \wedge \{(I) \Rightarrow [\neg(S)]\}\big) \Rightarrow \{\neg[(R) \Rightarrow (S)]\}.$$

Exercise 373 Determine whether the following derived rule holds. "If $(U) \Rightarrow (W)$ or $(V) \Rightarrow (W)$ is a theorem, then $[(U) \vee (V)] \Rightarrow (W)$ is also a theorem":

$$\{[(U) \Rightarrow (W)] \vee [(V) \Rightarrow (W)]\} \Rightarrow \{[(U) \vee (V)] \Rightarrow (W)\}.$$

Exercise 374 Determine whether the following derived rule holds. "If $(P) \Rightarrow (Q)$ or $(R) \Rightarrow (S)$ is a theorem, then $[(P) \vee (R)] \Rightarrow [(Q) \vee (S)]$ is also a theorem":

$$\{[(P) \Rightarrow (Q)] \vee [(R) \Rightarrow (S)]\} \Rightarrow \{[(P) \vee (R)] \Rightarrow [(Q) \vee (S)]\}.$$

Exercise 375 Determine whether the following derived rule holds. "If $(P) \Rightarrow (S)$ and $[\neg(Q)] \Rightarrow [\neg(S)]$ is a theorem, then $(P) \Rightarrow (Q)$ is also a theorem":

$$\big([(P) \Rightarrow (S)] \wedge \{[\neg(Q)] \Rightarrow [\neg(S)]\}\big) \Rightarrow [(P) \Rightarrow (Q)].$$

Exercise 376 Prove that if T is a theorem, then $(R) \Rightarrow [(T) \Rightarrow (R)]$ is a theorem.

Exercise 377 Prove that if T is a theorem, then $[(T) \Rightarrow (R)] \Rightarrow (R)$ is a theorem.

Exercise 378 Prove that if $\neg(F)$ is a theorem, then so is $\{[\neg(R)] \Rightarrow (F)\} \Rightarrow (R)$.

Exercise 379 Prove that if $\neg(F)$ is a theorem, then so is $(R) \Rightarrow \{[\neg(R)] \Rightarrow (F)\}$.

Exercise 380 Determine whether the following derived rule of inference holds. "If $\neg(F)$ and $(P) \wedge [\neg(Q)]\} \Rightarrow (F)$ are theorems, then $(P) \Rightarrow (Q)$ is also a theorem."

1.6 PREDICATE CALCULUS

1.6.1 Predicates, Quantifiers, Free or Bound Variables

The concept of **predicate** depends upon the theory under consideration. For instance, one version of set theory has only one primitive predicate ($\in$), which denotes set membership. A formula is then called **atomic** if and only if it is one of the formulae

$$\varnothing \in \varnothing, \quad \varnothing \in X, \quad X \in \varnothing, \quad X \in Y,$$

with any variables X and Y. If an atomic formula contains a variable, then it may, but need not, have a Truth value. For example, the formula $X \in Y$ has no Truth value, because different substitutions for X and Y can yield different Truth values. However some formulae may contain variables and yet have a Truth value.

Definition 223 A formula is **universally valid** if and only if it is True regardless of its variables. The notation $\models P$ indicates that P is universally valid [92, p. 163].

For the present exposition, the notation $\vdash P$ indicates that there exists a proof of the well-formed propositional form P relying solely on the axioms and rules of inference of the propositional calculus. In contrast, the notation $\models P$ indicates that there exists a proof of the well-formed logical formula P that may require not only the propositional calculus, but also the predicate calculus explained here. Optionally, and only for demonstration purposes, some examples may involve sets or binary arithmetic.

Example 224 In set theory the formula $X \in \varnothing$ is False for every X. Consequently, $\neg(X \in \varnothing)$ is universally valid. Thus, $\models \neg(X \in \varnothing)$.

Example 225 In the theory of well-formed sets the formula $X \in X$ is False for every X. Consequently, $\neg(X \in X)$ is universally valid. Thus, $\models \neg(X \in X)$.

Example 226 In binary arithmetic (example 15) the formula $0 * X = 0$ is True for every X. Consequently, $0 * X = 0$ is universally valid. Thus, $\models 0 * X = 0$.

Example 227 In binary arithmetic (example 15), $0 * X = 1$ is False for every X. Consequently, $\neg(0 * X = 1)$ is universally valid. Thus, $\models \neg(0 * X = 1)$.

Variables occurring in propositional forms, for instance, the variables P and Q in $(P) \Rightarrow [(Q) \Rightarrow (P)]$, describe patterns of deductive reasoning. In contrast, variables occurring in atomic formulae, for instance, X and Y in $X \in Y$, belong to applications of deductive reasoning. Different words reflect this difference between variables.

Definition 228 A variable is an **individual variable** if and only if it occurs in an atomic formula involving predicates. A variable is a **propositional variable** if and only if it occurs in a propositional form involving logical connectives.

Example 229 The variables X and Y in $X \in Y$ are individual variables because they occur with the predicate $\in$. The variables P and Q in $(P) \Rightarrow [(Q) \Rightarrow (P)]$ are propositional variables because they occur in a propositional form.

One method for a logical system to distinguish individual variables from propositional variables consists in maintaining two disjoint collections of variables, for instance, the Roman alphabet (with optional subscripts or superscripts) for propositional variables, and the Greek alphabet (with optional subscripts or superscripts) for individual variables. In the present section, $H, K, L, M, N, P, Q, R, S$ will be propositional variables, whereas W, X, Y, Z will be individual variables. Thus if X is an individual variable and if P is a logical formula, then in certain contexts

the formula $\forall X(P)$ means that "for each X the proposition P is True",
the formula $\exists X(P)$ means that "there exists X for which P is True".

In the logic presented here, only individual variables may appear immediately after either quantifier, $\forall$ (read "for each") or $\exists$ (read "there exists"). Because of this restriction, this logic is a **first order logic**. Logical systems allowing for propositional variables to appear immediately after a quantifier are of second order or higher order.

If a formula P is True regardless of X, but if P also contains another variable Z, then substituting Z for X can change the Truth value of P.

Counterexample 230 Consider any context with at least two different objects, for instance, two sets, or two binary numbers. Thus for each object X there exists a different object Z, whence $\exists Z[\neg(X = Z)]$ is True. Replacing Z by X in $\exists Z[\neg(X = Z)]$ gives $\exists X[\neg(X = X)]$, which is False, because each object equals itself. Thus, replacing Z by X in the True formula $\exists Z[\neg(X = Z)]$ yields the False formula $\exists X[\neg(X = X)]$.

Counterexample 230 shows that substitutions of a variable by another must obey certain rules, for instance, with the concepts introduced in definition 231.

Definition 231 For each individual variable X and for each logical formula P, an occurrence of the variable X is **bound** in the formula P if and only if in P that occurrence of the variable X immediately follows $\forall$ or $\exists$, or if it appears between either $\forall X($ or $\exists X($ and the corresponding right parenthesis $))$. An occurrence of the variable X is **free** in P if and only if that occurrence of X is not bound in P. A logical formula is **closed** if and only if it does not contain any free occurrence of any variable. A logical **sentence** is a closed logical formula.

Example 232

The variable X is bound in the formula $\exists X(\varnothing = \varnothing)$.
The formula $\exists X(\varnothing = \varnothing)$ is closed, hence it is a sentence.
The variable X is free in the formula $\forall Z(Z = X)$.
The formula $\forall Z(Z = X)$ is *not* closed (because X occurs freely).
Both occurrences of X are bound in the formula $\forall X[\neg(X \in \varnothing)]$.
The formula $\forall X[\neg(X \in \varnothing)]$ is closed, hence it is a sentence.
The first and second occurrences of X are bound and the third occurrence of X is free in the formula $[\exists X(X = \varnothing)] \vee (X = \varnothing)$.
The formula $[\exists X(X = \varnothing)] \vee (X = \varnothing)$ is *not* closed.

The following definition avoids the phenomenon exhibited in counterexample 230.

Definition 233 A **proper substitution** of a variable Z for each free occurrence of a *different* variable X in a logical formula P consists of the following three steps:

(1) Identify a variable that does *not* occur in P, for example, Y.

(2) In P, replace each *bound* occurrence of Z by Y.

(3) Then replace each *free* occurrence of X by Z.

The proper substitution of Z for each free occurrence of X in P is denoted by $\text{Subf}_Z^X(P)$ [79, p. 179]. For convenience, $\text{Subf}_X^X(P)$ is defined to be P, and if X does *not* occur freely in P, then $\text{Subf}_Z^X(P)$ is also defined to be P. The concept and notation for proper substitutions also apply to the substitution of a constant for a free variable. Because constants cannot appear immediately after a quantifier, they are not bound. Consequently, only the last step applies to the proper substitution of constants for free variables. Thus, $\text{Subf}_\varnothing^X(P)$ merely substitutes $\varnothing$ for every free occurrence of X in P.

Example 234 For P consider the formula $(\forall X\{\exists Z[\neg(X = Z)]\}) \vee (X = \varnothing)$.

(1) Verify that the variable Y does not occur in P.

(2) Replace the bound occurrences of Z by Y, which gives the formula $(\forall X\{\exists Y[\neg(X = Y)]\}) \vee (X = \varnothing)$.

(3) Replace the free occurrence of X by Z, which gives the formula $(\forall X\{\exists Y[\neg(X = Y)]\}) \vee (Z = \varnothing)$ for $\text{Subf}_Z^X(P)$.

In contrast, substituting the constant $\varnothing$ for every free occurrence of X in P yields $(\forall X\{\exists Z[\neg(X = Z)]\}) \vee (\varnothing = \varnothing)$ for $\text{Subf}_\varnothing^X(P)$.

Example 235 A situation like that in definition 233 occurs with computer algorithms to swap two variables X and Z, which typically use a third variable Y distinct from X and Z as a temporary storage. Firstly, the algorithm assigns X to Y, an operation denoted by $Y := X$. Secondly, the algorithm assigns Z to X, an operation denoted by $X := Z$. Finally, the algorithm assigns Y to Z, an operation denoted by $Z := Y$.

In contrast to substitutions of free variables, the following definition does not impose any restriction on the substitution of bound variables.

Definition 236 The formula $\text{Subb}_Z^X(P)$ results from substituting the variable Z for each bound occurrence of the variable X in P.

Definition 236 explicitly applies only to variables. In particular, it does *not* allow for substitutions of constants for bound variables.

1.6.2 Axioms and Rules for the Predicate Calculus

As the axioms of the propositional calculus reflect patterns of deductive reasoning with implications and negations, the axioms of predicate calculus reflect patterns of deductive reasoning with quantifiers. There also exist several choices of initial axioms for use with quantifiers, for instance, the following axioms [11, §30, p. 171–172].

Definition 237 The following **axioms of the predicate calculus** govern the use of the **universal quantifier** $\forall$ and the **existential quantifier** $\exists$.

Axiom Q0 Axioms of the propositional calculus are axioms of the predicate calculus.

Axiom Q1 (Specialization) $[\forall X(P)] \Rightarrow [\text{Subf}_Z^X(P)]$.

Axiom Q2 $\{\forall X[(P) \Rightarrow (Q)]\} \Rightarrow \{(P) \Rightarrow [\forall X(Q)]\}$ if P contains no free X .

Axiom Q3 $\{\neg[\exists X(P)]\} \Leftrightarrow \{\forall X[\neg(P)]\}$.

Axiom Q4 $\{\exists X[\neg(P)]\} \Leftrightarrow \{\neg[\forall X(P)]\}$.

The first axiom of the predicate calculus (Q0) and the rules of inference carry all the theorems from the propositional calculus over to theorems of the predicate calculus.

The second axiom (Q1) corresponds to the notion that if an individual variable X may occur in a formula P, and if P is True regardless of X, in other words, if $\forall X(P)$ is True, then P remains True with X replaced by any individual variable or constant Z. If X and Z are the same variable, then axiom Q1 gives $[\forall X(P)] \Rightarrow (P)$.

The third axiom (Q2) describes the relation between the universal quantifier ("there exists") and the logical connective of the implicational calculus ("if ... then").

The first axiom for the existential quantifier (Q3) states that a formula P is False for every X if and only if there does *not* exist any X for which P is True.

Similarly, the second axiom for the existential quantifier (Q4) states that there exists some X for which P is False if and only if it is False that P is True for every X. Axiom Q4 asserts the existence of an object. Consequently, axiom Q4 applies neither to "empty" theories where nothing exists, nor to logics that require not only existence but also the determination of which objects satisfy a formula.

Besides the axioms, the predicate calculus allows for proofs of theorems through the following **rules of inference**.

Definition 238 (Theorems and proofs) Within the predicate calculus, a formula is a **theorem** if and only if it is obtained by the following **rules of inference**.

Rule 239 (Axioms)

Every axiom of the predicate calculus is a theorem.

Rule 240 (Substitutions) The following generalization of the rule of substitution holds:

> For each component K of a theorem R of the propositional calculus,
> and for each well-formed propositional form *or atomic formula* L,
> the formula obtained by replacing in R every occurrence of K by L is again a theorem.

Rule 241 ("Modus Ponens" (abbreviated by M. P.), or "Detachment")

> If P is a theorem, and
> if $(P) \Rightarrow (Q)$ is a theorem,
> then Q is a theorem.

Rule 242 (Generalization)

> If P is a theorem,
> then $\forall X(P)$ is also a theorem.

A **proof** of a theorem R is a sequence of theorems $H, K, L, \ldots P, Q, R$ obtained from the rules of inference.

The following derived rule of inference will simplify proofs.

Theorem 243 *If X does not occur freely in P, and if $\forall X[(P) \Rightarrow (Q)]$ is a theorem, then $(P) \Rightarrow [\forall X(Q)]$ is also a theorem.*

Proof. Apply axiom Q2 and *Modus Ponens*:

$\models \forall X[(P) \Rightarrow (Q)]$	hypothesis,
$\models \{\forall X[(P) \Rightarrow (Q)]\} \Rightarrow \{(P) \Rightarrow [\forall X(Q)]\}$	axiom Q2,
$\models (P) \Rightarrow [\forall X(Q)]$	*Modus Ponens*.

□

1.6.3 Examples of Proofs with the Predicate Calculus

The following theorems provide examples of proofs in predicate calculus. The first two theorems allow for exchanging quantifiers by introducing negations.

Theorem 244 *For all X and P without restrictions,* $\models [\exists X(P)] \Leftrightarrow (\neg\{[\forall X[\neg(P)]\})$.

Proof. Apply the full propositional calculus and axiom Q3:

$\models \{\neg[\exists X(P)]\} \Leftrightarrow \{\forall X[\neg(P)]\}$	axiom Q3,
$\models (\neg\{\neg[\exists X(P)]\}) \Leftrightarrow (\neg\{\forall X[\neg(P)]\})$	contraposition and its converse,
$\models [\exists X(P)] \Leftrightarrow (\neg\{\forall X[\neg(P)]\})$	double negation and transitivity.

□

Theorem 245 *For all X and P without restrictions,* $\models [\forall X(P)] \Leftrightarrow (\neg\{[\exists X[\neg(P)]\})$.

Proof. Apply the full propositional calculus and axiom Q4:

$\models \{\neg[\forall X(P)]\} \Leftrightarrow \{\exists X[\neg(P)]\}$	axiom Q4,
$\models (\neg\{\neg[\forall X(P)]\}) \Leftrightarrow (\neg\{\exists X[\neg(P)]\})$	contraposition and its converse,
$\models [\forall X(P)] \Leftrightarrow (\neg\{\exists X[\neg(P)]\})$	double negation and transitivity.

□

The following two theorems relate the universal quantifier to implications.

Theorem 246 $\models \{\forall X[(P) \Rightarrow (Q)]\} \Rightarrow \{[\forall X(P)] \Rightarrow (Q)\}$.

Proof. Apply the full propositional calculus and axiom Q1:

$\models \{\forall X[(P) \Rightarrow (Q)]\} \Rightarrow [(P) \Rightarrow (Q)]$	specialization (Q1),
$\models [\forall X(P)] \Rightarrow (P)$	specialization (Q1),
$\models \{\forall X[(P) \Rightarrow (Q)]\} \Rightarrow \{[\forall X(P)] \Rightarrow (Q)\}$	transitivity (theorem 183).

□

Theorem 247 $\models \{\forall X[(P) \Rightarrow (Q)]\} \Rightarrow \{[\forall X(P)] \Rightarrow [\forall X(Q)]\}$.

Proof. All occurrences of X are bound in $\forall X(P)$ and in $\forall X[(P) \Rightarrow (Q)]$; hence

$\models \{\forall X[(P) \Rightarrow (Q)]\} \Rightarrow \{[\forall X(P)] \Rightarrow (Q)\}$	theorem 246,
$\models \{\forall X[(P) \Rightarrow (Q)]\} \Rightarrow (\forall X\{[\forall X(P)] \Rightarrow (Q)\})$	theorem 243,
$\models (\forall X\{[\forall X(P)] \Rightarrow (Q)\}) \Rightarrow \{[\forall X(P)] \Rightarrow [\forall X(Q)]\}$	axiom Q2,
$\models \{\forall X[(P) \Rightarrow (Q)]\} \Rightarrow \{[\forall X(P)] \Rightarrow [\forall X(Q)]\}$	theorem 175.

□

Counterexample 248 The converse of theorem 247, which would be

$$[\forall X(P) \Rightarrow \forall X(Q)] \Rightarrow \{\forall X[(P) \Rightarrow (Q)]\},$$

is *False* in contexts with two *different* objects Y and Z, so that $\neg(Y = Z)$ is True:

- $\forall X[(X = Y) \Rightarrow (X = Z)]$ is False, because substituting Y for X gives $[(Y = Y) \Rightarrow (Y = Z)]$, which is False, because of the True hypothesis $Y = Y$ and the False conclusion $Y = Z$.
- $\forall X(X = Y)$ is False, because substituting Z for X gives $(Z = Y)$, which is False by the assumption that $\neg(Y = Z)$.
- $[\forall X(X = Y)] \Rightarrow [\forall X(X = Z)]$ is True, because of its False hypothesis.
- $\{[\forall X(X = Y)] \Rightarrow [\forall X(X = Z)]\} \Rightarrow \{\forall X[(X = Y) \Rightarrow (X = Z)]\}$ is False, because of the True hypothesis and the False conclusion.

The next theorem involves the existential quantifier.

Theorem 249 *For all X and P without restrictions,* $\models [\text{Subf}_Z^X(P)] \Rightarrow [\exists X(P)]$.

Proof. Apply specialization, contraposition, double negations, and transitivity:

$\models \{\forall X[\neg(P)]\} \Rightarrow \{\text{Subf}_Z^X[\neg(P)]\}$	specialization (axiom Q1),
$\models \{\text{Subf}_Z^X[\neg(P)]\} \Rightarrow \{\neg[\text{Subf}_Z^X(P)]\}$	definition of Subf_Z^X,
$\models \{\forall X[\neg(P)]\} \Rightarrow \{\neg[\text{Subf}_Z^X(P)]\}$	transitivity (theorem 175),
$\models [\text{Subf}_Z^X(P)] \Rightarrow (\neg\{\forall X[\neg(P)]\})$	contraposition, double negation,
$\models (\neg\{\forall X[\neg(P)]\}) \Rightarrow (\neg\{\neg[\exists X(P)]\})$	axiom Q3,
$\models (\neg\{\forall X[\neg(P)]\}) \Rightarrow [\exists X(P)]$	double negation and transitivity,
$\models [\text{Subf}_Z^X(P)] \Rightarrow [\exists X(P)]$	transitivity (theorem 175).

□

1.6.4 Exercises

Each of the following ten exercises lists one formula P. Identify a formula that is logically equivalent to $\neg(P)$ among the same ten exercises.

Exercise 381 $\forall X[\exists Y(X \in Y)]$

Exercise 382 $\forall X[\exists Y(Y \in X)]$

Exercise 383 $\forall X[(X \in A) \Rightarrow (X \in B)]$

Exercise 384 $\forall X\{(X \in C) \Leftrightarrow [(X \in A) \wedge (X \in B)]\}$

Exercise 385 $\forall X\{(X \in C) \Leftrightarrow [(X \in A) \vee (X \in B)]\}$

Exercise 386 $\exists X(\{(X \in C) \wedge [\neg(X \in A)] \wedge [\neg(X \in B)]\} \vee \{[\neg(X \in C)] \wedge [(X \in A) \vee (X \in B)]\})$

Exercise 387 $\exists X\{(X \in A) \wedge [\neg(X \in B)]\}$

Exercise 388 $\exists X\{\forall Y[\neg(Y \in X)]\}$

Exercise 389 $\exists X([(X \in C) \wedge \{[\neg(X \in A)] \vee [\neg(X \in B)]\}] \vee \{[\neg(X \in C)] \wedge [(X \in A) \wedge (X \in B)]\})$

Exercise 390 $\exists X\{\forall Y[\neg(X \in Y)]\}$

1.7 INFERENCE WITH QUANTIFIERS

1.7.1 Proofs with Quantifiers

Some theorems from the propositional calculus correspond to analogous universally valid formulae in the predicate calculus, for instance, as in the following theorem.

Theorem 250 *For every individual variable X and for every* theorem $(P) \Rightarrow (Q)$, *the formula $[\forall X(P)] \Rightarrow [\forall X(Q)]$ is universally valid.*

Proof. Apply the rule of generalization, theorem 247, and *Modus Ponens*:

$\vdash (P) \Rightarrow (Q)$	hypothesis,
$\models \forall X[(P) \Rightarrow (Q)]$	generalization (rule 242),
$\models \{\forall X[(P) \Rightarrow (Q)]\} \Rightarrow \{[\forall X(P)] \Rightarrow [\forall X(Q)]\}$	theorem 247,
$\models [\forall X(P)] \Rightarrow [\forall X(Q)]$	*Modus Ponens*.

□

The following examples illustrate the preceding theorem.

Example 251 For all logical formulae P, Q, and individual variable X, the formulae

$$[\forall X(P)] \Rightarrow \{\forall X[(Q) \Rightarrow (P)]\}$$

$$[\forall X(P)] \Rightarrow \{[\forall X(Q)] \Rightarrow [\forall X(P)]\}$$

are universally valid:

$\vdash (P) \Rightarrow [(Q) \Rightarrow (P)]$	axiom P1,
$\models \forall X\{(P) \Rightarrow [(Q) \Rightarrow (P)]\}$	generalization,
$\models [\forall X(P)] \Rightarrow \{\forall X[(Q) \Rightarrow (P)]\}$	theorem 250,
$\models \{\forall X[(Q) \Rightarrow (P)]\} \Rightarrow \{[\forall X(Q)] \Rightarrow [\forall X(P)]\}$	theorem 247,
$\models [\forall X(P)] \Rightarrow \{[\forall X(Q)] \Rightarrow [\forall X(P)]\}$	theorem 175.

Example 252 For all logical formulae P, Q, and individual variable X, the formula

$$(\forall X\{[\neg(Q)] \Rightarrow [\neg(P)]\}) \Rightarrow \{[\forall X(P)] \Rightarrow [\forall X(Q)]\}$$

is universally valid:

$\vdash \{[\neg(Q)] \Rightarrow [\neg(P)]\} \Rightarrow [(P) \Rightarrow (Q)]$	axiom P3,
$\models (\forall X\{[\neg(Q)] \Rightarrow [\neg(P)]\}) \Rightarrow \{\forall X[(P) \Rightarrow (Q)]\}$	theorem 250,
$\models \{\forall X[(P) \Rightarrow (Q)]\} \Rightarrow \{[\forall X(P)] \Rightarrow [\forall X(Q)]\}$	theorem 247,
$\models (\forall X\{[\neg(Q)] \Rightarrow [\neg(P)]\}) \Rightarrow \{[\forall X(P)] \Rightarrow [\forall X(Q)]\}$	theorem 175.

The following theorem shows a similar pattern for equivalences.

Theorem 253 *For every variable X and for every* theorem $(P) \Leftrightarrow (Q)$*, the formula $[\forall X(P)] \Leftrightarrow [\forall X(Q)]$ is universally valid.*

Proof. Start from the theorem $(P) \Leftrightarrow (Q)$, and apply theorems 202, 250, and 175:

$\vdash [(P) \Leftrightarrow (Q)] \Rightarrow [(P) \Rightarrow (Q)]$	definition of $\Leftrightarrow$,
$\models \forall X\{[(P) \Leftrightarrow (Q)]\} \Rightarrow \{\forall X[(P) \Rightarrow (Q)]\}$	theorem 250,
$\models \{\forall X[(P) \Rightarrow (Q)]\} \Rightarrow \{[\forall X(P)] \Rightarrow [\forall X(Q)]\}$	theorem 247,
$\models \forall X\{[(P) \Leftrightarrow (Q)]\} \Rightarrow \{[\forall X(P)] \Rightarrow [\forall X(Q)]\}$	theorem 175,

$\models [(P) \Leftrightarrow (Q)] \Rightarrow [(Q) \Rightarrow (P)]$	definition of $\Leftrightarrow$,
$\models \forall X\{[(P) \Leftrightarrow (Q)]\} \Rightarrow \{\forall X[(Q) \Rightarrow (P)]\}$	theorem 250,
$\models \{\forall X[(Q) \Rightarrow (P)]\} \Rightarrow \{[\forall X(Q)] \Rightarrow [\forall X(P)]\}$	theorem 247,
$\models \forall X\{[(P) \Leftrightarrow (Q)]\} \Rightarrow \{[\forall X(Q)] \Rightarrow [\forall X(P)]\}$	theorem 175,

$\models \forall X\{[(P) \Leftrightarrow (Q)]\}$	
$\quad \Rightarrow (\{[\forall X(P)] \Rightarrow [\forall X(Q)]\} \wedge \{[\forall X(Q)] \Rightarrow [\forall X(P)]\})$	theorem 202,
$\models (\{[\forall X(P)] \Rightarrow [\forall X(Q)]\} \wedge \{[\forall X(Q)] \Rightarrow [\forall X(P)]\})$	
$\quad \Rightarrow \{[\forall X(Q)] \Leftrightarrow [\forall X(P)]\}$	definition of $\Leftrightarrow$,
$\models \forall X\{[(P) \Leftrightarrow (Q)]\} \Rightarrow \{[\forall X(Q)] \Leftrightarrow [\forall X(P)]\}$	theorem 175.

□

Example 254 For every variable X and for every logical formula P, the formula

$$[\forall X(P)] \Leftrightarrow (\forall X\{\neg[\neg(P)]\})$$

is universally valid. Indeed, apply theorem 253 to $(P) \Leftrightarrow \{\neg[\neg(P)]\}$.

The following two theorems involve both universal and existential quantifiers.

Theorem 255 *For all X, H, K, if X does not occur freely in K, then*

$$\models \{\forall X[(H) \Rightarrow (K)]\} \Rightarrow \{[\exists X(H)] \Rightarrow (K)\}.$$

Proof. Apply the full propositional calculus, axioms Q2, Q3, and generalization:

$\models [(H) \Rightarrow (K)] \Rightarrow \{[\neg(K)] \Rightarrow [\neg(H)]\}$	contraposition,
$\models \forall X([(H) \Rightarrow (K)] \Rightarrow \{[\neg(K)] \Rightarrow [\neg(H)]\})$	generalization,
$\models \{\forall X[(H) \Rightarrow (K)]\} \Rightarrow (\forall X\{[\neg(K)] \Rightarrow [\neg(H)]\})$	theorem 247,
$\models (\forall X\{[\neg(K)] \Rightarrow [\neg(H)]\}) \Rightarrow ([\neg(K)] \Rightarrow \{\forall X[\neg(H)]\})$	axiom Q2,
$\models \{\forall X[(H) \Rightarrow (K)]\} \Rightarrow ([\neg(K)] \Rightarrow \{\forall X[\neg(H)]\})$	theorem 175,
$\models \{\forall X[\neg(H)]\} \Rightarrow \{\neg[\exists X(H)]\}$	axiom Q3,
$\models \{\forall X[(H) \Rightarrow (K)]\} \Rightarrow ([\neg(K)] \Rightarrow \{\neg[\exists X(H)]\})$	theorem 184,
$\models ([\neg(K)] \Rightarrow \{\neg[\exists X(H)]\}) \Rightarrow \{[\exists X(H)] \Rightarrow (K)\}$	contraposition,
$\models \{\forall X[(H) \Rightarrow (K)]\} \Rightarrow \{[\exists X(H)] \Rightarrow (K)\}$	theorem 175.

□

Theorem 256 *For all X, H, K, without restrictions,*

$$\models \{\forall X[(P) \Rightarrow (Q)]\} \Rightarrow \{[\exists X(P)] \Rightarrow [\exists X(Q)]\}.$$

Proof. Apply the full propositional calculus and previous theorems:

$\models \{\forall X[(P) \Rightarrow (Q)]\} \Rightarrow (\forall X\{[\neg(Q)] \Rightarrow [\neg(P)]\})$	(194, 250),
$\models (\forall X\{[\neg(Q)] \Rightarrow [\neg(P)]\}) \Rightarrow (\{\forall X[\neg(Q)]\} \Rightarrow \{\forall X[\neg(P)]\})$	(247),
$\models \{\forall X[(P) \Rightarrow (Q)]\} \Rightarrow (\{\forall X[\neg(Q)]\} \Rightarrow \{\forall X[\neg(P)]\})$	(175),
$\models (\{\forall X[\neg(Q)]\} \Rightarrow \{\forall X[\neg(P)]\})$	
$\quad \Rightarrow [(\neg\{\forall X[\neg(P)]\}) \Rightarrow (\neg\{\forall X[\neg(Q)]\})]$	(194);
$\models [\exists X(P)] \Rightarrow (\neg\{\forall X[\neg(P)]\})$	(244),
$\models (\neg\{\forall X[\neg(Q)]\}) \Rightarrow [\exists X(Q)]$	(244),
$\models [(\neg\{\forall X[\neg(P)]\}) \Rightarrow (\neg\{\forall X[\neg(Q)]\})]$	
$\quad \Rightarrow \{[\exists X(P)] \Rightarrow [\exists X(Q)]\}$	(193),
$\models \{\forall X[(P) \Rightarrow (Q)]\} \Rightarrow \{[\exists X(P)] \Rightarrow [\exists X(Q)]\}$	(175).

□

1.7.2 Proofs With Universal Quantifiers and Other Connectives

The following two theorems prove the universally valid formula

$$\{\forall X[(P) \wedge (Q)]\} \Leftrightarrow \{[\forall X(P)] \wedge [\forall X(Q)]\}.$$

Theorem 257 $\models \{\forall X[(P) \wedge (Q)]\} \Rightarrow \{[\forall X(P)] \wedge [\forall X(Q)]\}$.

Proof. Apply theorems 200, 250, 199, 202:

$\vdash [(P) \wedge (Q)] \Rightarrow (P)$	theorem 200,
$\models \{\forall X[(P) \wedge (Q)]\} \Rightarrow \{\forall X(P)\}$	theorem 250,
$\vdash [(P) \wedge (Q)] \Rightarrow (Q)$	theorem 199,
$\models \{\forall X[(P) \wedge (Q)]\} \Rightarrow \{\forall X(Q)\}$	theorem 250,
$\models \{\forall X[(P) \wedge (Q)]\} \Rightarrow \{[\forall X(P)] \wedge [\forall X(Q)]$	theorem 202.

□

The converse implication forms the object of the following theorem.

Theorem 258 $\models \{[\forall X(P)] \wedge [\forall X(Q)]\} \Rightarrow \{\forall X[(P) \wedge (Q)]\}$.

Proof. Apply axiom Q1, theorems 219, 243, and generalization:

$\models [\forall X(P)] \Rightarrow (P)$	axiom Q1,
$\models [\forall X(Q)] \Rightarrow (Q)$	axiom Q1,
$\models \{[\forall X(P)] \wedge [\forall X(Q)]\} \Rightarrow [(P) \wedge (Q)]$	theorem 219,
$\models \forall X(\{[\forall X(P)] \wedge [\forall X(Q)]\} \Rightarrow [(P) \wedge (Q)])$	generalization,
$\models \{[\forall X(P)] \wedge [\forall X(Q)]\} \Rightarrow \{\forall X[(P) \wedge (Q)]\}$	theorem 243.

□

The following theorem relates $\forall$ to $\vee$ through an implication.

Theorem 259 $\models \{[\forall X(P)] \vee [\forall X(Q)]\} \Rightarrow \{\forall X[(P) \vee (Q)]\}$.

Proof. Apply axiom Q1, exercise 371, theorem 243, and generalization:

$\models [\forall X(P)] \Rightarrow (P)$	axiom Q1,
$\models [\forall X(Q)] \Rightarrow (Q)$	axiom Q1,
$\models \{[\forall X(P)] \vee [\forall X(Q)]\} \Rightarrow [(P) \vee (Q)]$	exercise 371,
$\models \forall X(\{[\forall X(P)] \vee [\forall X(Q)]\} \Rightarrow [(P) \vee (Q)])$	generalization,
$\models \{[\forall X(P)] \vee [\forall X(Q)]\} \Rightarrow \{\forall X[(P) \vee (Q)]\}$	theorem 243.

□

Counterexample 260 The converse of theorem 259, which would be

$$\{\forall X[(P) \vee (Q)]\} \Rightarrow \{[\forall X(P)] \vee [\forall X(Q)]\},$$

may be *False*. For instance, in every context with exactly *two different* objects V and W, consider the formulae $X = V$ for P and $X = W$ for Q:

$\models \forall X[(X = V) \vee (X = W)]$	because either $(X = V)$ or $(X = W)$;
$\forall X(X = V)$	is False if $X := W$;
$\forall X(X = W)$	is False if $X := V$;
$[\forall X(X = V)] \vee [\forall X(X = W)]$	is False by the preceding two lines;
$\{\forall X[(X = V) \vee (X = W)]\}$ $\Rightarrow \{[\forall X(X = V)] \vee [\forall X(X = W)]\}$	is False because $(T) \not\Rightarrow (F)$.

□

Nevertheless, the converse of theorem 247 holds under some additional hypotheses.

Theorem 261 *For every individual variable X and for all logical formulae P, Q, if $\forall X(Q)$ is universally valid, then so is $\forall X[(P) \Rightarrow (Q)]\}$. Moreover, if $\forall X(Q)$ and $\forall X(P)$ are both universally valid, then so is $\forall X[(P) \Leftrightarrow (Q)]\}$.*

Proof. The proof of the first assertion relies on example 251:

$\models \forall X(Q)$	hypothesis,
$\models [\forall X(Q)] \Rightarrow \{\forall X[(P) \Rightarrow (Q)]\}\}$	example 251,
$\models \forall X[(P) \Rightarrow (Q)]$	*Modus Ponens*.

Hence if $\forall X(P)$ is universally valid, then swapping P and Q yields $\models \forall X[(Q) \Rightarrow (P)]$. The proof of the second assertion then relies on theorems 201 and 258:

$\models \forall X(Q)$	hypothesis,
$\models \forall X[(P) \Rightarrow (Q)]$	just proved,
$\models \forall X(P)$	hypothesis,
$\models \forall X[(Q) \Rightarrow (P)]$	just proved,
$\models \{\forall X[(P) \Rightarrow (Q)]\} \wedge \{\forall X[(Q) \Rightarrow (P)]\}$	theorem 201,
$\models \forall X\{[(P) \Rightarrow (Q)] \wedge [(Q) \Rightarrow (P)]\}$	theorem 258 and *Modus Ponens*;
$\models \forall X[(P) \Leftrightarrow (Q)]$	definition of $\Leftrightarrow$.

□

1.7.3 Proofs with Quantifiers and Other Connectives

The following derived rule of inference relates existential quantifiers to equivalences.

Theorem 262 *If* $(P) \Leftrightarrow (Q)$ *is a theorem, then* $[\exists X(P)] \Leftrightarrow [\exists X(Q)]$ *is a theorem.*

Proof. Apply the full propositional calculus and theorem 253:

$$\begin{array}{rl}
\models (P) \Leftrightarrow (Q) & \\
\Updownarrow & \text{contraposition and transposition,} \\
[\neg(Q)] \Leftrightarrow [\neg(P)] & \\
\Downarrow & \text{theorem 253,} \\
\{\forall X[\neg(Q)]\} \Leftrightarrow \{\forall X[\neg(P)]\} & \\
\Updownarrow & \text{contraposition and transposition,} \\
(\neg\{\forall X[\neg(Q)]\}) \Leftrightarrow (\neg\{\forall X[\neg(P)]\}) & \\
\Updownarrow & \text{axiom Q3 and transitivity of } \Leftrightarrow, \\
(\neg\{\neg[\exists X(Q)]\}) \Leftrightarrow (\neg\{\neg[\exists X(P)]\}) & \\
\Updownarrow & \text{double negations.} \\
[\exists X(Q)] \Leftrightarrow [\exists X(P)] &
\end{array}$$

□

Example 263 Applied to the theorem $(P) \Leftrightarrow \{\neg[\neg(P)]\}$, theorem 262 shows that the formula $[\exists X(P)] \Leftrightarrow (\exists X\{\neg[\neg(P)]\})$ is universally valid.

The following proof demonstrates another use of theorem 262.

Theorem 264 $\models \{[\exists X(P)] \vee [\exists X(Q)]\} \Leftrightarrow \{[\exists X[(P) \vee (Q)]\}$.

Proof. Apply axiom Q4, theorems 257, 258, and 253:

$$\begin{array}{rl}
\models (\forall X\{[\neg(P)] \wedge [\neg(Q)]\}) \Leftrightarrow (\{\forall X[\neg(P)]\} \wedge \{\forall X[\neg(Q)]\}) & 257, 258, \\
\Updownarrow & \text{axiom P3,} \\
[\neg(\{\forall X[\neg(P)]\} \wedge \{\forall X[\neg(Q)]\})] \Leftrightarrow [\neg(\forall X\{[\neg(P)] \wedge [\neg(Q)]\})] & \\
\Updownarrow & 253, \\
[(\neg\{\forall X[\neg(P)]\}) \vee (\neg\{\forall X[\neg(Q)]\})] \Leftrightarrow [\neg(\forall X\{\neg[(P) \vee (Q)]\})] & \\
\Updownarrow & \text{axiom Q4,} \\
(\neg\{\neg[\exists X(P)]\}) \vee (\neg\{\neg[\exists X(Q)]\}) \Leftrightarrow [\neg(\neg\{\exists X[(P) \vee (Q)]\})] & \\
\Updownarrow & \text{negations.} \\
\{[\exists X(P)] \vee [\exists X(Q)]\} \Leftrightarrow \{\exists X[(P) \vee (Q)]\} &
\end{array}$$

□

1.7.4 Proofs with Restrictions On a Quantified Variable

The following theorems require a quantified variable not to occur freely.

Theorem 265 *If* P *has no free* X, *then* $\models (P) \Leftrightarrow [\forall X(P)]$.

Proof. Specialization (axiom Q1) yields $\models [\forall X(P)] \Rightarrow (P)$. For the converse,

$$\begin{array}{rl}
\vdash (P) \Rightarrow (P) & \text{theorem 173,} \\
\vdash \forall X[(P) \Rightarrow (P)] & \text{generalization,} \\
\vdash (P) \Rightarrow [\forall X(P)] & \text{theorem 243.}
\end{array}$$

□

Theorem 266 *If* P *has no free* X, *then* $\models (P) \Leftrightarrow [\exists X(P)]$.

Proof. Apply the complete law of double negation:

$$
\begin{array}{rl}
\exists X(P) & \\
\Updownarrow & \text{double negation and theorem 262,} \\
\exists X\{\neg[\neg(P)]\} & \\
\Updownarrow & \text{axiom Q4,} \\
\neg\{\forall X[\neg(P)]\} & \\
\Updownarrow & \text{theorem 265,} \\
\neg[\neg(P)] & \\
\Updownarrow & \text{double negation.} \\
(P) &
\end{array}
$$

□

The converse of axiom Q2 also holds.

Theorem 267 *If P has no free X, then* $\models \{(P) \Rightarrow [\forall X(Q)]\} \Rightarrow \{\forall X[(P) \Rightarrow (Q)]\}$.

Proof. Apply theorems 191, 243, specialization, and generalization:

$$
\begin{array}{ll}
\models \forall X(Q) \Rightarrow (Q) & \text{specialization,} \\
\models \{(P) \Rightarrow [\forall X(Q)]\} \Rightarrow [(P) \Rightarrow (Q)] & \text{theorem 191,} \\
\models \forall X\big(\{(P) \Rightarrow [\forall X(Q)]\} \Rightarrow [(P) \Rightarrow (Q)]\big) & \text{generalization,} \\
\models \{(P) \Rightarrow [\forall X(Q)]\} \Rightarrow \{\forall X[(P) \Rightarrow (Q)]\} & \text{theorem 243.}
\end{array}
$$

□

Theorem 268 *If P has no free X, then* $\models \{\forall X[(P) \vee (Q)]\} \Rightarrow \{(P) \vee [\forall X(Q)]\}$.

Proof. Apply the definition of $\vee$, theorem 253, and axiom Q2:

$$
\begin{array}{rl}
\forall X[(P) \vee (Q)] & \\
\Updownarrow & \text{definition of } (P) \vee (Q) \text{ and theorem 253,} \\
\forall X\{[\neg(P)] \Rightarrow (Q)\} & \\
\Downarrow & \text{axiom Q2,} \\
[\neg(P)] \Rightarrow [\forall X(Q)] & \\
\Updownarrow & \text{definition of } (R) \vee (S). \\
(P) \vee [\forall X(Q)] &
\end{array}
$$

□

Theorem 269 *If P has no free X, then* $\models \{\exists X[(P) \wedge (Q)]\} \Rightarrow \{(P) \wedge [\exists X(Q)]\}$.

Proof. Apply the full propositional calculus, theorem 262, and axiom Q4:

$$
\begin{array}{rl}
\exists X[(P) \wedge (Q)] & \\
\Updownarrow & \text{double negation,} \\
\exists X\big(\neg\{\neg[(P) \wedge (Q)]\}\big) & \\
\Updownarrow & \text{de Morgan's first law and theorem 262,} \\
\exists X\big(\neg\{[\neg(P)] \vee [\neg(Q)]\}\big) & \\
\Updownarrow & \text{axiom Q4,} \\
\neg\big(\forall X\{[\neg(P)] \vee [\neg(Q)]\}\big) & \\
\Downarrow & \text{preceding theorem,} \\
\neg\big([\neg(P)] \vee \{\forall X[\neg(Q)]\}\big) & \\
\Updownarrow & \text{de Morgan's second law and double negation,} \\
(P) \wedge \big(\neg\{\forall X[\neg(Q)]\}\big) & \\
\Updownarrow & \text{axiom Q4,} \\
(P) \wedge \big(\exists X \neg\{\neg[\neg(Q)]\}\big) & \\
\Updownarrow & \text{double negation and theorem 262.} \\
(P) \wedge [\exists X(Q)] &
\end{array}
$$

□

The converse also holds.

Theorem 270 *If P has no free X, then* $\models \{\exists X[(P) \wedge (Q)]\} \Leftrightarrow \{(P) \wedge [\exists X(Q)]\}$.

Proof. Apply the full propositional calculus, theorem 262, and axioms Q2, Q3, Q4:

$$\begin{array}{rl} \exists X[(P) \wedge (Q)] & \\ \Updownarrow & \text{double negation,} \\ \exists X(\neg\{\neg[(P) \wedge (Q)]\}) & \\ \Updownarrow & \text{de Morgan's first law and theorem 262,} \\ \exists X(\neg\{[\neg(P)] \vee [\neg(Q)]\}) & \\ \Updownarrow & \text{definition of } \vee, \\ \exists X(\neg\{(P) \Rightarrow [\neg(Q)]\}) & \\ \Updownarrow & \text{axioms Q4 and Q3,} \\ \neg(\forall X\{(P) \Rightarrow [\neg(Q)]\}) & \\ \Updownarrow & \text{axiom Q2 and theorem 267,} \\ \neg[(P) \Rightarrow \{\forall X[\neg(Q)]\}] & \\ \Updownarrow & \text{definition of } \vee, \\ \neg([\neg(P)] \vee \{\forall X[\neg(Q)]\}) & \\ \Updownarrow & \text{de Morgan's second law and double negation,} \\ (P) \wedge (\neg\{\forall X[\neg(Q)]\}) & \\ \Updownarrow & \text{axiom Q4,} \\ (P) \wedge (\exists X\neg\{\neg[\neg(Q)]\}) & \\ \Updownarrow & \text{double negation and theorem 262.} \\ (P) \wedge [\exists X(Q)] & \end{array}$$

□

1.7.5 Proofs with More than One Quantified Variable

The following theorems provide examples of proofs with two quantifiers. The first theorem allows for the deletion of a redundant universal quantifier.

Theorem 271 $\models [\forall X(Q)] \Leftrightarrow \{\forall X[\forall X(Q)]\}$.

Proof. Apply theorem 265 to $\forall X(Q)$, which has no free X. □

The second theorem allows for the swap of two consecutive universal quantifiers.

Theorem 272 $\models \{\forall X[\forall Y(P)]\} \Leftrightarrow \{\forall Y[\forall X(P)]\}$.

Proof. Apply axiom Q1, theorem 243, and generalization:

$$\begin{array}{ll} \models \{\forall X[\forall Y(P)]\} \Rightarrow [\forall Y(P)] & \text{axiom Q1,} \\ \models [\forall Y(P)] \Rightarrow (P) & \text{axiom Q1,} \\ \models \{\forall X[\forall Y(P)]\} \Rightarrow [\forall X(P)] & \text{theorem 243,} \\ \models \forall Y(\{\forall X[\forall Y(P)]\} \Rightarrow [\forall X(P)]) & \text{generalization,} \\ \models \{\forall X[\forall Y(P)]\} \Rightarrow \{\forall Y[\forall X(P)]\} & \text{theorem 243.} \end{array}$$

□

The third theorem allows for the swap of two consecutive existential quantifiers.

Theorem 273 $\models \{\exists X[\exists Y(P)]\} \Leftrightarrow \{\exists Y[\exists X(P)]\}$.

Proof. Apply the complete law of double negation, axiom Q3, and theorem 272:

$$
\begin{array}{rl}
\exists X[\exists Y(P)] & \\
\Updownarrow & \text{double negation,} \\
\neg(\neg\{\exists X[\exists Y(P)]\}) & \\
\Updownarrow & \text{axiom Q3,} \\
\neg(\forall X\{\neg[\exists Y(P)]\}) & \\
\Updownarrow & \text{axiom Q3,} \\
\neg(\forall X\{\forall Y[\neg(P)]\}) & \\
\Updownarrow & \text{theorem 272,} \\
\neg(\forall Y\{\forall X[\neg(P)]\}) & \\
\Updownarrow & \text{axiom Q3,} \\
\neg(\forall Y\{\forall X[\neg(P)]\}) & \\
\Updownarrow & \text{axiom Q3,} \\
\neg(\forall Y\{\neg[\exists X(P)]\}) & \\
\Updownarrow & \text{axiom Q3,} \\
\neg(\neg\{\exists Y[\exists X(P)]\}) & \\
\Updownarrow & \text{double negation.} \\
\exists Y[\exists X(P)] &
\end{array}
$$

□

The fourth theorem allows for the swap of different quantifiers in an implication.

Theorem 274 $\models \{\exists X[\forall Y(P)]\} \Rightarrow \{\forall Y[\exists X(P)]\}$.

Proof. Apply the full propositional calculus and theorems 266, 247:

$\models (P) \Rightarrow [\exists X(P)]$	theorem 266,
$\models \forall Y\{(P) \Rightarrow [\exists X(P)]\}$	generalization,
$\models (\forall Y\{(P) \Rightarrow [\exists X(P)]\}) \Rightarrow ([\forall Y(P)] \Rightarrow \{\forall Y[\exists X(P)]\})$	theorem 247,
$\models [\forall Y(P)] \Rightarrow \{\forall Y[\exists X(P)]\}$	*Modus Ponens.*
$\models (\neg\{\forall Y[\exists X(P)]\}) \Rightarrow \{\neg[\forall Y(P)]\}$	contraposition,
$\models \{\neg[\forall Y(P)]\} \Rightarrow (\forall X\{\neg[\forall Y(P)]\})$	generalization,
$\models (\neg\{\forall Y[\exists X(P)]\}) \Rightarrow (\forall X\{\neg[\forall Y(P)]\})$	theorem 175,
$\models [\neg(\forall X\{\neg[\forall Y(P)]\})] \Rightarrow [\neg(\neg\{\forall Y[\exists X(P)]\})]$	contraposition,
$\models [\exists X(\neg\{\neg[\forall Y(P)]\})] \Rightarrow [\neg(\neg\{\forall Y[\exists X(P)]\})]$	axiom Q4,
$\models \{\exists X[\forall Y(P)]\} \Rightarrow \{\forall Y[\exists X(P)]\}$	double negations.

□

Counterexample 275 The converse of theorem 274, which would be

$$\{\forall Y[\exists X(P)]\} \Rightarrow \{\exists X[\forall Y(P)]\},$$

can be *False*. For instance, in every context with at least two *different* objects V and W, consider the logical formula $X = Y$ for P.

$\models \forall Y[\exists X(X = Y)]$	for each Y, choose $X := Y$;
$\exists X[\forall Y(X = Y)]$	is False: no X equals V and W;
$\{\forall Y[\exists X(X = Y)]\} \Rightarrow \{\exists X[\forall Y(X = Y)]\}$	is False because $(T) \not\Rightarrow (F)$.

1.7.6 Exercises

For the following exercises, prove that the stated formulae are universally valid.

Exercise 391 $\{\exists X[(P) \vee (Q)]\} \Leftrightarrow \{\exists X[(Q) \vee (P)]\}$

Exercise 392 $\{\forall X[(P) \wedge (P)]\} \Leftrightarrow \{\forall X(P)\}$

Exercise 393 $\{\exists X[(P) \vee (P)]\} \Leftrightarrow \{\exists X(P)\}$

Exercise 394 $(\exists X\{[(P) \vee (Q)] \vee (R)\}) \Leftrightarrow (\exists X\{(P) \vee [(Q) \vee (R)]\})$

Exercise 395 $(\forall X\{[(P) \wedge (Q)] \vee (R)\}) \Leftrightarrow (\{\forall X[(P) \vee (R)]\} \wedge \{\forall X[(Q) \vee (R)]\})$

Exercise 396 $(\exists X\{[(P) \vee (Q)] \wedge (R)\}) \Leftrightarrow (\{\exists X[(P) \wedge (R)]\} \vee \{\exists[(Q) \wedge (R)]\})$

Exercise 397 $[\exists X(Q)] \Leftrightarrow [\exists X\,(\exists X(Q))]$

Exercise 398 If P has no free X, then $\{(P) \wedge [\forall X(Q)]\} \Leftrightarrow \{\forall X[(P) \wedge (Q)]\}$

Exercise 399 If P has no free X, then $\{(P) \vee [\forall X(Q)]\} \Leftrightarrow \{\forall X[(P) \vee (Q)]\}$

Exercise 400 If P has no free X, then $\{(P) \vee [\exists X(Q)]\} \Leftrightarrow \{\exists X[(P) \vee (Q)]\}$

1.8 FURTHER ISSUES IN LOGIC

A practically important issue pertains to the automation of logical processes.

In computer science, for example, McCluskey's algorithm automates the synthesis and simplification of logical circuits with specified Truth tables [76].

In classical logic, the Completeness Theorem presented in chapter 4 provides a finite algorithm to identify and prove each theorem of the propositional calculus.

For the predicate calculus, there exist algorithms to identify and prove each theorem within special classes of formulae [11, Ch. 46–47]. However, Alonzo Church has proved that the full predicate calculus lies beyond the realm of any algorithm [10], [9].

Another issue pertains to the selection of a logic close to common reasoning.

For instance, the Boolean and classical logics allow for reasoning by the converse law of contraposition, which is uncommon [73, p. 36]. There exist other logics that include the implicational calculus and the law of contraposition without its converse, but they cannot be described by Truth tables, as proved in chapter 4.

1.9 PROJECTS

Project 3 Raymond M. Smullyan published a method to verify tautologies with "tableaux" [88, Ch. II, p. 15–30]. To verify that a formula L is a tautology, Smullyan's method assumes that L is False and hence leads to the conclusion that at least one of its components P must be True and False. Thus $[\neg(L)] \Rightarrow \{[\neg(P)] \wedge (P)\}$, or, equivalently, $[\neg(L)] \Rightarrow \{\neg[(P) \Rightarrow (P)]\}$. Hence L is True, by *reductio ad absurdum*.

Verify the tautologies in this chapter by Smullyan's method of tableaux.

Example 276 For $(P) \Rightarrow \{[\neg(P)] \Rightarrow (Q)\}$, Smullyan's method proceeds as follows.

1. Here L has the pattern $(H) \Rightarrow (K)$, with P for H, and $[\neg(P)] \Rightarrow (Q)$ for K.
2. Assume $(H) \Rightarrow (K)$ False. Then there is only one possibility: H (which is P) is True, and K is False. Continue with the components that are False, here K.
3. Thus K, which is $[\neg(P)] \Rightarrow (Q)$ is False; hence Q is False and $[\neg(P)]$ is True, whence P is False.

Consequently P is True (step 2) and P is False (step 3); therefore, L is True.

Project 4 Investigate a predicate calculus with axiom Q4 replaced by the implication $\{\neg[\forall X(P)]\} \Leftarrow \{\exists X[\neg(P)]\}$.

Chapter 2

Set Theory

2.0 INTRODUCTION

Figure 2.1: Celestial bodies can be modelled in various ways with mathematical sets.

For some practical problems, features that are essential to their solutions can be specified in terms of **sets** or **collections** of objects.

Example 277 (Binary arithmetic) The binary arithmetic of computers relies on a *set of two symbols*, 0 and 1, which will be defined with yet other sets in this chapter.

Example 278 (Geometries) Geometries can be designed entirely with sets. Points are sets (of *sets of coordinates*), while lines, planes, and space are *sets of points*. Points, lines, planes, and space are "primitive" objects that may remain undefined, but relations between them are specified through axioms. For instance, the first axiom of incidence geometry specifies that through any two distinct points passes exactly one line [49, p. 3]. Likewise in this chapter, mathematical "sets" are "primitive" objects that remain undefined, while features of sets and relations between sets are specified by axioms.

Example 279 (Two-body problem) Johann Kepler's investigations of the orbit of Mars around the Sun pertained to the relative motions of a *set of two celestial objects*: Mars and the Sun. Such characteristics as the shapes and the inner compositions of Mars and the Sun did not matter for the investigation of their orbits.

Example 280 (Three-body problem) No simple description is available for the motion of three or more celestial bodies. Nevertheless, in 1907 K. F. Sundman proved that three objects in a rotating motion in space will never collide in a simultaneous triple collision [87], [94]. Rather than their shapes or internal composition, only the *set of three celestial objects* matters in their investigation of such collisions.

Example 281 (Geoid) From Kepler's elliptic planetary orbits, Isaac Newton (1642–1727) derived the law of gravity, according to which masses attract each other proportionally to the inverse of the square of the distance between them [80]. Representing a planet by a *set of many particles* attracting one another according to the law of gravity, Newton, Ivory, Huygens, Clairaut, and Laplace [68, Book III, §18] then proved that if a planet initially consisted of a rotating set of loose particles, then its shape could eventually stabilize as an **ellipsoid** rotating around its *shortest* axis [36, p. 172–175]. Yet from 1700 through 1733, three surveys in France suggested that the Earth was an ellipsoid rotating around its *largest* axis [8, p. 250–251]. Ordered by Louis XV, a survey in Lapland and a survey in Peru in 1735 finally confirmed that the shape of the earth — called the **geoid** — was an ellipsoid rotating around its *shortest* axis [8, p. 251–252].

The foregoing examples already demonstrate a major difficulty in using problems about "real" objects to illustrate logical and mathematical concepts: no exact answer might be available. For instance, electronic digital computers do not use anything like the symbols 0 and 1; indeed, they use two electrical potentials confined to two mutually exclusive ranges. Similarly, the earth is neither exactly spherical nor exactly ellipsoidal; rather, the earth is closer to an ellipsoid than to a sphere. Likewise, the orbit of any planet is neither circular nor elliptical; rather, it is closer to elliptical than circular. A precise answer would involve more advanced engineering, logic, mathematics, and physics. Therefore, the sets in the present exposition will not contain "real" objects; instead, all the following sets will contain only abstract objects defined by precise rules.

2.1 SETS AND SUBSETS

2.1.1 Equality and Extensionality

One of the major mathematical achievements around the beginning of the twentieth century was the realization that most of mathematics and computer science consists of logical relations between abstract objects called sets [74]. There is no definition of mathematical sets. Indeed, such a definition would have to define sets in terms of yet more foundational objects, but sets *are* the most foundational objects. Henceforth, in this text, *all mathematical objects are sets, and all quantified variables designate sets.*

Instead of a definition of sets, a few axioms specify certain characteristics of sets. The set theory presented here involves *one* undefined primitive binary relation, denoted by $\in$ and called "membership". The notation $X \in Y$ is read in various ways as "X is an element of Y", "X is a member of Y", or "X belongs to Y". For any set X and any set Y, the atomic formula $X \in Y$ has a Truth value, so that $X \in Y$ is either True (if X is an element of Y) or False (if X is not an element of Y). Consequently, the following generalization of the tautology $(B) \vee [\neg(B)]$ is universally valid:

$$\models \forall X\big(\forall Y\{(X \in Y) \vee [\neg(X \in Y)]\}\big).$$

The relation $\in$ of membership is the only foundational relation between sets. Consequently, the only characteristics of a set are its elements. Because set theory involves only one relation, every set A can relate to other sets X and Y in only four ways:

$$\begin{gathered} X \in A, \\ \neg(X \in A); \\ A \in Y, \\ \neg(A \in Y). \end{gathered}$$

Whether $A \in Y$ or $\neg(A \in Y)$ might also be considered as a characteristic of A. Consequently, rather than involving the elements of A (those sets X such that $X \in A$), another way to define the characteristics of A might involve the sets of which A is an element (those sets Y such that $A \in Y$). However, if the only characteristics of a set are its elements, then the two ways ought to be logically equivalent; this equivalence forms the essence of the axiom of extensionality.

Axiom S1 (Axiom of extensionality)

$$\models \forall A\{\forall B[(\{\forall X[(X \in A) \Leftrightarrow (X \in B)]\} \Leftrightarrow \{\forall Y[(A \in Y) \Leftrightarrow (B \in Y)]\})]\}.$$

In the axiom of extensionality (S1), the formula $\forall X[(X \in A) \Leftrightarrow (X \in B)]$ states that each set X is an element of A if and only if X is an element of B. The formula $\forall Y[(A \in Y) \Leftrightarrow (B \in Y)]$ states that A is an element of Y if and only if B is an element of Y. The axiom of extensionality states that these two formulae are logically equivalent.

Yet another way to state that two sets have exactly the same characteristics involves a derived binary relation (derived from $\in$) denoted by $=$ and called "equality". For each set A and each set B, the formula $A = B$ (read "A equals B") means that A and B have exactly the same characteristics. By the axiom of extensionality, the equality of two sets can be stated in two logically equivalent ways:

$$\models \forall A\{\forall B[(A = B) \Leftrightarrow \{\forall X[(X \in A) \Leftrightarrow (X \in B)]\}]\},$$

$$\models \forall A\{\forall B[(A = B) \Leftrightarrow \{\forall Y[(A \in Y) \Leftrightarrow (B \in Y)]\}]\}.$$

The notation $A = B$ is merely a shorthand to state that the following two formulae hold:

$$\forall X[(X \in A) \Leftrightarrow (X \in B)],$$

$$\forall Y[(A \in Y) \Leftrightarrow (B \in Y)],$$

and the axiom of extensionality states that these two formulae are logically equivalent.

There is another presentation of set theory with *two* undefined relations, equality ($=$) and membership ($\in$). Then the axiom of extensionality specifies that two sets are the "same" set if and only if they contain the "same" elements. In this exposition the distinction just made does not matter, because equality ($=$) serves only as a shorthand and all operations with sets pertain to elements of those sets. (See also the discussion by Bernays [3, p. 53] and its introduction by Fraenkel [3, p. 6–8].) For the negations of membership and equality, the following abbreviations prove convenient.

Definition 282 The symbols $\notin$ and $\neq$ denote the negations of $\in$ and $=$ so that

$$\models (X \notin Y) \Leftrightarrow [\neg(X \in Y)];$$

$$\models (A \neq B) \Leftrightarrow [\neg(A = B)].$$

The determination of the Truth value of an equality $A = B$ requires prior definitions of both sets A and B. In contrast, there exists a different use of the same concept of equality, denoted by $C := D$ (read "let C equal D") to specify a hitherto undefined set C in terms of an already defined set D [41, p. 8], [91, p. 5]. Alternatively, the notation $D =: C$ (also read "let C equal D") can also serve to specify C in terms of D, especially where a derivation leads to a lengthy formula D, which can thus be abbreviated by a shorter variable or string C [91, p. 271, p. 347].

Example 283 The notation $E := \varnothing$ specifies that E stands for $\varnothing$.

At the elementary stage of set theory, most formal logical proofs of relations between sets are straightforward, in the sense that they use only axioms and definitions to establish a sequence of equivalences between the objective of the proof and a universally valid formula. Such formal logical proofs are usually longer than "informal" proofs . To show a first example of a proof within set theory — a formal version and an informal version — the following theorem states that each set equals itself.

Remark 284 In *designing* a proof we may at any stage start from the conclusion — but we may *not* assume it as True — and then search for logically equivalent formulae that connect the conclusion to other formulae that we know how to prove. For instance, Smullyan's method of tableaux uses such an approach [88, Ch. II, p. 15–30].

Theorem 285 *Each set equals itself: the formula* $\forall S\,(S = S)$ *is universally valid.*

Proof. An informal proof can consist of the following statements.

- Every set X is an element of S if and only if X is an element of S;
- hence $S = S$ by definition of the equality of sets and extensionality (S1).

One method to *design* a formal proof transforms the objective, here the *yet unproved* formula $\forall S\,(S = S)$, into logically equivalent formulae, until one such equivalent formula appears that is True thanks to an axiom or to a previously proven theorem. For instance, substituting S for A and also S for B in the axiom of extensionality gives

$$\begin{array}{rl} S = S & \textbf{yet unproved,} \\ \Updownarrow & \text{definition of} = \\ \forall X[(X \in S) \Leftrightarrow (X \in S)] & \end{array}$$

where the logical formula $(X \in S) \Leftrightarrow (X \in S)$ has the pattern of the tautology $(P) \Leftrightarrow (P)$. Thus, a complete proof may proceed as follows:

$$\begin{array}{ll} \vdash (P) \Leftrightarrow (P) & \text{tautology from logic,} \\ \models (X \in S) \Leftrightarrow (X \in S) & \text{substitution in the tautology } (P) \Leftrightarrow (P), \\ \models \forall S\{\forall X[(X \in S) \Leftrightarrow (X \in S)]\} & \text{generalizations of the tautology,} \\ \models \forall S(S = S) & \text{definition of} = \text{and extensionality (S1).} \end{array}$$

The proof just presented relies on one of the two formulae for the axiom of extensionality: $\forall X[(X \in A) \Leftrightarrow (X \in B)]$. Another proof could rely on the other formula: $\forall Y[(A \in Y) \Leftrightarrow (B \in Y)]$. □

The axiom of extensionality merely provides two logically equivalent criteria to test whether sets have exactly the same characteristics. However, so far in this theory there is no "set" yet. The "existence" of at least one set — or, more accurately, a convention about an abstract concept of a specific set — requires a second axiom.

2.1.2 The Empty Set

The second axiom, called the **axiom of the empty set,** guarantees the existence of at least one set, denoted by $\varnothing$ or also by $\{\ \ \}$; this set contains no element.

Axiom S2 (Axiom of the empty set) $\models \forall X[\neg(X \in \varnothing)]$.

Set theory could dispense with the constant $\varnothing$ and state the axiom of the empty set in the form $\exists E[\forall X(X \notin E)]$. In either case, the following theorem shows that $E = \varnothing$.

Theorem 286 *If* $\models \forall X(X \notin \varnothing)$ *and* $\models \exists E[\forall X(X \notin E)]$, *then* $\models E = \varnothing$.

Proof. An informal proof can consist of the following statements.

- No set may belong to the empty set $\varnothing$, by the axiom of the empty set;
- no set may belong to the set E, by the hypothesis on E;
- thus $\varnothing$ and E have the same elements (in effect, none);
- therefore $\varnothing = E$, by the axiom of extensionality.

A formal proof can use the axioms of extensionality and the empty set, and theorem 261 (stating that if $\forall X(P)$ and $\forall X(Q)$ both hold, then so does $\forall X[(P) \Leftrightarrow (Q)]$):

$\models \forall X(X \notin \varnothing)$	axiom of the empty set (S2),
$\models \forall X(X \notin E)$	hypothesis on E,
$\models \forall X[(X \in \varnothing) \Leftrightarrow (X \in E)]$	theorem 261,
$\models E = \varnothing$	definition of = and extensionality (S1).

□

2.1.3 Subsets and Supersets

In some circumstances, only some of the elements of a set prove useful; the following definition then allows for the grouping of all such elements into a "subset".

Definition 287 (Subsets and supersets) For each set A, for each set B, the set A is a **subset** of the set B if and only if each element of A is also an element of B. Either notation $A \subseteq B$ or $A \subseteqq B$ indicates that "A is a subset of B"; thus,

$$\models \forall A\{\forall B[(A \subseteq B) \Leftrightarrow \{\forall X[(X \in A) \Rightarrow (X \in B)]\}]\}.$$

Similarly, a set B is a **superset** of a set A if and only if $A \subseteq B$, a relation also denoted by $B \supseteq A$ or $B \supseteqq A$.

In the definition of subsets, the equivalence ($\Leftrightarrow$) states that the relation of subset ($A \subseteq B$) is logically equivalent to the formula $\forall X[(X \in A) \Rightarrow (X \in B)]$. In this formula, for each set X the logical implication ($\Rightarrow$) states that *if* X is an element of A, *then* X is also an element of B.

There also exist symbols more specific than $A \subseteq B$. For instance, $A \subset B$ or $A \subsetneq B$ or $A \subsetneqq B$ indicate that A is a subset of B different from B; thus,

$$\models (A \subsetneqq B) \Leftrightarrow (A \subsetneq B) \Leftrightarrow (A \subset B) \Leftrightarrow [(A \subseteq B) \wedge (A \neq B)].$$

Similarly, $B \supset A$ or $B \supsetneq A$ or $B \supsetneqq A$ stand for $(B \supseteq A) \wedge (B \neq A)$.

The following three theorems provide further examples and practice with proofs in set theory. Moreover, they also demonstrate how to design a proof.

Theorem 288 *Each set is a subset of itself:* $\forall S(S \subseteq S)$ *is universally valid.*

Proof. An informal proof may consist of the single statement that each element of S is also an element of S, whence S is a subset of S, by definition 287. A formal proof can show how the definition of $\subseteq$ reveals that $S \subseteq S$ is logically equivalent to a tautology:

$$\begin{array}{rl} \forall S(S \subseteq S) & \textbf{yet unproved}, \\ \Updownarrow & \text{definition of } \subseteq \\ \forall S\,\{\forall X[(X \in S) \Rightarrow (X \in S)]\} & \end{array}$$

which has the form $\forall S\,\{\forall X[(P) \Rightarrow (P)]\}$, based on the tautology $(P) \Rightarrow (P)$, here with $X \in S$ instead of P. Thus a complete proof can proceed as follows.

$$\begin{array}{ll} \vdash (P) \Rightarrow (P) & \text{tautology from logic,} \\ \models (X \in S) \Rightarrow (X \in S) & \text{substitution in the tautology,} \\ \models \forall S\{\forall X[(X \in S) \Rightarrow (X \in S)]\} & \text{generalizations,} \\ \models \forall S(S \subseteq S) & \text{definition (287) of subsets.} \end{array}$$

□

Theorem 289 *The empty set is a subset of every set:* $\models \forall S(\varnothing \subseteq S)$.

Proof. An informal proof can use the converse law of contraposition (axiom P3):

- Every set not in S is also not in $\varnothing$, because no set belongs to $\varnothing$;
- the contraposition then means that every set in $\varnothing$ is also in S;
- hence $\varnothing$ is a subset of S, by the definition of subsets.

A formal proof can demonstrate how the definition of $\subseteq$ and the axiom of the empty set reveal that $\varnothing \subseteq S$ is logically equivalent to a tautology:

$$\begin{array}{rl} \forall S(\varnothing \subseteq S) & \textbf{yet unproved}, \\ \Updownarrow & \text{definition of } \subseteq \\ \forall S\,\{\forall X[(X \in \varnothing) \Rightarrow (X \in S)]\} & \end{array}$$

which is a tautology of the form $(F) \Rightarrow (P)$, with F False. Thus a complete proof may consist of the following steps.

$$\begin{array}{ll} \vdash (F) \Rightarrow (P) & \text{tautology with } F \text{ False,} \\ \models (X \in \varnothing) \Rightarrow (X \in S) & \text{substitutions in the tautology,} \\ \models \forall S\,\{\forall X[(X \in \varnothing) \Rightarrow (X \in S)]\} & \text{generalizations,} \\ \models \forall S(\varnothing \subseteq S) & \text{definition (287) of } \subseteq. \end{array}$$

□

Theorem 290 *Two sets are subsets of each other if and only if they equal each other:*

$$\models \forall A(\forall B\{(A = B) \Leftrightarrow [(A \subseteq B) \wedge (B \subseteq A)]\}).$$

Proof. An informal proof can proceed as follows.

- If each element of A is an element of B,
- and if each element of B is an element of A,
- then A and B have exactly the same elements;
- hence $A = B$ by extensionality, and conversely.

A formal proof can establish the following sequence of logical equivalences.

$(A \subseteq B) \wedge (B \subseteq A)$

$\Updownarrow$ definition of subset,

$\{\forall X[(X \in A) \Rightarrow (X \in B)]\} \wedge \{\forall X[(X \in B) \Rightarrow (X \in A)]\}$

$\Updownarrow$ thanks to $\models \{[\forall X(P)] \wedge [\forall X(Q)]\} \Leftrightarrow \{\forall X[(P) \wedge (Q)]\}$,

$\forall X\{[(X \in A) \Rightarrow (X \in B)] \wedge [(X \in B) \Rightarrow (X \in A)]\}$

$\Updownarrow$ definition of $\Leftrightarrow$,

$\forall X[(X \in A) \Leftrightarrow (X \in B)]$

$\Updownarrow$ axiom of extensionality (S1).

$A = B$

Inserting $\forall A$ and $\forall B$ before each step (generalizing) then completes the proof. □

The axioms of extensionality (S1) and of the empty set (S2) apply to all sets. Yet sets other than the empty set require further axioms, as explained in the next section.

2.1.4 Exercises

Exercise 401 Write a logical formula stating that a set S is not the empty set.

Exercise 402 Write a logical formula stating that a set A is not a subset of B.

Exercise 403 Write a logical formula stating that a set A is not a superset of B.

Exercise 404 Write a logical formula stating that a set A is not equal to a set B.

Exercise 405 Prove that $\varnothing \in \varnothing$ is False.

Exercise 406 Prove that $\varnothing \subseteq \varnothing$ is True.

Exercise 407 Prove that $\varnothing = \varnothing$ is True.

Exercise 408 Use the second formula, $\forall Y[(A \in Y) \Leftrightarrow (B \in Y)]$, in the axiom of extensionality to write a proof that $S = S$ for each set S.

Exercise 409 Prove that $\varnothing$ is the only subset of $\varnothing$: $\models \forall S[(S \subseteq \varnothing) \Rightarrow (S = \varnothing)]$.

Exercise 410 For each set S, prove that $S \supsetneq S$ is False.

Exercise 411 Prove that if a set S is the only subset of itself, then S is the empty set.

Exercise 412 Prove that there does *not* exist a set S such that $S \in Y$ for every set Y.

Exercise 413 Prove that if a set S is a subset of every set, then S is the empty set.

Exercise 414 For all sets A, B, and C, prove that if $A = B$ and $B = C$, then $A = C$.

Exercise 415 For all sets A, B, and C, prove that if $A \subseteq B$ and $B \subseteq C$, then $A \subseteq C$.

Exercise 416 For all sets A, B, and C, prove that if $A \subset B$ and $B \subset C$, then $A \subset C$.

Exercise 417 For all sets A and B, prove that $A = B$ if and only if for every set Z

$$(A \subseteq Z) \Leftrightarrow (B \subseteq Z).$$

Exercise 418 For all sets A and B, prove that $A \subseteq B$ if and only if for every set Z

$$(B \subseteq Z) \Rightarrow (A \subseteq Z).$$

Exercise 419 For all sets C and D, prove that $C \supseteq D$ if and only if for every set W

$$(D \supseteq W) \Rightarrow (C \supseteq W).$$

Exercise 420 For all sets R and S, prove that $R \supseteq S$ if and only if for every set W

$$(W \in S) \Rightarrow (W \in R).$$

2.2 PAIRING, POWER, AND SEPARATION

2.2.1 Pairing

A theory allowing for sets other than the empty set requires additional axioms. For instance, the **axiom of pairing** states that for every set H and every set K, there exists a set L, also denoted by $\{H, K\}$, which contains only the elements H and K.

Axiom S3 (Axiom of pairing)

$$\models \forall H \{\forall K [\exists L (\forall X \{(X \in L) \Leftrightarrow [(X = H) \vee (X = K)]\})]\}.$$

In the axiom of pairing (S3), the equivalence ($\Leftrightarrow$) states that a set X is an element of L if and only if X equals H or X equals K. Because the logical "or" is inclusive, the axiom of pairing thus allows $\{H, K\}$ to contain both H and K. Moreover, because the logical "or" commutes, the order in which H and K appear does not matter.

Theorem 291 *For each set H and for each set K, $\{H, K\} = \{K, H\}$.*

Proof. An informal proof can compare the elements of $\{H, K\}$ and $\{K, H\}$:

- The set $\{H, K\}$ contains the elements H and K, but no other element;
- the set $\{K, H\}$ contains the elements K and H, but no other element;
- thus $\{H, K\}$ and $\{K, H\}$ contain exactly the same elements;
- therefore $\{H, K\} = \{K, H\}$ by the axiom of extensionality (S1).

A formal proof can rely on the tautology $[(P) \vee (Q)] \Leftrightarrow [(Q) \vee (P)]$ to show that

$$[(X = H) \vee (X = K)] \Leftrightarrow [(X = K) \vee (X = H)]$$

is universally valid, and hence also

$$(X \in \{H, K\}) \Leftrightarrow [(X = H) \vee (X = K)] \Leftrightarrow [(X = K) \vee (X = H)] \Leftrightarrow (X \in \{K, H\}),$$

whence $\{H, K\} = \{K, H\}$ by the axiom of extensionality (S1). $\square$

With $H := S$ and $K := S$, the following theorem shows that for each set S, there exists a set, denoted by $\{S\}$, which contains only one element, S.

Theorem 292 *For each set S there exists a set L, also denoted by $\{S\}$, which contains only the element S; formally,* $\models \forall S(\exists L\,\{\forall X\,[(X \in L) \Leftrightarrow (X = S)]\})$.

Proof. An informal proof can use the axiom of pairing with $H := S$ and $K := S$:

- For each set S, the axiom of pairing yields a set $\{S, S\}$;
- by the axiom of pairing, $X \in \{S, S\}$ if and only if $X = S$ or $X = S$;
- yet "$X = S$ or $X = S$" is logically equivalent to "$X = S$";
- therefore $\{S, S\} = \{S\}$ by the axiom of extensionality.

A corresponding formal proof can rely on the tautology $(P) \Leftrightarrow [(P) \vee (P)]$:

$$\begin{array}{ll} \models \forall H\{\forall K[\exists L(\forall X\,\{(X \in L) \Leftrightarrow [(X = H) \vee (X = K)]\})]\} & \text{axiom S3,} \\ \models \exists L(\forall X\,\{(X \in L) \Leftrightarrow [(X = S) \vee (X = S)]\}) & \text{Subf}_S^H,\ \text{Subf}_S^K, \\ \models \forall S[\exists L(\forall X\,\{(X \in L) \Leftrightarrow [(X = S) \vee (X = S)]\})] & \text{generalization,} \\ \models \forall S(\exists L\{\forall X[(X \in L) \Leftrightarrow (X = S)]\}) & (P)\Leftrightarrow[(P)\vee(P)]. \end{array}$$

□

Definition 293 (Singleton) A **singleton** is a set containing exactly one element.

Example 294 In theorem 292, substituting $\varnothing$ for S gives the set $L = \{\varnothing\}$, which contains the single element $\varnothing$. In particular, $\varnothing \in \{\varnothing\}$, so that $\{\varnothing\}$ is *not* empty.

The sets $\varnothing$ and $\{\varnothing\}$ have *different* characteristics.

Theorem 295 *The sets $\varnothing$ and $\{\varnothing\}$ are two distinct sets:* $\varnothing \neq \{\varnothing\}$.

Proof. An informal proof can utilize substitutions in previous axioms and theorems:

- By definition of the empty set, $\varnothing \notin \varnothing$ (by a substitution in axiom S2);
- moreover, $\varnothing \in \{\varnothing\}$ (by a substitution in theorem 292);
- hence the two sets $\varnothing$ and $\{\varnothing\}$ have different elements: $\varnothing \in \{\varnothing\}$ but $\varnothing \notin \varnothing$;
- consequently, $\varnothing \neq \{\varnothing\}$, by the axiom of extensionality.

The following formal proof uses contraposition in the tautologies

$$\begin{array}{l} \vdash (P) \Rightarrow \{[(P) \Rightarrow (Q)] \Rightarrow (Q)\}, \\ \vdash (P) \Rightarrow ([\neg(Q)] \Rightarrow \{\neg[(P) \Rightarrow (Q)]\}), \end{array}$$

whence if P and $\neg(Q)$ are True then $\neg[(P) \Rightarrow (Q)]$ is True by *Modus Ponens* twice:

$$\begin{array}{ll} \models \neg(\varnothing \in \varnothing) & \neg(Q)\text{: substitution in axiom S2,} \\ \models \varnothing \in \{\varnothing\} & P\text{: substitution in theorem 292,} \\ \models \neg[(\varnothing \in \{\varnothing\}) \Rightarrow (\varnothing \in \varnothing)] & \neg[(P) \Rightarrow (Q)]\,, \\ \models \exists X\{\neg[(X \in \{\varnothing\}) \Rightarrow (X \in \varnothing)]\} & \text{substitution of } \varnothing \text{ for } X\ (\text{Subf}_\varnothing^X), \\ \models \neg\{\forall X[(X \in \{\varnothing\}) \Rightarrow (X \in \varnothing)]\} & \{\exists X[\neg(P)]\} \Leftrightarrow \{\neg[\forall X(P)]\}, \\ \models \neg(\{\varnothing\} \subseteq \varnothing) & \text{definition of subsets,} \\ \models \varnothing \neq \{\varnothing\} & \text{contraposition of theorem 290}\,. \end{array}$$

□

The distinction between $\varnothing$ and $\{\varnothing\}$ allows for the formation of various other sets.

Example 296 Substituting $H := \varnothing$ and $K := \{\varnothing\}$ in the axiom of pairing gives the set $L = \{H, K\} = \{\varnothing, \{\varnothing\}\}$. This set L has *two* elements, because $\varnothing \neq \{\varnothing\}$.

Example 297 Substituting $S := \{\varnothing\}$ in theorem 292 gives the set $L = \{S\} = \{\ \{\varnothing\}\ \}$. This set L has *one* element, in effect $\{\varnothing\}$; thus, $\{\varnothing\} \in \{\ \{\varnothing\}\ \}$.

2.2.2 Power Sets

For sets with more than one or two elements, a new axiom becomes necessary. For instance, the **axiom of the power set** states that for each set A, the collection of all subsets of A forms a new set, denoted by $\mathcal{P}$ or by $\mathcal{P}(A)$ and called the **power set** of A:

Axiom S4 (Axiom of the power set)

$$\models \forall A\big(\exists \mathcal{P}\,\{\forall S[(S \in \mathcal{P}) \Leftrightarrow (S \subseteq A)]\}\big).$$

In the axiom of the power set (S4), the equivalence ($\Leftrightarrow$) states that a set S is an *element* of the power set $\mathcal{P}(A)$ if and only if S is a *subset* of the set A.

Example 298 The empty set $\varnothing$ is the only subset of itself. Hence its power set has only one element, $\varnothing$, so that $\mathcal{P}(\varnothing) = \{\varnothing\}$.

A set A and its power set $\mathcal{P}(A)$ might have no elements in common. Yet the next theorem shows that for every element X of A the singleton $\{X\}$ is an element of $\mathcal{P}(A)$.

Theorem 299 *For all sets A and X, if $X \in A$, then $\{X\} \in \mathcal{P}(A)$, and conversely.*

Proof. An informal proof can rely on the definitions of subsets and power sets.

- $X \in A$ if and only if $\{X\} \subseteq A$, by definitions of $\{X\}$ and subsets;
- $\{X\} \subseteq A$ if and only if $\{X\} \in \mathcal{P}(A)$ by definition of power sets.

A formal proof carries out similar verification from the formal definitions.

$$\begin{aligned}
&\forall A\big(\forall X\{(X \in A) \Leftrightarrow [\{X\} \in \mathcal{P}(A)]\}\big) \quad \textbf{yet unproved,}\\
&\quad\Updownarrow \quad \text{definition of power sets,}\\
&\forall A\big(\forall X\{(X \in A) \Leftrightarrow [\{X\} \subseteq A]\}\big)\\
&\quad\Updownarrow \quad \text{definition of subsets,}\\
&\forall A\big(\forall X\{(X \in A) \Leftrightarrow [\forall Z(Z \in \{X\}) \Rightarrow (Z \in A)]\}\big)\\
&\quad\Updownarrow \quad \text{definition of } \{X\} \text{ (pairing),}\\
&\forall A\big(\forall X\{(X \in A) \Leftrightarrow [\forall Z(Z = X) \Rightarrow (Z \in A)]\}\big)\\
&\quad\Updownarrow \quad \text{extensionality,}\\
&\forall A\Big[\forall X\big\{(X \in A) \Leftrightarrow \big(\forall Z\big[\{\forall Y[(Z \in Y) \Leftrightarrow (X \in Y)]\} \Rightarrow (Z \in A)\big]\big)\big\}\Big]\\
&\quad\Updownarrow \quad \text{Subf}_X^Z \text{ and Subf}_A^Y,\\
&\forall A\Big[\forall X\big\{(X \in A) \Leftrightarrow \big[\{[(X \in A) \Leftrightarrow (X \in A)]\} \Rightarrow (X \in A)\big]\big\}\Big]
\end{aligned}$$

which holds by Peirce's law $\{[(P) \Rightarrow (Q)] \Rightarrow (P)\} \Rightarrow (P)$ (exercise 347, page 67) and its converse (exercise 367, page 77). □

The next theorem shows that the power set of a singleton has exactly two elements.

Theorem 300 *For each set H, $\mathcal{P}(\{H\}) = \{\varnothing, \{H\}\}$.*

Proof. An informal proof can list the subsets of $\{H\}$ by cases:

- $\{H\} \subseteq \{H\}$ by theorem 288;
- $\varnothing \subseteq \{H\}$ by theorem 289;
- if $B \subseteq \{H\}$, then B has no element or B has the single element H;
- hence $\{H\}$ has no subets other than $\varnothing$ and $\{H\}$; thus $\mathcal{P}(\{H\}) = \{\varnothing, \{H\}\}$.

A formal proof can proceed through two cases, using the tautology (theorem 221)

$$([(H) \Rightarrow (K)] \wedge \{[\neg(H)] \Rightarrow (K)\}) \Rightarrow (K)$$

with H for $B = \varnothing$ and K for $B \in \{\varnothing, \{\varnothing\}\}$. In the first case, $(B = \varnothing) \Rightarrow (B \in \{\varnothing, \{\varnothing\}\})$ by substituting $\{\varnothing, \{\varnothing\}\}$ for Y in the axiom of extensionality (S1). In the second case, the definitions of $B \neq \varnothing$ and $B \subseteq \{H\}$ give

$$\begin{array}{rl}
(B \neq \varnothing) \wedge (B \subseteq \{H\}) & \\
\Updownarrow & \text{definitions of } \varnothing \text{ and } \subseteq, \\
[\exists X(X \in B)] \wedge \{\forall Z[(Z \in B) \Rightarrow (Z \in \{H\})]\} & \\
\Updownarrow & \text{definitions of } \{H\}, \\
[\exists X(X \in B)] \wedge \{\forall Z[(Z \in B) \Rightarrow (Z = H)]\} & \\
\Downarrow & (Z{=}H) \Rightarrow \{\forall Y[(Z \in Y) \Leftrightarrow (H \in Y)]\} \\
\exists X\big[(X \in B) \wedge \{\forall Z[(Z \in B) \Rightarrow (H \in B)]\}\big] & \\
\Downarrow & \text{specialization } \mathrm{Subf}_X^Z. \\
\exists X\{(X \in B) \wedge [(X \in B) \Rightarrow (H \in B)]\} & \\
\Downarrow & \{(P) \wedge [(P) \Rightarrow (Q)]\} \Rightarrow (Q), \\
H \in B &
\end{array}$$

Thus $B \subseteq \{H\}$, but $H \in B$ whence $\{H\} \subseteq B$, and hence $B = \{H\}$. □

Example 301 The singleton $\{\varnothing\}$ has two subsets: $\{\varnothing\}$ by theorem 288, and $\varnothing$ by theorem 289. Morevoer, $\{\varnothing\}$ has no other subsets. Hence, $\mathcal{P}(\{\varnothing\}) = \{\varnothing, \{\varnothing\}\}$.

Counterexample 302 The set $A := \{\{\varnothing\}\}$ has *no* element in common with its power set $\mathcal{P}(\{\{\varnothing\}\}) = \Big\{ \varnothing, \{\{\varnothing\}\} \Big\}$. In particular, A is *not* a subset of $\mathcal{P}(A)$.

The axioms of the empty set (S2), of pairing (S3), and of the power set (S4) allow for increasingly large sets, for instance,

$$\begin{array}{rcl}
\varnothing, & & \\
\mathcal{P}(\varnothing) & = & \{\varnothing\}, \\
\mathcal{P}(\{\varnothing\}) & = & \{\varnothing, \{\varnothing\}\}, \\
\mathcal{P}(\{\varnothing, \{\varnothing\}\}) & = & \Big\{\varnothing, \{\varnothing\}, \{\{\varnothing\}\}, \{\varnothing, \{\varnothing\}\}\Big\}. \\
 & \vdots &
\end{array}$$

However, the "selection" or "separation" of subsets requires yet another axiom.

2.2.3 Separation of Sets

For each set A and for each logical formula P, the **axiom schema of separation** "separates" from A a subset S consisting of all the elements X of A for which P is True. The logical formula P may *not* involve any free occurrence of the variable S, but it may involve free occurrences of X and of other variables, as indicated in the following axiom by the ellipsis $\ldots$, which may represent other free variables. Such a "separation rule" to form subsets differs from other axioms, because the logic used here allows for the quantification of the elements, with the symbols $\forall X$, but this logic has no provision for the quantification of formulae: it does not allow for expressions like $\forall P$ with P standing for formulae. Thus, the "separation rule" provides a schema for infinitely many axioms, in effect one axiom for each logical formula, whence its name.

Axiom S5 (Axiom schema of separation) For each set A, and for each logical formula P that does *not* contain any free occurrence of the variable S, there exists a subset $S \subseteq A$ that consists of only those elements $X \in A$ for which P is True:

$$\models \forall A\big[\exists S\big(\forall X\,\{(X \in S) \Leftrightarrow [(X \in A) \wedge (P)]\}\big)\big].$$

Common notations for the subset S have the formats

$$\begin{aligned} S &= \{X : (X \in A) \wedge (P)\}, \\ S &= \{X \in A : P\}. \end{aligned}$$

An alternative notation replaces the colon (:) by a vertical bar (|), but in the context of further mathematics such vertical bars become difficult to recognize against absolute values, norms, and other similar symbols.

Example 303 Consider the set

$$\begin{aligned} A &:= \mathcal{P}(\{\varnothing, \{\varnothing\}\}) \\ &= \{\,\varnothing,\ \{\varnothing\},\ \{\{\varnothing\}\},\ \{\varnothing, \{\varnothing\}\}\,\}, \end{aligned}$$

and let P be the formula $\neg(X = \varnothing)$. Then

$$\begin{aligned} A &= \{\,\varnothing,\ \{\varnothing\},\ \{\{\varnothing\}\},\ \{\varnothing, \{\varnothing\}\}\,\}, \\ (P) &\Leftrightarrow [\neg(X = \varnothing)], \\ S &= \{X : (X \in A) \wedge [\neg(X = \varnothing)]\}, \\ S &= \{X \in A : \neg(X = \varnothing)\}, \\ S &= \{\,\{\varnothing\},\ \{\{\varnothing\}\},\ \{\varnothing, \{\varnothing\}\}\,\}. \end{aligned}$$

The existence of this set S would not have followed from the previous axioms.

The following examples demonstrate the axiom schema of separation with various instances of the formula P. Additional examples appear in the next section.

Example 304 For each set A and for each set B, let P be $(X \in B)$. Then

$$S = \{X : (X \in A) \wedge (X \in B)\},$$

which is usually denoted by $S = A \cap B$ and called the **intersection** of A and B.

Example 305 For each set A and for each set B, let P be $\neg(X \in B)$. Then

$$S = \{X : (X \in A) \wedge [\neg(X \in B)]\},$$

which is usually denoted by $S = A \setminus B$ and called the **difference** of A and B.

The symbol $\setminus$ adopted here for the difference of two sets aims at avoiding confusions with the arithmetic difference of sets $A - B$ in such further branches of mathematics as convexity, linear algebra, and functional analysis.

Example 306 For each set A, and for all subsets $B \subseteq A$ and $C \subseteq A$, let P be the formula $(X \in B) \vee (X \in C)$. Then

$$S = \{X : (X \in A) \wedge [(X \in B) \vee (X \in C)]\},$$

which is usually denoted by $S = B \cup C$ and called the **union** of B and C.

Unions and intersections of more than two sets form the subject of the next section.

2.2.4 Exercises

Exercise 421 Give examples of sets X, Y, Z with $X \in Y$ and $Y \in Z$ but $X \notin Z$.

Exercise 422 Provide examples of sets X, Y such that $X \in Y$ but $X \subsetneqq Y$.

Exercise 423 Provide examples of sets X, Y such that $X \subseteq Y$ but $X \notin Y$.

Exercise 424 Provide examples of sets X, Y such that $X \in Y$ and $X \subseteq Y$.

Exercise 425 Provide examples of sets A, X such that $X \in A$ but $X \notin \mathcal{P}(A)$.

Exercise 426 Provide examples of sets A, X such that $X \subseteq A$ but $X \subsetneqq \mathcal{P}(A)$.

Exercise 427 Prove that $\{\varnothing\} \neq \{\{\varnothing\}\}$.

Exercise 428 Prove that $\{\varnothing\} \neq \{\varnothing, \{\varnothing\}\}$.

Exercise 429 Prove that $\{\{\varnothing\}\} \neq \{\varnothing, \{\varnothing\}\}$.

Exercise 430 Determine whether $\{\varnothing\} \subseteq \{\{\varnothing\}\}$.

Exercise 431 Determine whether $\{\{\varnothing\}\} \subseteq \{\{\varnothing\}\}$.

Exercise 432 Determine whether $\{\varnothing\} \subseteq \{\varnothing, \{\varnothing\}\}$.

Exercise 433 Determine whether $\{\{\varnothing\}\} \subseteq \{\varnothing, \{\varnothing\}\}$.

Exercise 434 Prove that replacing $\vee$ by $\wedge$ in axiom S3 yields the *False* formula

$$\forall H\{\forall K[\exists L(\forall X\{(X \in L) \Leftrightarrow [(X = H) \wedge (X = K)]\})]\}.$$

Exercise 435 For theorem 291, explain how the word "and" in the informal proof corresponds to the logical connective $\vee$ in the formal proof.

Exercise 436 For each set S, prove that $\{\varnothing, S\} \subseteq \mathcal{P}(S)$.

Exercise 437 Prove that two sets equal each other if and only if they have the same power sets: $\vdash \forall A(\forall B\{(A = B) \Leftrightarrow [\mathcal{P}(A) = \mathcal{P}(B)]\})$.

Exercise 438 Identify the set $\{\varnothing\} \cap \{\varnothing, \{\varnothing\}\}$.

Exercise 439 Identify the set $\{\{\varnothing\}\} \cap \{\varnothing, \{\varnothing\}\}$.

Exercise 440 Identify the set $\{\varnothing\} \cap \{\ \{\varnothing\}\ \}$.

Exercise 441 Identify the set $\{\varnothing\} \cup \{\ \varnothing,\ \{\varnothing\}\ \}$ and identify one of its supersets.

Exercise 442 Identify the set $\{\ \{\varnothing\}\ \} \cup \{\ \varnothing,\ \{\varnothing\}\ \}$ and identify one of its supersets.

Exercise 443 For each set S, identify the set $S \cap \varnothing$.

Exercise 444 For each set S, identify the set $S \cup \varnothing$.

Exercise 445 Prove that $\models \forall A[(A \setminus \varnothing) = A]$.

Exercise 446 Prove that $\models \forall A[(A \setminus A) = \varnothing]$.

Exercise 447 Prove that $\models \forall A\big(\forall B\{[(A \setminus B) = \varnothing] \Leftrightarrow (A \subseteq B)\}\big)$.

Exercise 448 Prove that $\models \forall A\big(\forall B\{[(A \setminus B) = A] \Leftrightarrow [(A \cap B) = \varnothing)]\}\big)$.

Exercise 449 Prove or disprove $\forall A\{\forall B[(A \setminus B) = (B \setminus A)]\}$.

Exercise 450 Prove or disprove $\forall A\big[\forall B\big(\forall C\{[(A \setminus B) \setminus C] = [A \setminus (B \setminus C)]\}\big)\big]$.

2.3 UNIONS AND INTERSECTIONS OF SETS

2.3.1 Unions of Sets

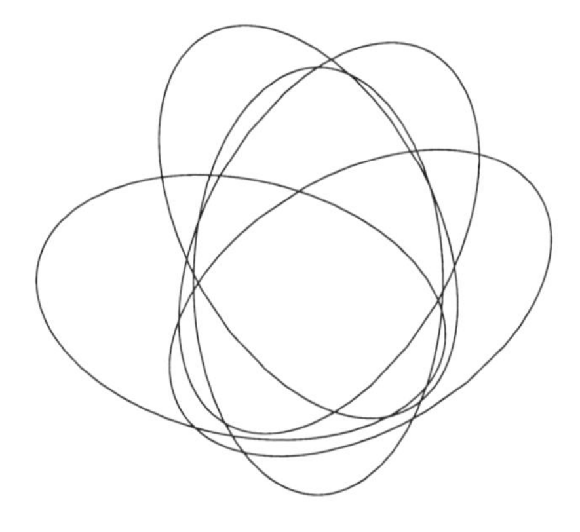

Figure 2.2: Venn diagram for the unions and intersections of five sets [39], [40].

Many mathematical situations involve unions and intersections of any "collection" or "family" of sets, or, in other words, of any set of sets. For instance, the **Venn diagram** in figure 2.2 (adapted from Hamburger and Pippert's [39], [40]) illustrates all the possible unions and intersections for a set of five sets. For the existence of such general unions and intersections, however, new axioms become necessary. Thus, the **axiom of union** asserts that for each set $\mathcal{F}$ there exists a new set U, denoted by $\bigcup \mathcal{F}$ and called the **union** of $\mathcal{F}$, which consists of all the elements X of all the elements S of $\mathcal{F}$:

Axiom S6 (Axiom of union) For each set $\mathcal{F}$, there exists a set U, also denoted by $\bigcup \mathcal{F}$, which consists of all the elements that belong to any element of $\mathcal{F}$:

$$\models \forall \mathcal{F}\big(\exists U\big\{\forall X\big[(X \in U) \Leftrightarrow \{\exists S[(S \in \mathcal{F}) \wedge (X \in S)]\}\big]\big\}\big).$$

In the axiom of union (S6), the equivalence ($\Leftrightarrow$) states that a set X is an element of the union $U = \bigcup \mathcal{F}$ if and only if there exists an element S of $\mathcal{F}$ such that $X \in S$.

Example 307 If $\mathcal{F} = \varnothing$, then $(\bigcup \varnothing) = \varnothing$, because in the axiom of union (S6) the condition $(S \in \mathcal{F})$ is False for every set S, and then the equivalent condition $X \in (\bigcup \varnothing)$ is False for every set X. Thus $\bigcup \varnothing$ contains no element.

For the union of two sets, a special notation proves convenient.

Definition 308 The notation $A \cup B$ stands for $\bigcup\{A, B\}$.

With only two sets in $\mathcal{F}$, the definition of $\bigcup \mathcal{F}$ simplifies considerably.

Theorem 309 *For all sets A and B, if $\mathcal{F} = \{A, B\}$, then*

$$\forall X \left\{ \left[X \in \bigcup\{A, B\} \right] \Leftrightarrow [(X \in A) \vee (X \in B)] \right\}.$$

Proof. For $\mathcal{F} = \{A, B\}$, the axiom of pairing (S3) shows that

$$(S \in \mathcal{F}) \Leftrightarrow [(S = A) \vee (S = B)],$$

whence the axiom of union (S6) gives the following condition for $X \in (A \cup B)$:

$$\begin{array}{rcl} [X \in (A \cup B)] & \Leftrightarrow & \{\exists S[(S \in \{A, B\}) \wedge (X \in S)]\} \\ & \Leftrightarrow & (\exists S\{[(S = A) \vee (S = B)] \wedge (X \in S)\}) \\ & \Leftrightarrow & (\exists S\{[(S = A) \wedge (X \in S)] \vee [(S = B) \wedge (X \in S)]\}) \\ & \Leftrightarrow & \{\exists S[(X \in A) \vee (X \in B)]\} \\ & \Leftrightarrow & [(X \in A) \vee (X \in B)]. \end{array}$$

The last equivalences result from extensionality, which gives

$$\begin{array}{rcl} [(S = A) \wedge (X \in S)] & \Leftrightarrow & (X \in A), \\ [(S = B) \wedge (X \in S)] & \Leftrightarrow & (X \in B). \end{array}$$

Moreover, S does not occur freely in $(X \in B) \vee (X \in B)$, which makes $\exists S$ superfluous, so that $\{\exists S[(X \in B) \vee (X \in B)]\} \Leftrightarrow [(X \in B) \vee (X \in B)]$. □

Example 310 For $A := \{\varnothing, \{\varnothing\}\}$ and $B := \left\{ \{\varnothing, \{\varnothing\}\} \right\}$, the union $A \cup B$ consists of the elements that belong to A (these are $\varnothing$ and $\{\varnothing\}$) or that belong to B (where there is only one element, $\{\varnothing, \{\varnothing\}\}$). Therefore, $A \cup B = \left\{ \varnothing,\ \{\varnothing\},\ \{\varnothing, \{\varnothing\}\} \right\}$.

Example 311 Some abbreviations are common [73, p. 98], [74, p. 453], [95, p. 129]:

$$\begin{aligned}
0 &:= \varnothing, \\
1 &:= 0 \cup \{0\} = \varnothing \cup \{\varnothing\} = \{\varnothing\}, \\
2 &:= 1 \cup \{1\} = \{\varnothing\} \cup \{\{\varnothing\}\} = \{\varnothing, \{\varnothing\}\}, \\
3 &:= 2 \cup \{2\} = \{\varnothing, \{\varnothing\}\} \cup \Big\{\{\varnothing, \{\varnothing\}\}\Big\} = \Big\{\varnothing, \{\varnothing\}, \{\varnothing, \{\varnothing\}\}\Big\}, \\
4 &:= 3 \cup \{3\} = \left\{ \varnothing,\ \{\varnothing\},\ \{\varnothing, \{\varnothing\}\},\ \Big\{\varnothing, \{\varnothing\}, \{\varnothing, \{\varnothing\}\}\Big\} \right\}, \\
5 &:= 4 \cup \{4\}, \\
6 &:= 5 \cup \{5\}, \\
7 &:= 6 \cup \{6\}, \\
8 &:= 7 \cup \{7\}, \\
9 &:= 8 \cup \{8\}.
\end{aligned}$$

The sets $0, 1$ are the **binary digits**; $0, 1, 2, 3, 4, 5, 6, 7, 8, 9$ are the **decimal digits**.

Example 312 For all sets A, B, and C, there exists a set V whose only elements are A, B, and C. Indeed, the axiom of pairing yields two sets $\{A, B\}$ and $\{B, C\}$. With

$$\mathcal{F} = \{\ \{A, B\},\ \{B, C\}\ \},$$

the axiom of union produces a set $\bigcup \mathcal{F} = \{A, B\} \cup \{B, C\}$ such that for each set X,

$$\begin{aligned}
[X \in (\{A, B\} \cup \{B, C\})] &\Leftrightarrow [(X \in \{A, B\}) \vee (X \in \{B, C\})] \\
&\Leftrightarrow \{[(X = A) \vee (X = B)] \vee [(X = B) \vee (X = C)]\} \\
&\Leftrightarrow [(X = A) \vee (X = B) \vee (X = C)].
\end{aligned}$$

The common notation $\{A, B, C\}$ can replace $\{A, B\} \cup \{B, C\}$, so that

$$\{A, B, C\} = \{A, B\} \cup \{B, C\}.$$

Example 313 If the set $\mathcal{F}$ contains only three sets A, B, and C, so that $\mathcal{F} = \{A, B, C\}$, then $\bigcup\{A, B, C\}$ is also denoted by $A \cup B \cup C$, so that

$$A \cup B \cup C = \bigcup\{A, B, C\} = \bigcup \mathcal{F}$$

consists of all the elements that are elements of at least one of A, B, or C.

The following theorems show that unions and intersections of sets have features similar to — and derived from — those of logical disjunctions and conjunctions.

Theorem 314 *The union of sets commutes:* $\models \forall A\, \{\forall B[(A \cup B) = (B \cup A)]\}$.

Proof. An informal proof may consist of the statements that a set belongs to $A \cup B$ if and only if it belongs to A or to B, which also means that it belongs B or A and hence to $B \cup A$, whence $A \cup B = B \cup A$ by the axiom of extensionality. A formal proof reveals that the commutativity of the union $\cup$ corresponds to the commutativity of the logical connective $\vee$ in the proof of theorem 291, which states that $\{A, B\} = \{B, A\}$:

$$A \cup B = \bigcup\{A, B\} = \bigcup\{B, A\} = B \cup A.$$

□

Theorem 315 *The union of sets is associative:*

$$\vdash \forall A \forall B \big[\forall C(\{[(A \cup B) \cup C] = [A \cup (B \cup C)]\})\big].$$

Proof. An informal proof can merely point out that a set "A or B, or C" is equivalent to "A, or B or C." A formal proof reveals that the associativity of the union $\cup$ corresponds to the associativity of the logical connective $\vee$ in a translation of $[(A \cup B) \cup C] = [A \cup (B \cup C)]$ into a tautology with atomic formulae and connectives:

$$\begin{array}{rl} [(A \cup B) \cup C] = [A \cup (B \cup C)] & \textbf{yet unproved,} \\ \Updownarrow & \text{axiom of extensionality (S1),} \\ \forall X\, (\{X \in [(A \cup B) \cup C]\} & \\ \Leftrightarrow \{X \in [A \cup (B \cup C)]\}) & \\ \Updownarrow & \text{theorem 309 twice,} \\ \forall X\, (\{[(X \in A) \vee (X \in B)] \vee (X \in C)\} & \\ \Leftrightarrow \{(X \in A) \vee [(X \in B) \vee (X \in C)]\}) & \end{array}$$

which is uniformly valid associativity of the logical connective $\vee$:

$$\{[(P) \vee (Q)] \vee (R)\} \Leftrightarrow \{(P) \vee [(Q) \vee (R)]\}$$

□

The following theorem shows that if a set B is an *element* of a set $\mathcal{F}$, then B is a *subset* of the union $\bigcup \mathcal{F}$.

Theorem 316 *For each set $\mathcal{F}$ and for each set B, if $B \in \mathcal{F}$, then $B \subseteq (\bigcup \mathcal{F})$:*

$$\models \forall \mathcal{F} \left(\forall B \left\{(B \in \mathcal{F}) \Rightarrow \left[B \subseteq \left(\bigcup \mathcal{F}\right)\right]\right\}\right).$$

Proof. An informal proof can substitute B for S in the axiom of union (S6):

- If $B \in \mathcal{F}$, then for each $X \in B$ there exists $S \in \mathcal{F}$ with $X \in S$, namely $S = B$;
- hence $X \in \bigcup \mathcal{F}$ for every $X \in B$, by definition of $\bigcup \mathcal{F}$ (axiom S6);
- consequently $B \subseteq \bigcup \mathcal{F}$ by definition of subsets (definition 287).

A formal proof consists in transforming the proposed formula until a tautology appears. Because two steps involve not an equivalence but an implication, however, the final proof reorders the investigative five steps in their reverse order:

(5) $\forall \mathcal{F}(\forall B\{[B \in \mathcal{F}] \Rightarrow [B \subseteq (\bigcup \mathcal{F})]\})$ **yet unproved,**

$\Updownarrow$ definitions of $\subseteq$,

(4) $\forall \mathcal{F}\Big[\forall B\Big([B \in \mathcal{F}] \Rightarrow \{\forall X\{(X \in B) \Rightarrow [X \in (\bigcup \mathcal{F})]\}\}\Big)\Big]$

$\Updownarrow$ definitions of $\bigcup$,

(3) $\forall \mathcal{F}\Big[\forall B\Big([B \in \mathcal{F}] \Rightarrow \{\forall X[(X \in B) \Rightarrow \{\exists A[(A \in \mathcal{F}) \wedge (X \in A)]\}]\}\Big)\Big]$

$\Uparrow$ no X in $(B \in \mathcal{F})$, hence $\{\forall X[(P) \Rightarrow (Q)]\} \Rightarrow \{(P) \Rightarrow [\forall X(Q)]\}$,

(2) $\forall \mathcal{F}\Big\{\forall B(\forall X\{[B \in \mathcal{F}] \Rightarrow [(X \in B) \Rightarrow \{\exists A[(A \in \mathcal{F}) \wedge (X \in A)]\}]\})\Big\}$

$\Uparrow \models [\text{Subf}_B^A(P)] \Rightarrow [\exists A(P)]$,

(1) $\forall \mathcal{F}(\forall B\{\forall X[(B \in \mathcal{F}) \Rightarrow \{(X \in B) \Rightarrow [(B \in \mathcal{F}) \wedge (X \in B)]\}]\})$

which holds thanks to the tautology $(P) \Rightarrow \{(Q) \Rightarrow [(P) \wedge (Q)]\}$ (theorem 201). □

2.3.2 Intersections of Sets

Based on the axiom of union and the axiom of separation, the following definition specifies for each *non-empty* set $\mathcal{F}$ a new set, denoted by $\bigcap \mathcal{F}$ and called the **intersection** of $\mathcal{F}$, which consists of every element X that is an element of every element of $\mathcal{F}$.

Definition 317 (Intersection of sets) For each set $\mathcal{F}$, apply the axiom of union to define $\mathcal{A} := \bigcup \mathcal{F}$, and apply the axiom of separation to the set $\mathcal{A}$ and to the formula

$$\forall Y[(Y \in \mathcal{F}) \Rightarrow (X \in Y)].$$

Then define the **intersection** of $\mathcal{F}$, a set denoted by $\bigcap \mathcal{F}$, through the formula

$$\bigcap \mathcal{F} := \Big\{X \in \bigcup \mathcal{F} : \forall Y[(Y \in \mathcal{F}) \Rightarrow (X \in Y)]\Big\},$$

so that

$$\forall X\{(X \in \bigcap \mathcal{F}) \Leftrightarrow [(X \in \bigcup \mathcal{F}) \wedge \{\forall Y[(Y \in \mathcal{F}) \Rightarrow (X \in Y)]\}]\}.$$

The definition of the intersection $\bigcap \mathcal{F}$ of a set of sets $\mathcal{F}$ states that a set X is an element of $\bigcap \mathcal{F}$ if and only if $\forall Y[(Y \in \mathcal{F}) \Rightarrow (X \in Y)]$ is True, which occurs if and only if X is an element of every element Y of $\mathcal{F}$. This definition also holds if $\mathcal{F}$ is empty because of the requirement that $\bigcap \mathcal{F}$ first be a subset of the union $\bigcup \mathcal{F}$. (This definition of $\bigcap$ in terms of $\bigcup$ conforms to Bernays's [3, p. 14].) If the set $\mathcal{F}$ contains only two elements, then the definition of $\bigcap \mathcal{F}$ simplifies considerably.

Theorem 318 *For all sets A and B, if $\mathcal{F} = \{A, B\}$ then*

$$\bigcap \{A, B\} = \{X : [X \in (A \cup B)] \wedge [(X \in A) \wedge (X \in B)]\}.$$

Proof. Apply theorem 202, $[(H) \Rightarrow (K)] \Rightarrow \big([(H) \Rightarrow (L)] \Rightarrow \{(H) \Rightarrow [(K) \wedge (L)]\}\big)$:

$\models (\forall Y\{(Y \in \{A,B\}) \Rightarrow (X \in Y)\})$
$\quad \Rightarrow [(A \in \{A,B\}) \Rightarrow (X \in A)]$ specialization Subf_A^Y,
$\models A \in \{A,B\}$ axiom of pairing (S3),
$\models X \in A$ *Modus Ponens*;
$\models (\forall Y\{(Y \in \{A,B\}) \Rightarrow (X \in Y)\})$
$\quad \Rightarrow [(B \in \{A,B\}) \Rightarrow (X \in B)]$ specialization Subf_B^Y,
$\models B \in \{A,B\}$ axiom of pairing (S3),
$\models X \in B$ *Modus Ponens*;
$\models (\forall Y\{(Y \in \{A,B\}) \Rightarrow (X \in Y)\})$
$\quad \Rightarrow [(X \in A) \wedge (X \in B)]$ theorem 202;

$\models (X \in A) \Rightarrow [(A \in \{A,B\}) \Rightarrow (X \in A)]$ from axiom P1,
$\models (X \in B) \Rightarrow [(B \in \{A,B\}) \Rightarrow (X \in B)]$ from axiom P1,
$\models [(X \in A) \wedge (X \in B)]$
$\quad \Rightarrow (\forall Y\{(Y \in \{A,B\}) \Rightarrow (X \in Y)\})$

which is universally valid thanks to the tautology (theorem 219)

$$\{[(P) \Rightarrow (Q)] \wedge [(R) \Rightarrow (S)]\} \Rightarrow \{[(P) \wedge (R)] \Rightarrow [(Q) \wedge (S)]\}.$$

□

For the intersection of two sets, a special notation proves convenient.

Definition 319 The notation $A \cap B$ stands for $\bigcap\{A,B\}$.

Example 320 For the sets

$$\begin{aligned} A &:= \{\,\varnothing, \{\varnothing\}, \{\{\varnothing\}\}\,\}, \\ B &:= \{\,\{\varnothing\}, \{\{\varnothing\}\}, \{\{\{\varnothing\}\}\}\,\}, \end{aligned}$$

$A \cap B$ contains only the elements belonging simultaneously to both A and B:

$$\begin{aligned} A \cap B &= \{\,\varnothing, \{\varnothing\}, \{\{\varnothing\}\}\,\} \bigcap \{\,\{\varnothing\}, \{\{\varnothing\}\}, \{\{\{\varnothing\}\}\}\,\} \\ &= \{\,\{\varnothing\}, \{\{\varnothing\}\}\,\}. \end{aligned}$$

Because the definition of the intersection of sets relies upon the conjunction $\wedge$, the intersection has formal features similar to the logical features of the conjunction, for instance, commutativity and associativity, as demonstrated in the following theorems.

Theorem 321 *The intersection of sets commutes:*

$$\models \forall A\,\{\forall B[(A \cap B) = (B \cap A)]\}.$$

Proof. An informal proof can state that a set X is an element of A and B if and only if X is an element of B and A. A formal proof shows that the commutativity of the intersection $\cap$ corresponds to the commutativity of the logical connective $\wedge$:

$(A \cap B) = (B \cap A)$ **yet unproved**,
$\Updownarrow$ axiom S1,
$\forall X\,\{[X \in (A \cap B)] \Leftrightarrow [X \in (B \cap A)]\}$
$\Updownarrow$ theorem 318,
$\forall X\,\{[(X \in A) \wedge (X \in B)] \Leftrightarrow [(X \in B) \wedge (X \in A)]\}$

which holds thanks to the tautology $[(P) \wedge (Q)] \Leftrightarrow [(Q) \wedge (P)]$ (theorem 198). □

Theorem 322 *The intersection of sets is associative:*

$$\models \forall A\big[\forall B\big(\forall C\{[(A \cap B) \cap C] = [A \cap (B \cap C)]\}\big)\big].$$

Proof. An informal proof can rely on the equivalence of "A and B, and C" with "A, and B and C." A formal proof shows that the associativity of the intersection $\cap$ corresponds to the associativity of the logical connective $\wedge$.

$$\begin{array}{rl} [(A \cap B) \cap C] = [A \cap (B \cap C)] & \textbf{yet unproved}, \\ \Updownarrow & \text{axiom of extensionality (S1)}, \\ \forall X\big(\{X \in [(A \cap B) \cap C]\} & \\ \Leftrightarrow \{X \in [[A \cap (B \cap C)]\}\big) & \\ \Updownarrow & \text{theorem 318 twice}, \\ \forall X\big(\{[(X \in A) \wedge (X \in B)] \wedge (X \in C)\} & \\ \Leftrightarrow \{(X \in A) \wedge [(X \in B) \wedge (X \in C)]\}\big) & \end{array}$$

which holds thanks to the associativity of $\wedge$ (theorem 207):

$$\{[(P) \wedge (Q)] \wedge (R)\} \Leftrightarrow \{(P) \wedge [(Q) \wedge (R)]\}.$$

□

The following theorem shows that the intersection $\bigcap \mathcal{F}$ of a set of sets $\mathcal{F}$ is a subset of every element of $\mathcal{F}$. In other terms, for each set B, if $B \in \mathcal{F}$, then $(\bigcap \mathcal{F}) \subseteq B$.

Theorem 323 *For each (non-empty) set $\mathcal{F}$ and for each $B \in \mathcal{F}$, $(\bigcap \mathcal{F}) \subseteq B$:*

$$\models \forall \mathcal{F}\left(\forall B\left\{(B \in \mathcal{F}) \Rightarrow \left[\left(\bigcap \mathcal{F}\right) \subseteq B\right]\right\}\right).$$

Proof. An informal proof can consist of the following statements:

- If $X \in \bigcap \mathcal{F}$, then X is an element of every $Y \in \mathcal{F}$, and hence of $B \in \mathcal{F}$;
- consequently, $(\bigcap \mathcal{F}) \subseteq B$, by definition of subsets.

A formal proof can consist in transforming the proposed formula until a universally valid formula appears.

$$\begin{array}{rl} \forall \mathcal{F}\left(\forall B\left\{(B \in \mathcal{F}) \Rightarrow [(\bigcap \mathcal{F}) \subseteq B]\right\}\right) & \textbf{yet unproved}, \\ \Updownarrow & \text{definition of } \subseteq, \\ \forall \mathcal{F}\Big\{\forall B\Big[(B \in \mathcal{F}) \Rightarrow \{\forall X\,[(X \in \bigcap \mathcal{F}) \Rightarrow (X \in B)]\}\Big]\Big\} & \\ \Updownarrow & \text{definition of } \bigcap, \\ \forall \mathcal{F}\Big\{\forall B\Big[(B \in \mathcal{F}) \Rightarrow & \\ \{\forall X\,[\{\forall Y[(Y \in \mathcal{F}) \Rightarrow (X \in Y)]\} \Rightarrow (X \in B)]\}\Big]\Big\} & \\ \Uparrow & \text{no } X \text{ in } B \in \mathcal{F}, \\ \forall \mathcal{F}\Big\{\forall B\Big[\forall X\{(B \in \mathcal{F}) \Rightarrow & \\ [\{\forall Y[(Y \in \mathcal{F}) \Rightarrow (X \in Y)]\} \Rightarrow (X \in B)]\}\Big]\Big\} & \\ \Updownarrow & \text{theorem 185}, \\ \forall \mathcal{F}\Big\{\forall B\Big[\forall X(\overbrace{\{\forall Y[(Y \in \mathcal{F}) \Rightarrow (X \in Y)]\}}^{P} & \\ \Rightarrow [\underbrace{(B \in \mathcal{F})}_{Q} \Rightarrow \underbrace{(X \in B)}_{R}])\Big]\Big\} & \end{array}$$

which holds by specialization (Subf_B^Y) and the law of commutation (theorem 185):

$$\big[(P) \Rightarrow \{(Q) \Rightarrow (R)\}\big] \Leftrightarrow \big[(Q) \Rightarrow \{(P) \Rightarrow (R)\}\big].$$

□

Definition 324 (Disjoint sets) Two sets A and B are **disjoint** if and only if $A \cap B = \varnothing$. Similarly, a set of sets $\mathcal{F}$ is **pairwise disjoint** if and only if either $A = B$ or $A \cap B = \varnothing$ for all elements A and B of $\mathcal{F}$.

For the union of disjoint sets, a special notation proves convenient.

Definition 325 (Disjoint unions) A union $A \cup B$ is **disjoint** if and only if $A \cap B = \varnothing$; only for disjoint sets, the notation $A \dot{\cup} B$ stands for $A \cup B$. Similarly, a union $\bigcup \mathcal{F}$ is **pairwise disjoint** if and only if either $A = B$ or $A \cap B = \varnothing$ for all elements A and B of $\mathcal{F}$; only for pairwise disjoint sets, the notation $\dot{\bigcup} \mathcal{F}$ stands for $\bigcup \mathcal{F}$.

2.3.3 Unions and Intersections of Sets

In some contexts yet another notation for unions and intersections of a set $\mathcal{G}$ proves convenient, especially where the set $\mathcal{G}$ relates in a specific way to a second set $\mathcal{F}$. The new notation may then "index" the elements of $\mathcal{G}$ with the corresponding elements of $\mathcal{F}$.

Example 326 For each set $\mathcal{F} \subseteq \mathcal{P}(U)$ of subsets of a set U, consider the set $\mathcal{G}$ of all the complements $U \setminus S$ of all the elements $S \in \mathcal{F}$; thus,

$$\mathcal{G} := \{B \in \mathcal{P}(U) : \exists S\,[(S \in \mathcal{F}) \wedge (B = U \setminus S)]\}.$$

The indexed notation then denotes the union and the intersection of $\mathcal{G}$ as follows:

$$\bigcup_{S \in \mathcal{F}} (U \setminus S) := \bigcup \mathcal{G},$$

$$\bigcap_{S \in \mathcal{F}} (U \setminus S) := \bigcap \mathcal{G}.$$

The notation on the left-hand sides avoids the need to write a formula for the set $\mathcal{G}$.

Because the definitions of the intersection and union of sets rely on conjunction and disjunction, the union and intersection have formal features similar to the logical features of the conjunction and disjunction, for instance, distributivity.

Theorem 327 (De Morgan's Laws) *For each set U, and for each set $\mathcal{F} \subseteq \mathcal{P}(U)$ of subsets of U, the complement of the union equals the intersection of the complements,*

$$U \setminus \left(\bigcup \mathcal{F}\right) = \bigcap_{A \in \mathcal{F}} (U \setminus A),$$

whereas for each set U, and for each non-empty family $\mathcal{F} \subseteq \mathcal{P}(U)$ of subsets of U, the complement of the intersection equals the union of the complements,

$$U \setminus \left(\bigcap \mathcal{F}\right) = \bigcup_{A \in \mathcal{F}} (U \setminus A).$$

Proof. For the complement of the intersection, an informal proof can proceed as follows.

- For each set X, $X \in U \setminus (\bigcap \mathcal{F})$ if and only if $X \in U$ but $X \notin (\bigcap \mathcal{F})$;
- by definition, $X \notin (\bigcap \mathcal{F})$ if and only if there exists $A \in \mathcal{F}$ with $X \notin A$;
- hence $X \in U \setminus (\bigcap \mathcal{F})$ if and only if $X \in U$ and there exists $A \in \mathcal{F}$ with $X \notin A$;
- equivalently, $X \in U \setminus (\bigcap \mathcal{F})$ if and only if there exists $A \in \mathcal{F}$ with $X \in (U \setminus A)$;
- hence $X \in U \setminus (\bigcap \mathcal{F})$ if and only if $X \in \bigcup_{A \in \mathcal{F}} (U \setminus A)$.

The foregoing informal proof does not justify the permutation of the two statements "$X \in U$" and "there exists $A \in \mathcal{F}$" but the following formal proof justifies such a permutation by the absence of any free occurrence of A in $X \in U$.

$$\begin{array}{rl}
U \setminus (\bigcap \mathcal{F}) = \bigcup_{A \in \mathcal{F}} (U \setminus A) & \\
\Updownarrow & \text{(axiom S1)}, \\
\forall X \{[X \in \{U \setminus (\bigcap \mathcal{F})\}] \Leftrightarrow [X \in \bigcup_{A \in \mathcal{F}} (U \setminus A)]\} & \\
\Updownarrow & (\setminus), \\
\forall X \{[(X \in U) \wedge \{\neg (X \in \bigcap \mathcal{F})\}] \Leftrightarrow [X \in \bigcup_{A \in \mathcal{F}} (U \setminus A)]\} & \\
\Updownarrow & \bigcup, \bigcap, \\
\forall X \{[(X \in U) \wedge (\neg \{\forall A [(A \in \mathcal{F}) \Rightarrow (X \in A)]\})] & \\
\Leftrightarrow \{\exists A [(A \in \mathcal{F}) \wedge \{(X \in U) \wedge [\neg (X \in A)]\}]\}\} & \\
\Updownarrow & \begin{array}{l}\neg[\forall A(P)], \\ \exists A[\neg(P)],\end{array} \\
\forall X \{[(\exists A \{(X \in U) \wedge \{\neg \{[\{\neg (A \in \mathcal{F})\} \vee (X \in A)]\}\}\})] & \\
\Leftrightarrow [\exists A \{(A \in \mathcal{F}) \wedge [(X \in U) \wedge [\neg (X \in A)]]\}]\} & \\
\Updownarrow & \text{de Morgan}, \\
\forall X \{[(\exists A \{(X \in U) \wedge \{(A \in \mathcal{F}) \wedge [\neg (X \in A)]\}\})] & \\
\Leftrightarrow \{\exists A [(A \in \mathcal{F}) \wedge \{(X \in U) \wedge [\neg (X \in A)]\}]\}\} &
\end{array}$$

which is universally valid thanks to $\vdash \{(P) \wedge [(Q) \wedge (R)] \Leftrightarrow \{[(P) \wedge (Q)] \wedge (R)\}$.

For the complement of the union, an informal proof can proceed as follows.

- For each set X, $X \in U \setminus (\bigcup \mathcal{F})$ if and only if $X \in U$ but $X \notin (\bigcup \mathcal{F})$;
- by definition, $X \notin (\bigcup \mathcal{F})$ if and only if $X \notin A$ for every $A \in \mathcal{F}$;
- hence $X \in U \setminus (\bigcup \mathcal{F})$ if and only if $X \in U$ and $X \notin A$ for every $A \in \mathcal{F}$;
- equivalently, $X \in U \setminus (\bigcup \mathcal{F})$ if and only if $X \in (U \setminus A)$ for every $A \in \mathcal{F}$;
- hence $X \in U \setminus (\bigcup \mathcal{F})$ if and only if $X \in \bigcap_{A \in \mathcal{F}} (U \setminus A)$.

The foregoing informal proof does not justify the permutation of the two statements "$X \in U$" and "for every $A \in \mathcal{F}$" because it hides the permutation by placing the quantifier at the end of the statements. The following formal proof justifies such a permutation by the absence of any free occurrence of A in $X \in U$.

$U \setminus (\bigcup \mathcal{F}) = \bigcap_{A \in \mathcal{F}} (U \setminus A)$

$\Updownarrow$ (S1),

$\forall X \{[X \in U \setminus (\bigcup \mathcal{F})] \Leftrightarrow [X \in \bigcap_{A \in \mathcal{F}} (U \setminus A)]\}$

$\Updownarrow (\setminus)$,

$\forall X \{[(X \in U) \wedge \{\neg (X \in \bigcup \mathcal{F})\}] \Leftrightarrow [X \in \bigcap_{A \in \mathcal{F}} (U \setminus A)]\}$

$\Updownarrow \bigcup, \bigcap$,

$\forall X \{[(X \in U) \wedge \{\neg (\exists A \{(A \in \mathcal{F}) \wedge (X \in A)\})\}]$

$\Leftrightarrow [(X \in \bigcup_{A \in \mathcal{F}} (U \setminus A)) \wedge \forall A \{(A \in \mathcal{F}) \Rightarrow \{(X \in U) \wedge [\neg (X \in A)]\}\}\}$

$\Updownarrow$

$\forall X \{[(\forall A \{(X \in U) \wedge [\neg ((A \in \mathcal{F}) \wedge (X \in A))]\})]$

$\Leftrightarrow [(X \in U \setminus \bigcap \mathcal{F}) \wedge \forall A \{[\neg (A \in \mathcal{F})] \vee \{(X \in U) \wedge [\neg (X \in A)]\}\}]\}$

$\Updownarrow$

$\forall X \{[(\forall A \{(X \in U) \wedge [([\neg (A \in \mathcal{F})] \vee [\neg (X \in A)])]\})]$

$\Leftrightarrow ([X \in U] \wedge \exists B [(B \in \mathcal{F}) \wedge \{\neg (X \in B)\}])$

$\wedge (\forall A \{[\neg (A \in \mathcal{F})] \vee \{(X \in U) \wedge [\neg (X \in A)]\}\})\}$

$\Uparrow$

$\forall X \{[(\forall A \{(X \in U) \wedge [([\neg (A \in \mathcal{F})] \vee [\neg (X \in A)])]\})]$

$\Leftrightarrow [(X \in U) \wedge \forall A \{[\neg (A \in \mathcal{F})] \vee \{(X \in U) \wedge [\neg (X \in A)]\}\}]\}$

where the first implication, $\Uparrow$, used the assumption that $\mathcal{F} \neq \varnothing$ for the formula beginning with $\exists B \ldots$; the last line follows from the distributivity of $\wedge$ over $\vee$. □

Theorem 328 *For all sets $\mathcal{F}$ and $\mathcal{G}$, intersection distributes over their unions:*

$$\models \left[\left(\bigcup \mathcal{F}\right) \cap \left(\bigcup \mathcal{G}\right)\right] = \left[\bigcup \{A \cap B : (A \in \mathcal{F}) \wedge (B \in \mathcal{G})\}\right].$$

Proof. An informal proof can proceed as follows.

- For each set X, $X \in [(\bigcup \mathcal{F}) \cap (\bigcup \mathcal{G})]$ if and only if $X \in (\bigcup \mathcal{F})$ and $X \in (\bigcup \mathcal{G})$;
- equivalently, X belongs to some $A \in \mathcal{F}$ and X belongs to some $B \in \mathcal{G}$,
- equivalently, $X \in (A \cap B)$ for some elements $A \in \mathcal{F}$ and $B \in \mathcal{G}$;
- equivalently, $X \in [\bigcup \{A \cap B : (A \in \mathcal{F}) \wedge (B \in \mathcal{G})\}]$.

A formal proof can proceed as follows.

$$
\begin{array}{rl}
[(\bigcup \mathcal{F}) \cap (\bigcup \mathcal{G})] & \\
= [\bigcup \{A \cap B : (A \in \mathcal{F}) \wedge (B \in \mathcal{G})\}] & \textbf{yet unproved}, \\
\Updownarrow & \text{extensionality}, \\
\forall X\, (\{X \in [(\bigcup \mathcal{F}) \cap (\bigcup \mathcal{G})]\} & \\
\Leftrightarrow \{X \in [\bigcup \{A \cap B : (A \in \mathcal{F}) \wedge (B \in \mathcal{G})\}]\}) & \\
\Updownarrow & \text{definitions: } \cap, \bigcup, \\
\forall X\, (\{[X \in (\bigcup \mathcal{F})] \wedge [X \in (\bigcup \mathcal{G})]\} & \\
\Leftrightarrow \left[\exists A \left(\exists B \{(A \in \mathcal{F}) \wedge (B \in \mathcal{G}) \wedge [X \in (A \cap B)]\}\right)\right]) & \\
\Updownarrow & \text{definition of } \bigcup, \\
[\forall X\, (\{\exists A[(A \in \mathcal{F}) \wedge (X \in A)]\} \wedge \{\exists B[(B \in \mathcal{G}) \wedge (X \in B)]\})] & \\
\Leftrightarrow \left[\exists A \left(\exists B \{(A \in \mathcal{F}) \wedge (B \in \mathcal{G}) \wedge [X \in (A \cap B)]\}\right)\right] & \\
\Updownarrow & \text{no free } A \text{ or } B, \\
[\forall X\, (\exists A \{\exists B[(A \in \mathcal{F}) \wedge (B \in \mathcal{G}) \wedge (X \in A) \wedge (X \in B)]\})] & \\
\Leftrightarrow \left[\exists A \left(\exists B \{(A \in \mathcal{F}) \wedge (B \in \mathcal{G}) \wedge [X \in (A \cap B)]\}\right)\right] &
\end{array}
$$

which is True by definition of $A \cap B$. □

Theorem 329 *For all sets $\mathcal{F}$ and $\mathcal{G}$, union distributes over their intersections:*

$$\models \left[\left(\bigcap \mathcal{F}\right) \cup \left(\bigcap \mathcal{G}\right)\right] = \left[\bigcap \{A \cup B : (A \in \mathcal{F}) \wedge (B \in \mathcal{G})\}\right].$$

Proof. An informal proof can proceed as follows.

- For each set X, $X \in [(\bigcap \mathcal{F}) \cup (\bigcap \mathcal{G})]$ if and only if $X \in (\bigcap \mathcal{F})$ or $X \in (\bigcap \mathcal{G})$;
- equivalently, X belongs to every $A \in \mathcal{F}$ or X belongs to every $B \in \mathcal{G}$,
- equivalently, $X \in (A \cup B)$ for all elements $A \in \mathcal{F}$ and $B \in \mathcal{G}$;
- equivalently, $X \in [\bigcap \{A \cup B : (A \in \mathcal{F}) \wedge (B \in \mathcal{G})\}]$.

A formal proof can proceed as follows.

$$\begin{aligned}
&[(\bigcap \mathcal{F}) \cup (\bigcap \mathcal{G})] \\
&= [\bigcap\{A \cup B : (A \in \mathcal{F}) \wedge (B \in \mathcal{G})\}] && \textbf{unproved}, \\
&\Updownarrow && \text{(S1)}, \\
&\forall X \, (\{X \in [(\bigcap \mathcal{F}) \cup (\bigcap \mathcal{G})]\} \\
&\Leftrightarrow \{X \in [\bigcap\{A \cup B : (A \in \mathcal{F}) \wedge (B \in \mathcal{G})\}]\}) \\
&\Updownarrow && \text{def. } \cup, \bigcap, \\
&\forall X \, (\{[X \in (\bigcap \mathcal{F})] \vee [X \in (\bigcap \mathcal{G})]\} \\
&\Leftrightarrow \big[\forall A \big(\forall B\{[(A \in \mathcal{F}) \wedge (B \in \mathcal{G})] \Rightarrow [X \in (A \cup B)]\}\big)\big]\big) \\
&\Updownarrow && \text{definition: } \bigcap, \\
&[\forall X \, (\{\forall A[(A \in \mathcal{F}) \Rightarrow (X \in A)]\} \vee \{\forall B[(B \in \mathcal{G}) \Rightarrow (X \in B)]\})] \\
&\Leftrightarrow \big[\forall A \big(\forall B\{[(A \in \mathcal{F}) \wedge (B \in \mathcal{G})] \Rightarrow [X \in (A \cup B)]\}\big)\big] \\
&\Updownarrow && \text{no free } A, B, \\
&\{\forall X \, [\forall A \, (\forall B\{[(A \in \mathcal{F}) \Rightarrow (X \in A)] \vee [(B \in \mathcal{G}) \Rightarrow (X \in B)]\})]\} \\
&\Leftrightarrow \big[\forall A \big(\forall B\{(A \in \mathcal{F}) \wedge (B \in \mathcal{G}) \wedge [X \in (A \cup B)]\}\big)\big] \\
&\Updownarrow && [(P) \wedge (Q)] \Rightarrow (P), \\
&\{\forall X \, [\forall A \, (\forall B\{[(A \in \mathcal{F}) \Rightarrow (X \in A)] \vee [(B \in \mathcal{G}) \Rightarrow (X \in B)]\})]\} \\
&\Leftrightarrow \big[\forall A \big(\forall B\{[(A \in \mathcal{F}) \wedge (B \in \mathcal{G})] \Rightarrow [(X \in A) \vee (X \in B)]\}\big)\big] \\
&\Updownarrow && \text{definition: } \cup, \\
&\{\forall X \, [\forall A \, (\forall B\{[(A \in \mathcal{F}) \Rightarrow (X \in A)] \vee [(B \in \mathcal{G}) \Rightarrow (X \in B)]\})]\} \\
&\Leftrightarrow \big[\forall A \big(\forall B\{[(A \in \mathcal{F}) \wedge (B \in \mathcal{G})] \Rightarrow [X \in (A \cup B)]\}\big)\big]
\end{aligned}$$

which is True by definition of $A \cup B$. □

Besides union, intersection, and difference, another operation with sets is useful.

Definition 330 (Symmetric difference) The **symmetric difference** of any sets A and B is denoted by $A \bigtriangleup B$ and defined by

$$A \bigtriangleup B := (A \cup B) \setminus (A \cap B).$$

2.3.4 Exercises

Exercise 451 List all the elements of $\{2, 3, 7\} \cup \{3, 5, 7\}$.

Exercise 452 List all the elements of $\{4, 6, 8\} \cup \{4, 8, 9\}$.

Exercise 453 List all the elements of $\{2, 3, 7\} \cap \{3, 5, 7\}$.

Exercise 454 List all the elements of $\{4, 6, 8\} \cap \{4, 8, 9\}$.

Exercise 455 List all the elements of $\{2, 3, 7\}\Delta\{3, 5, 7\}$.

Exercise 456 List all the elements of $\{4, 6, 8\}\Delta\{4, 8, 9\}$.

Exercise 457 Provide an example of a set $\mathcal{F}$ such that $\bigcup \mathcal{F} = \mathcal{F}$.

Exercise 458 Provide an example of a set $\mathcal{G}$ such that $\bigcap \mathcal{G} = \mathcal{G}$.

Exercise 459 Provide an example of a set $\mathcal{F}$ such that $\bigcup \mathcal{F} \neq \mathcal{F}$.

Exercise 460 Provide an example of a set $\mathcal{G}$ such that $\bigcap \mathcal{G} \neq \mathcal{G}$.

Exercise 461 Provide examples of sets B and $\mathcal{F}$ such that $B \in \mathcal{F}$ but $B \notin \bigcup \mathcal{F}$.

Exercise 462 Provide examples of sets B and $\mathcal{G}$ such that $B \in \mathcal{G}$ but $\bigcap \mathcal{G} \notin B$.

Exercise 463 Provide examples of sets A, B, X, Y such that $X \in A$ and $Y \in B$ but $(X \cup Y) \notin (A \cup B)$.

Exercise 464 Provide examples of sets A, B, X, Y such that $X \in A$ and $Y \in B$ but $(X \cap Y) \notin (A \cap B)$.

Exercise 465 For each set A, prove that $\bigcup \{A\} = A$.

Exercise 466 For each set A, prove that $\bigcup \mathcal{P}(A) = A$.

Exercise 467 Prove that $\forall A \forall B \forall C \{[(A \cup B) \cap C] = [(A \cap C) \cup (B \cap C)]\}$.

Exercise 468 Prove that $\forall A \forall B \forall C \{[(A \cap B) \cup C] = [(A \cup C) \cap (B \cup C)]\}$.

Exercise 469 Prove that $\forall A [(A \cup \varnothing) = A]$.

Exercise 470 Prove that $\forall A [(A \cap \varnothing) = \varnothing]$.

Exercise 471 Prove that $\forall A [(A \cup A) = A]$.

Exercise 472 Prove that $\forall A [(A \cap A) = A]$.

Prove the following formulae for all subsets A and B of each set U.

Exercise 473 $[U \setminus (A \cap B)] = (U \setminus A) \cup (U \setminus B)$

Exercise 474 $[U \setminus (A \cup B)] = (U \setminus A) \cap (U \setminus B)$

Exercise 475 $[(A \setminus B) \setminus U] = (A \setminus U) \setminus (B \setminus U)$

Exercise 476 $[U \setminus (A \setminus B)] = (U \setminus A) \cup (U \cap B)$

Exercise 477 $[(A \cup B) \setminus U] = (A \setminus U) \cup (B \setminus U)$

Exercise 478 $[(A \cap B) \setminus U] = (A \setminus U) \cap (B \setminus U)$

Exercise 479 $[U \cap (A \setminus B)] = (U \setminus B) \setminus (U \setminus A)$

Exercise 480 Prove or disprove that $([\mathcal{P}(A)] \cup [\mathcal{P}(B)]) \subseteq \mathcal{P}(A \cup B)$.

Exercise 481 Prove or disprove that $([\mathcal{P}(A)] \cap [\mathcal{P}(B)]) \supseteq \mathcal{P}(A \cap B)$.

Exercise 482 Prove or disprove that $([\mathcal{P}(A)] \cap [\mathcal{P}(B)]) \subseteq \mathcal{P}(A \cap B)$.

Exercise 483 Prove or disprove that $([\mathcal{P}(A)] \cup [\mathcal{P}(B)]) \supseteq \mathcal{P}(A \cup B)$.

Exercise 484 Prove or disprove that $([\mathcal{P}(A)] \setminus [\mathcal{P}(B)]) \subseteq \mathcal{P}(A \setminus B)$.

Exercise 485 Prove or disprove that $([\mathcal{P}(A)] \setminus [\mathcal{P}(B)]) \supseteq \mathcal{P}(A \setminus B)$.

Exercise 486 For each non-empty set $\mathcal{F} \subseteq \mathcal{P}(U)$, and for each $B \subseteq U$, prove that

$$\left(\bigcap \mathcal{F}\right) \cup B = \bigcap_{A \in \mathcal{F}} (A \cup B).$$

Exercise 487 For each set $\mathcal{F} \subseteq \mathcal{P}(U)$, and for each subset $B \subseteq U$, prove that

$$\left(\bigcup \mathcal{F}\right) \cap B = \bigcup_{A \in \mathcal{F}} (A \cap B).$$

Exercise 488 Prove that $A \triangle \varnothing = A$.

Exercise 489 Prove that $A \triangle A = \varnothing$.

Exercise 490 Prove that the symmetric difference is associative: $(A \triangle B) \triangle C = A \triangle (B \triangle C)$.

Exercise 491 Prove that the symmetric difference commutes: $A \triangle B = B \triangle A$.

Exercise 492 Prove or disprove that $[(A\Delta C) \cup (B\Delta C)] \subseteq [(A \cup B)\Delta C]$.

Exercise 493 Prove or disprove that $[(A\Delta C) \cup (B\Delta C)] \supseteq [(A \cup B)\Delta C]$.

Exercise 494 Prove or disprove that $[(A\Delta C) \cap (B\Delta C)] \subseteq [(A \cap B)\Delta C]$.

Exercise 495 Prove or disprove that $[(A\Delta C) \cap (B\Delta C)] \supseteq [(A \cap B)\Delta C]$.

Exercise 496 Prove or disprove that $[(A\Delta C) \setminus (B\Delta C)] \subseteq [(A \setminus B)\Delta C]$.

Exercise 497 Prove or disprove that $[(A\Delta C) \setminus (B\Delta C)] \supseteq [(A \setminus B)\Delta C]$.

Exercise 498 Prove or disprove that $[(A \cup C)\Delta(B \cup C)] \subseteq [(A\Delta B) \cup C]$.

Exercise 499 Prove or disprove that $[(A \cup C)\Delta(B \cup C)] \supseteq [(A\Delta B) \cup C]$.

Exercise 500 Prove or disprove that $[(A \cap C)\Delta(B \cap C)] \subseteq [(A\Delta B) \cap C]$.

Exercise 501 Prove or disprove that $[(A \cap C)\Delta(B \cap C)] \supseteq [(A\Delta B) \cap C]$.

Exercise 502 Prove or disprove that $[(A \setminus C)\Delta(B \setminus C)] \subseteq [(A\Delta B) \setminus C]$.

Exercise 503 Prove or disprove that $[(A \setminus C)\Delta(B \setminus C)] \supseteq [(A\Delta B) \setminus C]$.

Exercise 504 Prove or disprove that $[(C \setminus A)\Delta(C \setminus B)] \subseteq [C \setminus (A\Delta B)]$.

Exercise 505 Prove or disprove that $[(C \setminus A)\Delta(C \setminus B)] \supseteq [C \setminus (A\Delta B)]$.

Exercise 506 Prove or disprove that $([\mathcal{P}(A)]\Delta[\mathcal{P}(B)]) \subseteq \mathcal{P}(A\Delta B)$.

Exercise 507 Prove or disprove that $([\mathcal{P}(A)]\Delta[\mathcal{P}(B)]) \supseteq \mathcal{P}(A\Delta B)$.

Exercise 508 Prove that $A \cup B = (A \setminus B)\dot{\cup}(B \setminus A)$.

Exercise 509 Prove that $A \cup B = (A \triangle B)\dot{\cup}(A \cap B)$.

Exercise 510 Prove that $A \triangle B = (A \setminus B)\dot{\cup}(B \setminus A)$.

2.4 CARTESIAN PRODUCTS AND RELATIONS

2.4.1 Cartesian Products of Sets

Beyond logic and sets, much of mathematics consists of connections between types of sets called Cartesian products, mathematical functions, and mathematical relations. These types of sets allow for mathematical specifications, analysis, synthesis, and processing of such concepts as graphs, maps, algorithms, and rankings. To such ends, Cartesian products contain certain sets with two elements. Whereas $\{X, Y\} = \{Y, X\}$ for all sets X and Y, however, some situations require a method for listing the elements of a set in a specific order by means of "ordered pairs" or otherwise, for example, in geography and in navigation as in figure 2.3. The following definition (attributed to Wiener and Kuratowski [74, p. 455]) and theorem 333 derive such ordered pairs from sets, which shows that the concept of ordered pairs does not require any additional axiom.

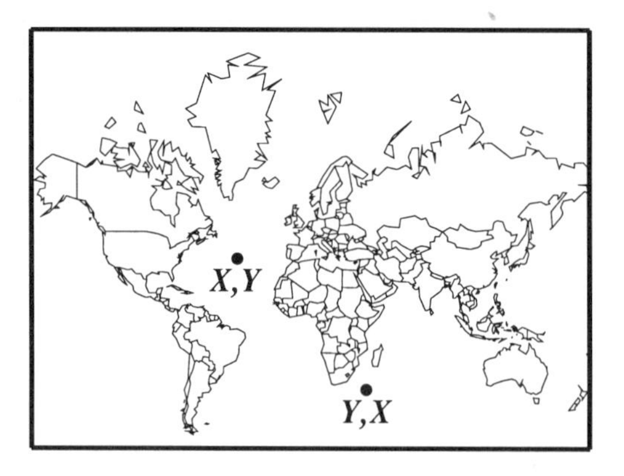

Figure 2.3: If $X \neq Y$, then $(X, Y) \neq (Y, X)$.

Definition 331 (Ordered pairs) For all sets X and Y, the **ordered pair** (X, Y) is the set defined by three applications of the axiom of pairing as

$$(X, Y) = \{\{X\}, \{X, Y\}\}.$$

X is the **first coordinate** of (X, Y), and Y is the **second coordinate** of (X, Y).

Example 332

- If $X = 0$ and $Y = 1$, then $(X, Y) = \{\{0\}, \{0, 1\}\}$.
- If $X = 1$ and $Y = 0$, then $(X, Y) = \{\{1\}, \{1, 0\}\} = \{\{1\}, \{0, 1\}\}$.
- If $X = 0$ and $Y = 0$, then $(X, Y) = \{\{0\}, \{0, 0\}\} = \{\{0\}, \{0\}\} = \{\{0\}\}$.

The following theorem confirms that, in contrast to sets with two elements, ordered pairs record the order of their coordinates.

Theorem 333 *For all sets X and Y, if $X \neq Y$, then $(X, Y) \neq (Y, X)$.*

Proof. An informal proof can consist of showing that the two sets (X, Y) and (Y, X) contain different elements, whence $(X, Y) \neq (Y, X)$.

- For all sets X and Y, $X \in \{X\}$ and $Y \in \{Y\}$ by pairing (S3);
- if $X \neq Y$, then $X \notin \{Y\}$ and $Y \notin \{X\}$, by pairing (S3);
- hence if $X \neq Y$, then $\{X\} \neq \{Y\}$ by extensionality (S1);
- from $X \notin \{Y\}$ and $X \in \{Y, X\}$ it follows that $\{X\} \neq \{Y, X\}$ by (S1);
- from $\{X\} \neq \{Y\}$, $\{X\} \neq \{Y, X\}$ follows $\{X\} \notin \{\{Y\}, \{Y, X\}\} = (Y, X)$;
- yet $\{X\} \in \{\{X\}, \{X, Y\}\} = (X, Y)$;
- from $\{X\} \in (X, Y)$ and $\{X\} \notin (Y, X)$ follows $(X, Y) \neq (Y, X)$, by (S1).

A formal proof can parallel the same reasoning.

(1) $\models \forall Z[(Z \in \{X\}) \Leftrightarrow (Z = X)]$	axiom of pairing (S3),
(2) $\models \forall Z[(Z \neq X) \Leftrightarrow (Z \notin \{X\})]$	contraposition and transposition,
(3) $\models (Y \neq X) \Leftrightarrow (Y \notin \{X\})$	specialization Subf_Y^Z,
(4) $\models Y \neq X$	hypothesis,
(5) $\models Y \notin \{X\}$	*Modus Ponens*;
(6) $\models Y \in \{X, Y\}$	axiom of pairing (S3),
(7) $\models \{X\} \neq \{X, Y\}$	(5), (6), and extensionality (S1);
(8) $\models Y \in \{Y\}$	axiom of pairing (S3),
(9) $\models \{X\} \neq \{Y\}$	(5), (8), and extensionality (S1);
(10) $\models \{X\} \notin \{\{Y\}, \{Y, X\}\} = (Y, X)$	(7), (9), and extensionality (S1);
(11) $\models \{X\} \in \{\{X\}, \{X, Y\}\} = (X, Y)$	pairing (S3),
(12) $\models (X, Y) \neq (Y, X)$	(10), (11), and extensionality (S1).

□

The definition of the ordered pair (X, Y) holds for all sets X and Y, in particular, for all elements X and Y of two sets A and B. The following theorem shows that all these ordered pairs (X, Y) are themselves elements of a set.

Theorem 334 *For all sets A and B, for each $X \in A$ and for each $Y \in B$, the pair (X, Y) belongs to $\mathcal{P}[\mathcal{P}(A \cup B)]$.*

Proof. An informal proof can trace back the definition of the ordered pair (X, Y):

- From $X \in A$ follows $X \in A \cup B$, because $A \subseteq A \cup B$, whence $\{X\} \in \mathcal{P}(A \cup B)$;
- from $Y \in B$ it follows that $Y \in A \cup B$, because $B \subseteq A \cup B$;
- from $X \in A \cup B$ and $Y \in A \cup B$ it follows that $\{X, Y\} \in \mathcal{P}(A \cup B)$;
- from $\{X\} \in \mathcal{P}(A \cup B)$ and $\{X, Y\} \in \mathcal{P}(A \cup B)$, it follows that $\{\{X\}, \{X, Y\}\} \in \mathcal{P}[\mathcal{P}(A \cup B)]$.

A formal proof can parallel the foregoing argument.

$\models X \in A$	hypothesis,
$\models A \subseteq (A \cup B)$	theorem 316 ,
$\models X \in (A \cup B)$	definition of subsets and specialization,
$\models \{X\} \subseteq (A \cup B)$	definitions of subsets and singletons,
$\models \{X\} \in \mathcal{P}(A \cup B)$	definition of power sets.
$\models Y \in B$	hypothesis,
$\models Y \in (A \cup B)$	as for $X \in (A \cup B)$,
$\models \{X, Y\} \in \mathcal{P}(A \cup B)$	$X \in (A \cup B)$ and $Y \in (A \cup B)$,
$\models \{\{X\}, \{X, Y\}\} \subseteq \mathcal{P}(A \cup B)$	$\{X\} \in \mathcal{P}(A \cup B)$, $\{X, Y\} \in \mathcal{P}(A \cup B)$,
$\models \{\{X\}, \{X, Y\}\} \in \mathcal{P}[\mathcal{P}(A \cup B)]$	definition of power set.

□

Because every ordered pair (X, Y) belongs to the set $\mathcal{P}[\mathcal{P}(A \cup B)]$, the axiom of separation guarantees that the collection of all such ordered pairs is a set.

Definition 335 (Cartesian product) For all sets A and B, the **Cartesian product** of A and B is the set $A \times B$ (read "A cross B"), consisting of all ordered pairs (X, Y) with $X \in A$ and $Y \in B$. Thus,

$$\begin{aligned} & A \times B \\ & = \left\{C \in \mathcal{P}\left[\mathcal{P}(A \cup B)\right] : \exists X\big(\exists Y\{(X \in A) \wedge (Y \in B) \wedge [C = (X,Y)]\}\big)\right\} \\ & = \{(X,Y) : (X \in A) \wedge (Y \in B)\} \end{aligned}$$

with $\exists X\big(\exists Y\{(X \in A) \wedge (Y \in B) \wedge [C = (X,Y)]\}\big)$ for P in the axiom of separation (S5). The sets A and B are the **factors** of the Cartesian product $A \times B$.

A common graphical representation of a Cartesian product $A \times B$ lists all the elements of A along a horizontal axis, and all the elements of B along a vertical axis, so that the element (X, Y) of $A \times B$ appears directly above X and across from Y.

Example 336 For the sets

$$\begin{array}{rcl} A & := & \{0,1,2\}, \\ B & := & \{0,1\}, \end{array}$$

the Cartesian product $A \times B$ takes the form

$$A \times B = \{(0,0),(1,0),(2,0),(0,1),(1,1),(2,1)\},$$

with the following graphical representation:

$$\begin{array}{ccccc} B & & & & A \times B \\ 1 & (0,1) & (1,1) & (2,1) & \\ 0 & (0,0) & (1,0) & (2,0) & \\ & 0 & 1 & 2 & A \end{array}$$

For practice and for future use, the following theorems establish relations between Cartesian products and other operations with sets.

Theorem 337 *For all sets A, B, C, and D,*

$$[(A \cap C) \times (B \cap D)] = [(A \times B) \cap (C \times D)].$$

Proof. An informal proof can establish that $[(A \cap C) \times (B \cap D)]$ and $[(A \times B) \cap (C \times D)]$ have exactly the same elements. These two sets are Cartesian products, and, consequently their elements are ordered pairs.

- An ordered pair (X, Y) is an element of $(A \cap C) \times (B \cap D)$ if and only if $X \in (A \cap C)$ and $Y \in (B \cap D)$;
- hence $(X, Y) \in [(A \cap C) \times (B \cap D)]$ if and only if $X \in A$ and $X \in C$, and $Y \in B$ and $Y \in D$,

- which is equivalent to $X \in A$ and $Y \in B$, and $X \in C$ and $Y \in D$,
- which is equivalent to $(X,Y) \in (A \times B)$ and $(X,Y) \in (C \times D)$,
- which is thus equivalent to $(X,Y) \in [(A \times B) \cap (C \times D)]$.

As the preceding informal proof swapped $X \in C$ and $Y \in B$, a formal proof can rely on the commutativity $[(P) \wedge (Q)] \Leftrightarrow [(Q) \wedge (P)]$ of $\wedge$ by theorem 198:

$$\begin{array}{rl}
(X,Y) \in [(A \cap C) \times (B \cap D)] & \\
\Updownarrow & \text{definition of Cartesian products,} \\
[X \in (A \cap C)] \wedge [Y \in (B \cap D)] & \\
\Updownarrow & \text{definition of intersection,} \\
[(X \in A) \wedge (X \in C)] \wedge [(Y \in B) \wedge (Y \in D)] & \\
\Updownarrow & \text{commutativity and associativity of } \wedge, \\
[(X \in A) \wedge (Y \in B)] \wedge [(X \in C) \wedge (Y \in D)] & \\
\Updownarrow & \text{definition of Cartesian products,} \\
[(X,Y) \in (A \times B)] \wedge [(X,Y) \in (C \times D)] & \\
\Updownarrow & \text{definition of intersection.} \\
(X,Y) \in [(A \times B) \cap (C \times D)] &
\end{array}$$

□

Theorem 338 *For all sets A, B, C, and D,*

$$[(A \times B) \cup (C \times D) \cup (A \times D) \cup (C \times B)] = [(A \cup C) \times (B \cup D)].$$

Proof. An informal proof can establish that $[(A \times B) \cup (C \times D) \cup (A \times D) \cup (C \times B)]$ and $[(A \cup C) \times (B \cup D)]$ have exactly the same elements.

- An ordered pair (X,Y) is an element of $(A \times B) \cup (C \times D) \cup (A \times D) \cup (C \times B)$ if and only if $(X,Y) \in (A \times B)$, or $(X,Y) \in (C \times D)$, or $(X,Y) \in (A \times D)$, or $(X,Y) \in (C \times B)$,
- which is equivalent to $X \in A$ and $Y \in B$, or $X \in C$ and $Y \in D$, or $X \in A$ and $Y \in D$, or $X \in C$ and $Y \in B$,
- which is equivalent to $X \in A$ or $X \in C$, and $Y \in B$ or $Y \in D$,
- which is equivalent to $(X,Y) \in [(A \cup C) \times (B \cup D)]$.

A formal proof translates the alleged equivalences into the logical tautology

$$\{[(P) \wedge (Q)] \vee [(P) \wedge (S)] \vee [(R) \wedge (Q)] \vee [(R) \wedge (S)]\} \Leftrightarrow \{[(P) \vee (R)] \wedge [(Q) \vee (S)]\},$$

which follows from the distributivity of $\wedge$ over $\vee$ (theorem 215), with

$$\begin{array}{rcl}
(P) & \Leftrightarrow & (X \in A), \\
(Q) & \Leftrightarrow & (Y \in B), \\
(R) & \Leftrightarrow & (X \in C), \\
(S) & \Leftrightarrow & (Y \in D).
\end{array}$$

Thus,

$$
\begin{array}{rl}
(X,Y) \in [(A \times B) \cup (C \times D) \cup (A \times D) \cup (C \times B)] & \\
\Updownarrow & \text{definition of } \bigcup, \\
[(X,Y) \in (A \times B)] \vee [(X,Y) \in (C \times D)] & \\
\vee[(X,Y) \in (A \times D)] \vee [(X,Y) \in (C \times B)] & \\
\Updownarrow & \text{definition of } \times, \\
[(X \in A) \wedge (Y \in B)] \vee [(X \in C) \wedge (Y \in D)] & \\
\vee[(X \in A) \wedge (Y \in D)] \vee [(X \in C) \wedge (Y \in B)] \vee & \\
\Updownarrow & \text{tautology}, \\
[(X \in A) \vee (X \in C)] \wedge [(Y \in B) \vee (Y \in D)] & \\
\Updownarrow & \text{definition of } \bigcup, \\
[X \in (A \cup C)] \wedge [Y \in (B \cup D)] & \\
\Updownarrow & \text{definition of } \times. \\
(X,Y) \in [(A \cup C) \times (B \cup D)] &
\end{array}
$$

□

2.4.2 Cartesian Products of Unions and Intersections

The foregoing theorems generalize to Cartesian products of unions or intersections of any sets of sets.

Theorem 339 *For all sets of sets $\mathcal{F}$ and $\mathcal{G}$,*

$$\left(\bigcup \mathcal{F}\right) \times \left(\bigcup \mathcal{G}\right) = \bigcup_{(A,B) \in \mathcal{F} \times \mathcal{G}} (A \times B).$$

Proof. An informal proof can show that $(\bigcup \mathcal{F}) \times (\bigcup \mathcal{G})$ and $\bigcup_{(A,B) \in \mathcal{F} \times \mathcal{G}} (A \times B)$ have exactly the same elements.

- A pair (X,Y) is an element of $(\bigcup \mathcal{F}) \times (\bigcup \mathcal{G})$ if and only if $X \in (\bigcup \mathcal{F})$ and $Y \in (\bigcup \mathcal{G})$,
- which is equivalent to $X \in A$ for some $A \in \mathcal{F}$ and $Y \in B$ for some $B \in \mathcal{G}$,
- which is equivalent to $(X,Y) \in (A \times B)$ for some $(A,B) \in (\mathcal{F} \times \mathcal{G})$.

A formal proof can parallel the foregoing reasoning.

$$
\begin{array}{rl}
(X,Y) \in (\bigcup \mathcal{F}) \times (\bigcup \mathcal{G}) & \\
\Updownarrow & \text{definition of } \times, \\
[X \in (\bigcup \mathcal{F})] \wedge [Y \in (\bigcup \mathcal{G})] & \\
\Updownarrow & \text{definition of } \bigcup, \\
\{\exists A[(A \in \mathcal{F}) \wedge (X \in A)]\} \wedge \{\exists B[(B \in \mathcal{G}) \wedge (Y \in B)]\} & \\
\Updownarrow & \text{no } B, \text{ no } A, \\
\exists A(\exists B\{[(A \in \mathcal{F}) \wedge (X \in A)] \wedge [(B \in \mathcal{G}) \wedge (Y \in B)]\}) & \\
\Updownarrow & \text{properties of } \wedge, \\
\exists A(\exists B\{[(A \in \mathcal{F}) \wedge (B \in \mathcal{G})] \wedge [(X \in A) \wedge (Y \in B)]\}) & \\
\Updownarrow & \text{definition of } \times, \\
\exists A(\exists B\{[(A \times B) \in (\mathcal{F} \times \mathcal{G})] \wedge [(X,Y) \in (A \times B)]\}) & \\
\Updownarrow & \text{definition of } \bigcup. \\
(X,Y) \in \bigcup_{(A \times B) \in (\mathcal{F} \times \mathcal{G})} (A \times B) &
\end{array}
$$

□

Theorem 340 *For all sets of sets $\mathcal{F}$ and $\mathcal{G}$,*

$$\left(\bigcap \mathcal{F}\right) \times \left(\bigcap \mathcal{G}\right) = \bigcap_{(A,B)\in\mathcal{F}\times\mathcal{G}} (A \times B).$$

Proof. An informal proof can show that $(\bigcap \mathcal{F}) \times (\bigcap \mathcal{G})$ and $\bigcap_{(A,B)\in\mathcal{F}\times\mathcal{G}}(A \times B)$ have exactly the same elements.

- A pair (X, Y) is an element of $(\bigcap \mathcal{F}) \times (\bigcap \mathcal{G})$ if and only if $X \in (\bigcap \mathcal{F})$ and $Y \in (\bigcap \mathcal{G})$,
- which is equivalent to $X \in A$ for every $A \in \mathcal{F}$ and $Y \in B$ for every $B \in \mathcal{G}$,
- which is equivalent to $(X, Y) \in (A \times B)$ for every $(A, B) \in (\mathcal{F} \times \mathcal{G})$.

The foregoing informal proof glosses over the case in which at least one of $\mathcal{F}$ or $\mathcal{G}$ is empty, which would require invoking the corresponding unions as supersets of the intersections, because of the definition of intersections. One remedy could consist in proving such cases separately. Indeed, $\mathcal{F} = \varnothing$ or $\mathcal{G} = \varnothing$ if and only if $(\mathcal{F} \times \mathcal{G}) = \varnothing$. A formal proof can parallel the foregoing reasoning with the tautology (theorem 219)

$$\vdash \{[(P) \Rightarrow (Q)] \wedge [(R) \Rightarrow (S)]\} \Rightarrow \{[(P) \wedge (R)] \Rightarrow [(Q) \wedge (S)]\}$$

to prove that

$$\left(\bigcap \mathcal{F}\right) \times \left(\bigcap \mathcal{G}\right) \subseteq \bigcap_{(A,B)\in\mathcal{F}\times\mathcal{G}} (A \times B),$$

with the additional hypotheses $C \in \mathcal{F}$ for P and $D \in \mathcal{G}$ for R for the converse, so that

$$\vdash \{[(P) \Rightarrow (Q)] \wedge [(R) \Rightarrow (S)]\} \Leftarrow \big([(P) \wedge (R)] \wedge \{[(P) \wedge (R)] \Rightarrow [(Q) \wedge (S)]\}\big).$$

Thus,

$$\begin{array}{rl}
(X, Y) \in (\bigcap \mathcal{F}) \times (\bigcap \mathcal{G}) & \\
\Updownarrow & \text{definition of } \times, \\
[X \in (\bigcap \mathcal{F})] \wedge [Y \in (\bigcap \mathcal{G})] & \\
\Updownarrow & \bigcap \mathcal{H} \subseteq \bigcup \mathcal{H}, \\
\big(\{\forall A[(A \in \mathcal{F}) \Rightarrow (X \in A)]\} \wedge \{\forall B[(B \in \mathcal{G}) \Rightarrow (Y \in B)]\}\big) & \\
\wedge\big(\{\exists C[(C \in \mathcal{F}) \wedge (X \in C)]\} \wedge \{\exists D[(D \in \mathcal{G}) \wedge (Y \in D)]\}\big) & \\
\Updownarrow & \text{no } B, \text{ no } A, \\
\big[\forall A\big(\forall B\{[(A \in \mathcal{F}) \Rightarrow (X \in A)] \wedge [(B \in \mathcal{G}) \Rightarrow (Y \in B)]\}\big)\big] & \\
\wedge\big[\exists C\big(\exists D\{[(C \in \mathcal{F}) \wedge (X \in C) \wedge (D \in \mathcal{G}) \wedge (Y \in D)]\}\big)\big] & \\
\Updownarrow & \text{tautologies}, \\
\big[\forall A\big(\forall B\{[(A \in \mathcal{F}) \wedge (B \in \mathcal{G})] \Rightarrow [(X \in A) \wedge (Y \in B)]\}\big)\big] & \\
\wedge\big[\exists C\big(\exists D\{[(C \in \mathcal{F}) \wedge (X \in C) \wedge (D \in \mathcal{G}) \wedge (Y \in D)]\}\big)\big] & \\
\Updownarrow & \text{definition of } \times, \\
\big[\forall A\big(\forall B\{[(A, B) \in (\mathcal{F} \times \mathcal{G})] \Rightarrow [(X, Y) \in (A \times B)]\}\big)\big] & \\
\wedge\big[\exists C\big(\exists D\{[(C, D) \in (\mathcal{F} \times \mathcal{G})] \wedge [(X, Y) \in (C \times D)]\}\big)\big] & \\
\Updownarrow & \text{definition of } \bigcap. \\
(X, Y) \in \bigcap_{(A\times B)\in(\mathcal{F}\times\mathcal{G})}(A \times B)\} &
\end{array}$$

□

2.4.3 Mathematical Relations

The Cartesian product $A \times B$ provides a means to draw connections, or, in other words, to specify relations, between elements of the two sets A and B.

Definition 341 (Relation) A **relation** between elements of sets A and B is a subset $R \subseteq A \times B$ of their Cartesian product.

- Two elements $X \in A$ and $Y \in B$ are **related** with respect to the relation R if and only if $(X,Y) \in R$.
- For each relation $R \subseteq A \times B$, the **domain** $\mathcal{D}(R) \subseteq A$ of the relation R consists of those elements of A related by R to at least one element of B:

$$\mathcal{D}(R) = \{X \in A : \exists Y\{(Y \in B) \wedge [(X,Y) \in R]\}\}.$$

- Similarly, the **range** $\mathcal{R}(R) \subseteq B$ of the relation R consists of those elements of B related by R to at least one element of A:

$$\mathcal{R}(R) := \{Y \in B : \exists X\{(X \in A) \wedge [(X,Y) \in R]\}\}.$$

Another common notation for $(X,Y) \in R$ is XRY, especially if such a special symbol as $\subseteq$, $\subsetneqq$, or $=$ denotes the relation.

Example 342 For each set A, the **diagonal** $\Delta_A \subseteq A \times A$ is the subset

$$\Delta_A = \{(X,X) \in A \times A : X \in A\}.$$

Thus, the diagonal corresponds to the relation called "equality", or, equivalently, "identity": a pair of elements (X,Y) lies on the diagonal Δ_A if and only if $X = Y$. Because $X = X$, and hence $(X,X) \in \Delta_A$, for every $X \in A$, it follows that the diagonal Δ_A relates every element of A to itself, whence $\mathcal{D}(R) = A = \mathcal{R}(R)$. For example, if

$$A := \{\, \varnothing,\ \{\varnothing\}\,\},$$

then

$$A \times A = \left\{ \begin{matrix} (\varnothing,\{\varnothing\}) & (\{\varnothing\},\{\varnothing\}) \\ (\varnothing,\varnothing) & (\{\varnothing\},\varnothing) \end{matrix} \right\},$$

$$\Delta_A = \left\{ \begin{matrix} & (\{\varnothing\},\{\varnothing\}) \\ (\varnothing,\varnothing) & \end{matrix} \right\},$$

because the only pairs (X,Y) with $X = Y$ are the pairs $(\varnothing,\varnothing)$ and $(\{\varnothing\},\{\varnothing\})$.

Example 343 For all sets H and K, consider their respective power sets $A := \mathcal{P}(H)$ and $B := \mathcal{P}(K)$. The relation $R \subseteq \mathcal{P}(H) \times \mathcal{P}(K)$ of **inclusion** consists of all, but only those, pairs (V,W) of subsets $V \subseteq H$ and $W \subseteq K$ such that $V \subseteq W$:

$$R = \{(V,W) : (V \subseteq H) \wedge (W \subseteq K) \wedge (V \subseteq W)\}.$$

The domain of the relation $\subseteq$ consists of all subsets of A included as a subset in at least one subset of B. Similarly, the range of the relation $\subseteq$ consists of all subsets of B that contain as a subset at least one subset of A.

Example 344 For all sets H and K, let $A := \mathcal{P}(H)$ and $B := \mathcal{P}(K)$. The relation $S \subseteq \mathcal{P}(H) \times \mathcal{P}(K)$ of **strict inclusion** consists of all, but only those, pairs (V, W) of subsets $V \subseteq H$ and $W \subseteq K$ such that $V \subseteq W$ but $V \neq W$:

$$S = \{(V, W) : (V \subseteq H) \wedge (W \subseteq K) \wedge (V \subseteq W) \wedge (V \neq W)\}.$$

The strict inclusion $(V \subseteq W) \wedge (V \neq W)$, is also denoted by $V \subset W$ or by $V \subsetneqq W$:

$$\models [(V \subseteq W) \wedge (V \neq W)] \Leftrightarrow (V \subset W) \Leftrightarrow (V \subsetneqq W).$$

The domain of the relation of strict inclusion consists of all subsets of A included in, but not equal to, at least one subset of B; its range consists of all subsets of B that contain, but do not coincide with, at least one subset of A.

Example 345 For all sets A and B, the relation $\in$ relates every element $X \in A$ that is an element of some $Y \in B$. Let E be this relation. (The symbol $\in$ cannot denote a subset of $A \times B$ because $\in$ is not a variable.) Thus, $E \subseteq (A \times B)$ is defined by

$$\models [(X, Y) \in E] \Leftrightarrow [(X \in A) \wedge (X \in Y) \wedge (Y \in B)].$$

For example, if

$$\begin{aligned} A &:= \{\varnothing, \{\varnothing\}\}, \\ B &:= \{\varnothing, \{\varnothing\}\}, \end{aligned}$$

then

$$\begin{aligned} A \times B &= \left\{ \begin{array}{cc} (\varnothing, \{\varnothing\}) & (\{\varnothing\}, \{\varnothing\}) \\ (\varnothing, \varnothing) & (\{\varnothing\}, \varnothing) \end{array} \right\}, \\ E &= \left\{ \begin{array}{cc} (\varnothing, \{\varnothing\}) & \end{array} \right\}, \end{aligned}$$

because in $A \times B$ the only pair (X, Y) such that $X \in Y$ is the pair $(\varnothing, \{\varnothing\})$.

Example 346 For the sets

$$\begin{aligned} A &:= \{\varnothing, \{\varnothing\}\}, \\ B &:= \{\varnothing, \{\varnothing\}, \{\varnothing, \{\varnothing\}\}\}, \end{aligned}$$

the Cartesian product $A \times B$ and the relation E take the forms

$$\begin{aligned} A \times B &= \left\{ \begin{array}{cc} (\varnothing, \{\varnothing, \{\varnothing\}\}) & (\{\varnothing\}, \{\varnothing, \{\varnothing\}\}) \\ (\varnothing, \{\varnothing\}) & (\{\varnothing\}, \{\varnothing\}) \\ (\varnothing, \varnothing) & (\{\varnothing\}, \varnothing) \end{array} \right\}, \\ E &= \left\{ \begin{array}{cc} (\varnothing, \{\varnothing, \{\varnothing\}\}) & (\{\varnothing\}, \{\varnothing, \{\varnothing\}\}) \\ (\varnothing, \{\varnothing\}) & \end{array} \right\}, \end{aligned}$$

because in $A \times B$ the only pairs (X, Y) such that $X \in Y$ are those just displayed.

In some contexts a relation $R \subseteq A \times B$ may also prove useful with B and A listed in the reverse order. Then the "inverse" relation $R^{\circ -1} \subseteq B \times A$ contains similar pairs as R does but with coordinates listed in the reverse order. Some texts denote the inverse relation by R^{-1}, which can cause confusion because the same notation also represents reciprocals in arithmetic and algebra. The notation adopted here for the inverse relation, $R^{\circ -1}$, conforms to [58].

Definition 347 (Inverse relation) The **inverse** of a relation $R \subseteq A \times B$ between two sets A and B is the relation $R^{\circ -1} \subseteq B \times A$, defined by

$$R^{\circ -1} = \{(Y, X) \in B \times A : (X, Y) \in R\}.$$

Example 348 The inverse of the diagonal is the same diagonal.

Example 349 The inverse of the subset relation $\subseteq$ is the superset relation $\supseteq$.

Definition 350 (Composite relation) The **composition** of two relations $R \subseteq A \times B$ and $S \subseteq B \times C$ is the relation $S \circ R$ on $A \times C$ defined as follows. The **composite** relation $S \circ R$ relates an element $U \in A$ to an element $W \in C$ if and only if R relates U to some element $V \in B$ and S relates the same element V to $W \in C$:

$$(S \circ R) = \{(U, W) \in (A \times C) : \exists V([V \in B] \wedge [(U, V) \in R] \wedge [(V, W) \in S]).$$

Example 351 For each relation $R \subseteq A \times B$ between any sets A and B, the composition $R \circ R^{\circ -1}$ contains the diagonal $\triangle_{\mathcal{R}(R)}$ for the range of R. Indeed, by definition of its range, R relates every $Y \in \mathcal{R}(R)$ to an element $X \in A$ so that $(X, Y) \in R$; then the inverse $R^{\circ -1}$ relates Y to X so that $(Y, X) \in R^{\circ -1}$. From $(Y, X) \in R^{\circ -1}$ and $(X, Y) \in R$ it follows that $(Y, Y) \in (R \circ R^{\circ -1})$ for every $(Y, Y) \in \triangle_{\mathcal{R}(R)}$. Therefore $\triangle_{\mathcal{R}(R)} \subseteq (R \circ R^{\circ -1})$. Similarly, $R^{\circ -1} \circ R$ contains the diagonal $\triangle_{\mathcal{D}(R)}$ for the domain of R. Indeed, by definition of its domain, R relates every $X \in \mathcal{D}(R)$ to an element $Y \in B$ so that $(X, Y) \in R$; then the inverse $R^{\circ -1}$ relates Y to X so that $(Y, X) \in R^{\circ -1}$. From $(X, Y) \in R$ and $(Y, X) \in R^{\circ -1}$ it follows that $(X, X) \in (R^{\circ -1} \circ R)$ for every $(X, X) \in \triangle_{\mathcal{D}(R)}$. Therefore $\triangle_{\mathcal{D}(R)} \subseteq (R^{\circ -1} \circ R)$.

Because every relation R between sets A and B is a subset $R \subseteq (A \times B)$, operations with sets apply to all relations.

Definition 352 (Unions and intersections of relations) For all sets A, B, C, E, and for all relations $R \subseteq A \times B$ and $T \subseteq C \times E$, the **union** of the relations R and T is the relation $R \cup T$ between $A \cup C$ and $B \cup E$, so that

$$(R \cup T) := \{(X, Y) \in [(A \cup C) \times (B \cup E)] : [(X, Y) \in R] \vee [(X, Y) \in T]\}.$$

Similarly, the **intersection** of the relations R and T is the relation $R \cap T$ between $A \cup C$ and $B \cup E$, so that

$$(R \cap T) = \{(X, Y) \in [(A \cap C) \times (B \cap E)] : [(X, Y) \in R] \wedge [(X, Y) \in T]\}.$$

A particular instance of intersections of relations $R \subseteq (A \times B)$ and $T \subseteq (A \times B)$ consists of the intersection of R with a subset $S \subseteq A$ and its Cartesian product with B. The "restriction" of R to S is then the relation $R \cap T$ with $T = (S \times B)$. The concept of the "restriction" of a relation to a subset is useful if the subset has characteristics that are useful for the purpose at hand while the complement of the subset does not.

Definition 353 (Restricted relation) For each relation $R \subseteq A \times B$, and for each subset $S \subseteq A$ of A, the **restriction of R to S** is the relation $R|_S \subseteq S \times B$ defined by

$$\begin{aligned} R|_S &= R \cap (S \times B) \\ &= \{(X, Y) \in R : X \in S\}. \end{aligned}$$

Thus, $R|_S$ restricts its first coordinates only to those elements of S.

There also exists a similar instance of intersections of relations $R \subseteq (A \times B)$ and $T \subseteq (A \times B)$ as the intersection of R with a subset $V \subseteq B$ and its Cartesian product with A. The "restriction" of R to V is then the relation $R \cap T$ with $T = (A \times V)$. Similarly, the restriction of R to $S \subseteq A$ and $V \subseteq B$ is the relation $R \cap (S \times V)$.

2.4.4 Exercises

Exercise 511 Determine whether the Cartesian product is associative.

Exercise 512 For each set A, prove that $A \times \varnothing = \varnothing$, and that $\varnothing \times A = \varnothing$.

Exercise 513 Prove that $A \times B = \varnothing$ if and only if $A = \varnothing$ or $B = \varnothing$.

Exercise 514 Prove that $\times$ distributes over $\cap$: $[(A \times B) \cap (C \times B)] = [(A \cap C) \times B]$.

Exercise 515 Prove that $\times$ distributes over $\cup$: $[(A \times B) \cup (C \times B)] = [(A \cup C) \times B]$.

Exercise 516 Prove that $[(A \setminus C) \times B] = [(A \times B) \setminus (C \times B)]$.

Exercise 517 Prove that $[A \times (B \setminus D)] = [(A \times B) \setminus (A \times D)]$.

Exercise 518 Prove or disprove that $[(A \setminus C) \times (B \setminus D)] \subseteq [(A \times B) \setminus (C \times D)]$.

Exercise 519 Prove or disprove that $[(A \setminus C) \times (B \setminus D)] \supseteq [(A \times B) \setminus (C \times D)]$.

Exercise 520 Prove or disprove that $[(A \Delta C) \times (B \Delta D)] \subseteq [(A \times B) \Delta (C \times D)]$.

Exercise 521 Prove or disprove that $[(A \Delta C) \times (B \Delta D)] \supseteq [(A \times B) \Delta (C \times D)]$.

Exercise 522 Prove or disprove that $([\mathcal{P}(A)] \times [\mathcal{P}(B)]) \subseteq \mathcal{P}(A \times B)$.

Exercise 523 Prove or disprove that $([\mathcal{P}(A)] \times [\mathcal{P}(B)]) \supseteq \mathcal{P}(A \times B)$.

Exercise 524 Provide a formula for the inverse of the relation of inclusion.

Exercise 525 Provide a formula for the inverse of the relation of strict inclusion.

Exercise 526 For each relation $R \subseteq A \times B$, and for each subset $S \subseteq A$ of A, prove that $R|_S = R \cap (S \times B)$.

Exercise 527 For all sets A and B, prove that $\varnothing$ is an element of the domain $\mathcal{D}$ of the relation $\subseteq$ on $\mathcal{P}(A) \times \mathcal{P}(B)$.

Exercise 528 For all sets A and B, prove that $\mathcal{P}(B)$ is the range $\mathcal{R}$ of the relation $\subseteq$ on $\mathcal{P}(A) \times \mathcal{P}(B)$.

Exercise 529 Provide examples of sets A and B, such that $\mathcal{P}(A)$ is the domain $\mathcal{D}$ of the relation $\subseteq$ on $\mathcal{P}(A) \times \mathcal{P}(B)$.

Exercise 530 Provide examples of sets A and B, such that $\mathcal{P}(A)$ is *not* the domain $\mathcal{D}$ of the relation $\subseteq$ on $\mathcal{P}(A) \times \mathcal{P}(B)$, so that $\mathcal{D} \subsetneqq \mathcal{P}(A)$.

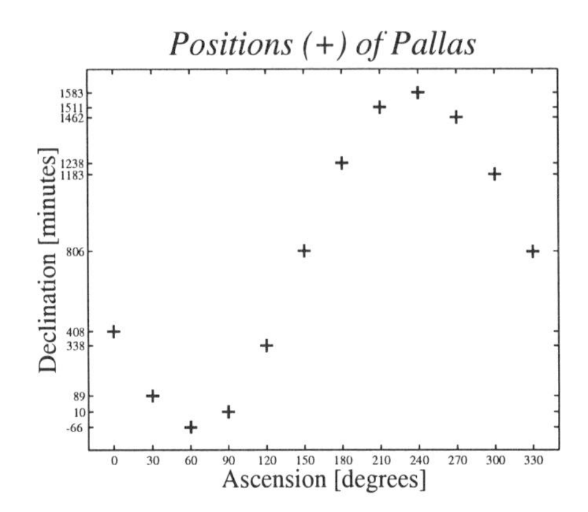

Figure 2.4: Positions include at most one declination for each ascension.

2.5 MATHEMATICAL FUNCTIONS

2.5.1 Mathematical Functions

In some applications of relations, the domain and the range can contain measurements.

Example 354 Results from astronomical observations can consist of a relation between two coordinates of position, with ordered pairs (X, Y) where X is the observed ascension (elevation) and Y is the observed declination (azimuth) of an asteroid. For example, the following pairs (X, Y) record the ascension X and the declination Y of the asteroid Pallas, measured by Baron von Zach about 1800 A.D. [7, p. 5]:

$$\begin{array}{cccccc} (0,408) & (30,89) & (60,-66) & (90,10) & (120,338) & (150,807) \\ (180,1238) & (210,1511) & (240,1583) & (270,1462) & (300,1183) & (330,804) \end{array}$$

with the corresponding graphical representation in figure 2.4.

Results from astronomical observations can also consist of a relation between time and position, with ordered pairs (T, Y) where Y is the observed position (declination or azimuth) of a planet at time T. Such a relation has the following properties.

- If no observation was made at some time T, then the results do not contain any ordered pair with T as their first coordinate.
- Each observation yields only one position Y at any time T.
- Observations can yield the same position Y at several times, for instance, if the motion is periodic.

Mathematical "functions" are relations corresponding to such applications.

Definition 355 A **function** from a set A to a set B is a relation $F \subseteq (A \times B)$ such that for each $X \in A$ there exists *at most one* $Y \in B$ for which $(X, Y) \in F$.

The **domain** of F is the subset $\mathcal{D}(F) \subseteq A$ consisting of every $X \in A$ for which there exists some $Y \in B$ such that $(X, Y) \in F$.

$$\mathcal{D}(F) = \{X \in A : \exists Y[(Y \in B) \wedge \{(X, Y) \in F\}]\}.$$

The **range** of F, denoted by $\mathcal{R}(F)$, consists of every $Y \in B$ such that there exists some $X \in A$ such that $(X, Y) \in F$:

$$\mathcal{R}(F) = \{Y \in B : \exists X[(X \in A) \wedge \{(X,Y) \in F\}]\}.$$

Moreover, B is called the **co-domain** of F.

For each $X \in \mathcal{D}(F)$, the unique Y such that $(X, Y) \in F$ is called the **value of** F **at** X, or also the **image of** X **by** F. The same Y is denoted by $F(X)$ (read "F of X"); thus, $Y = F(X)$ if and only if $(X, Y) \in F$. The element X is called the **argument** of F in the expression $F(X)$. The notation

$$F : A \to B$$

(read "F maps A to B") means that F is a function from A to B; the notation $X \mapsto F(X)$ (read "F maps X to $F(X)$") may specify $F(X)$ by a formula or otherwise.

Remark 356 A variant of the concept of a function from a set A to a set B is a subset $F \subseteq (A \times B)$ such that for each $X \in A$ there exists *exactly one* $Y \in B$ for which $(X, Y) \in F$, and then the notation $F : A \to B$ implies that $\mathcal{D}(F) = A$.

The requirement that $\mathcal{D}(F) = A$ remains harmless with simple examples of functions, but it presents unnecessary obstacles with more realistic examples of functions, whose complexity can make the domain difficult or impossible to identify, especially if the identification of the domain is irrelevant to the task at hand.

The specification with *"at most"* in definition 355 is common in set theory [95, p. 58 & 86], algebraic geometry (especially with "rational" functions) [103, p. 34–35], complex analysis (especially with "meromorphic" functions) [1, p. 128], and functional analysis [97, p. 4 & 18], in instances where several functions called "operators" have different domains but arise from a relation common to all of them.

Example 357 The pairs in example 354 define a function $F : A \to B$ from the set of ascensions $A := \{0, 30, 60, 90, 120, 150, 180, 210, 240, 270, 300, 330\}$ to the set of declinations $B := \{-66, 10, 89, 338, 408, 804, 807, 1183, 1238, 1462, 1511, 1583\}$. This function $F : A \to B$ has domain A, codomain B, and range B.

Example 358 With $A := \{6, 7, 8, 9\}$ and $B := \{1, 2, 3, 4\}$, let

$$F := \{(6,2), (7,1), (8,2), (9,3)\}.$$

Then $F : A \to B$ is a function with domain A, codomain B, and range $\{1, 2, 3\}$.

Example 359 For all non-empty sets A and B, and for any element $Z \in B$, there exists a **constant function**

$$\begin{aligned} C_Z \quad : \quad & A \to B, \\ & X \mapsto Z. \end{aligned}$$

Thus the constant function C_Z maps every element $X \in A$ to the same value Z, so that

$$C_Z := \{(X, Z) : X \in A\}.$$

In particular, $\mathcal{D}(C_Z) = A$ for the domain of C_Z, and $\mathcal{R}(C_Z) = \{Z\}$ for its range.

Example 360 For each set A, there exists an **identity function**

$$\begin{aligned} I_A \quad : \quad & A \to A, \\ & X \mapsto X. \end{aligned}$$

Thus the identity function I_A maps every element to itself, so that

$$I_A := \{(X,X) : X \in A\}.$$

In particular, $\mathcal{D}(I_A) = A$ and $\mathcal{R}(I_A) = A$, because $(X,X) \in I_A$ for *every* $X \in A$. The function I_A is also denoted by Δ_A and is called the the "diagonal" of $A \times A$.

Example 361 For all sets A and B, the **canonical projection functions** from the Cartesian product $A \times B$ into its factors A and B are the functions P_A and P_B with

$$\begin{aligned} P_A \quad : \quad & (A \times B) \to A, \\ & (X,Y) \mapsto X; \end{aligned}$$

$$\begin{aligned} P_B \quad : \quad & (A \times B) \to B, \\ & (X,Y) \mapsto Y. \end{aligned}$$

Thus, P_A maps (X,Y) to its first coordinate X in A, whereas P_B maps (X,Y) to its second coordinate Y in B.

Example 362 For all sets A and B, for any subset $V \subseteq A$ and any element $Z \in B$, the **slice function** $S_{V,Z}$ maps each element $X \in V$ to $(X,Z) \in (A \times B)$:

$$\begin{aligned} S_{V,Z} \quad : \quad & V \to (A \times B), \\ & X \mapsto (X,Z). \end{aligned}$$

Thus, $S_{V,Z}$ maps its domain V to the **slice** $V \times \{Z\}$ in $A \times B$.

Example 363 For each set A and each subset $S \subseteq A$, the **characteristic function** χ_S — denoted by the Greek letter χ (read "chi") — maps every element of the subset S to 1, and every element outside the subset S to 0, (with $0 = \varnothing$, $1 = \{\varnothing\}$, and $2 = \{0,1\}$):

$$\begin{aligned} \chi_S \quad : \quad & A \to 2, \\ & X \mapsto \begin{cases} 1 & \text{if } X \in S, \\ 0 & \text{if } X \notin S. \end{cases} \end{aligned}$$

The following theorem provides a means to compare two functions to each other.

Theorem 364 *Two functions $F : A \to B$ and $G : A \to B$ equal each other, $F = G$, if and only if $F(X) = G(X)$ for every $X \in A$.*

Proof. This proof rewrites $F(X) = G(X)$ in terms of the definition of functions:

$$\begin{aligned} F &= G \\ &\Updownarrow \quad \text{extensionality,} \\ \forall X\big(\forall Y \{[(X,Y) \in F] &\Leftrightarrow [(X,Y) \in G]\}\big) \\ &\Updownarrow \quad \text{functional notation.} \\ \forall X\big(\forall Y \{[Y = F(X)] &\Leftrightarrow [Y = G(X)]\}\big) \end{aligned}$$

□

Some situations involve only parts of a function, or combinations of several functions, for instance, as defined by the concepts introduced here.

Definition 365 (Restriction of functions) For each function $F : A \to B$, and for each subset $S \subseteq A$ of A, the **restriction of** F **to** S is the function $F|_S \subseteq S \times B$ defined by

$$\begin{aligned} F|_S \quad : \quad & S \to B, \\ & X \mapsto F(X). \end{aligned}$$

Thus, $F|_S(X) = F(X)$, but $F|_S$ maps only the elements of $S \cap \mathcal{D}(F)$.

Example 366 With $A := \{6, 7, 8, 9\}$ and $B := \{1, 2, 3, 4\}$, let

$$F := \{(6, 2), (7, 1), (8, 2), (9, 3)\}.$$

Then the restriction of F to the subset $S := \{6, 8\}$ is $F|_S = \{(6, 2), (8, 2)\}$.

Example 367 For each set B, and for each subset $W \subseteq B$, the **inclusion function** — denoted by the Greek letter ι ("iota") — is the restriction of the identity function to W:

$$\begin{aligned} \iota_W \ : \ & W \to B, \\ & X \mapsto X. \end{aligned}$$

Thus $\iota_W : W \to B$ is the restriction of $I_B : B \to B$ to the subset $W \subseteq B$.

Definition 368 (Union of functions) For all *disjoint* sets A and C, so that $A \cap C = \varnothing$, for all sets B and E, and for all functions $F : A \to B$ and $G : C \to E$, the **union** of the functions F and G is the union $F\dot{\cup}G$ of the two sets $F \subseteq A \times B$ and $G \subseteq C \times E$:

$$\begin{aligned} F\dot{\cup}G \ : \ & (A\dot{\cup}C) \to (B \cup E), \\ & X \mapsto \begin{cases} F(X) & \text{if } X \in \mathcal{D}(F) \subseteq A, \\ G(X) & \text{if } X \in \mathcal{D}(G) \subseteq C. \end{cases} \end{aligned}$$

Definition 369 (Intersection of functions) For all sets A, B, C, E, and all functions $F : A \to B$ and $G : C \to E$, the **intersection** of the functions F and G is the intersection $F \cap G$ of the two sets $F \subseteq A \times B$ and $G \subseteq C \times E$, so that

$$\begin{aligned} F \cap G \ : \ & D \to (B \cap E), \\ & X \mapsto F(X) = G(X), \end{aligned}$$

with domain $D := \mathcal{D}(F \cap G) = \{X \in A \cap C : F(X) = G(X)\}$.

2.5.2 Images and Inverse Images of Sets by Functions

Some situations involve the images by functions of not only single elements in the domain, but also subsets of the domain, as defined here.

Definition 370 For each function $F : A \to B$ and each subset $V \subseteq A$, the **image of** V **by** F is the subset $F``(V)$ consisting of all images of every $X \in V \cap \mathcal{D}(F)$ by F:

$$\begin{aligned} F``(V) &= \{Y \in B : \exists X\, [(X \in V) \wedge (F(X) = Y)]\} \\ &= \{F(X) : X \in V \cap \mathcal{D}(F)\}. \end{aligned}$$

For each function $F : A \to B$, $F``$ is a function of subsets: $F`` : \mathcal{P}(A) \to \mathcal{P}(B)$.

Remark 371 The notation $F``(V)$ adopted in definition 370 is common in set theory [64, p. 14], [95, p. 65]. Informal mathematical usage employs the notation $F(V)$ for the image of a subset $V \subseteq A$ by a function $F : A \to B$, but this usage is ambiguous. For example, consider the set $A := \{\ \varnothing,\ \{\varnothing\}\ \}$ and the constant function

$$\begin{aligned} C_\varnothing \ : \ & \{\ \varnothing,\ \{\varnothing\}\ \} \to \{\ \varnothing,\ \{\varnothing\}\ \}, \\ & X \mapsto \varnothing. \end{aligned}$$

The set $\{\varnothing\}$ is an element of A, whence $C_{\varnothing}(\{\varnothing\}) = \varnothing$ because $C_{\varnothing}(X) = \varnothing$ for every $X \in A$. Yet $\{\varnothing\}$ is also a subset of A, containing the single element $\varnothing \in \{\varnothing\}$; because $C_{\varnothing}(\varnothing) = \varnothing$ it follows that $C"_{\varnothing}(\{\varnothing\}) = \{\varnothing\}$ as the image of a subset. Thus,

$$\begin{aligned} C_{\varnothing}(\{\varnothing\}) &= \varnothing, \\ C"_{\varnothing}(\{\varnothing\}) &= \{\varnothing\}. \end{aligned}$$

The common informal notation leads to the contradiction

$$\begin{aligned} C_{\varnothing}(\{\varnothing\}) &= \varnothing, \\ C_{\varnothing}(\{\varnothing\}) &= \{\varnothing\}. \end{aligned}$$

In a formal theory containing this contradiction, every proposition would be True.

Example 372 If $C_Z : A \to B$ is a constant function that maps each $X \in A$ to the same $Z \in B$, then $C_Z"(V) = \{Z\}$ for each *non-empty* subset $V \subseteq A$.

Example 373 If $I_A : A \to A$, $X \mapsto X$ is an identity function, so that $I_A(X) = X$. for each $X \in A$ then $I_A"(V) = V$ for each subset $V \subseteq A$. Thus, $I_A"$ is the identity function $I_A" = I_{\mathcal{P}(A)} : \mathcal{P}(A) \to \mathcal{P}(A)$, with $V \mapsto V$ for every $V \in \mathcal{P}(A)$.

Besides images of subsets, such problems as the solution of equations involve the identification of a subset, called a pre-image, mapped to a specified image.

Definition 374 (Pre-images) For each function $F : A \to B$, and for each element $Y \in B$, the **pre-image of** Y **by** F is the *subset* of A denoted by $F^{\circ -1}"(\{Y\})$ and consisting of all elements $X \in A$ such that $F(X) = Y$:

$$F^{\circ -1}"(\{Y\}) = \{X \in A : F(X) = Y\}.$$

For each subset $W \subseteq B$, the **inverse image**, or **pre-image, of** W **by** F is the subset of A denoted by $F^{\circ -1}"(W)$ and consisting of all pre-images of all elements of V by F:

$$F^{\circ -1}"(W) = \{X \in A : F(X) \in W\}.$$

Example 375 If $C_Z : A \to B$ is a constant function that maps each $X \in A$ to the same $Z \in B$, then $C_Z^{\circ -1}"(\{Z\}) = A$.

Also, $C_Z^{\circ -1}"(W) = A$ for each subset $W \subseteq B$ for which $Z \in W$.

In contrast, $C_Z^{\circ -1}"(S) = \varnothing$ for each subset $S \subseteq B$ for which $Z \notin S$. Thus,

$$C_Z^{\circ -1}"(W) = \begin{cases} A & \text{if } Z \in W, \\ \varnothing & \text{if } Z \notin W. \end{cases}$$

Example 376 If $I_A : A \to A$ is an identity function, with $I_A(X) = X$ for each $X \in A$, then $I_A^{\circ -1}"(W) = W$ for each subset $W \subseteq A$.

The following theorem relates images and pre-images to unions and intersections.

Theorem 377 *For each function $F : A \to B$, for each set $\mathcal{F}$ of subsets of A, and for each set $\mathcal{G}$ of subsets of B, the following relations hold.*

$$\begin{aligned} F^{\circ -1}"\left(\bigcup \mathcal{G}\right) &= \bigcup_{W \in \mathcal{G}} F^{\circ -1}"(W) \\ F^{\circ -1}"\left(\bigcap \mathcal{G}\right) &= \bigcap_{W \in \mathcal{G}} F^{\circ -1}"(W) \\ F"\left(\bigcup \mathcal{F}\right) &= \bigcup_{V \in \mathcal{F}} F"(V) \\ F"\left(\bigcap \mathcal{F}\right) &\subseteq \bigcap_{V \in \mathcal{F}} F"(V). \end{aligned}$$

Proof. Apply the definitions of union and intersection. For $F^{\circ-1}{}^{"}\left(\bigcup \mathcal{G}\right)$,

$$\begin{aligned}
F^{\circ-1}{}^{"}\left(\bigcup \mathcal{G}\right) = \bigcup_{W\in\mathcal{G}} F^{\circ-1}{}^{"}(W) & \quad \textbf{yet unproved,}\\
\Updownarrow & \quad \text{extensionality (S1),}\\
\forall X\left\{\left[X \in F^{\circ-1}{}^{"}\left(\bigcup \mathcal{G}\right)\right]\right. &\\
\left.\Leftrightarrow \left[X \in \bigcup_{W\in\mathcal{G}} F^{\circ-1}{}^{"}(W)\right]\right\} &\\
\Updownarrow & \quad \text{definitions of } F^{\circ-1}{}^{"} \text{ and } \bigcup,\\
\forall X\left\{\left[F(X) \in \left(\bigcup \mathcal{G}\right)\right]\right. &\\
\left.\Leftrightarrow \left(\exists W\left\{(W \in \mathcal{G}) \wedge [F(X) \in W]\right\}\right)\right\} &\\
\Updownarrow & \quad \text{definition of } \bigcup,\\
\forall X\left\{\left(\exists W\left\{(W \in \mathcal{G}) \wedge [F(X) \in W]\right\}\right)\right. &\\
\left.\Leftrightarrow \left(\exists W\left\{(W \in \mathcal{G}) \wedge [F(X) \in W]\right\}\right)\right\} &
\end{aligned}$$

which is universally valid thanks to the tautology $(P) \Leftrightarrow (P)$. For $F^{\circ-1}{}^{"}\left(\bigcap \mathcal{G}\right)$,

$$\begin{aligned}
F^{\circ-1}{}^{"}\left(\bigcap \mathcal{G}\right) = \bigcap_{W\in\mathcal{G}} F^{\circ-1}{}^{"}(W) & \quad \textbf{yet unproved,}\\
\Updownarrow & \quad \text{extensionality (S1),}\\
\forall X\left\{\left[X \in F^{\circ-1}{}^{"}\left(\bigcap \mathcal{G}\right)\right]\right. &\\
\left.\Leftrightarrow \left[X \in \bigcap_{W\in\mathcal{G}} F^{\circ-1}{}^{"}(W)\right]\right\} &\\
\Updownarrow & \quad \text{definitions of } F^{\circ-1} \text{ and } \bigcap,\\
\forall X\left\{\left[F(X) \in \left(\bigcap \mathcal{G}\right)\right] \Leftrightarrow\right. &\\
\left.\left(\forall W\left\{(W \in \mathcal{G}) \Rightarrow [F(X) \in W]\right\}\right)\right\} &\\
\Updownarrow & \quad \text{definitions of } \bigcap,\\
\forall X\left\{\left(\forall W\left\{(W \in \mathcal{G}) \Rightarrow [F(X) \in W]\right\}\right)\right. &\\
\left.\Leftrightarrow \left(\forall W\left\{(W \in \mathcal{G}) \Rightarrow [F(X) \in W]\right\}\right)\right\} &
\end{aligned}$$

which is universally valid thanks to the tautology $(P) \Leftrightarrow (P)$. For $F"\left(\bigcup \mathcal{F}\right)$,

$$\begin{aligned}
F"\left(\bigcup \mathcal{F}\right) = \bigcup_{V\in\mathcal{F}} F"(V) & \quad \textbf{yet unproved,}\\
\Updownarrow & \quad \text{extensionality (S1),}\\
\forall Y\left\{\left[Y \in F"\left(\bigcup \mathcal{F}\right)\right] \Leftrightarrow \left[Y \in \bigcup_{V\in\mathcal{F}} F"(V)\right]\right\} &\\
\Updownarrow & \quad \text{definitions: } F", \bigcup,\\
\forall Y\left[\left(\exists X\left\{\left[X \in \left(\bigcup \mathcal{F}\right)\right] \wedge [Y = F(X)]\right\}\right)\right. &\\
\left.\Leftrightarrow \left(\exists V\left\{[V \in \mathcal{F}] \wedge [Y \in F"(V)]\right\}\right)\right] &\\
\Updownarrow & \quad \text{definitions: } \bigcup, F",\\
\forall Y\left[\left(\exists X\left\{\exists V[(V \in \mathcal{F}) \wedge (X \in V)]\right\} \wedge [Y = F(X)]\right)\right. &\\
\left.\Leftrightarrow \left\{\exists V\left[(V \in \mathcal{F}) \wedge \left(\exists X\{(X \in V) \wedge [Y = F(X)]\}\right)\right]\right\}\right] &\\
\Updownarrow & \quad \text{no } X \text{ in } V \in \mathcal{F},\\
\forall Y\left[\left(\exists X\left\{\exists V[(V \in \mathcal{F}) \wedge (X \in V)]\right\} \wedge [Y = F(X)]\right)\right. &\\
\left.\Leftrightarrow \left(\exists V\left\{\exists X\left[(V \in \mathcal{F}) \wedge \{(X \in V) \wedge [Y = F(X)]\}\right]\right\}\right)\right] &
\end{aligned}$$

which is universally valid thanks to the associativity of $\wedge$. For $F"\left(\bigcap \mathcal{F}\right)$,

$$
\begin{aligned}
F\text{"}\left(\bigcap \mathcal{F}\right) \subseteq \bigcap_{V\in\mathcal{F}} F\text{"}(V) &\quad \textbf{yet unproved,}\\
\Updownarrow &\quad \text{definition 287,}\\
\forall Y \left\{[Y \in F\text{"}\left(\bigcap \mathcal{F}\right)] \Rightarrow \left[Y \in \bigcap_{V\in\mathcal{F}} F\text{"}(V)\right]\right\} &\\
\Updownarrow &\quad \text{definitions: } F\text{"}, \bigcap,\\
\forall Y \left[\{\exists X \left[X \in \left(\bigcap \mathcal{F}\right)\right] \wedge [Y = F(X)]\}\right. &\\
\left.\Rightarrow (\forall V \{(V \in \mathcal{F}) \Rightarrow [Y \in F\text{"}(V)]\})\right] &\\
\Updownarrow &\quad \text{definitions: } \bigcap, F\text{"},\\
\forall Y \Big(\{\exists X [\forall V ([V \in \mathcal{F}] \Rightarrow [X \in V])] \wedge [Y = F(X)]\} &\\
\Rightarrow \{\forall V [(V \in \mathcal{F}) \Rightarrow (\exists X\{(X \in V) \wedge [Y = F(X)]\})]\}\Big) &\\
\Updownarrow &\quad \text{no } X \text{ in } (V \in \mathcal{F}),\\
\forall Y \Big(\{\exists X [\forall V ([V \in \mathcal{F}] \Rightarrow [X \in V])] \wedge [Y = F(X)]\} &\\
\Rightarrow \{\forall V [\exists X(V \in \mathcal{F}) \Rightarrow (\{(X \in V) \wedge [Y = F(X)]\})]\}\Big) &
\end{aligned}
$$

which is universally valid thanks to the universally valid formulae

$$
\begin{gathered}
\{[(P) \Rightarrow (Q)] \wedge (V)\} \Rightarrow \{(P) \Rightarrow [(Q) \wedge (V)]\},\\
\{\exists X[\forall R(W)]\} \Rightarrow \{\forall R[\exists X(W)]\},\\
[(W) \Rightarrow (V)] \Rightarrow \{[\exists X(W)] \Rightarrow [\exists X(V)]\}.
\end{gathered}
$$

□

2.5.3 Exercises

Exercise 531 Determine whether $F := \{(0,1),(2,3),(1,2),(0,4)\}$ is a function.

Exercise 532 Determine whether $G := \{(9,2),(7,3),(8,2),(6,1)\}$ is a function.

Exercise 533 Determine whether the following relation is a function:

$$R := \{(0,1),(1,2),(2,4),(3,8),(4,8),(5,4),(6,2),(7,1),(8,0)\}.$$

Exercise 534 Determine whether the following relation is a function:

$$S := \{(0,2),(1,5),(2,7),(3,5),(4,2),(5,5),(6,7),(7,5),(8,2)\}.$$

Exercise 535 Determine whether the following relation is a function:

$$Z := \{(0,0),(1,0),(2,0),(3,0),(4,0),(5,0),(6,0),(7,0),(8,0),(9,0)\}.$$

Exercise 536 Prove that exactly one function exists from $A := \varnothing$ to $B := \varnothing$.

Exercise 537 Investigate whether a function $F : \varnothing \to B$ exists from $\varnothing$ to any set B.

Exercise 538 Investigate whether a function $F : A \to \varnothing$ exists from any set A to $\varnothing$.

Exercise 539 For each set A let $\mathbf{1}_A$ denote the constant function with value 1, so that

$$\mathbf{1}_A : A \to \{1\}, \quad \mathbf{1}_A(a) = 1.$$

For each subset $B \subseteq A$, prove that $\mathbf{1}_A|_B$ coincides with χ_B.

Exercise 540 For all subsets V, W of each set A, investigate whether the characteristic function of the intersection, $\chi_{V\cap W} : A \to 2$, is the intersection of the two characteristic functions $\chi_V : A \to 2$ and $\chi_W : A \to 2$, so that $\chi_{V\cap W} = \chi_V \cap \chi_W$.

Exercise 541 For all subsets V, W of each set A, investigate whether the characteristic function of the union, $\chi_{V\cup W} : A \to 2$, is the union of the two characteristic functions $\chi_V : A \to 2$ and $\chi_W : A \to 2$, so that $\chi_{V\cup W} = \chi_V \cup \chi_W$.

Exercise 542 Prove that for each function $F : A \to B$, for all subsets $R, S \subseteq A$, and for all subsets $V, W \subseteq B$, the following relations hold.

$$\begin{array}{rcl} F^{\circ -1}{}''(V \cup W) & = & F^{\circ -1}{}''(V) \cup F^{\circ -1}{}''(W) \\ F^{\circ -1}{}''(V \cap W) & = & F^{\circ -1}{}''(V) \cap F^{\circ -1}{}''(W) \\ F''(R \cup S) & = & F''(R) \cup F''(S) \\ F''(R \cap S) & \subseteq & F''(R) \cap F''(S). \end{array}$$

Exercise 543 Provide an example for which $F''(R \cap S) \subsetneqq F''(R) \cap F''(S)$.

Exercise 544 For each function $F : A \to B$ and for all subsets $V, W \subseteq B$, prove that $F^{\circ -1}{}''(W \setminus V) = [F^{\circ -1}{}''(W)] \setminus [F^{\circ -1}{}''(V)]$.

Exercise 545 For each function $F : A \to B$ and for all subsets $H, K \subseteq A$, investigate whether inclusion or equality holds for $F''(K \setminus H)$ and $[F''(K)] \setminus [F''(H)]$.

Exercise 546 For each function $F : A \to B$, prove that $F''(V) = \mathcal{R}(F|_V)$.

Exercise 547 For each function $F : A \to B$, prove that $F'' : \mathcal{P}(A) \to \mathcal{P}(B)$ contains all the information about F, in the sense that $\{F(X)\} = F''(\{X\})$ for every $X \in A$.

Exercise 548 For each function $\mathcal{F} : \mathcal{P}(A) \to \mathcal{P}(B)$, investigate whether there exists a function $F : A \to B$ such that $F'' = \mathcal{F}$.

Exercise 549 Consider the function $F : A \to B$ defined by

$$\begin{array}{rcl} A & := & 3 = \{0, 1, 2\} = \Big\{ \varnothing,\ \{\varnothing\},\ \{\varnothing,\ \{\varnothing\}\} \Big\}, \\ B & := & A, \\ F & := & \{(0, 2), (1, 2), (2, 0)\}. \end{array}$$

Recall that the superscript " indicates images of *subsets* (rather than of elements).

(549.1) $\{\varnothing,\ \{\varnothing\}\}$ is a *subset* of A. Find its image: $F''(\{\varnothing,\ \{\varnothing\}\})$.

(549.2) $\{\varnothing,\ \{\varnothing\}\}$ is a *subset* of B. Find its inverse image $F^{\circ -1}{}''(\{\varnothing,\ \{\varnothing\}\})$.

(549.3) $\{\varnothing,\ \{\varnothing\}\}$ is an *element* of A. Find its image $F(\{\varnothing,\ \{\varnothing\}\})$.

(549.4) $\{\varnothing,\ \{\varnothing\}\}$ is an *element* of B. Find $F^{\circ -1}{}''\Big(\big\{\{\varnothing,\ \{\varnothing\}\}\big\}\Big)$.

Exercise 550 Consider the function $G : C \to D$ defined by

$$\begin{array}{rcl} C & := & 4 = \{0, 1, 2, 3\} \\ & = & \Big\{ \varnothing,\ \{\varnothing\},\ \{\varnothing,\ \{\varnothing\}\},\ \Big\{ \varnothing,\ \{\varnothing\},\ \{\varnothing,\ \{\varnothing\}\} \Big\} \Big\}, \\ D & := & C, \\ G & := & \{(0, 3), (1, 0), (2, 3), (3, 1)\}. \end{array}$$

Recall that the superscript " indicates images of *subsets* (rather than of elements).

(550.1) 3 is a *subset* of A. Find its image: $G"(3)$.

(550.2) 3 is a *subset* of B. Find its inverse image $G^{\circ-1}"(3)$.

(550.3) 3 is an *element* of A. Find its image $G(3)$.

(550.4) 3 is an *element* of B. Find $G^{\circ-1}"(\{3\})$.

2.6 COMPOSITE AND INVERSE FUNCTIONS

2.6.1 Compositions of Functions

Some situations involve sequences of operations corresponding to sequences of functions. For instance, if a first function consists of pairs (T, X) with the ascension (elevation) X of a planet at time T, and if a second function consists of pairs (X, Y) with the declination (azimuth) Y of the planet at ascension X, then the composition of the two functions consists of pairs (T, Y) with the declination Y of the planet at time T.

Definition 378 (Composition of functions) For all functions $F : A \to B$ and $G : B \to C$, the **composite function** $G \circ F$ (read "G preceded by F" or "F followed by G" or "the composition of G and F") is the function $G \circ F : A \to C$ defined by

$$(G \circ F)(X) := G[F(X)]$$

for each $X \in F^{\circ-1}"[\mathcal{D}(G)]$. Thus,

$$[(X, Z) \in (G \circ F)] \Leftrightarrow \{[\exists Y (Y \in B)] \wedge [(X, Y) \in F] \wedge [(Y, Z) \in G]\}.$$

Example 379 Consider the following functions F and G:

$$\begin{array}{rcl} A & = & \{0, 1\}, \\ B & = & \{0, 1, 2\}, \\ F{:}\, A & \to & B, \\ F & = & \{(0, 0), (1, 2)\}, \end{array} \qquad \begin{array}{rcl} B & = & \{0, 1, 2\}, \\ C & = & \{0, 1\}, \\ G{:}\, B & \to & C, \\ G & = & \{(0, 1), (1, 0), (2, 1)\}, \end{array}$$

Their composition

$$(G \circ F){:}\, A \to C,$$

has values

$$\begin{array}{rcccccl} (G \circ F)(0) & = & G[F(0)] & = & G[0] & = & 1, \\ (G \circ F)(1) & = & G[F(1)] & = & G[2] & = & 1, \end{array}$$

so that

$$(G \circ F) = \{(0, 1), (1, 1)\}.$$

Theorem 380 *The composition of functions is associative: For all functions $F : A \to B$, $G : B \to C$, and $H : C \to D$,*

$$[H \circ (G \circ F)] = [(H \circ G) \circ F].$$

Proof. For each $X \in F^{\circ-1}"\{G^{\circ-1}"[\mathcal{D}(H)]\}$, apply the definition of $\circ$ repeatedly:

$$\begin{array}{rcl} [H \circ (G \circ F)](X) & = & H\{(G \circ F)(X)\} \\ & = & H\{G[F(X)]\} \\ & = & [H \circ G][F(X)] \\ & = & ([H \circ G] \circ F)(X). \end{array}$$

□

In contrast to its associativity, the composition of functions is *not* commutative.

Counterexample 381 Consider the following functions F and G:

$$\begin{array}{rcl} A & = & \{0,1\}, \\ B & = & \{0,1,2\}, \\ F\colon A & \to & B, \\ F & = & \{(0,0),(1,2)\}, \end{array} \qquad \begin{array}{rcl} B & = & \{0,1,2\}, \\ C & = & \{0,1\}, \\ G\colon B & \to & C, \\ G & = & \{(0,1),(1,0),(2,1)\}, \end{array}$$

Their composition

$$(F \circ G)\colon C \to A,$$

has values

$$\begin{array}{rcccccl} (F \circ G)(0) & = & F[G(0)] & = & F[1] & = & 2, \\ (F \circ G)(1) & = & F[G(1)] & = & F[0] & = & 0, \end{array}$$

so that

$$(F \circ G) = \{(0,2),(1,0)\}.$$

In contrast,

$$(G \circ F) = \{(0,1),(1,1)\}$$

from example 379, which confirms that $(F \circ G) \neq (G \circ F)$.

2.6.2 Injective, Surjective, Bijective, and Inverse Functions

Such problems as the solution of equations involve the determination of whether an equation has no solution, exactly one solution, or more than one solution, which correspond to the features of functions introduced here.

Definition 382 (Injectivity) A function $F : A \to B$ is **injective** if and only if for all $W \in \mathcal{D}(F)$ and $X \in \mathcal{D}(F)$, if $W \neq X$, then $F(W) \neq F(X)$:

$$\forall W \big[\forall X (\{[W \in \mathcal{D}(F)] \wedge [X \in \mathcal{D}(F)]\} \Rightarrow \{[(W \neq X) \Rightarrow [F(X) \neq F(W)]\})\big].$$

By contraposition, the condition just stated is equivalent to the following alternative condition: for all $W \in \mathcal{D}(F)$ and $X \in \mathcal{D}(F)$, if $F(W) = F(X)$, then $W = X$:

$$\forall W \big[\forall X (\{[W \in \mathcal{D}(F)] \wedge [X \in \mathcal{D}(F)]\} \Rightarrow \{[F(X) = F(W)] \Rightarrow (W = X)\})\big].$$

The notation $F : A \hookrightarrow B$ indicates that F is an injection.

Another common mathematical usage consists of saying that F maps only **one** X **to** each **one** Y; yet this alternative terminology fails to indicate which "one" it emphasizes (the *first* "one"), which leads to confusion, and, therefore, will not be adopted here.

Example 383 With $A := \{0,1,2\}$ and $B := \{0,1,2,3,4,5,6,7,8,9\}$, the function $G := \{(0,1),(1,3),(2,9)\}$ is injective.

Example 384 With $A := \{4,6,8,9\}$ and $B := \{2,3\}$, the function

$$H := \{(4,2),(6,3),(8,2),(9,3)\}$$

is *not* injective, because $4 \neq 8$ but $H(4) = 2 = H(8)$.

Definition 385 (Surjectivity) A function $F : A \to B$ is **surjective** if and only if for each $Y \in B$, there exists some $X \in A$ for which $Y = F(X)$. In other words, the condition just stated means that the range of F consists of all of the co-domain B:

$$\forall Y\{(Y \in B) \Rightarrow (\exists X\{(X \in A) \wedge [F(X) = Y]\})\}.$$

The notation $F : A \twoheadrightarrow B$ indicates that F is a surjection.

Another common mathematical usage consists of saying that F maps A **onto** B.

Example 386 With $A := \{0, 1, 2\}$ and $B := \{0, 1, 2, 3, 4, 5, 6, 7, 8, 9\}$, the function $G := \{(0, 1), (1, 3), (2, 9)\}$ is *not* surjective: there is *no* $X \in A$ with $G(X) = 6$.

Example 387 With $A := \{4, 6, 8, 9\}$ and $B := \{2, 3\}$, the function

$$H := \{(4, 2), (6, 3), (8, 2), (9, 3)\}$$

is surjective. Indeed, $H(4) = 2$ for $Y := 2$, and $H(6) = 3$ for $Y := 3$,

Example 388 With $A := \{6, 7, 8, 9\}$ and $B := \{1, 2, 3, 4\}$, the function

$$F := \{(6, 2), (7, 1), (8, 2), (9, 3)\}$$

is *neither* injective, because $F(6) = 2 = F(8)$ with $6 \neq 8$, *nor* surjective, because there does not exist any $X \in A$ with $F(X) = 4$.

Definition 389 (Bijectivity) A function $F : A \to B$ is **bijective** if and only if F is both injective and surjective, which can be denoted by $F : A \rightleftarrows B$ or $F : A \rightleftharpoons B$.

Example 390 With $A := \{0, 1, 2, 3\}$ and $B := \{1, 2, 4, 8\}$, the function $P : A \to B$ defined by $P := \{(0, 1), (1, 2), (2, 4), (3, 8)\}$ is bijective.

Example 391 For each set A the identity function $I_A : A \to A$, $X \mapsto X$ is bijective. Indeed, I_A is injective, because if $I_A(W) = I_A(X)$ then $W = I_A(W) = I_A(X) = X$. Similarly, I_A is surjective, because for each $Y \in A$ there exists $X \in A$, in effect $X := Y$, with $Y = I_A(Y)$.

Example 392 For each *proper* subset $S \subsetneq A$, the inclusion function $\iota : S \to A$, $X \mapsto X$, is injective but not surjective. Indeed, ι is injective, for if $W, X \in S$ and $W \neq X$, then $\iota(W) = W \neq X = \iota(X)$, whence $\iota(W) \neq \iota(X)$. However, ι is not surjective: because S is a proper subset of A, there exists some element Z in $A \setminus S$; in particular, $Z \neq X$ for each $X \in S$, and, consequently, $\iota(X) = X \neq Z$, which means that ι is not surjective. Thus, ι is not bijective either.

Example 393 For each non-empty set A and for each set B containing more than one element, the canonical projection $P_A : (A \times B) \to A$ is surjective but not injective. Indeed, B contains at least one element $Y \in B$, because $B \neq \varnothing$; hence, $X = P_A(X, Y)$ for each $X \in A$. However, B also contains some $Z \in B$ such that $Y \neq Z$. Consequently, $(X, Y) \neq (X, Z)$, and yet $P_A(X, Y) = X = P_A(X, Z)$, which means that P_A is not injective. Thus, P_A is not bijective either.

Theorem 394 *For all functions $F : A \to B$ and $G : B \to C$,*

- *if F and G are both injective, then $G \circ F$ is also injective;*

- *if F and G are both surjective, then $G \circ F$ is also surjective;*
- *if F and G are both bijective, then $G \circ F$ is also bijective;*
- *if $G \circ F$ is injective, then F is injective;*
- *if $G \circ F$ is surjective, then G is surjective;*
- *if $G \circ F$ is bijective, then F is injective and G is surjective.*

Proof. Assume that F and G are both injective. For all distinct elements $W \neq X$ in $F^{\circ -1}{}''[\mathcal{D}(G)]$, the injectivity of F ensures that $F(W) \neq F(X)$. Hence $G(F(W)) \neq G(F(X))$ by the injectivity of G. Hence, $(G \circ F)(W) = G(F(W)) \neq G(F(X)) = (G \circ F)(X)$, whence $(G \circ F)(W) \neq (G \circ F)(X)$, so that $G \circ F$ is injective.

Assume that F and G are both surjective. For each $Z \in C$, the surjectivity of G ensures the existence of an element $Y \in B$ such that $G(Y) = Z$. Hence, by the surjectivity of F, there exists an element $X \in A$ for which $F(X) = Y$. Therefore, $(G \circ F)(X) = G(F(X)) = G(Y) = Z$, which means that $(G \circ F)$ is surjective.

In particular, if F and G are both bijective, then $G \circ F$ is also bijective.

Assume that $G \circ F$ is injective. For all distinct elements $W \neq X$ in $F^{\circ -1}{}''[\mathcal{D}(G)]$, the injectivity of $G \circ F$ ensures that $(G \circ F)(W) \neq (G \circ F)(X)$, whence $G(F(W)) \neq G(F(X))$. Because G is a function, G cannot take different values at the same argument, whence it follows that $F(W) \neq F(X)$, so that F is injective.

Assume that $G \circ F$ is surjective. For each element $Z \in C$, the surjectivity of $G \circ F$ ensures the existence of an element $X \in A$ such that $(G \circ F)(X) = Z$. Hence, letting $Y := F(X)$ demonstrates the existence of an element $Y \in B$ for which $G(Y) = G(F(X)) = (G \circ F)(X) = Z$, which means that G is surjective.

In particular, if $G \circ F$ is bijective, then F is injective and G is surjective. □

Definition 395 (Invertibility) A function $F : A \to B$ is **invertible** if and only if there exists a function $G : B \to A$ for which $G \circ F = I_{\mathcal{D}(F)}$ and $F \circ G = I_{\mathcal{D}(G)}$. Such a function G is denoted by $F^{\circ -1}$ and called the **inverse function** of F. Thus,

$$\begin{aligned} F^{\circ -1} \circ F &= I_{\mathcal{D}(F)}, \\ F \circ F^{\circ -1} &= I_{\mathcal{D}(G)}. \end{aligned}$$

Example 396 With $A := \{0, 1, 2, 3\}$ and $B := \{1, 2, 4, 8\}$, the function

$$P := \{(0, 1), (1, 2), (2, 4), (3, 8)\}$$

is invertible, with inverse $P^{\circ -1} := \{(1, 0), (2, 1), (4, 2), (8, 3)\}$.

Theorem 397 *Each function $F : A \to B$ admits at most one inverse function. Moreover, if $G : B \to A$ is an inverse function for F, then G consists of all pairs obtained by swapping the coordinates in each pair of F.*

Proof. Assume that G is an inverse function for F, which means that $G \circ F = I_{\mathcal{D}(F)}$ and $F \circ G = I_{\mathcal{D}(G)}$, and consider the set

$$H := \{(Y, X) : (X, Y) \in F\}.$$

If $(X, Y) \in F$, then $X \in \mathcal{D}(F)$. Because $G \circ F = I_{\mathcal{D}(F)}$, it follows that there exists some $Z \in B$ such that $(X, Z) \in F$ and $(Z, X) \in G$. With $(X, Y) \in F$ and $(X, Z) \in F$, it follows that $Y = Z$, because F is a function. This shows that if $(Y, X) \in H$, so that $(X, Y) \in F$, then $(Y, X) = (Z, X) \in G$, whence $G \subseteq H$.

Conversely, If $(Z,X) \in G$, then $Z \in \mathcal{D}(G)$. Because $F \circ G = I_{\mathcal{D}(G)}$ it follows that there exists some $W \in \mathcal{D}(F)$ such that $(Z,W) \in G$ and $(W,Z) \in F$. With $(Z,X) \in G$ and $(Z,W) \in G$, it follows that $W = X$, because G is a function. This shows that if $(Z,X) \in G$ then $(X,Z) \in F$, whence $(Z,X) \in H$, so that $G \subseteq H$.

Finally, $G = H$, which shows that if F has an inverse function G, then the only possibility is $G = H$. Thus, $G = H = F^{\circ -1} = \{(Y,X) : (X,Y) \in F\}$. □

Theorem 398 *For each function $F : A \to B$ with $\mathcal{D}(F) = A$, the function F is invertible if and only if F is bijective.*

Proof. Assume that F is invertible, with inverse $G := F^{\circ -1}$. Because $F \circ F^{\circ -1} = I_{\mathcal{D}(G)}$ is surjective, it follows from theorem 394 that F is surjective. Similarly, because $F^{\circ -1} \circ F = I_{\mathcal{D}(F)}$ is injective, it follows from theorem 394 that F is injective.

Conversely, assume that F is bijective. Construct an inverse function by means of the set G defined by swapping both coordinates in each pair of the function F:

$$G := \{(Y,X) : (X,Y) \in F\}.$$

Then verify that G is the inverse function of F.

Firstly, the injectivity of F ensures that G is a function: if $(Y,X) \in G$ and $(Y,W) \in G$, then $(X,Y) \in F$ and $(W,Y) \in F$, whence $X = W$ by injectivity.

Secondly, $G \circ F = I_{\mathcal{D}(F)}$. Indeed, if $(X,Z) \in (G \circ F)$, then there exists some $Y \in B$ for which $(X,Y) \in F$ and $(Y,Z) \in G$. Consequently, $(Z,Y) \in F$, and again the injectivity of F shows that $X = Z$, whence $(X,Z) = (X,X)$. Thus, $(G \circ F) \subseteq I_{\mathcal{D}(F)}$. Because the foregoing reasoning holds for each $X \in \mathcal{D}(F)$, however, it also follows that $I_{\mathcal{D}(F)} \subseteq (G \circ F)$, and thus $(G \circ F) = I_{\mathcal{D}(F)}$.

Finally, $F \circ G = I_{\mathcal{D}(G)}$. Indeed, if $(Y,W) \in (F \circ G)$, then there exists some $X \in \mathcal{D}(G)$ with $(Y,X) \in G$ and $(X,W) \in F$. From $(Y,X) \in G$, it follows that $(X,Y) \in F$. Because F is a function, it also follows that $Y = W$, whence $(Y,W) = (Y,Y)$. Thus, $(F \circ G) \subseteq I_{\mathcal{D}(G)}$. Then the surjectivity of F guarantees that for each $Y \in \mathcal{D}(G)$ there exists some $X \in A$ with $(X,Y) \in F$. Hence, $(Y,X) \in G$ and then $(Y,Y) \in (F \circ G)$, so that $I_B \subseteq (F \circ G)$. Therefore, $(F \circ G) = I_B$. □

Some situations involve a concept more general than invertibility.

Definition 399 (Left or right invertibility) A function $F : A \to B$ is **invertible on its left**, or **left invertible**, if and only if there exists a function $G : B \to A$ for which $G \circ F = I_{\mathcal{D}(F)}$. Such a function G is called a **left inverse function** for F.

Similarly, a function $F : A \to B$ is **invertible on its right**, or **right invertible**, if and only if there exists a function $G : B \to A$ for which $F \circ G = I_{\mathcal{D}(G)}$. Such a function G is called a **right inverse function** for F.

Example 400 With $A := \{0,1,2\}$ and $B := \{0,1,2,3,4,5,6,7,8,9\}$, the function $F : A \to B$ defined by $F := \{(0,1),(1,3),(2,9)\}$ has a left inverse function

$$G := \{(1,0),(2,0),(3,1),(4,0),(5,0),(6,0),(7,0),(8,0),(9,2)\}.$$

Example 401 For $A := \{4,6,8,9\}$ and $B := \{2,3\}$, the function $F : A \to B$ with $F := \{(4,2),(6,3),(8,2),(9,3)\}$ has a right inverse $G := \{(2,4),(3,6)\}$

Example 402 Perspectives to draw a picture of space A on a flat screen $B \subset A$ can be represented by a function $F : A \to B$, mapping each point X in space A to its image $F(X)$ on the screen B. For such perspectives, each point $Y \in B$ on the screen B is its own image, so that $F(Y) = Y$. Thus the inclusion function $\iota_B : B \to A$ is a right inverse for F, because

$(F \circ \iota_B)(Y) = F[\iota_B(Y)] = F(Y) = Y$ for every $Y \in B$, so that $F \circ \iota_B = I_B$. In contrast, such a perspective F has no left inverse, because F maps many points in space to the same image on the screen.

The existence of a *left*-inverse $G : B \to A$ for a function $F : A \to B$ indicates that for each $Y \in B$ the equation $F(X) = Y$ has *at most* one solution. In contrast, the existence of a *right*-inverse $G : B \to A$ for a function $F : A \to B$ indicates that for each $Y \in B$ the equation $F(X) = Y$ has *at least* one solution.

2.6.3 Exercises

Exercise 551 For each function $F : A \to B$ and $I_B : B \to B$, prove that $I_B \circ F = F$.

Exercise 552 For each function $F : A \to B$ and $I_A : A \to A$, prove that $F \circ I_A = F$.

Exercise 553 For each function $F : A \to B$ and $\varnothing : \varnothing \to A$, prove that $F \circ \varnothing = \varnothing$.

Exercise 554 Prove that if a function has a left inverse, then it is injective.

Exercise 555 Prove that if a function is injective, then it has a left inverse.

Exercise 556 Prove that if a function has a right inverse, then it is surjective.

Exercise 557 Provide an example of a function with a right inverse but no left inverse.

Exercise 558 Provide an example of a function with a left inverse but no right inverse.

Exercise 559 Provide an example of a function that has more than one left inverse.

Exercise 560 Provide an example of a function that has more than one right inverse.

2.7 EQUIVALENCE RELATIONS

2.7.1 Reflexive, Symmetric, Transitive, or Anti-Symmetric Relations

Besides functions, mathematics contains several other types of relations. For instance, ordering relations define orders or rankings, whereas equivalence relations define equivalences relative to certain criteria. Such various types of relations can be defined by combinations of several features called reflexivity, symmetry, and transitivity.

Definition 403 (Reflexivity) For each set A, a relation $R \subseteq A \times A$ is **reflexive** if and only if $(X, X) \in R$ for each $X \in A$:

$$\forall X\{(X \in A) \Rightarrow [(X, X) \in R]\}.$$

Theorem 404 *A relation $\mathcal{R} \subseteq A \times A$ is reflexive if and only if $\Delta_A \subseteq \mathcal{R}$.*

Proof. A relation $\mathcal{R} \subseteq A \times A$ is reflexive if and only if $(X, X) \in \mathcal{R}$, for every $X \in A$, in other words, if and only if $\mathcal{R}$ contains the set $\{(X, X) : X \in A\} = \Delta_A$. □

Example 405 For each set A, the diagonal Δ_A is a reflexive relation.

Example 406 For each set A and its power set $\mathcal{P}(A)$, the relation $\subseteq$ on $\mathcal{P}(A)$ is reflexive. Indeed, $B \subseteq B$ for each $B \in \mathcal{P}(A)$.

Counterexample 407 The relation of membership $\in$ is *not* reflexive. For instance, if $A = \{\varnothing\}$, then $\varnothing \notin \varnothing$. Consequently, $\in$ is *not* reflexive on $A \times A$.

Definition 408 (Symmetry) For each set A, a relation $R \subseteq A \times A$ is **symmetric** if and only if $(X, Y) \in R$ is equivalent to $(Y, X) \in R$ for all $X \in A$ and $Y \in A$:

$$\forall X \forall Y(\{(X \in A) \wedge (Y \in A)\} \Rightarrow \{[(X, Y) \in R] \Leftrightarrow [(Y, X) \in R]\}).$$

Example 409 For each set A, the diagonal Δ_A is a symmetric relation.

Counterexample 410 The relation of membership $\in$ is *not* symmetric in general: if $A := \{\, \varnothing,\ \{\varnothing\}\, \}$, then $\varnothing \in \{\varnothing\}$ but $\{\varnothing\} \notin \varnothing$; thus $\in$ is *not* symmetric.

Definition 411 (Transitivity) For each set A, a relation $R \subseteq A \times A$ is **transitive** if and only if $(X, Y) \in R$ and $(Y, Z) \in R$ imply $(X, Z) \in R$ for all $X, Y, Z \in A$:

$$\begin{aligned}&\forall X \forall Y \forall Z\\ &\{[(X \in A) \wedge (Y \in A) \wedge (Z \in A)]\\ &\Rightarrow (\{[(X, Y) \in R] \wedge [(Y, Z) \in R]\} \Rightarrow [(X, Z) \in R])\}\end{aligned}$$

Example 412 For each set A, the diagonal Δ_A is a transitive relation.

Example 413 For each set A, the relation $\subseteq$ on $\mathcal{P}(A)$ is transitive: for all $U \in \mathcal{P}(A)$, $V \in \mathcal{P}(A)$, $W \in \mathcal{P}(A)$, if $U \subseteq V$ and $V \subseteq W$, then $U \subseteq W$.

Counterexample 414 The relation of membership $\in$ is *not* transitive in general: if $A = \left\{\, \varnothing,\ \{\varnothing\},\ \{\{\varnothing\}\}\, \right\}$, then $\varnothing \in \{\varnothing\}$ and $\{\varnothing\} \in \{\{\varnothing\}\}$, but $\varnothing \notin \{\{\varnothing\}\}$.

2.7.2 Partitions and Equivalence Relations

The concept of an equivalence relation on a set corresponds to a "partition" of that set into a union of disjoint subsets called "equivalence" classes.

Definition 415 (Equivalence relations) For each set A, a relation $R \subseteq A \times A$ is an **equivalence relation** if and only if R is *reflexive, symmetric, and transitive*.

Example 416 For each set A, the diagonal Δ_A is an equivalence relation.

Example 417 The set $A := \{0, 1, 2, 3\}$ admits an equivalence relation

$$\begin{aligned}\mathcal{R} &:= \{(0,0), (1,1), (2,2), (3,3), (0,2), (2,0), (1,3), (3,1)\}\\ &= \Delta_A \cup \{(0,2), (2,0), (1,3), (3,1)\}.\end{aligned}$$

- The relation $\mathcal{R}$ is reflexive because it contains the diagonal

$$\Delta_A = \{(0,0), (1,1), (2,2), (3,3)\}.$$

- The relation $\mathcal{R}$ is symmetric: it contains $(0,2)$, $(2,0)$, as well as $(1,3)$, $(3,1)$.
- The relation $\mathcal{R}$ is transitive because if it contains (X, Y) and (Y, Z), then it contains (X, Z), for instance, $(0,2)$ and $(2,0)$ hence $(0,0)$; $(2,0)$ and $(0,2)$ hence $(2,2)$, as well as $(1,3)$ and $(3,1)$ hence $(1,1)$; $(3,1)$ and $(1,3)$ hence $(3,3)$.

Counterexample 418 The relation of membership $\in$ is *not* an equivalence relation in general, because it is *not* symmetric and *not* transitive.

Definition 419 (Partition) A **partition** of a set A is a set $\mathcal{F} \subseteq \mathcal{P}(A)$ of subsets of A with all of the following properties.

- No member of $\mathcal{F}$ is empty: $\forall V\{(V \in \mathcal{F}) \Rightarrow (V \neq \varnothing)\}$.
- The unions of all the members of $\mathcal{F}$ "covers" A, which means that $A = (\bigcup \mathcal{F})$.
- All pairs of distinct members of $\mathcal{F}$ are disjoint:

$$\forall V\big(\forall W\{[(V \in \mathcal{F}) \wedge (W \in \mathcal{F}) \wedge (V \neq W)] \Rightarrow [(V \cap W) = \varnothing]\}\big).$$

Example 420 The empty set $\varnothing$ admits only one partition: the empty set $\mathcal{F} = \varnothing$.

Example 421 Each non-empty set S has a partition: the singleton $\mathcal{F} = \{S\}$.

Example 422 The set $A := \{0, 1, 2, 3\}$ admits a partition $\mathcal{F}$ into two disjoint sets, $\mathcal{F} := \{\ \{0, 2\},\ \{1, 3\}\ \}$:

- No member of $\mathcal{F}$ is empty: $\{0, 2\} \neq \varnothing$ and $\{1, 3\} \neq \varnothing$.
- The unions of the members of $\mathcal{F}$ covers A, so that $\{0, 1, 2, 3\} = \{0, 2\} \cup \{1, 3\}$.
- Distinct members of $\mathcal{F}$ are disjoint: $\{0, 2\} \cap \{1, 3\} = \varnothing$.

Definition 423 (Relations from partitions) For each partition $\mathcal{F} \subseteq \mathcal{P}(A)$ of each set A, define a relation $R_{\mathcal{F}}$ on A so that R relates elements $X \in A$ and $Y \in A$ if and only if X and Y belong to the same element of the partition $\mathcal{F}$:

$$[(X, Y) \in R_{\mathcal{F}}] \Leftrightarrow \{\exists B[(B \in \mathcal{F}) \wedge (X \in B) \wedge (Y \in B)]\}.$$

Theorem 424 (Equivalence relations from partitions) *For each set A and for each partition $\mathcal{F}$ of A, the relation $R_{\mathcal{F}}$ is an equivalence relation.*

Proof. The relation $R_{\mathcal{F}}$ is reflexive: for each $X \in A$, the partition $\mathcal{F}$ has an element $B \in \mathcal{F}$ that contains X, because $\mathcal{F}$ covers A. Thus X and X both belong to the same $B \in \mathcal{F}$, whence $(X, X) \in R_{\mathcal{F}}$. Symbolically,

$$\begin{array}{rl} \forall X\{(X \in A) \Rightarrow [(X, X) \in R_{\mathcal{F}}]\} & \textbf{yet unproved,} \\ \Updownarrow & \text{definition 423 for } R_{\mathcal{F}}, \\ \forall X\big(\{X \in A\} \Rightarrow & \\ \{\exists B[(B \in \mathcal{F}) \wedge (X \in B) \wedge (X \in B)]\}\big) & \\ \Updownarrow & \text{tautology } [(P) \wedge (P)] \Leftrightarrow (P), \\ \forall X\big(\{X \in A\} \Rightarrow & \\ \{\exists B[(B \in \mathcal{F}) \wedge (X \in B)]\}\big) & \\ \Updownarrow & \text{definition of } \bigcup. \\ A \subseteq (\bigcup \mathcal{F}) & \end{array}$$

which is universally valid by definition of a partition.

The relation $R_{\mathcal{F}}$ is symmetric: $(X, Y) \in R_{\mathcal{F}}$ if and only if the partition $\mathcal{F}$ has an element $B \in \mathcal{F}$ that contains X and Y, whence B contains Y and X, which means that $(Y, X) \in R_{\mathcal{F}}$. Symbolically, from the tautology $[(P) \wedge (Q)] \Leftrightarrow [(Q) \wedge (P)]$,

$$\begin{array}{rl} \models \forall X \forall Y\Big\{[(X \in A) \wedge (Y \in A)] \Rightarrow & \\ \big(\{\exists B[(B \in \mathcal{F}) \wedge (X \in B) \wedge (Y \in B)]\} \Leftrightarrow & \\ \{\exists B[(B \in \mathcal{F}) \wedge (Y \in B) \wedge (X \in B)]\}\big)\Big\} & \\ \Updownarrow & \text{definition 423 for } R_{\mathcal{F}}. \\ \forall X\big[\forall Y\big([(X \in A) \wedge (Y \in A)] \Rightarrow & \\ \{[(X, Y) \in R_{\mathcal{F}}] \Leftrightarrow [(Y, X) \in R_{\mathcal{F}}]\}\big)\big] & \end{array}$$

The relation $R_{\mathcal{F}}$ is transitive: if $(X,Y) \in R_{\mathcal{F}}$, then the partition $\mathcal{F}$ has an element $B \in \mathcal{F}$ that contains X and Y. If also $(Y,Z) \in R_{\mathcal{F}}$, then the partition $\mathcal{F}$ has an element $C \in \mathcal{F}$ that contains Y and Z. However, from $X \in B$ and $Y \in C$ it follows that $X \in (B \cap C)$, whence B and C are not disjoint and hence $B = C$, by definition of a partition. Consequently, $X \in B$ and $Z \in B$, and, therefore, $(X,Z) \in R_{\mathcal{F}}$. □

Definition 425 (Equivalence classes) For each equivalence relation $R \subseteq (A \times A)$, and for each element $X \in A$, define the subset $[X]_R$ of all the elements $Y \in A$ equivalent to X with respect to R, called the **equivalence class of** X:

$$[X]_R := \{Y \in A : (X,Y) \in R\}.$$

Then let $\mathcal{F}_R$ consist of all such equivalence classes:

$$\mathcal{F}_R := \{[X]_R \in \mathcal{P}(A) : X \in A\}.$$

Another common notation for $\mathcal{F}_R$ is A/R, so that $\mathcal{F}_R = A/R = \{[X]_R : X \in A\}$. The set $\mathcal{F}_R = A/R$ of all equivalence classes is also called the **quotient** of the set A by the relation R.

Example 426 For each set A, the equivalence classes of the "diagonal" equivalence relation Δ_A consist of every singleton $[X]_{\Delta_A} = \{X\}$ for every $X \in A$. Thus $A/\Delta_A = \{\,\{X\} : X \in A\}$.

Example 427 For the set $A := \{0,1,2,3\}$, the equivalence relation

$$\begin{aligned} \mathcal{R} &:= \{(0,0),(1,1),(2,2),(3,3),(0,2),(2,0),(1,3),(3,1)\} \\ &= \Delta_A \cup \{(0,2),(2,0),(1,3),(3,1)\} \end{aligned}$$

corresponds to the equivalence classes

$$\begin{aligned} [0]_{\mathcal{R}} &= \{0,2\}, \\ [1]_{\mathcal{R}} &= \{1,3\}. \end{aligned}$$

Thus $A/\mathcal{R} = \{[0]_{\mathcal{R}},\ [1]_{\mathcal{R}}\}$.

Theorem 428 (Partitions from equivalence relations) *For each equivalence relation $R \subseteq (A \times A)$, the set of subsets $\mathcal{F}_R \subseteq \mathcal{P}(A)$ is a partition of A.*

Proof. The partition $\mathcal{F}_R$ covers A: the reflexivity of R guarantees that $(X,X) \in R$ for each $X \in A$, whence $X \in [X]_R$, and hence $X \in (\bigcup_{Z \in A} B_Z) = (\bigcup \mathcal{F}_R)$. Thus $A \subseteq (\bigcup \mathcal{F})$. The reverse inclusion follows from $(\bigcup \mathcal{F}) \subseteq [\bigcup \mathcal{P}(A)] = A$:

$$\begin{array}{rl} \models \forall X\{(X \in A) \Rightarrow [(X,X) \in R]\} & \text{reflexivity of } R, \\ \Updownarrow & \text{definition of } [X]_R, \\ \forall X\{(X \in A) \Rightarrow (X \in [X]_R)\} & \\ \Updownarrow & \text{definition of } \mathcal{F}_R, \\ \forall X\{(X \in A) \Rightarrow [([X]_R \in \mathcal{F}_R) \wedge (X \in [X]_R)]\} & \\ \Updownarrow & \text{definition of } \exists, \\ \forall X[(X \in A) \Rightarrow \{\exists B[(B \in \mathcal{F}_R) \wedge (X \in B)]\}] & \\ \Updownarrow & \text{definition of } \bigcup, \\ \forall X\{(X \in A) \Rightarrow [X \in (\bigcup \mathcal{F})]\} & \\ \Updownarrow & \text{definition of } \subseteq. \\ A \subseteq (\bigcup \mathcal{F})] & \end{array}$$

Any two distinct elements of $\mathcal{F}_R$ are disjoint: if two members B and C of $\mathcal{F}_R$ are not disjoint, then their intersection $B \cap C$ contains an element $X \in A$; by definition of $\mathcal{F}_R$, however, every element of B is equivalent to X, and so is every element of C, whence $B = [X]_R = C$, which is a negation of the distinctness of B and C. Finally, $\mathcal{F}_R$ does not contain any empty element. Indeed, if $A = \varnothing$, then $R \subseteq A \times A$, whence $R = \varnothing$ and $\mathcal{F}_R = \varnothing$, which does not contain any element, and hence no empty element. If $A \neq \varnothing$, then $\mathcal{F}_R = \{[X]_R : X \in A\}$ where $X \in [X]_R$, whence $[X]_R \neq \varnothing$. □

Example 429 For each equivalence relation $R \subseteq A \times A$ on a set A, the **canonical map,** also called the **quotient map,** is the function $P : A \to A/R$ that maps each element $X \in A$ to its equivalence class $[X]_R = \{Y \in A : (X, Y) \in R\}$:

$$\begin{aligned} P\colon A &\to A/R, \\ X &\mapsto [X]_R. \end{aligned}$$

2.7.3 Exercises

Exercise 561 Prove that the empty relation $\varnothing \subseteq A \times A$ is reflexive, symmetric, and transitive.

Exercise 562 Prove that for each set A the relation of strict inclusion $\subsetneqq$ on $\mathcal{P}(A)$ is *not* reflexive.

Exercise 563 Prove that for each set A the relation of strict inclusion $\subsetneqq$ on $\mathcal{P}(A)$ is antisymmetric.

Exercise 564 Prove that for each set A the relation of strict inclusion $\subsetneqq$ on $\mathcal{P}(A)$ is transitive.

Exercise 565 For the set $A := \{0, 1, 2, 3, 4, 5\}$, verify that the following relation $\mathcal{R}$ is an equivalence relation, and list all its equivalence classes:

$$\mathcal{R} := \left\{ \begin{array}{cccccc} & (1,5) & & (3,5) & & (5,5) \\ (0,4) & & (2,4) & & (4,4) & \\ & (1,3) & & (3,3) & & (5,3) \\ (0,2) & & (2,2) & & (4,2) & \\ & (1,1) & & (3,1) & & (5,1) \\ (0,0) & & (2,0) & & (4,0) & \end{array} \right\}.$$

Exercise 566 For the set $A := \{0, 1, 2, 3, 4, 5\}$, verify that the following relation $\mathcal{S}$ is an equivalence relation, and list all its equivalence classes:

$$\mathcal{S} := \left\{ \begin{array}{cccccc} & & (5,2) & & & (5,5) \\ & (1,4) & & & (4,4) & \\ (0,3) & & & (3,3) & & \\ & & (2,2) & & & (2,5) \\ & (1,1) & & & (4,1) & \\ (0,0) & & & (3,0) & & \end{array} \right\}.$$

Exercise 567 For the set $B := \{0, 1, 2, 3, 4, 5, 6, 7\}$, verify that the following set $\mathcal{F}$ of subsets is a partition, and list the corresponding equivalence relation:

$$\mathcal{F} := \{ \{0, 2, 4, 6\}, \{1, 3, 5, 7\} \}.$$

Exercise 568 For the set $B := \{0, 1, 2, 3, 4, 5, 6, 7\}$, verify that the following set $\mathcal{G}$ of subsets is a partition, and list the corresponding equivalence relation:

$$\mathcal{G} := \{ \{0,4\}, \{1,5\}, \{2,6\}, \{3,7\} \}.$$

Exercise 569 Prove that $R_{(\mathcal{F}_R)} = R$ for each equivalence relation R.

Exercise 570 Prove that $\mathcal{F}_{(R_\mathcal{F})} = \mathcal{F}$ for each partition $\mathcal{F}$.

2.8 ORDERING RELATIONS

2.8.1 Preorders and Partial Orders

Besides the reflexivity, symmetry, and transitivity used to define equivalence relations, such other types of relations as rankings also require different variations of these concepts, as introduced here (with the terminology of Suppes [95, §3.2, p. 69]).

Definition 430 (Strict, irreflexivity) For each set A, a relation $R \subseteq A \times A$ is **irreflexive**, or, equivalently, **strict**, if and only if R does *not* relate any element of A to itself:

$$\forall X\{(X \in A) \Rightarrow [(X,X) \notin R]\}.$$

Example 431 The empty relation $\varnothing$ is strict, because the conclusion $(X, X) \notin \varnothing$ is universally valid.

Example 432 For each set A, the relation of strict inclusion $\subset$ is strict on $\mathcal{P}(A)$. Indeed, for all subsets $V \subseteq A$ and $W \subseteq A$, the definition of $V \subset W$ includes the requirement that $V \neq W$, so that $(V \subset W) \Rightarrow (V \neq W)$. Contraposition then confirms that $(V = W) \Rightarrow (V \not\subset W)$, so that $V \not\subset V$.

One method to specify a ranking or direction on a set removes the requirement of symmetry from the concept of equivalence, which gives the following concept of preorder.

Definition 433 (Preorder or quasi-order) For each set A, a relation $R \subseteq A \times A$ is a **preorder** or a **quasi-order**, if and only if R is *reflexive and transitive*. It is a **strict preorder** if and only if R is *irreflexive and transitive*.

Example 434 Consider the set $A := \{0, 1, 2\}$.
The relation $\mathcal{Q} := \{(0,0), (0,1), (1,0), (1,1), (2,2)\}$ is a preorder.
The relation $\mathcal{R} := \{(0,0), (0,1), (1,1), (2,2)\}$ is a preorder.
The relation $\mathcal{S} := \{(0,1)\}$ is a strict preorder.

Example 435 For each set A, the diagonal Δ_A is a preorder. If $A \neq \varnothing$, then Δ_A is *not* strict, because there exists $X \in A$ and then $(X, X) \in \Delta_A$.

Example 436 For each set A the relation $\subseteq$ is a preorder on $\mathcal{P}(A)$. The relation $\subseteq$ is *not* strict because $\varnothing \in \mathcal{P}(A)$ and $\varnothing \subseteq \varnothing$.

Example 437 For each set A, the relation $\subset$ is a *strict* preorder on $\mathcal{P}(A)$.

The concept of a preorder R allows for "circular" rankings, with $(X,Y) \in R$ and $(Y,X) \in R$ even though $X \neq Y$. To specify different types of rankings, a different concept — anti-symmetry — becomes necessary.

Definition 438 (Anti-Symmetry) For each set A, a relation $R \subseteq A \times A$ is **anti-symmetric** if and only if $(X, Y) \in R$ and $(Y, X) \in R$ imply $X = Y$ for all $X, Y \in A$:

$$\forall X \left(\forall Y \left[\{[(X,Y) \in R] \wedge [(Y,X) \in R]\} \Rightarrow (X = Y)\right]\right).$$

Example 439 For each set A, the diagonal Δ_A is an anti-symmetric relation.

Example 440 For each set A, the relation $\subseteq$ on $\mathcal{P}(A)$ is antisymmetric. Indeed, for each $B \in \mathcal{P}(A)$ and for each $C \in \mathcal{P}(A)$ if $B \subseteq C$ and $C \subseteq B$, then $B = C$.

Similar features can also be defined through the following concept of asymmetry.

Definition 441 (Asymmetry) For each set A, a relation $R \subseteq A \times A$ is **asymmetric** if and only if $(X, Y) \in R$ implies $(Y, X) \notin R$ for each $X \in A$ and for each $Y \in A$:

$$\forall X \left(\forall Y \{[(X,Y) \in R] \Rightarrow [(Y,X) \notin R]\}\right).$$

Example 442 Consider the set $A := \{0, 1, 2\}$.

The relation $\mathcal{Q} := \{(0,0), (0,1), (1,0), (1,1), (2,2)\}$ is *not* asymmetric: $\mathcal{Q}$ contains $(0,1)$ and $(1,0)$.

The relation $\mathcal{R} := \{(0,0), (0,1), (1,1), (2,2)\}$ is *not* asymmetric: $\mathcal{R}$ contains $(0,0)$, $(1,1)$, and $(2,2)$.

The relation $\mathcal{S} := \{(0,1)\}$ is asymmetric.

In contrast to preorders, "partial orders" do *not* allow for circular rankings.

Definition 443 (Partial order) For each set A, a relation $R \subseteq A \times A$ is a **partial order** if and only if R is *reflexive, anti-symmetric, and transitive*. It is a **strict partial order** if and only if R is *irreflexive and transitive*.

Example 444 Consider the set $A := \{0, 1, 2\}$.

The relation $\mathcal{Q} := \{(0,0), (0,1), (1,0), (1,1), (2,2)\}$ is *not* a partial order, because it is *not* anti-symmetric: $\mathcal{Q}$ contains $(0,1)$ and $(1,0)$ but $0 \neq 1$.

The relation $\mathcal{R} := \{(0,0), (0,1), (1,1), (2,2)\}$ is a partial order.

The relation $\mathcal{S} := \{(0,1)\}$ is a strict partial order.

Example 445 For each set A, the diagonal Δ_A is a partial order.

Example 446 For each set A the relation $\subseteq$ is a partial order on $\mathcal{P}(A)$.

Example 447 For each set A the relation $\subsetneq$ is a *strict* partial order on $\mathcal{P}(A)$.

As their names might suggest, neither preorders nor partial orders need relate all elements to one another. Relations that do so are called strongly connected.

Definition 448 (Strong connectivity) For each set A, a relation $R \subseteq A \times A$ is **strongly connected** if and only if $(X, Y) \in R$ or $(Y, X) \in R$ (including both possibilities) for each $X \in A$ and for each $Y \in A$:

$$\forall X \left[\forall Y \left([(X \in A) \wedge (Y \in A)] \Rightarrow \{[(X,Y) \in R] \vee [(Y,X) \in R]\}\right)\right].$$

Strict rankings replace strong connectivity by connectivity.

Definition 449 (Connectivity) For each set A, a relation $R \subseteq A \times A$ is **connected** if and only if $(X,Y) \in R$ or $(Y,X) \in R$ (including both possibilities) for all *distinct* $X \in A$ and $Y \in A$ (such that $X \neq Y$):

$$\forall X \left[\forall Y \left([(X \in A) \wedge (Y \in A) \wedge (X \neq Y)] \Rightarrow \{[(X,Y) \in R] \vee [(Y,X) \in R]\}\right)\right].$$

Example 450 Consider the set $A := \{0, 1, 2\}$.

The relation $\mathcal{Q} := \{(0,0),(0,1),(1,0),(1,1),(2,2)\}$ is neither connected nor strongly connected, because it contains neither $(0,2)$ nor $(2,0)$.

The relation $\mathcal{T} := \{(0,0),(0,1),(0,2),(1,1),(1,2),(2,2)\}$ is strongly connected.

2.8.2 Total Orders and Well-Orderings

The geometric "direction along a line" corresponds to a **total order**, also called **linear order**, **complete order** [59, p. 14], or **simple order** [59, p. 14], [95, p. 69].

Definition 451 (Total order) For each set A, a relation $R \subseteq A \times A$ is a **total order,** or **total ordering,** if and only if R is a *strongly connected partial order* (strongly connected, reflexive, anti-symmetric, and transitive). It is a **strict total order** if and only if R is *connected, irreflexive, and transitive*.

Example 452 The empty relation on the empty set is a *strict* total order.

Example 453 For each set S and the singleton $A = \{S\}$, the relation $\subseteq$ on $\mathcal{P}(A) = \{\varnothing,\ \{S\}\}$, is a total order. Indeed, $\varnothing \subseteq \varnothing$, $\varnothing \subseteq \{S\}$, and $\{S\} \subseteq \{S\}$.

Counterexample 454 The relation $\subset$ is *not* a total order in general. For example, if

$$A := \{\ \varnothing,\ \{\varnothing\}\ \},$$

then $\subset$ is *not* a total order on the power set

$$\mathcal{P}(A) = \Big\{\ \varnothing,\ \{\varnothing\},\ \{\ \{\varnothing\}\ \},\ \{\ \varnothing,\ \{\varnothing\}\ \}\ \Big\},$$

because

$$\begin{aligned} \{\varnothing\} &\not\subset \{\ \{\varnothing\}\ \}, \\ \{\ \{\varnothing\}\ \} &\not\subset \{\varnothing\}. \end{aligned}$$

Thus the relation $\subset$ contains neither the pair $(\ \{\varnothing\},\ \{\ \{\varnothing\}\ \}\ )$ nor the pair $(\ \{\ \{\varnothing\}\ \},\ \{\varnothing\}\ )$. Instead, for this set A the relation $\subset$ takes the following form:

$$\begin{array}{ccc} \{\ \{\varnothing\}\ \} & \subset & \{\ \varnothing,\ \{\varnothing\}\ \} \\ \cup & & \cup \\ \varnothing & \subset & \{\varnothing\} \end{array}$$

Some relations that are not total orders can restrict to total orders on subsets.

Definition 455 (Chain) For each set A, and for each relation $R \subseteq A \times A$, a subset $B \subseteq A$ is a **chain** if and only if the restriction $R|_B$ is a total order on B.

In particular, for each total order R on each set A, the set A is a chain relative to R.

In partially ordered sets, a subset might have a first element, which precedes every element of that subset, or a last element, which follows every element of that subset.

Definition 456 (First or last element) For each set A, for each subset $B \subseteq A$, and for each partial order $R \subseteq A \times A$, an element $X \in B$ is a **first, or smallest, element in** B if and only if $(X, Y) \in R$ for each $Y \in B$. Similarly, an element $Z \in B$ is a **last, or largest, element in** B if and only if $(Y, Z) \in R$ for each $Y \in B$.

Example 457 For each set A and for the relation $\subseteq$ on $\mathcal{P}(A)$, the element $\varnothing \in \mathcal{P}(A)$ is a first element of $\mathcal{P}(A)$. Indeed, $\varnothing \subseteq C$ for each $C \in \mathcal{P}(A)$. Also, the element $A \in \mathcal{P}(A)$ is a last element of $\mathcal{P}(A)$. Indeed, $C \subseteq A$ for each $C \in \mathcal{P}(A)$.

Definition 458 (Well-ordering) For each set A and for each partial order $R \subseteq A \times A$, the set A is **well-ordered** by R if and only if each non-empty subset $B \subseteq A$ has a first element. A relation R is called a **well-ordering** (for the lack of a grammatical and logically equivalent terminology) if and only if A is well-ordered by R.

Remark 459 Instead of requiring that a well-ordering be a partial order [19, p. 31], some texts impose the stronger requirement that a well-ordering be a total order [59, p. 29]. Yet the insistence on a total order is redundant; indeed, if each non-empty subset has a first element, then a partial order is automatically a total order [95, p. 74–76]. □

Example 460 The set $\{\varnothing\} = \mathcal{P}(\varnothing)$ is well-ordered by the relation of inclusion $\subseteq$. Indeed, the only non-empty subset of $\{\varnothing\} = \mathcal{P}(\varnothing)$ is $\{\varnothing\}$, and it has a first element, namely $\varnothing$, because $\varnothing \subseteq C$ for each $C \in \{\varnothing\}$.

Definition 461 (Upper or lower bound) For each set A, for each subset $B \subseteq A$, and for each partial order $R \subseteq A \times A$, an element $Z \in A$ is an **upper bound for** B if and only if $(Y, Z) \in R$ for each $Y \in B$. Similarly an element $X \in A$ is a **lower bound for** B if and only if $(X, Y) \in R$ for each $Y \in B$.

Thus a last element can differ from an upper bound because an upper bound need not belong to the same subset. Similarly a first element can differ from a lower bound because a lower bound need not belong to the same subset.

Definition 462 (Maximal element) For each set A, for each subset $B \subseteq A$, and for each partial order $R \subseteq A \times A$, an element $Z \in B$ is a **maximal element of** B if and only if $[(Z, Y) \in R] \Rightarrow [(Y, Z) \in R]$ for each $Y \in B$. In other words, $Z \in B$ is a maximal element of B if and only if every element $Y \in B$ that follows Z also precedes Z (which allows for the possibility that no $Y \in B$ follows Z).

Thus, either a maximal element precedes no element, or it follows every element that it precedes, but in either case it belongs to the same subset. For instance, a last element is a maximal element. However, a maximal element need not be a last element.

Example 463 Consider the set $A := \{0, 1, 2\}$ and the relation

$$\mathcal{Q} := \{(0,0), (0,1), (1,0), (1,1), (2,2)\}.$$

The element $Z := 2$ is maximal in A, because the hypothesis $(Z, Y) \in \mathcal{Q}$ in $[(Z, Y) \in \mathcal{Q}] \Rightarrow [(Y, Z) \in \mathcal{Q}]$ is False for each $Y \neq 2$ in A. Yet 2 is not a last element in A, because 2 does not follow every element: $(0, 2) \notin \mathcal{Q}$ and $(1, 2) \notin \mathcal{Q}$.

Such notions as the geometric direction on an "infinite line, plane, or space" require another axiom, the axiom of infinity, which forms the subject of chapter 3.

Axiom S7 (Axiom of infinity) There exists a set H, such that $\varnothing \in H$, and for each element $C \in H$ its successor $C \cup \{C\}$ is also an element of H:

$$\models \exists H \left[(\varnothing \in H) \wedge \left(\forall C \left\{ (C \in H) \Rightarrow [(C \cup \{C\}) \in H] \right\} \right) \right].$$

The concept of well-ordered sets allows for many logically equivalent statements of the last axiom of Zermelo-Frankel set theory [19, Ch. 1, §9, p. 23, #9.2(3)].

Axiom S8 (Axiom of choice) For each set $\mathcal{F}$ of *non-empty* sets ($A \neq \varnothing$ for each $A \in \mathcal{F}$), there is a "choice" function $F : \mathcal{F} \to \bigcup \mathcal{F}$ with $F(A) \in A$ for each $A \in \mathcal{F}$.

The axiom of choice is logically equivalent to each of the following two theorems. Because neither theorem is needed here, this text omits their proof [19, p. 31–35].

Theorem 464 (Zermelo's Theorem) Every set is well-ordered by some relation.

Theorem 465 (Zorn's Lemma) In a pre-ordered set A, if every chain in A has an upper bound, then A has a maximal element.

2.8.3 Exercises

For the following exercises, determine all of the characteristics — among connectivity, strong connectivity, reflexivity, irreflexivity, symmetry, anti-symmetry, asymmetry, transitivity — of the given relation on the set $A := \{0, 1, 2, 3, 4, 5, 6, 7, 8, 9\}$.

Exercise 571

$$\mathcal{Q} := \left\{ \begin{array}{lllllll}
(0,9) & (1,9) & & (3,9) & & & (6,9) \\
(0,8) & (1,8) & (2,8) & & (4,8) & & (6,8) \\
(0,7) & (1,7) & & & & & \\
(0,6) & (1,6) & (2,6) & (3,6) & (4,6) & & \\
(0,5) & (1,5) & & & & & \\
(0,4) & (1,4) & (2,4) & & & & \\
(0,3) & (1,3) & & & & & \\
(0,2) & (1,2) & & & & & \\
(0,1) & & & & & &
\end{array} \right\}.$$

Exercise 572

$$\mathcal{R} := \left\{ \begin{array}{llllllllll}
(0,9) & (1,9) & & (3,9) & & & (6,9) & & & (9,9) \\
(0,8) & (1,8) & (2,8) & & (4,8) & & (6,8) & & (8,8) & \\
(0,7) & (1,7) & & & & & & (7,7) & & \\
(0,6) & (1,6) & (2,6) & (3,6) & (4,6) & & (6,6) & & & \\
(0,5) & (1,5) & & & & (5,5) & & & & \\
(0,4) & (1,4) & (2,4) & & (4,4) & & & & & \\
(0,3) & (1,3) & & (3,3) & & & & & & \\
(0,2) & (1,2) & (2,2) & & & & & & & \\
(0,1) & (1,1) & & & & & & & & \\
(0,0) & & & & & & & & &
\end{array} \right\}.$$

Exercise 573

$$\mathcal{S} := \left\{ \begin{array}{cccccccccc}
(0,9) & (1,9) & & (3,9) & & & (6,9) & & & \\
(0,8) & (1,8) & (2,8) & & (4,8) & & (6,8) & & & \\
(0,7) & (1,7) & & & & & & & & \\
(0,6) & (1,6) & (2,6) & (3,6) & (4,6) & & & & (8,6) & (9,6) \\
(0,5) & (1,5) & & & & & & & & \\
(0,4) & (1,4) & (2,4) & & & & (6,4) & & (8,4) & \\
(0,3) & (1,3) & & & & & (6,3) & & & (9,3) \\
(0,2) & (1,2) & & & (4,2) & & (6,2) & & (8,2) & \\
(0,1) & & (2,1) & (3,1) & (4,1) & (5,1) & (6,1) & (7,1) & (8,1) & (9,1) \\
& (1,0) & (2,0) & (3,0) & (4,0) & (5,0) & (6,0) & (7,0) & (8,0) & (9,0)
\end{array} \right\}.$$

Exercise 574

$$\mathcal{U} := \left\{ \begin{array}{cccccccccc}
(0,9) & (1,9) & & (3,9) & & & (6,9) & & & (9,9) \\
(0,8) & (1,8) & (2,8) & & (4,8) & & (6,8) & & (8,8) & \\
(0,7) & (1,7) & & & & & & (7,7) & & \\
(0,6) & (1,6) & (2,6) & (3,6) & (4,6) & & (6,6) & & (8,6) & (9,6) \\
(0,5) & (1,5) & & & & (5,5) & & & & \\
(0,4) & (1,4) & (2,4) & & (4,4) & & (6,4) & & (8,4) & \\
(0,3) & (1,3) & & (3,3) & & & (6,3) & & & (9,3) \\
(0,2) & (1,2) & (2,2) & & (4,2) & & (6,2) & & (8,2) & \\
(0,1) & (1,1) & (2,1) & (3,1) & (4,1) & (5,1) & (6,1) & (7,1) & (8,1) & (9,1) \\
(0,0) & (1,0) & (2,0) & (3,0) & (4,0) & (5,0) & (6,0) & (7,0) & (8,0) & (9,0)
\end{array} \right\}.$$

Exercise 575

$$\mathcal{V} := \left\{ \begin{array}{cccccccccc}
& (1,9) & & & & & & & & (9,9) \\
& (1,8) & & & & & & & (8,8) & \\
& (1,7) & & & & & & (7,7) & & \\
& (1,6) & & & & & (6,6) & & & \\
& (1,5) & & & & (5,5) & & & & \\
& (1,4) & & & (4,4) & & & & & \\
& (1,3) & & (3,3) & & & & & & (9,3) \\
& (1,2) & (2,2) & & (4,2) & & & & (8,2) & \\
(0,1) & (1,1) & & & & & & & & \\
& (1,0) & & & & & & & &
\end{array} \right\}.$$

Exercise 576

$$\mathcal{W} := \left\{ \begin{array}{cccccccccc}
& (1,9) & & (3,9) & & & & & & (9,9) \\
& (1,8) & (2,8) & & & & & & (8,8) & \\
& (1,7) & & & & & & (7,7) & & \\
& (1,6) & & & & & (6,6) & & & \\
& (1,5) & & & & (5,5) & & & & \\
& (1,4) & (2,4) & & (4,4) & & & & & \\
& (1,3) & & (3,3) & & & & & & (9,3) \\
& (1,2) & (2,2) & & (4,2) & & & & (8,2) & \\
(0,1) & (1,1) & (2,1) & (3,1) & (4,1) & (5,1) & (6,1) & (7,1) & (8,1) & (9,1) \\
& (1,0) & & & & & & & &
\end{array} \right\}.$$

Exercise 577 Prove that for each set A the empty relation $\varnothing \subseteq A \times A$ is a partial order.

Exercise 578 Prove that a relation $\mathcal{R}$ is *symmetric* if and only if $\mathcal{R} = \mathcal{R}^{\circ -1}$.

Exercise 579 Prove that a relation $\mathcal{R}$ on A is *irreflexive* if and only if $\mathcal{R} \cap \Delta_A = \varnothing$.

Exercise 580 Prove that a relation $\mathcal{R}$ on a set A is *anti-symmetric* if and only if $\mathcal{R} \cap \mathcal{R}^{\circ -1} \subseteq \Delta_A$.

Exercise 581 Prove that every strict partial order is asymmetric

Exercise 582 Prove that if a relation is irreflexive and anti-symmetric, then it is also asymmetric

Exercise 583 Prove that every asymmetric relation is also anti-symmetric.

Exercise 584 Prove that every asymmetric relation is also irreflexive.

Exercise 585 Prove that a relation is a strict partial order if and only if it is asymmetric and transitive

Exercise 586 Provide an example of an anti-symmetric but not asymmetric relation.

Exercise 587 Provide an example of an asymmetric and strongly connected relation.

Exercise 588 Exhibit an example of a set A for which the relation $\subseteq$ on $\mathcal{P}(A)$ is *not* a total order: supply subsets B and C such that neither $B \subseteq C$ nor $C \subseteq B$.

Exercise 589 Exhibit an example of a set A and a subset $S \subseteq \mathcal{P}(A)$ that is a chain with respect to the relation $\subseteq$ on $\mathcal{P}(A)$.

Exercise 590 Prove that if a function is surjective, then it has a right inverse.

2.9 PROJECTS

Project 5 Document instances of claims that something could not be done, but for which a subsequent proof revealed that the same thing could be done.

For example, in 1880 "Venn Said It Couldn't Be Done" about Venn diagrams for five sets; yet in 1997 Peter Hamburger and Raymond E. Pippert produced Venn diagrams for five sets ([39], [40]), for instance, as in Figure 2.2 on page 110.

Explain how such proofs rely on the "one half" $\{\exists X[\neg(P)]\} \Rightarrow \{\neg[\forall X(P)]\}$ of axiom Q4 from the predicate calculus, $\{\exists X[\neg(P)]\} \Leftrightarrow \{\neg[\forall X(P)]\}$.

Project 6 Read and supply some of the elementary logical steps in proofs of the mutual logical equivalence of the axiom of choice (S8), Zermelo's Theorem (464), and Zorn's Lemma (465), for instance, as in [19, p. 31–35].

Chapter 3

Induction, Recursion, Arithmetic, Cardinality

3.0 INTRODUCTION

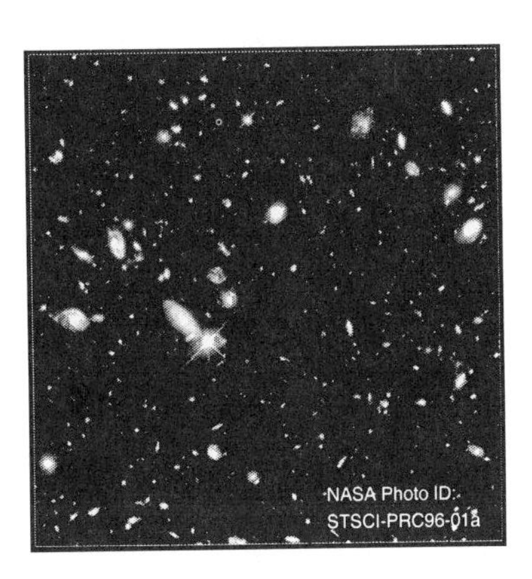

Figure 3.1: The number of particles in the universe may determine its fate.

The concepts of "numbers" and "counting" allow for the determination and comparison of the "sizes" of various objects. The same concepts also lead to precise definitions of "infinite" sets, which then reveals an "infinite" variety of "infinite" sets. For example, the number of particles in the universe (see figure 3.1) is related to its mass, volume, and density, which quantities might affect the fate of the universe, in particular, whether the universe will remain finite or will expand to "infinity" [57, p. 445].

Mathematically, this chapter shows how the concepts of "numbers", "arithmetic", "counting", and "infinity" fit within set theory, without any new axiomatic system.

3.1 MATHEMATICAL INDUCTION

3.1.1 The Axiom of Infinity

The Principle of Mathematical Induction forms the theoretical basis for such algorithms as counting and arithmetic, by means of *sequences* of definitions, computations, and verifications, where the length of the sequence depends on the situation. One framework to allow for sequences of yet unspecified lengths consists in embedding all such sequences into one set $\mathbb{N}$ that already allows for sequences of all lengths. To this end, with a construction attributed to John von Neumann [3, p. 22], [95, p. 129], [102], for each element $X \in \mathbb{N}$, the set $\mathbb{N}$ also contains a "next" element $X \cup \{X\}$.

Definition 466 For each set X the **successor** of X is $X \cup \{X\}$.

Example 467 The successor of the empty set $\varnothing$ is the set $\varnothing \cup \{\varnothing\} = \{\varnothing\}$.

A subsequent theorem will verify that for each set $X \in \mathbb{N}$ the successor $X \cup \{X\}$ is strictly larger than X, so that $X \subsetneq (X \cup \{X\})$. Within the theory presented so far, however, nothing guarantees the existence a set containing the successor of each of its elements. To this end, a new axiom becomes necessary [3, p. 21]. The following version is identical to that of axiom S7 already mentioned in chapter 2, here with $\mathbb{I}$ instead of H.

Axiom S7 (Axiom of infinity) There exists a set $\mathbb{I}$, such that $\varnothing \in \mathbb{I}$, and for each element $C \in \mathbb{I}$ its successor $C \cup \{C\}$ is also an element of $\mathbb{I}$:

$$\models \exists \mathbb{I}\left[(\varnothing \in \mathbb{I}) \wedge \left(\forall C\,\{(C \in \mathbb{I}) \Rightarrow [(C \cup \{C\}) \in \mathbb{I}]\}\right)\right].$$

Digits or other symbols can abbreviate the set notation for the elements of this set.

Example 468 The set $\mathbb{I}$ described in axiom S7 contains the following elements.

$$\begin{aligned}
0 &:= \varnothing, \\
1 &:= 0 \cup \{0\} = \varnothing \cup \{\varnothing\} = \{\varnothing\}, \\
2 &:= 1 \cup \{1\} = \{\varnothing\} \cup \{\{\varnothing\}\} = \{\varnothing, \{\varnothing\}\}, \\
3 &:= 2 \cup \{2\} = \{\varnothing, \{\varnothing\}\} \cup \Big\{\{\varnothing, \{\varnothing\}\}\Big\} = \Big\{\varnothing, \{\varnothing\}, \{\varnothing, \{\varnothing\}\}\Big\}, \\
4 &:= 3 \cup \{3\} = \left\{\varnothing,\ \{\varnothing\},\ \{\varnothing, \{\varnothing\}\},\ \Big\{\varnothing, \{\varnothing\}, \{\varnothing, \{\varnothing\}\}\Big\}\right\}, \\
&\ \ \vdots
\end{aligned}$$

Such elements are called "natural numbers" (see remark 472 about 0).

The axiom of infinity does *not* restrict $\mathbb{I}$ to contain only the elements listed in example 468, so that the set $\mathbb{I}$ might also contain other elements. Therefore, a further construction with the preceding axioms becomes necessary to form a set containing *only* the natural numbers. The construction selected here forms the intersection of subsets of $\mathbb{I}$, using a method common to several parts of mathematics [19, p. 66], [27, p. 21], [42, p. 132]. Specifically, denote by P the logical formula defined by

$$(P) \Leftrightarrow \left[(\varnothing \in A) \wedge \left(\forall X\,\{(X \in A) \Rightarrow [(X \cup \{X\}) \in A]\}\right)\right].$$

Thus the formula P asserts that a set A contains among its elements the empty set and the successor of every one of its elements. The Axiom of Infinity asserts that there exists a set $\mathbb{I}$ for which $\mathrm{Subf}_{\mathbb{I}}^{A}(P)$ is True:

$$\models \exists \mathbb{I}\left[(\varnothing \in \mathbb{I}) \wedge \left(\forall X\,\{(X \in \mathbb{I}) \Rightarrow [(X \cup \{X\}) \in \mathbb{I}]\}\right)\right].$$

The following construction forms a subset $\mathbb{N} \subseteq \mathbb{I}$ containing only elements such as those in example 468. To this end, consider the set $\mathcal{F} \subseteq \mathcal{P}(\mathbb{I})$ of all subsets $B \subseteq \mathbb{I}$ for which $\text{Subf}_B^A(P)$ is also True:

$$\begin{aligned}\mathcal{F} &:= \{B : (B \subseteq \mathbb{I}) \wedge [\text{Subf}_B^A(P)]\} \\ &= \Big\{B \in \mathcal{P}(\mathbb{I}) : \\ &\quad [(\varnothing \in B) \wedge (\forall X \{(X \in B) \Rightarrow [(X \cup \{X\}) \in B]\})]\Big\}.\end{aligned}$$

The following definition and theorems will confirm that $\mathbb{N}$ can be defined as $\bigcap \mathcal{F}$.

Theorem 469 *The set* $\mathcal{F} = \{B \in \mathcal{P}(\mathbb{I}) : \mathit{Subf}_B^A(P)\}$ *is not empty:* $\mathbb{I} \in \mathcal{F}$.

Proof. By the Axiom of Infinity (S7), the formula P is True for $\mathbb{I}$, so that $\text{Subf}_{\mathbb{I}}^A(P)$ is True. Moreover, $\mathbb{I} \subseteq \mathbb{I}$. Therefore, $\mathbb{I}$ is a subset of $\mathbb{I}$ for which P is True, so that $(\mathbb{I} \subseteq \mathbb{I}) \wedge [\text{Subf}_{\mathbb{I}}^A(P)]$ holds, which means that $\mathbb{I} \in \mathcal{F}$. □

From $\mathbb{I}$ the following definition extracts $\mathbb{N}$ as the smallest subset for which P holds.

Definition 470 (Natural numbers) The set of **natural numbers**, denoted by $\mathbb{N}$, is the intersection of all the elements of $\mathcal{F}$:

$$\mathbb{N} := \bigcap \mathcal{F}$$

A **natural number** is an element of $\mathbb{N}$. Also, define $\mathbb{N}^* := \mathbb{N} \setminus \{\varnothing\}$.

Remark 471 In the present theory, every natural number is a *set*.

Remark 472 Some texts exclude the element $\varnothing$ from the set $\mathbb{N}$. The two definitions differ only in their terminology and lead to the same theory. Yet the definition of $\mathbb{N}$ with $\varnothing \in \mathbb{N}$ has proved more convenient than without it in the context of set theory [95, p. 121], ordinals [64], and in such situations as Kurt Gödel's work [34] on logic and mathematics. Therefore, the definition adopted here includes $\varnothing \in \mathbb{N}$.

The following theorems verify that P also holds for $\mathbb{N}$.

Theorem 473 *The set* $\mathbb{N} = \bigcap \mathcal{F}$ *is not empty; indeed,* $\varnothing \in \mathbb{N}$.

Proof. Firstly, $\varnothing \in \mathbb{I}$ and $\mathbb{I} \in \mathcal{F}$ by theorem 469, whence $\varnothing \in \bigcup \mathcal{F}$. Secondly, for each $B \in \mathcal{F}$ the formula $\text{Subf}_B^A(P)$ is True, hence $\varnothing \in B$; consequently, $\varnothing \in \bigcap \mathcal{F} = \mathbb{N}$. □

Theorem 474 *The formula P is True for* $\mathbb{N}$: $\forall X \{(X \in \mathbb{N}) \Rightarrow [(X \cup \{X\}) \in \mathbb{N}]\}$.

Proof. For each X, if $X \in \mathbb{N} = \bigcap \mathcal{F}$, then $X \in B$ for each $B \in \mathcal{F}$, but then $(X \cup \{X\}) \in B$ for each $B \in \mathcal{F}$, whence $(X \cup \{X\}) \in \bigcap \mathcal{F} = \mathbb{N}$. □

Theorem 475 *The set* $\mathbb{N}$ *is an element of* $\mathcal{F}$.

Proof. Theorems 473 and 474 show that $\varnothing \in \mathbb{N}$ and $\text{Subf}_{\mathbb{N}}^A(P)$ are True. Moreover, $\mathbb{N} = \bigcap \mathcal{F} \subseteq \mathbb{I}$. Thus, $(\mathbb{N} \subseteq \mathbb{I}) \wedge [\text{Subf}_{\mathbb{N}}^A(P)]$ holds, whence $\mathbb{N} \in \mathcal{F}$. □

The concept of "successor" can serve to specify functions defined on $\mathbb{N}$.

Example 476 The "successor" function is defined by

$$G := \{(X,Y) \in (\mathbb{N} \times \mathbb{N}) : Y = (X \cup \{X\})\},$$

so that

$$\begin{array}{rcl} G : \mathbb{N} & \rightarrow & \mathbb{N}, \\ X & \mapsto & X \cup \{X\}. \end{array}$$

Definition 477 (Sequence) For each set E, a **sequence** in E is a function $F : \mathbb{N} \rightarrow E$. For each $N \in \mathbb{N}$, the value $F(N)$ can also be denoted by F_N, and then the function F can also be denoted by (F_N).

Example 478 The "successor" function in example 476 is a sequence in $\mathbb{N}$.

For convenience, mathematical usage abbreviates the successor $N \cup \{N\}$ as $N + 1$.

Definition 479 For each natural number N, let $N + 1 := N \cup \{N\}$.

In general, however, the specification of functions defined on $\mathbb{N}$ requires a method known as the Principle of Mathematical Induction, as described in the next subsections.

3.1.2 The Principle of Mathematical Induction

The following theorem forms the theoretical basis for the methods of proof by induction and recursive computation. Specifically, the "Principle of Mathematical Induction" shows that if a subset $S \subseteq \mathbb{N}$ contains $\varnothing$, and if S also contains the successor $X \cup \{X\}$ of each of its elements X, then S contains all the natural numbers: $S = \mathbb{N}$.

Theorem 480 (Principle of Mathematical Induction) *For each subset $S \subseteq \mathbb{N}$, if S contains the empty set and the successor of every one of its elements, then $S = \mathbb{N}$. Thus,* $\models \forall S \{(S \subseteq \mathbb{N}) \wedge [\text{Subf}^A_S(P)] \Rightarrow (S = \mathbb{N})\}$*; or, with* $\text{Subf}^A_S(P)$ *spelled out:*

$$\models \forall S[(S \subseteq \mathbb{N}) \wedge (\varnothing \in S) \wedge (\forall X \{(X \in S) \Rightarrow [(X \cup \{X\}) \in S]\}) \Rightarrow (S = \mathbb{N})].$$

Proof. If $S \subseteq \mathbb{N}$ then $S \subseteq \mathbb{N} = \bigcap \mathcal{F} \subseteq \mathbb{I}$. If moreover $\text{Subf}^A_S(P)$ is True, then $S \in \mathcal{F}$ by definition of $\mathcal{F}$. From $S \in \mathcal{F}$, it then follows that $\bigcap \mathcal{F} \subseteq S$. Thus, $\bigcap \mathcal{F} \subseteq S \subseteq \bigcap \mathcal{F}$, whence (by theorem 290) equality holds: $S = \bigcap \mathcal{F} = \mathbb{N}$. □

The following example demonstrates a pattern amenable to induction.

Example 481 The elements of $\mathbb{N}$ listed in example 468 reveal the following pattern:

- Every *element* of 0 is also a *subset* of 0, so that $\models \forall X[(X \in 0) \Rightarrow (X \subseteq 0)]$. The hypothesis $X \in 0$ is False, whence the implication is universally valid.
- Every *element* of $1 = \{\varnothing\}$ is also a *subset* of 1, because $\varnothing \in 1$ and $\varnothing \subseteq 1$.
- Every *element* of $2 = \{\varnothing, \{\varnothing\}\}$ is also a *subset* of 2. Indeed $\varnothing \in 2$ and $\varnothing \subseteq 2$; also, $\{\varnothing\} \in 2$ and $\{\varnothing\} \subseteq 2$.
- Similarly, every *element* of $3 = \left\{\varnothing, \{\varnothing\}, \{\varnothing, \{\varnothing\}\}\right\}$ is also a *subset* of 3. Indeed $\varnothing \in 3$ and $\varnothing \subseteq 3$; moreover, $\{\varnothing\} \in 3$ and $\{\varnothing\} \subseteq 3$; furthermore, $\{\varnothing, \{\varnothing\}\} \in 3$ and $\{\varnothing, \{\varnothing\}\} \subseteq 3$.

The proof of the following theorem demonstrates the use of the Principle of Mathematical Induction to verify the pattern of example 481 for all natural numbers.

Theorem 482 *For all $L \in \mathbb{N}$ and $N \in \mathbb{N}$, if $L \in N$ then $L \subseteq N$.*

Proof. This proof proceeds by induction with N. Define a set $S \subseteq \mathbb{N}$ by

$$S := \{N \in \mathbb{N} : \forall L[(L \in N) \Rightarrow (L \subseteq N)]\}.$$

Initial step If $N := \varnothing$, then $L \in \varnothing$ is False, whence $(L \in \varnothing) \Rightarrow (L \subseteq \varnothing)$ is universally valid; thus $\varnothing \in S$, by definition of S.

Inductive hypothesis Assume $K \in S$; thus $\forall L[(L \in K) \Rightarrow (L \subseteq K)]$ is True.

Inductive step To verify the theorem for $N := K + 1$, assume that $L \in (K+1) = (K \cup \{K\})$; then $L \in K$ or $L \in \{K\}$, by definition of the union $K \cup \{K\}$.

In the case $L \in K$, it follows from the inductive hypothesis that $L \subseteq K$, whence $L \subseteq K \subseteq (K \cup \{K\}) = (K+1)$ and then $L \subseteq (K+1)$.

In the case $L \in \{K\}$, it follows that $L = K$, whence $L = K \subseteq (K \cup \{K\}) = (K+1)$ and then $L \subseteq (K+1)$.

Consequently $L \subseteq (K+1)$ in either case. Because $L \subseteq (K+1)$ for every $L \in K$, it follows that $(K+1) \in S$.

Completion of the proof by induction From the initial step $\varnothing \in S$, and from the inductive step, if $K \in S$ then $(K+1) \in S$; it follows from the Principle of Mathematical Induction (theorem 480) that $S = \mathbb{N}$. Thus $N \in S$ for every $N \in \mathbb{N}$. This means that for every $N \in \mathbb{N}$, if $L \in N$ then $L \subseteq N$. □

Theorem 482 holds for all natural numbers, but it can fail for other sets.

Counterexample 483 If $X = \{\varnothing\}$ and $Y = \{\{\varnothing\}\}$, then $X \in Y$ but $X \not\subseteq Y$, because $\varnothing \in X$ but $\varnothing \notin Y$. Theorem 482 does not apply to Y because $Y \notin \mathbb{N}$.

The following subsection demonstrates how to use the Principle of Mathematical Induction to prove the "existence" of certain functions.

3.1.3 Definitions by Mathematical Induction or Recursion

The preceding discussion has introduced the Principle of Mathematical Induction as a method to *prove* theorems and verify formulae involving natural numbers. In addition, the following theorem shows how the Principle of Mathematical Induction, and the concept of unions of functions, also form the foundation of a method to *define* by induction the values of some functions. The same method is also known as **recursion** [11, p. 322, n. 526], [53, p. 10]; the resulting functions are also called **recursive functions**, with a terminology attributed [35, p. 167] to Kurt Gödel [31], [34, p. 46].

Theorem 484 (Definition by Induction or Recursion) *For each non-empty set C, for each element $A \in C$, and for each function $G : C \to C$, there exists* exactly one *function $F : \mathbb{N} \to C$ that satisfies the following two conditions:*

(DMI.0) $F(0) = A$.

(DMI.1) $F(N+1) = G[F(N)]$ *for each* $N \in \mathbb{N}$.

Proof. This proof verifies by induction that there exists a sequence of functions (F_N), with each function $F_N \subseteq (\mathbb{N} \times C)$ defined on the set $S_N := (N \cup \{N\})$ so that the following two conditions hold:

(N.0) $F_N(0) = A$.

(N.1) $F_N(I+1) = G[F_N(I)]$ for each $I \in S_N \setminus \{N\} = N$.

Then the proof ends by verifying that the function F defined by $F := \bigcup_{N \in \mathbb{N}} F_N$, so that $F(N) = F_N(N)$, satisfies all the requirements.

Initial step For $N = 0$, consider the set $S_0 = \{0\}$, and consider the function

$$\begin{aligned} F_0 : \{0\} &\to C, \\ 0 &\mapsto A. \end{aligned}$$

Then the function F_0 just defined satisfies the following two conditions:

(0.0) $F_0(0) = A$.

(0.1) $F_0(I+1) = G[F_0(I)]$ for each $I \in S_0 \setminus \{0\} = \varnothing$.

Moreover, there exists only one function of singletons $F_0 : \{0\} \to \{A\}$.

Induction hypothesis Assume that there exists a natural number $J \in \mathbb{N}$ such that the theorem holds for $N := J$, so that for each $L \in S_J = J \cup \{J\}$ there exists exactly one function F_L satisfying the two conditions $(N.0)$ and $(N.1)$ for $N := L$.

Induction step Let $S \subseteq \mathbb{N}$ consist of every $N \in \mathbb{N}$ for which there exists exactly one function F_N satisfying $(N.0)$ and $(N.1)$.

There exists exactly one function of singletons $H_J : \{J+1\} \to \{G[F_J(J)]\}$, because there exists exactly one function F_J and hence exactly one value $F_J(J)$.

Define $F_{J+1} := F_J \dot{\cup} H_J$, which is a function because of the disjoint domains $S_J \cap \{J+1\} = \varnothing$ (by definition 368 in chapter 2).

For $I := J+1$, the definition of F_{J+1} gives $F_{J+1}(J+1) = H_J(J+1) = G[F_J(J)] = G[F_{J+1}(J)]$. For each $L \in S_J \setminus \{J\} = J$, the definition of F_{J+1} gives $F_{J+1}(L+1) = F_J(L+1) = G[F_J(L)] = G[F_{J+1}(L)]$ so that the following two conditions hold:

$(J+1.0)$ $F_{J+1}(0) = F_J(0) = A$,

$(J+1.1)$ $F_{J+1}(L+1) = G[F_{J+1}(L)]$ for every $L \in S_{J+1} \setminus \{J+1\}$.

Moreover, there exists exactly one such function F_{J+1} because there exists exactly one restriction $F_J|_{S_J}$, namely F_J, and exactly one value for $F_{J+1}(J+1) = G[F_J(J)]$.

Completion of the proof of the theorem Having constructed the sequence of functions (F_N), define a function $F : \mathbb{N} \to C$ by $F(N) := F_N(N)$. Specifically, let

$$F := \bigcup_{N \in \mathbb{N}} F_N.$$

Then verify that

(DMI.1) $F(0) = F_0(0) = 0$.

(DMI.2) $F(N+1) = F_{N+1}(N+1) = G[F_{N+1}(N)] = G[F_N(N)] = G[F(N)]$.

The uniqueness of F follows from the uniqueness of each restriction $F|_{N+1} = F_N$. □

Because the proof of the validity of the method of recursion (theorem 484) relies on mathematical induction (theorem 480), it follows that recursion is a logical consequence of induction. A different version of the Principle of Mathematical Induction replaces the function $G : C \to C$ by a function $G : C^{\mathbb{N}} \to C$ defined on the set $C^{\mathbb{N}}$ of all functions from $\mathbb{N}$ to C. Such a version will follow from the Principle of Transfinite Induction proved in chapter 4. Meanwhile, the following sections show how recursion suffices to define arithmetic with natural numbers.

3.1.4 Exercises

Exercise 591 With $5 := 4 \cup \{4\}$, write 5 in terms of sets, as in example 468.

Exercise 592 With $6 := 5 \cup \{5\}$, write 6 in terms of sets, as in example 468.

Exercise 593 Prove that every set is an *element* and a *subset* of its successor.

Exercise 594 Provide an example of a non-empty proper subset $S \subsetneqq \mathbb{N}$ that contains the successor of every one of its elements.

Prove or disprove each of the following statements for all sets A and B.

Exercise 595 Prove or disprove that if $A \subseteq B$, then $(A \cup \{A\}) \subseteq (B \cup \{B\})$.

Exercise 596 Prove or disprove that $(A \cup \{A\}) \cup (B \cup \{B\}) = (A \cup B) \cup \{A \cup B\}$.

Exercise 597 Prove or disprove that $(A \cup \{A\}) \cap (B \cup \{B\}) = (A \cap B) \cup \{A \cap B\}$.

Exercise 598 Prove or disprove that $(A \cup \{A\}) \setminus (B \cup \{B\}) = (A \setminus B) \cup \{A \setminus B\}$.

Prove or disprove each of the following statements for all $K, L, M, \in \mathbb{N}$.

Exercise 599 Prove that if $K \in M$ and $L \in M$, then $K \cup L \in M$.

Exercise 600 Prove that if $K \in M$ and $L \in M$, then $K \cap L \in M$.

Exercise 601 Outline a formal proof of theorem 469.

Exercise 602 Outline a formal proof of theorem 473.

Exercise 603 Outline a formal proof of theorem 474.

Exercise 604 Outline a formal proof of theorem 475.

Exercise 605 Outline a formal proof of theorem 480.

Exercise 606 Prove that $\bigcup \mathbb{N} = \mathbb{N}$.

Exercise 607 Prove that $\bigcap \mathbb{N} = \varnothing$.

Exercise 608 For the successor function, prove that $N = G^{\circ N}(\varnothing)$ for each $N \in \mathbb{N}^*$.

Exercise 609 (Twice-Shifted Induction.) Prove that if a subset $V \subseteq \mathbb{N}$ contains 2, and if V also contains the "successor" $X \cup \{X\}$ of each of its element X, then V contains all the natural numbers except 0 and 1: $V = \mathbb{N} \setminus \{\varnothing, \{\varnothing\}\}$. In other words, replace $\varnothing$ by $\{\varnothing, \{\varnothing\}\}$ in P to obtain the logical formula R defined by

$$(R) \Leftrightarrow \left[(\{\varnothing, \{\varnothing\}\} \in A) \wedge \forall X \{(X \in A) \Rightarrow [(X \cup \{X\}) \in A]\}\right].$$

Prove that $\models \forall V \left[\{(V \subseteq \mathbb{N}) \wedge [\text{Subf}_V^A(R)]\} \Rightarrow (V = \mathbb{N} \setminus \{\varnothing, \{\varnothing\}\})\right]$.

Exercise 610 (Shifted Induction.) Prove that if a subset $U \subseteq \mathbb{N}$ contains 1, and if U also contains the "successor" $X \cup \{X\}$ of each of its elements X, then U contains all the *positive* natural numbers: $U = \mathbb{N} \setminus \{\varnothing\}$. In other words, replace $\varnothing$ by $\{\varnothing\}$ in P to obtain the logical formula Q defined by

$$(Q) \Leftrightarrow \left[(\{\varnothing\} \in A) \wedge \forall X \{(X \in A) \Rightarrow [(X \cup \{X\}) \in A]\}\right].$$

Prove that $\models \forall U \left[\{(U \subseteq \mathbb{N}) \wedge [\text{Subf}_U^A(Q)]\} \Rightarrow (U = \mathbb{N} \setminus \{\varnothing\})\right]$.

3.2 ARITHMETIC WITH NATURAL NUMBERS

3.2.1 Addition with Natural Numbers

This subsection defines the addition of natural numbers and establishes some of its properties, all by induction. Besides explaining the foundations of integer arithmetic, the following considerations also provide examples of proofs by induction.

Definition 485 (Addition) For every $M \in \mathbb{N}$, define

$$\begin{array}{rcll} M+0 & := & M, & (A0) \\ M+1 & := & M \cup \{M\}. & (A1) \end{array}$$

Then for every $M \in \mathbb{N}$, define an addition function by induction, so that for every $N \in \mathbb{N}$,

$$M+(N+1) \quad := \quad (M+N)+1. \quad (A2)$$

Remark 486 According to definition 485, adding N amounts to adding 1 repeatedly N times. With an alternative but logically equivalent notation, for each $M \in \mathbb{N}$ definition 485 uses theorem 484 and the successor function

$$\begin{array}{rcl} G: \mathbb{N} & \to & \mathbb{N}, \\ N & \mapsto & N+1, \end{array}$$

to specify by induction (recursively) an addition function $F^{(M)}: \mathbb{N} \to \mathbb{N}$ such that

$$\begin{array}{rcll} F^{(M)}(0) & := & M, & (A0) \\ G(L) & := & L+1, & (A1) \\ F^{(M)}(N+1) & := & G\left[F^{(M)}(N)\right]. & (A2) \end{array}$$

The following theorem shows that addition is associative.

Theorem 487 *For all $P, Q, R \in \mathbb{N}$, $(P+Q)+R = P+(Q+R)$.*

Proof. For each P and each Q, proceed by induction with R. For $R := 0$,

$$\begin{array}{rcll} (P+Q)+0 & = & P+Q & \text{by (A0) with } M := (P+Q), \\ & = & P+(Q+0) & \text{by (A0) with } M := Q. \end{array}$$

Secondly, *assume* that there exists some $K \in \mathbb{N}$ such that the theorem holds for $R : K$, so that $(P+Q)+K = P+(Q+K)$ for each P and each Q; then

$$\begin{array}{rcll} (P+Q)+(K+1) & = & [(P+Q)+K]+1 & \text{(A2), } M := (P+Q), N := K, \\ & = & [P+(Q+K)]+1 & \text{induction hypothesis,} \\ & = & P+[(Q+K)+1] & \text{(A2), } M := P, N := (Q+K), \\ & = & P+[Q+(K+1)] & \text{(A2), } M := Q, N := K. \end{array}$$

□

The following three theorems show that addition commutes. The first theorem shows that adding 0 commutes.

Theorem 488 *For each $N \in \mathbb{N}$, $0+N = N$.*

Proof. Proceed by induction with N. Firstly, establish the conclusion for $N := 0$:

$$0 + 0 = 0 \quad \text{by (A0) with } M := 0.$$

Secondly, *assume* that $0 + K = K$ for some $K \in \mathbb{N}$, so that the theorem holds for $N := K$; then

$$\begin{aligned} 0 + (K + 1) &= (0 + K) + 1 && \text{by (A2),} \\ &= K + 1 && \text{induction hypothesis.} \end{aligned}$$

□

The second theorem shows that adding 1 commutes.

Theorem 489 *For each $P \in \mathbb{N}$, $1 + P = P + 1$.*

Proof. Proceed by induction with P. Firstly, for $P := 0$,

$$\begin{aligned} 1 + 0 &= 1 && \text{by (A0) with } M := 1, \\ &= 0 + 1 && \text{by theorem 488.} \end{aligned}$$

Secondly, *assume* that $1 + K = K + 1$ for some $K \in \mathbb{N}$, so that the theorem holds for $P := K$; then

$$\begin{aligned} 1 + (K + 1) &= (1 + K) + 1 && \text{(A2) with } M := 1 \text{ and } N := K, \\ &= (K + 1) + 1 && \text{induction hypothesis.} \end{aligned}$$

□

Finally, the third theorem shows that addition commutes.

Theorem 490 *For all natural numbers P and Q, $P + Q = Q + P$.*

Proof. For each $P \in \mathbb{N}$, proceed by induction with Q. Firstly,

$$\begin{aligned} P + 0 &= P && \text{by (A0),} \\ &= 0 + P && \text{by theorem 488.} \end{aligned}$$

Secondly, *assume* that there exists $K \in \mathbb{N}$ such that the theorem holds for $Q := K$, so that $R + K = K + R$ for each $R \in \mathbb{N}$; then for each $P \in \mathbb{N}$,

$$\begin{aligned} P + (K + 1) &= P + (1 + K) && \text{induction hypothesis with } R := 1, \\ &= (P + 1) + K && \text{theorem 487,} \\ &= K + (P + 1) && \text{induction hypothesis with } R := (P + 1), \\ &= K + (1 + P) && \text{theorem 489,} \\ &= (K + 1) + P && \text{theorem 487.} \end{aligned}$$

□

3.2.2 Multiplication with Natural Numbers

This subsection defines by induction the multiplication of natural numbers and establishes some of its properties, as well as some relations between addition and multiplication. Besides a continuation of the foundations of integer arithmetic, the following considerations also provide further examples of proofs by induction.

Definition 491 (Multiplication) For every non-negative integer $M \in \mathbb{N}$, define a multiplication function by induction based on the following specifications:

$$\begin{aligned} M * 0 &:= 0, && (M0) \\ M * (N + 1) &:= (M * N) + M. && (M1) \end{aligned}$$

Remark 492 According to definition 491, multiplying M and N amounts to starting from 0 and adding M repeatedly N times. With an alternative but logically equivalent notation, for each $M \in \mathbb{N}$, definition 491 uses a recursive definition with G (in theorem 484) replaced by the addition function $F^{(M)}$ from remark 486,

$$\begin{array}{rcl} G := F^{(M)} : \mathbb{N} & \to & \mathbb{N}, \\ N & \mapsto & N + M, \end{array}$$

to specify by induction a multiplication function $H^{(M)} : \mathbb{N} \to \mathbb{N}$ such that

$$\begin{array}{rcll} G(L) := F^{(M)}(L) & = & L + M, & \\ H^{(M)}(0) & := & 0, & (M0) \\ H^{(M)}(N+1) & := & G\left[H^{(M)}(N)\right]. & (M1) \end{array}$$

The following theorem shows that multiplication distributes over addition on the right-hand side.

Theorem 493 *For all $P, Q, R \in \mathbb{N}$, $(P + Q) * R = (P * R) + (Q * R)$.*

Proof. For each P and each Q, proceed by induction with R. For $R := 0$,

$$\begin{array}{rcll} (P+Q) * 0 & = & 0 & \text{(M0) with } M := (P+Q), \\ & = & 0 + 0 & \text{(A0) with } M := 0, \\ & = & (P * 0) + (Q * 0) & \text{(M0) with } M := P \text{ and (M0) with } M := Q. \end{array}$$

Secondly, *assume* that there exists $K \in \mathbb{N}$ such that the theorem holds for $R := K$, so that $(P + Q) * K = (P * K) + (Q * K)$ for each P and each Q; then

$$\begin{array}{rcll} (P+Q) * (K+1) & = & [(P+Q) * K] + (P+Q) & \text{(M1), } M := P + Q, \\ & = & [(P * K) + (Q * K)] + (P + Q) & \text{induction hypothesis,} \\ & = & (P * K) + [(Q * K) + (P + Q)] & \text{theorem 487,} \\ & = & (P * K) + [\{(Q * K) + P\} + Q] & \text{theorem 487,} \\ & = & (P * K) + [\{P + (Q * K)\} + Q] & \text{theorem 490,} \\ & = & (P * K) + [P + \{(Q * K) + Q\}] & \text{theorem 487,} \\ & = & [(P * K) + P] + [(Q * K) + Q] & \text{theorem 487,} \\ & = & [P * (K+1)] + [Q * (K+1)] & \text{(M1) twice.} \end{array}$$

□

The following three theorems show that multiplication commutes. The first theorem shows that multiplication by 0 commutes.

Theorem 494 *For each natural number $N \in \mathbb{N}$, $0 * N = 0$.*

Proof. Proceed by induction with N. Firstly, establish the conclusion for $N := 0$:

$$0 * 0 \quad = \quad 0 \quad (M0).$$

Secondly, *assume* that there exists $K \in \mathbb{N}$ such that the theorem holds for $N := K$, so that $0 * K = 0$; then

$$\begin{array}{rcll} 0 * (K+1) & = & (0 * K) + 0 & (M1), \\ & = & 0 + 0 & \text{induction hypothesis,} \\ & = & 0 & (A0). \end{array}$$

□

The second theorem shows that multiplication by 1 commutes.

Theorem 495 *For each non-negative integer $N \in \mathbb{N}$, $1 * N = N$.*

Proof. Proceed by induction with N. Firstly, establish the conclusion for $N := 0$:

$$1 * 0 = 0 \quad (M0).$$

Secondly, *assume* that there exists $K \in \mathbb{N}$ such that the theorem holds for $N := K$, so that $1 * K = K$; then

$$\begin{aligned} 1 * (K+1) &= (1 * K) + 1 && (M1), \\ &= K + 1 && \text{induction hypothesis.} \end{aligned}$$

□

Finally, the third theorem shows that multiplication commutes.

Theorem 496 *For all non-negative integers $P, Q \in \mathbb{N}$, $P * Q = Q * P$.*

Proof. For each $P \in \mathbb{N}$, proceed by induction with Q. Firstly, for $Q := 0$,

$$\begin{aligned} P * 0 &= 0 && (M0), \\ &= 0 * P && \text{theorem 494.} \end{aligned}$$

Secondly, *assume* that there exists $K \in \mathbb{N}$ such that the theorem holds for $Q := K$, so that $P * K = K * P$ for each $P \in \mathbb{N}$; then

$$\begin{aligned} P * (K+1) &= (P * K) + P && (M1), \\ &= (K * P) + P && \text{induction hypothesis,} \\ &= (K * P) + (1 * P) && \text{theorem 495,} \\ &= (K+1) * P && \text{theorem 493.} \end{aligned}$$

□

The next theorem shows that multiplication distributes over addition also on the left-hand side.

Theorem 497 *For all $P, Q, R \in \mathbb{N}$, $P * (Q + R) = (P * Q) + (P * R)$.*

Proof. Use commutativity and distributivity on the right-hand side (theorems 496, 493):

$$\begin{aligned} P * (Q + R) &= (Q + R) * P && \text{theorem 496,} \\ &= (Q * P) + (R * P) && \text{theorem 493,} \\ &= (P * Q) + (P * R) && \text{theorem 496.} \end{aligned}$$

□

The following theorem shows that multiplication is associative.

Theorem 498 *For all $P, Q, R \in \mathbb{N}$, $P * (Q * R) = (P * Q) * R$.*

Proof. For all $P, Q \in \mathbb{N}$, proceed by induction with R. For $R := 0$,

$$\begin{aligned} P * (Q * 0) &= P * 0 && (M0), \\ &= 0 && (M0), \\ &= (P * Q) * 0 && \text{(M0) with } M := (P * Q). \end{aligned}$$

Secondly, *assume* that there exists $K \in \mathbb{N}$ such that the theorem holds for $R := K$, so that $P * (Q * K) = (P * Q) * K$ for all $P, Q \in \mathbb{N}$; then

$$\begin{array}{rcll} (P * Q) * (K + 1)] & = & [(P * Q) * K] + (P * Q) & \text{(M1) with } M := (P * Q), \\ & = & [P * (Q * K)] + (P * Q) & \text{induction hypothesis,} \\ & = & P * [(Q * K) + Q] & \text{theorem 497,} \\ & = & P * [Q * (K + 1)] & (M1). \end{array}$$

□

There are other arithmetic operations with natural numbers, such as the factorial.

Definition 499 For each $N \in \mathbb{N}$, define $N!$ (read "N **factorial**") recursively by

$$\begin{array}{rcl} 0! & := & 1, \\ (N + 1)! & := & (N!) * (N + 1). \end{array}$$

3.2.3 Exercises

The following exercises involve the following sets (also defined in example 311):

$$\begin{array}{rcl} 0 & := & \varnothing, \\ 1 & := & \{0\}, \\ 2 & := & \{0, 1\}, \\ 3 & := & \{0, 1, 2\}, \\ 4 & := & \{0, 1, 2, 3\}, \\ 5 & := & \{0, 1, 2, 3, 4\}, \\ 6 & := & \{0, 1, 2, 3, 4, 5\}, \\ 7 & := & \{0, 1, 2, 3, 4, 5, 6\}, \\ 8 & := & \{0, 1, 2, 3, 4, 5, 6, 7\}, \\ 9 & := & \{0, 1, 2, 3, 4, 5, 6, 7, 8\}. \end{array}$$

Exercise 611 Prove that $1 + 1 = 2$.

Exercise 612 Prove that $2 + 1 = 3$.

Exercise 613 Prove that $3 + 1 = 4$.

Exercise 614 Prove that $4 + 1 = 5$.

Exercise 615 Prove that $5 + 1 = 6$.

Exercise 616 Prove that $6 + 1 = 7$.

Exercise 617 Prove that $7 + 1 = 8$.

Exercise 618 Prove that $8 + 1 = 9$.

Exercise 619 Prove that $2 + 2 = 4$.

Exercise 620 Prove that $3 + 2 = 5$.

Exercise 621 Prove that $4 + 2 = 6$.

Exercise 622 Prove that $5 + 2 = 7$.

Exercise 623 Prove that $6 + 2 = 8$.

Exercise 624 Prove that $7 + 2 = 9$.

Exercise 625 Prove that $3 + 3 = 6$.

Exercise 626 Prove that $4 + 3 = 7$.

Exercise 627 Prove that $5 + 3 = 8$.

Exercise 628 Prove that $6 + 3 = 9$.

Exercise 629 Prove that $4 + 4 = 8$.

Exercise 630 Prove that $5 + 4 = 9$.

Exercise 631 Prove that $2 * 2 = 4$.

Exercise 632 Prove that $3 * 2 = 6$.

Exercise 633 Prove that $4 * 2 = 8$.

Exercise 634 Prove that $3 * 3 = 9$.

Exercise 635 Prove or disprove that addition distributes over multiplication on the left.

Exercise 636 Prove or disprove that addition distributes over multiplication on the right.

The following exercises refer to the factorial specified by definition 499.

Exercise 637 Identify a set C and functions F and G that fit theorem 484 to justify definition 499.

Exercise 638 Calculate $0!$, $1!$, $2!$, $3!$.

Exercise 639 Prove or disprove that $(P + Q)! = (P!) + (Q!)$ for all $P, Q \in \mathbb{N}$.

Exercise 640 Prove or disprove that $(P * Q)! = (P!) * (Q!)$ for all $P, Q \in \mathbb{N}$.

3.3 ORDERS AND CANCELLATIONS

3.3.1 Orders on the Natural Numbers

The set of natural numbers can serve to model geometric concepts, for instance, a direction from left to right, and also algebraic concepts, for instance, increasing magnitudes:

$$0 < 1 < 2 < 3 < 4 < 5 < 6 < 7 < 8 < 9 < \dots$$

For both types of concepts — geometric and algebraic — it suffices to introduce an ordering relation on the natural numbers. To this end, this subsection shows that the natural numbers are well-ordered by a relation $\leq$ defined by

$$(M \leq N) \Leftrightarrow [(M = N) \vee (M \in N)].$$

From this well-ordering relation will result the laws of arithmetic cancellations, which will also allow for the solutions of certain equations. The first results define a strict order $<$ in terms of the foundational relation $\in$ of set membership.

Definition 500 For all $M, N \in \mathbb{N}$, define $M < N$ (read "M is less than N") by

$$(M < N) \Leftrightarrow (M \in N).$$

Example 501 $0 < 1$ because $0 \in 1$; indeed, $0 = \varnothing$ and $1 = \{\varnothing\}$, whence $\varnothing \in \{\varnothing\}$.
$1 < 2$ because $1 \in 2$; $1 = \{\varnothing\}$, $2 = \big\{\varnothing, \{\varnothing\}\big\}$, whence $\{\varnothing\} \in \big\{\varnothing, \{\varnothing\}\big\}$.

The following theorem establishes the *transitivity* of the relation $\in$ on $\mathbb{N}$.

Theorem 502 *For all $L, M, N \in \mathbb{N}$, if $L \in M$ and $M \in N$, then $L \in N$:*

$$\forall L \forall M \forall N \{[(L \in \mathbb{N}) \wedge (M \in \mathbb{N}) \wedge (N \in \mathbb{N})]$$

$$\Rightarrow [\{(L \in M) \wedge (M \in N)\} \Rightarrow (L \in N)]\}.$$

Proof. Proceed by induction with N.

Initial step If $N := 0 = \varnothing$, then for all non-negative integers $L, M \in \mathbb{N}$ the proposition $M \in \varnothing$ is False, whence so is the conjunction $(L \in M) \wedge (M \in 0)$, and hence the implication $[(L \in M) \wedge (M \in 0)] \Rightarrow (L \in 0)$ is True.

Induction hypothesis Assume that there exists some $K \in \mathbb{N}$ such that the theorem holds for $N := K$, so that $[(L \in M) \wedge (M \in K)] \Rightarrow (L \in K)$ for all $L, M \in \mathbb{N}$.

Induction step If $L \in M$ and $M \in K + 1 = K \cup \{K\}$, then two cases arise: $M \in \{K\}$ or $M \in K$.

In the first case, $M \in \{K\}$, whence $M = K$, and the hypothesis $L \in M$ then yields $L \in M = K \subseteq K \cup \{K\}$, so that $L \in K + 1$.

In the second case, $M \in K$, whence the induction hypothesis yields $L \in K \subset K \cup \{K\} = K + 1$, and then again $L \in K + 1$. □

The transitivity of the relation $\in$ holds on the set $\mathbb{N}$, but it can fail on other sets.

Counterexample 503 If $X := \varnothing$, $Y := \{\varnothing\}$, and $Z := \big\{\{\varnothing\}\big\}$, then $X \in Y$ and $Y \in Z$, but $X \notin Z$. Theorem 502 fails for $Z = \big\{\{\varnothing\}\big\}$ because $\big\{\{\varnothing\}\big\} \notin \mathbb{N}$.

The following theorem shows that on the natural numbers the relation $\in$ is *neither* reflexive *nor* symmetric, but, instead, $\in$ is both *irreflexive* and *asymmetric*.

Theorem 504 *For all $I, L, M \in \mathbb{N}$, $M \notin M$, and $(I \notin L) \vee (L \notin I)$.*

Proof. This proof of the first result ($M \notin M$) proceeds by induction with M.

Initial step If $M = 0$, then $M = \varnothing$. From $\varnothing \notin \varnothing$ it follows that $M \notin M$.

Induction hypothesis Assume that there exists some $K \in \mathbb{N}$ such that the theorem holds for $M := K$, so that $K \notin K$.

Induction step This step of the proof proceeds by contraposition. If $(K \cup \{K\}) \in (K \cup \{K\})$, then two cases can arise: $(K \cup \{K\}) \in K$ or $(K \cup \{K\}) \in \{K\}$.

Firstly, if $(K \cup \{K\}) \in K$, then the transitivity of the relation $\in$ (theorem 502) and $K \in (K \cup \{K\})$ give $K \in K$, which contradicts the induction hypothesis.

Secondly, if $(K \cup \{K\}) \in \{K\}$, then $(K \cup \{K\}) = K$, whence $K \in \{K\} \subseteq (K \cup \{K\}) = K$ gives $K \in K$, which contradicts the induction hypothesis.

The proof of the second result proceeds by contraposition. For all $I \in \mathbb{N}$ and $L \in \mathbb{N}$, the conjunction $(I \in L) \wedge (L \in I)$ and the transitivity of $\in$ (theorem 502) give $I \in I$, which contradicts the result just proved. Therefore, $\neg[(I \in L) \wedge (L \in I)]$. □

The following theorem shows that every natural number differs from its successor.

Theorem 505 *For each natural number $N \in \mathbb{N}$, $N \neq N+1$.*

Proof. From $N \in \{N\} \subseteq (N \cup \{N\}) = N+1$ it follows that $N \in (N+1)$. Yet $N \notin N$ by theorem 504. Therefore $N \neq N+1$ by the axiom of extensionality (S1). □

The following theorem shows that adding 1 to both sides of a valid inequality gives a valid inequality, so that if $M < N$ then $(M+1) < (N+1)$.

Theorem 506 *For all $M, N \in \mathbb{N}$, if $M \in N$, then $(M+1) \in (N+1)$:*

$$\models \forall M \forall N\, ([(M \in \mathbb{N}) \wedge (N \in \mathbb{N})] \Rightarrow \{(M \in N) \Rightarrow [(M+1) \in (N+1)]\}).$$

Proof. Apply induction with N. For $N := 0 = \varnothing$, and for each $M \in \mathbb{N}$, the proposition $M \in \varnothing$ is False; hence the implication $(M \in \varnothing) \Rightarrow [(M+1) \in (N+1)]$ is True.

Next, *assume* that there exists $K \in \mathbb{N}$ such that the theorem holds for $N := K$, so that $(M \in K) \Rightarrow [(M+1) \in (N+1)]$ holds for every $M \in \mathbb{N}$. If $M \in K+1 = K \cup \{K\}$, then two cases occur.

In the first case, $M \in \{K\}$, and then $M = K$, whence $M+1 = K+1 \in (K+1) \cup \{K+1\} = (K+1)+1$.

In the second case, $M \in K$, whence $M+1 \in K+1$ by induction hypothesis; with $K+1 \subseteq (K+1) \cup \{K+1\} = (K+1)+1$, it follows that $M+1 \in (K+1)+1$. □

The next theorem shows that $0 = \varnothing$ is the smallest natural number.

Theorem 507 *For each natural number $M \in \mathbb{N}$, if $M \neq 0$, then $0 \in M$.*

Proof. This proof proceeds by induction with M.

For $M := 0$, the proposition $M \neq 0$ is False, whence the implication $(M \neq 0) \Rightarrow (0 \in M)$ is True.

Next, *assume* that there exists $K \in \mathbb{N}$ such that the theorem holds for $M := K$, so that $(K \neq 0) \Rightarrow (0 \in K)$. Two cases arise: $K = 0$ or $K \neq 0$.

In the first case, if $K = 0$, then $0 = \varnothing \in (\varnothing \cup \{\varnothing\}) = K \cup \{K\} = K+1$; thus, $0 \in K+1$ is True.

In the second case, if $K \neq 0$, then $0 \in K$, by the induction hypothesis, whence $0 \in K \subseteq K \cup \{K\} = K+1$, and hence $0 \in K+1$. □

The following theorem shows that the relation $\in$ is *connected* on the natural numbers: if $M \neq N$, then either $M \in N$ or $N \in M$.

Theorem 508 *For all natural numbers $M, N \in \mathbb{N}$, exactly one of the following three formulae is True, while the other two are False:*

$$\begin{aligned} &M \in N, \\ &M = N, \\ &N \in M. \end{aligned}$$

Proof. Firstly, observe that at most one of the three formulae may be True. Indeed, $M \notin M$ by theorem 504, whence $(M \in N) \wedge (M = N)$ is False, and $(M = N) \wedge (N \in M)$ is False. Similarly, also by theorem 504, $(M \in N) \wedge (N \in M)$ is also False. Secondly, at least one of the three formulae must be True. This proof uses induction with N.

For $N := 0$, and for each $M \in \mathbb{N}$ either $M = 0 = N$ is True, or $M \neq 0$, and then $0 \in M$ is True by theorem 507. To complete the induction, assume that there exists $K \in \mathbb{N}$ such that the theorem holds for $N := K$, so that $(M \in K) \vee (M = K) \vee (K \in M)$ for each $M \in \mathbb{N}$, and examine all three formulae.

If $M \in K$, then $M \in K \subseteq K + 1$, whence $M \in K + 1$.

If $M = K$, then $M = K \in \{K\} \subseteq K \cup \{K\}$, whence $M \in K + 1$.

If $K \in M$, then $K + 1 \in M + 1 = M \cup \{M\}$, (theorem 506); two cases occur.

In the first case, $K + 1 \in \{M\}$, whence $K + 1 = M$. In the second case, $K + 1 \in M$ already. Thus, $(M \in [K + 1]) \vee (M = [K + 1]) \vee ([K + 1] \in M)$ is True. □

The foregoing result completes the proof that $<$ is a strict total order (irreflexive, asymmetric, and transitive) on the natural numbers. The following theorem shows that the natural numbers are well-ordered by the relation $\leq$.

Theorem 509 (Well-Ordering Principle) *Each non-empty subset of the natural numbers has a smallest element.*

Proof. By contraposition, this proof establishes that every subset of the natural numbers without a smallest element (definition 456) is empty. To this end, assume that a subset $E \subseteq \mathbb{N}$ has no smallest element. Thus every $N \in \mathbb{N}$ is *not* the smallest element of E, which means that $N \notin E$ or that E contains an element M such that $M < N$.

To prove that $E = \varnothing$, the proof proceeds by induction with the set $S \subseteq \mathbb{N}$ consisting of every $N \in \mathbb{N}$ such that the complement $\mathbb{N} \setminus E$ contains every $M \leq N$:

$$S := \{N \in \mathbb{N} : (N \cup \{N\}) \subseteq (\mathbb{N} \setminus E)\}.$$

Initial step The set $S = \mathbb{N} \setminus E$ contains 0; indeed, contraposition shows that if E contained 0, then 0 would be the smallest element of E, because $0 \leq N$ for every $N \in \mathbb{N}$ by theorem 507 and hence also for every $N \in E$.

Induction step Assume that $K \in S$ for some $K \in \mathbb{N}$; thus, every $M \leq K$ belongs to the complement of E. Contraposition again confirms that if E contained $K+1$, then $K+1$ would be the smallest element of E; consequently, $K + 1 \notin E$, whence $K + 1 \in S$. Therefore, $S = \mathbb{N}$, whence $E = \mathbb{N} \setminus S = \varnothing$. □

Definition 510 (Minimum) For each non-empty subset $E \subseteq \mathbb{N}$, the **minimum** of E is the smallest element of E and it is denoted by $\min(E)$.

In contrast to the Well-Ordering Principle, some non-empty subsets of the natural numbers have no maximum. Yet every non-empty subset of the natural numbers that has an upper bound (definition 461) also has a largest element (definition 456).

Theorem 511 *Each non-empty subset of the natural numbers with an upper bound in the natural numbers has a largest element.*

Proof. If there exists an upper bound $M \in \mathbb{N}$ for a non-empty subset $E \subseteq \mathbb{N}$, then $I \leq M$ for every $I \in E$. Let $S \subseteq \mathbb{N}$ be the set of all the upper bounds for E:

$$S := \{N \in \mathbb{N} : \forall I[(I \in E) \Rightarrow (I \leq N)]\}.$$

Then $M \in S$ by hypothesis on E; in particular, $S \neq \varnothing$. Consequently, by the Well-Ordering Principle (theorem 509) S has a smallest element $K := \min(S)$. Because $K \in S$, it follows that $I \leq K$ for every $I \in E$.

Moreover, $K \in E$, as proved by the following induction. If $K = 0$, then $E = \{0\}$ because $K = 0 < I$ for every $I \in \mathbb{N} \setminus \{0\}$; hence $K = 0 \in E$.

Suppose that there exists $L \in \mathbb{N}$ such that the theorem holds for $K := L$, so that for each $E \subset \mathbb{N}$ if $L = \min(S)$, then $L \in E$. If $\min(S) = L + 1$, then $L + 1 \in E$, for otherwise $I \leq L$ for every $I \in E$ and then $L \in S$; thus the theorem holds for $L + 1$. □

Definition 512 (Maximum) For each non-empty subset $E \subseteq \mathbb{N}$ with an upper bound in $\mathbb{N}$, the **maximum** of E is the largest element of E and it is denoted by $\max(E)$.

The following abbreviations occasionally prove convenient.

Definition 513 For all $I \in \mathbb{N}$ and $N \in \mathbb{N}$ define a set $\{I, \ldots, N\}$ by

$$\{I, \ldots, N\} := \{K \in \mathbb{N} : (I \leq K) \wedge (K \leq N)\},$$

which thus consists of all the natural numbers from I through N . Similarly,

$$\{I, \ldots\} := \{I, I+1, I+2, \ldots\} := \{K \in \mathbb{N} : I \leq K\},$$

which thus consists of all the natural numbers larger than or equal to I .

Example 514 If $E := \{2, \ldots, 7\}$, then $E = \{2, 3, 4, 5, 6, 7\}$. Also, $\min(E) = 2$ and $\max(E) = 7$.

Example 515 If $E := \{2, 3, \ldots\}$, then $\min(E) = 2$ but E has no maximum.

3.3.2 Laws of Arithmetic Cancellations

This subsection establishes rules to cancel terms in equations with additions or multiplications. The material also provides more practice with proofs by mathematical induction. The first rule shows how to solve equations of the form $M + 1 = L + 1$.

Theorem 516 *For all* $L, M \in \mathbb{N}$, *if* $M + 1 = L + 1$, *then* $M = L$.

Proof. If $M + 1 = L + 1$, then $M \cup \{M\} = L \cup \{L\}$, and the sets on both sides have the same elements. In particular, $L \in (L \cup \{L\})$ and thus $L \in (M \cup \{M\})$, whence two cases arise: $L \in \{M\}$, or $L \in M$.

In the first case, $L \in \{M\}$, and then $L = M$ indeed.

In the second case, $L \in M$; then $M \notin \{L\}$, for otherwise $M = L$ and $L \in L$ would contradict theorem 504. However, $M \in L \cup \{L\}$, whence $M \in L$, but that would also contradict theorem 504. Therefore, this second case does not occur. □

The following theorem allows for the cancellation of an additive term N common to both sides of an equation of the type $M + N = L + N$.

Theorem 517 *For all* $M, N, L \in \mathbb{N}$, *if* $M + N = L + N$, *then* $M = L$.

Proof. For all natural numbers $L, M \in \mathbb{N}$, proceed by induction with N. For $N := 0$, if $M + 0 = L + 0$, then $M = M + 0 = L + 0 = L$, by hypothesis and by (A0).

Secondly, *assume* that there exists $K \in \mathbb{N}$ such that the theorem holds for $N := K$, so that for all natural numbers $L, M \in \mathbb{N}$, if $M + K = L + K$ then $M = L$.

If $M + (K + 1) = L + (K + 1)$, then the associativity and the commutativity of addition lead to $(M + 1) + K = (L + 1) + K$, whence $M + 1 = L + 1$ by induction hypothesis, whence finally $M = L$ by theorem 516. □

The following theorem shows that addition preserves the direction of the ordering.

Theorem 518 *For all $L, M, N \in \mathbb{N}$, if $L < M$ then $L + N < M + N$.*

Proof. For all L and M, proceed by induction with N.

For $N := 0$, and for all L and M, if $L + 0 = M + 0$, then $L = L + 0 = M + 0 = M$.

To complete the induction, assume that there exists $K \in \mathbb{N}$ such that the theorem holds for $N := K$, so that for all L and M, if $L < M$ then $L + K < M + K$. Hence, if $L < M$ then $L + 1 < M + 1$ by theorem 506, whence

$$\begin{aligned} L + (K + 1) &= L + (1 + K) \\ &= (L + 1) + K \\ &< (M + 1) + K \\ &= M + (1 + K) \\ &= M + (K + 1). \end{aligned}$$

□

Theorem 519 *For all $M, N \in \mathbb{N}$, if $M \neq 0$, then $N < N + M$.*

Proof. Set $L := 0$ in theorem 518. If $M \neq 0$ then $0 < M$ by theorem 507, whence $L < M$. Hence $N = 0 + N < M + N$ by theorem 518. □

The following theorem forms the basis for the concept of subtraction of a natural number from a larger natural number.

Theorem 520 *For all $M, N \in \mathbb{N}$, if $M < N$, then there exists* exactly one *natural number $L \in \mathbb{N}$ such that $M + L = N$.*

Proof. This proof establishes the existence and the uniqueness separately.

Existence Firstly, establish the existence of such a number L. For each $M \in \mathbb{N}$, proceed by induction with N.

If $N := 0$, then $(M < 0)$ is False, because $(M < 0) \Leftrightarrow (M \in \varnothing)$. Therefore

$$\models \forall M \big([(M \in \mathbb{N}) \wedge (M < 0)] \Rightarrow \{\exists L[(L \in \mathbb{N}) \wedge (M + L = N)]\} \big)$$

is universally valid, and hence the theorem holds for $N = 0$.

For each $M \in \mathbb{N}$, assume that there exists $K \in \mathbb{N}$ such that the theorem holds for $N := K$, so that for each $M \in \mathbb{N}$, if $M < K$ then there exists $L \in \mathbb{N}$ such that $M + L = K$. If $M < (K + 1) = K \cup \{K\}$, then two cases arise.

In the first case, $M \in \{K\}$, whence $M = K$, and then $M+1 = K+1$, so that $L = 1$. In the second case, $M \in K$, that is, $M < K$, and the induction hypothesis yields some $L \in \mathbb{N}$ for which $M + L = K$; hence $M + (L + 1) = (M + L) + 1 = K + 1$.

Uniqueness Secondly, verify the uniqueness of L, which results from the theorem that if $M + L = N = M + K$, then $L = K$ (theorem 517). □

The following definition specifies the concept of subtraction of a natural number from a larger natural number.

Definition 521 (Subtraction) For all $L, M, N, \in \mathbb{N}$, if $M < N$, then $N - M := L$ is the natural number L defined in theorem 520 such that $M + L = N$

Example 522 $8 - 5 = 3$ because $5 + 3 = 8$ by exercise 627.

The following theorem shows that multiplication by a *non-zero* natural number preserves the ordering.

Theorem 523 *For all $L, M, N \in \mathbb{N}$, if $M < N$ and $0 < L$, then $L * M < L * N$.*

Proof. For all non-negative integers $M, N \in \mathbb{N}$, proceed by induction with L, the smallest non-zero value of L being 1. For $L := 1$, if $M < N$, then $1 * M = M < N = 1 * N$.

Assume that there exists $K \in \mathbb{N}$ with $0 < K$ such that the theorem holds for $L := K$, so that for all natural numbers $M, N \in \mathbb{N}$, if $M < N$, then $K * M < K * N$. Thus, if $M < N$, then apply theorem 518 twice:

$$(K + 1) * M = (K * M) + M < (K * M) + N < (K * N) + N = (K + 1) * N.$$

□

The following theorem allows for the cancellation of a non-zero multiplicative term on both sides of an equation, which forms the basis for the division of a natural number by a non-zero natural number, and the solution of equations of the type $L * M = L * N$.

Theorem 524 *For all $L, M, N \in \mathbb{N}$, if $0 < L$ and $L * M = L * N$, then $M = N$.*

Proof. Proceed by contraposition.

If $M \neq N$, then either $M < N$ or $N < M$.

If $M < N$, then $L * M < L * N$ and $L * M \neq L * N$.

Similarly, if $N < M$, then $L * N < L * M$ and $L * M \neq L * N$. □

Definition 525 (Division) For all $K, M, N \in \mathbb{N}$, if $0 < L$ and $K * L = N$, then define N/L by $N/L := K$, as defined uniquely by theorem 524.

Example 526 $8/4 = 2$ because $2 * 4 = 8$ by exercise 633.

3.3.3 Exercises

The following exercises involve the natural numbers (sets) $0, 1, 2, 3, 4, 5, 6, 7, 8, 9$ defined in example 311, page 111, and reviewed in subsection 3.2.3.

Exercise 641 Prove that $2 < 5$.

Exercise 642 Prove that $5 < 7$.

Exercise 643 Prove that $2 < 7$.

Exercise 644 Prove that $3 < 5$.

Exercise 645 Prove that $3 < 7$.

Exercise 646 Prove that $4 < 9$.

Exercise 647 Prove that $2 < 8$.

Exercise 648 Prove that there are no natural numbers $K > 1, L > 1$ with $K * L = 2$.

Exercise 649 Prove that there are no natural numbers $K > 1, L > 1$ with $K * L = 3$.

Exercise 650 Prove that there are no natural numbers $K > 1, L > 1$ with $K * L = 5$.

Exercise 651 Prove that there are no natural numbers $K > 1$, $L > 1$ with $K * L = 7$.

Exercise 652 Determine all the natural numbers $K > 1$ and $L > 1$ such that $K * L = 6$ and prove that there are no other such natural numbers.

Exercise 653 Determine all the natural numbers $K > 1$ and $L > 1$ such that $K * L = 9$ and prove that there are no other such natural numbers.

Exercise 654 Prove that the relation $\leq$ is a total ordering on $\mathbb{N}$, with

$$(M \leq N) \Leftrightarrow [(M = N) \vee (M \in N)].$$

Exercise 655 Prove that for each non-empty $S \subseteq \mathbb{N}$ there exists $I \in S$ with $I \cap S = \varnothing$.

Exercise 656 Prove that for each $L \in \mathbb{N}^*$ there exists $I \in L$ such that $I \cap L = \varnothing$.

Exercise 657 Prove that $(I \notin K) \vee (K \notin L) \vee (L \notin I)$ for all $I, K, L \in \mathbb{N}$.

Exercise 658 Prove that $1 < N$ for each $N \in (\mathbb{N} \setminus \{0, 1\})$.

Exercise 659 Prove that $\mathbb{N} \notin \mathbb{N}$.

Exercise 660 Prove that $\mathbb{N} \neq (\mathbb{N} \cup \{\mathbb{N}\})$.

3.4 INTEGERS

3.4.1 Negative Integers

Several operations remain undefined within the natural numbers. For instance, the ordering relation does not include any element smaller than the empty set, which precludes the use of natural numbers to model relations extending in two opposite directions. Also, the arithmetic of natural numbers does not contain any provision to record the "difference" from a larger natural number to a smaller one. Finally, the arithmetic of natural numbers does not include any concept of division other than special cases.

There exist several methods to extend the ordering and the arithmetic of natural numbers, to allow for elements "smaller" than zero, and for differences of any two elements. Such methods begin with the specification of a larger set of "integers" $\mathbb{Z}$ [from the German "die Zahl(en)" for "the number(s)"].

One method of defining a larger set, outlined by Kunen [64, p. 35], is sufficiently general to produce not only the integers, but also the rational numbers and the real numbers. Essentially, the method consists in defining the new numbers in terms of equivalence classes. For the integers specifically, the strategy consists in introducing the concept of the "difference" between two natural numbers M and N by means of the pair (M, N). Then relate (M, N) to every other pair (P, Q) with the same "difference" between P and Q. Because the concept of "difference" has not yet been defined for all pairs of natural numbers, however, a precise definition uses sums instead.

Two cases arise: either $N \leq M$, or $N > M$. In the first case, if $N \leq M$, then definition 521 specifies their difference $J := M - N$. If also (M, N) and (P, Q) represent the same difference, then $P - Q = J = M - N$. An equivalent statement without subtractions results from adding J to both extremes:

$$\begin{array}{ccccccccccc} M & - & N & = & J, & \quad & M & = & N & + & J; \\ P & - & Q & = & J, & \quad & P & = & Q & + & J. \end{array}$$

In the second case, if $N > M$ with $J := N - M$ and also $Q - P = J = N - M$, then

$$\begin{array}{ccccccccccc} N & - & M & = & J, & N & = & M & + & J; \\ Q & - & P & = & J, & Q & = & P & + & J. \end{array}$$

In either case, the statement that (M, N) and (P, Q) represent the same "difference" can be reworded without subtractions but with sums instead: there exists $J \in \mathbb{N}$ with

- either $M = N + J$ and $P = Q + J$ (if $M > N$ and $P > Q$),

$$\begin{array}{ccccccccccccc} & & & & & M & & & & N & & & \\ \cdots & \bullet & \bullet & \bullet & \bullet & \bullet & \bullet & \bullet & \bullet & \bullet & \bullet & \cdots \\ & & & P & & & & Q & & & & & \end{array}$$

- or $N = M + J$ and $Q = P + J$ (if $M < N$ and $P < Q$),

$$\begin{array}{ccccccccccccc} & & & & & N & & & & M & & & \\ \cdots & \bullet & \bullet & \bullet & \bullet & \bullet & \bullet & \bullet & \bullet & \bullet & \bullet & \cdots \\ & & & Q & & & & P & & & & & \end{array}$$

Swapping coordinates, from (M, N) to (N, M), will amount geometrically to reversing an order or a direction, which will amount algebraically to passing from a positive to a negative integer. Such considerations lead to the following definition.

Definition 527 Define a relation $\doteqdot$ on the set $A := \mathbb{N} \times \mathbb{N}$ for all $M, N, P, Q \in \mathbb{N}$ by

$$\begin{gathered} (M, N) \doteqdot (P, Q) \\ \Updownarrow \\ \exists J \big[(J \in \mathbb{N}) \wedge \{ [(M = N + J) \wedge (P = Q + J)] \vee [(N = M + J) \wedge (Q = P + J)] \} \big] : \end{gathered}$$

there is a $J \in \mathbb{N}$ with $M = N + J$ and $P = Q + J$, or $N = M + J$ and $Q = P + J$.

Example 528 $(3, 6) \doteqdot (5, 8)$ because $J := \mathbf{3}$ confirms that $6 = 3 + \mathbf{3}$ and $8 = 5 + \mathbf{3}$, by exercises 625 and 627.

The following theorem provides an equivalent formulation of this relation.

Theorem 529 *For all pairs of natural numbers,*

$$\begin{array}{rcl} (M, N) & \doteqdot & (P, Q) \\ & \Updownarrow & \\ M + Q & = & N + P. \end{array}$$

Proof. This proof establishes the two implications ($\Rightarrow$ and $\Leftarrow$) separately.

Assume first that $(M, N) \doteqdot (P, Q)$. Then two cases can arise.

If there exists $I \in \mathbb{N}$ with $M = N + I$ and $P = Q + I$, then

$$M + Q = (N + I) + Q = N + (I + Q) = N + (Q + I) = N + P.$$

If there exists $I \in \mathbb{N}$ with $N = M + I$ and $Q = P + I$, then

$$M + Q = M + (P + I) = (M + (I + P) = (M + I) + P = N + P.$$

For the converse, assume that $M + Q = N + P$. Then two cases can arise.

If $M \leq N$, then there exists $I \in \mathbb{N}$ such that $N = M + I$. Hence

$$M + Q = N + P = (M + I) + P = M + (P + I)$$

whence (by theorem 517) cancelling M yields $Q = P + I$.

If $M > N$, then there exists $I \in \mathbb{N}^*$ such that $M = N + I$. Hence

$$N + P = M + Q = (N + I) + Q = N + (Q + I)$$

whence (by theorem 517) cancelling N yields $P = Q + I$. □

Example 530 $(5, 4) \doteq (2, 1)$ because $5 + 1 = 4 + 2$, by exercises 615 and 621.

Instead of definition 527 the derivations can hence utilize theorem 529. Thus the following theorem shows that the relation $\doteq$ is an equivalence relation (definition 415).

Theorem 531 *The relation $\doteq$ is an equivalence relation on $\mathbb{N} \times \mathbb{N}$.*

Proof. This proof verifies that the relation $\doteq$ is reflexive, symmetric, and transitive.

Reflexivity For each pair $(M, N) \in (\mathbb{N} \times \mathbb{N})$, the equality $M + N = N + M$ holds by the commutativity of addition, whence $(M, N) \doteq (M, N)$.

Symmetry If $(M, N) \doteq (P, Q)$, then $M + Q = N + P$, whence $Q + M = P + N$, which means that $(P, Q) \doteq (M, N)$.

Transitivity If $(K, L) \doteq (M, N)$, then $K + N = L + M$. If also $(M, N) \doteq (P, Q)$, then $M + Q = N + P$. Consequently,

$$K + (M + Q) = K + (N + P) = (K + N) + P = (L + M) + P = L + (M + P)$$

whence (theorem 517) cancelling M yields $K + Q = L + P$, and $(K, L) \doteq (P, Q)$. □

Definition 532 (Kunen's definition of the integers) The set of integers is the set $\mathbb{Z} := (\mathbb{N} \times \mathbb{N}) / \doteq$ of all equivalence classes $[(M, N)]_{\doteq}$ for the relation $\doteq$ [64, p. 35].

Example 533 Setting $I := 3$ shows that $(0, 3) \doteq (1, 4) \doteq (2, 5) \doteq (3, 6) \doteq (4, 7)$. Thus the pairs $(0, 3), (1, 4), (2, 5), (3, 6), (4, 7)$ are elements of the equivalence class $[(0, 3)]_{\doteq}$. Similarly, $(3, 0) \doteq (4, 1) \doteq (5, 2) \doteq (6, 3) \doteq (7, 4)$. Thus the pairs $(3, 0), (4, 1), (5, 2), (6, 3), (7, 4)$ are elements of the equivalence class $[(3, 0)]_{\doteq}$.

Each pair $(I, J) \in \mathbb{N} \times \mathbb{N}$ is equivalent to a pair of the type $(K, 0)$ or $(0, K)$.

Theorem 534 *If $I > J$, then $(I, J) \doteq (I - J, 0)$.*
If $I \leq J$, then $(I, J) \doteq (0, J - I)$.

Proof. If $I > J$, then $I = (I - J) + J$ and $J = 0 + J$ whence $(I, J) \doteq (I - J, 0)$.

If $I \leq J$, then $I = 0 + I$ and $J = (J - I) + I$ whence $(I, J) \doteq (0, J - I)$. □

The following diagram shows a few elements from three equivalence classes: $[(0, 2)]_{\doteq}$, $[(0, 0)]_{\doteq}$, and $[\mathbf{(3, 0)}]_{\doteq}$ relative to the relation $\doteq$ on $A := \mathbb{N} \times \mathbb{N}$:

(1, 3) $(3, 3)$ **(6, 3)**
(0, 2) $(2, 2)$ **(5, 2)**
$(1, 1)$ **(4, 1)**
$(0, 0)$ **(3, 0)**

3.4.2 Arithmetic with Integers

From the preceding definition of integers in terms of equivalent pairs of natural numbers — which represent equivalent differences — follows a definition of arithmetic in terms of such pairs.

Definition 535 (Kunen's definition of integer arithmetic) Define an arithmetic with pairs of natural numbers as follows:

$$\begin{aligned}(M,N)+(P,Q) &:= (M+P,N+Q),\\ (M,N)*(P,Q) &:= ([M*P]+[N*Q],\ [M*Q]+[N*P]).\end{aligned}$$

The following theorem verifies that addition and multiplication of pairs commute.

Theorem 536 *For all $M,N,P,Q\in\mathbb{N}$,*

$$\begin{aligned}(M,N)+(P,Q) &= (P,Q)+(M,N),\\ (M,N)*(P,Q) &= (P,Q)*(M,N).\end{aligned}$$

Proof. Apply the commutativity of addition and multiplication with natural numbers:

$$\begin{aligned}(M,N)+(P,Q) &= (M+P,N+Q)\\ &= (P+M,Q+N)\\ &= (P,Q)+(M,N);\end{aligned}$$

$$\begin{aligned}(M,N)*(P,Q) &= ([M*P]+[N*Q],[M*Q]+[N*P])\\ &= ([P*M]+[Q*N],[Q*M]+[P*N])\\ &= ([P*M]+[Q*N],[P*N]+[Q*M])\\ &= (P,Q)*(M,N).\end{aligned}$$

The next theorem checks that arithmetic with equivalent pairs yields equivalent results: different pairs representing the same difference yield the same sum or product.

Theorem 537 *For all $I,J,K,L,M,N,P,Q\in\mathbb{N}$, if*

$$\begin{aligned}(I,J) &\simeq (K,L),\\ (M,N) &\simeq (P,Q),\end{aligned}$$

then

$$\begin{aligned}(I,J)+(M,N) &\simeq (K,L)+(P,Q),\\ (I,J)*(M,N) &\simeq (K,L)*(P,Q).\end{aligned}$$

Proof. If $(I,J)\simeq(K,L)$, then $I+L=J+K$. If also $(M,N)\simeq(P,Q)$, then $M+Q=N+P$. Moreover,

$$\begin{aligned}(I,J)+(M,N) &= (I+M,J+N),\\ (K,L)+(P,Q) &= (K+P,L+Q),\end{aligned}$$

whence

$$\begin{aligned}(I+M)+(L+Q) &= (I+L)+(M+Q)\\ &= (J+K)+(N+P)\\ &= (J+N)+(K+P),\end{aligned}$$

which means that $(I, J) + (M, N) \simeq (K, L) + (P, Q)$. For the multiplication,

$$\begin{aligned}(I, J) * (M, N) &= ([I * M] + [J * N], [I * N] + [J * M]),\\(K, L) * (P, Q) &= ([K * P] + [L * Q], [K * Q] + [L * P]),\end{aligned}$$

so that $(I, J) * (M, N) \simeq (K, L) * (P, Q)$ if and only if

$$([I * M] + [J * N]) + ([K * Q] + [L * P]) = ([I * N] + [J * M]) + ([K * P] + [L * Q]).$$

Additions and cancellations (theorem 517) give

$$\begin{aligned}
I * M + J * N + K * Q + L * P &= I * N + J * M + K * P + L * Q\\
&\Updownarrow\\
I * M + I * Q + J * N + K * Q + L * P &= I * N + I * Q + J * M + K * P + L * Q\\
&\Updownarrow\\
I * (M + Q) + J * N + K * Q + L * P &= I * N + J * M + K * P + (I + L) * Q\\
&\Updownarrow\\
I * (N + P) + J * N + K * Q + L * P &= I * N + J * M + K * P + (J + K) * Q\\
&\Updownarrow\\
I * P + J * N + L * P &= J * M + K * P + J * Q\\
&\Updownarrow\\
I * P + J * N + J * P + L * P &= J * M + K * P + J * P + J * Q\\
&\Updownarrow\\
I * P + J * (N + P) + L * P &= J * M + (K + J) * P + J * Q\\
&\Updownarrow\\
I * P + J * (M + Q) + L * P &= J * M + (I + L) * P + J * Q\\
&\Updownarrow\\
0 &= 0,
\end{aligned}$$

which is universally valid, whence $(I, J) * (M, N) \simeq (K, L) * (P, Q)$. □

Hence the following definition specifies an arithmetic with equivalence classes based on the arithmetic of representative pairs.

Definition 538 (Integer addition and multiplication)

$$\begin{aligned}([(M, N)]_{\simeq}) + ([(P, Q)]_{\simeq}) &:= ([(M + P,\ N + Q)]_{\simeq}),\\([(M, N)]_{\simeq}) * ([(P, Q)]_{\simeq}) &:= ([(M * P + N * Q,\ M * Q + N * P)]_{\simeq}).\end{aligned}$$

Theorem 539 *For $\mathbb{Z}$, and for all $K, L, M, N, P, Q \in \mathbb{N}$,*
addition commutes,
addition is associative,
$[(0, 0)]_{\simeq}$ *is the additive unit:* $[(M, N)]_{\simeq} + [(0, 0)]_{\simeq} = [(M, N)]_{\simeq}$,
$[(N, M)]_{\simeq}$ *is the additive inverse:* $[(M, N)]_{\simeq} + [(N, M)]_{\simeq} = [(0, 0)]_{\simeq}$,
multiplication commutes,
multiplication is associative,
multiplication distributes over addition, and
$[(1, 0)]_{\simeq}$ *is a multiplicative unit* $[(M, N)]_{\simeq} * [(1, 0)]_{\simeq} = [(M, N)]_{\simeq}$.

Proof. The proof forms the object of exercises. □

The following definition specifies a subtraction.

Definition 540 (Integer subtraction) Define a subtraction of integers by

$$([(M,N)]_{\simeq}) - ([(P,Q)]_{\simeq}) \quad := \quad ([(M,N)]_{\simeq}) + ([(Q,P)]_{\simeq}),$$

which is a binary operation $-$ from $\mathbb{Z} \times \mathbb{Z}$ to $\mathbb{Z}$. Define an "opposite" by

$$-([(P,Q)]_{\simeq}) \quad := \quad ([(Q,P)]_{\simeq}),$$

which is a unary operation (unfortunately also denoted by $-$) from $\mathbb{Z}$ to $\mathbb{Z}$.

Example 541 Reducing every pair (I,J) to either form $(I-J,0)$ or $(0,J-I)$ can facilitate calculations:

$$\begin{aligned}
[(2,3)]_{\simeq} + [(5,1)]_{\simeq} &= [(0,1)]_{\simeq} + [(4,0)]_{\simeq} \\
&= [(0+4,1+0)]_{\simeq} \\
&= [(4,1)]_{\simeq} \\
&= [(3,0)]_{\simeq};
\end{aligned}$$

$$\begin{aligned}
[(2,1)]_{\simeq} - [(7,3)]_{\simeq} &= [(2,1)]_{\simeq} + [(3,7)]_{\simeq} \\
&= [(1,0)]_{\simeq} + [(0,4)]_{\simeq} \\
&= [(1,4)]_{\simeq} \\
&= [(0,3)]_{\simeq} \\
&= -[(3,0)]_{\simeq};
\end{aligned}$$

$$\begin{aligned}
[(3,1)]_{\simeq} + [(2,5)]_{\simeq} &= [(2,0)]_{\simeq} + [(0,3)]_{\simeq} \\
&= [(2*0+0*3, 2*3+0*0)]_{\simeq} \\
&= [(0,6)]_{\simeq} \\
&= -[(6,0)]_{\simeq}.
\end{aligned}$$

3.4.3 Order on the Integers

The order $<$ on $\mathbb{N}$ extends to $\mathbb{Z}$ by a definition of "positive" and "negative" on $\mathbb{Z}$. By theorem 534, every integer $[(M,N)]_{\simeq}$ is the equivalence class of a pair $(I,0)$ or $(0,I)$ for some $I \in \mathbb{N}$. In the first case $M - N = I - 0 \geq 0$, so that $M - N$ is a non-negative difference. In the second case $N - M = I - 0 \geq 0$, so that the opposite, $N - M$, is a non-negative difference, and then $M - N$ represents a non-positive difference.

Definition 542 An integer $[(M,N)]_{\simeq}$ is **positive** if and only if $M > N$; an integer $[(M,N)]_{\simeq}$ is **negative** if and only if $M < N$.

Definition 543 Define the sets of non-zero integers ($\mathbb{Z}^*$), negative integers ($\mathbb{Z}^*_-$), positive integers ($\mathbb{Z}^*_+$), non-positive integers ($\mathbb{Z}_-$), and non-negative integers ($\mathbb{Z}_+$) by

$$\mathbb{Z}^* \quad := \quad \mathbb{Z} \setminus \{[(0,0)]_{\simeq}\},$$

$$\begin{aligned}
\mathbb{Z}_- &:= \{[(M,N)]_{\simeq} \in \mathbb{Z} : M \leq N\}, \\
&:= \{[(M,N)]_{\simeq} \in \mathbb{Z} : \exists I\{(I \in \mathbb{N}) \wedge [(M,N) \simeq (0,I)]\}\},
\end{aligned}$$

$$\begin{aligned}
\mathbb{Z}_+ &:= \{[(M,N)]_{\simeq} \in \mathbb{Z} : M \geq N\}, \\
&:= \{[(M,N)]_{\simeq} \in \mathbb{Z} : \exists I\{(I \in \mathbb{N}) \wedge [(M,N) \simeq (I,0)]\}\},
\end{aligned}$$

$$\begin{aligned}
\mathbb{Z}^*_- &:= \{[(M,N)]_{\simeq} \in \mathbb{Z} : M < N\}, \\
&:= \{[(M,N)]_{\simeq} \in \mathbb{Z} : \exists I\{(I \in \mathbb{N}^*) \wedge [(M,N) \simeq (0,I)]\}\},
\end{aligned}$$

$$\begin{aligned}
\mathbb{Z}^*_+ &:= \{[(M,N)]_{\simeq} \in \mathbb{Z} : M > N\} \\
&:= \{[(M,N)]_{\simeq} \in \mathbb{Z} : \exists I\{(I \in \mathbb{N}^*) \wedge [(M,N) \simeq (I,0)]\}\}.
\end{aligned}$$

The following relation on pairs will lead to an order on equivalence classes.

Definition 544 Define a relation $<$ on $\mathbb{N} \times \mathbb{N}$ as follows:

$$\begin{array}{rcl} (M,N) & < & (0,0) \\ & \Updownarrow & \\ M & < & N \end{array}$$

and

$$\begin{array}{rcl} (M,N) & < & (P,Q) \\ & \Updownarrow & \\ (M,N)+(Q,P) & < & (0,0) \\ & \Updownarrow & \\ M+Q & < & N+P. \end{array}$$

The following theorem verifies that the ordering does not depend on the choice of the pairs representing the equivalence classes.

Theorem 545 *For all $I, J, K, L, M, N, P, Q \in \mathbb{N}$, if*

$$\begin{array}{rcl} (I,J) & \doteqdot & (K,L), \\ (M,N) & \doteqdot & (P,Q), \end{array}$$

then

$$\begin{array}{rcl} (I,J) & < & (M,N) \\ & \Updownarrow & \\ (K,L) & < & (P,Q). \end{array}$$

Proof. If $(I,J) \doteqdot (K,L)$, then $I+L = J+K$. If also $(M,N) \doteqdot (P,Q)$, then also $M+Q = N+P$. Consequently, adding $L+P$ to each side of the inequalities yields

$$\begin{array}{rcl} (I,J) & < & (M,N) \\ & \Updownarrow & \\ I+N & < & J+M \\ & \Updownarrow & \\ (I+L)+(N+P) & < & (L+P)+(J+M) \\ & \Updownarrow & \\ (J+K)+(M+Q) & < & (L+P)+(J+M) \\ & \Updownarrow & \\ (J+M)+(K+Q) & < & (L+P)+(J+M) \\ & \Updownarrow & \\ K+Q & < & L+P \\ & \Updownarrow & \\ (K,L) & < & (P,Q). \end{array}$$

□

A definition of an order on $\mathbb{Z}$ can thus use any pair from an equivalence class.

Definition 546 Define an order $<$ on $\mathbb{Z}$ by

$$\begin{array}{rcl} [(M,N)]_{\doteqdot} & < & [(P,Q)]_{\doteqdot} \\ & \Updownarrow & \\ (M,N) & < & (P,Q) \\ & \Updownarrow & \\ M+Q & < & N+P. \end{array}$$

The next theorem shows that the relation $<$ is a strict total order (definition 451).

Theorem 547 *The relation $<$ is a strict total order on $\mathbb{Z} \times \mathbb{Z}$. In particular, if*

$$[(K, L)]_{\simeq} < [(M, N)]_{\simeq}$$

and

$$[(M, N)]_{\simeq} < [(P, Q)]_{\simeq},$$

then

$$[(K, L)]_{\simeq} < [(P, Q)]_{\simeq}.$$

Proof. This proof verifies that $<$ is irreflexive, connected, and transitive.

Irreflexivity (The relation does not relate any element to itself.) For each pair $(M, N) \in \mathbb{N} \times \mathbb{N}$, $M + N \not< N + M$, whence $([(M, N)]_{\simeq}) \not< ([(M, N)]_{\simeq})$.

Connectedness (Each element is related to every *different* element.) If $([(M, N)]_{\simeq}) \neq ([(P, Q)]_{\simeq})$, then $M + Q \neq N + P$ whence either $M + Q < N + P$, and then $([(M, N)]_{\simeq}) < ([(P, Q)]_{\simeq})$, or $M + Q > N + P$, and then $([(M, N)]_{\simeq}) > ([(P, Q)]_{\simeq})$.

Transitivity If $([(K, L)]_{\simeq}) < ([(M, N)]_{\simeq})$, then

$$K + N < L + M;$$

if also $([(M, N)]_{\simeq}) < ([(P, Q)]_{\simeq})$, then also

$$M + Q < N + P.$$

Consequently, adding the preceding two inequalities gives

$$(K + N) + (M + Q) < (L + M) + (N + P)$$

whence the commutativity and associativity of addition yields

$$K + [(M + N) + Q] < L + [(M + N) + P]$$

whence (by theorem 518) cancelling $M + N$ yields

$$K + Q < L + P$$

so that $[(K, L)]_{\simeq} < [(P, Q)]_{\simeq}$. □

The next theorem shows that multiplication by a positive integer keeps the order.

Theorem 548 *If*

$$[(M, N)]_{\simeq} \quad < \quad [(P, Q)]_{\simeq},$$

$$[(0, 0)]_{\simeq} \quad < \quad [(I, J)]_{\simeq} \quad < \quad [(K, L)]_{\simeq},$$

then

$$([(I, J)]_{\simeq} * [(M, N)]_{\simeq}) \quad < \quad ([(I, J)]_{\simeq} * [(P, Q)]_{\simeq}),$$

whence

$$([(I, J)]_{\simeq} * [(M, N)]_{\simeq}) \quad < \quad ([(K, L)]_{\simeq} * [(P, Q)]_{\simeq}).$$

Proof. The proof uses the equivalence $(I,J) \doteq (I-J,0)$. Firstly, from $[(M,N)]_{\doteq} < [(P,Q)]_{\doteq}$ it follows that $M+Q < N+P$ by definition of $<$ on the pairs. Hence a multiplication throughout by $I-J>0$ yields

$$(I-J)*(M+Q) < (I-J)*(N+P)$$

whence

$$\begin{array}{rcl} (I,J)*(M,N) & \doteq & (I-J,0)*(M,N) \\ & = & ([I-J]*M,[I-J]*N) \\ & < & ([I-J]*P,[I-J]*Q) \\ & = & (I-J,0)*(P,Q) \\ & \doteq & (I,J)*(P,Q). \end{array}$$

Swapping the roles of (I,J) with (K,L), and (M,N) with (P,Q), then yields

$$(I,J)*(P,Q) = (P,Q)*(I,J) < (P,Q)*(K,L) = (K,L)*(P,Q).$$

Consequently,

$$(I,J)*(M,N) < (I,J)*(P,Q) < (K,L)*(P,Q).$$

□

The following theorem shows that the square of every integer is non-negative.

Theorem 549 *For every $X \in \mathbb{Z}$, $X*X \geq 0$.*

Proof. For every $X \in \mathbb{Z}$ there exists $K \in \mathbb{N}$ such that $X = [(K,0)]_{\doteq}$ or $X = [(0,K)]_{\doteq}$, by theorem 534. However,

$$\begin{array}{rcl} (K,0)*(K,0) & = & (K*K+0*0, K*0+0*K) \\ & = & (K*K,0) \\ & = & (0*0+K*K, 0*K+K*0) \\ & = & (0,K)*(0,K) \end{array}$$

and in either case $X*X = [(K*K,0)]_{\doteq} \geq [(0,0)]_{\doteq}$ because $K*K+0 \geq 0+0$. □

Example 550 $(-[(1,0)]_{\doteq})*(-[(1,0)]_{\doteq}) = ([(0,1)]_{\doteq})*([(0,1)]_{\doteq}) = [(1,0)]_{\doteq}$.

An alternative method to define all the integers specifies from $\mathbb{N}$ a similar but disjoint set $\mathbb{Z}_-^*$ for all the "negative" integers, and then defines $\mathbb{Z} := \mathbb{Z}_-^* \cup \mathbb{N}$. For instance, apply the axioms of the power set, pairing, and separation to define

$$\begin{array}{rcl} \mathbb{Z}_-^* & := & \{K \in \mathcal{P}(\mathbb{N}) : \exists N\,[(N \in \mathbb{N}) \wedge (N \neq 0) \wedge (K = \{N\})]\} \\ & = & \{\,\{N\} : N \in \mathbb{N}^*\,\} \\ & = & \{\ldots, \{\{\varnothing,\{\varnothing\}\}, \{\{\varnothing\}\}\} \\ & = & \{\ldots, \{3\}, \{2\}, \{1\}\}. \end{array}$$

Then change the notation to define $-N := \{N\}$ for each $N \in \mathbb{N}^*$. In particular, $\mathbb{Z}_-^* \cap \mathbb{N} = \varnothing$. Indeed, $\varnothing \in N$ by theorem 507 and $\varnothing \notin \{N\}$ for each positive natural number $N \in \mathbb{N}\setminus\{\varnothing\}$. Consequently, $\{N\} \notin \mathbb{N}$ for every $\{N\}\mathbb{Z}_-^*$.

Definition 551 (Landau's definition of integer arithmetic) The set $\mathbb{Z} := \mathbb{Z}_-^* \cup \mathbb{N}$ is the set of of **integers**. The arithmetic operations extend from $\mathbb{N}$ to $\mathbb{Z}$ by setting

$$\begin{array}{rcll} (-M)+(-N) & := & -(M+N), & \\ (-N)+M := M+(-N) & := & -(N-M) & \text{if } M < N, \\ (-N)+M := M+(-N) & := & M-N & \text{if } N < M. \end{array}$$

Similarly,

$$\begin{aligned} (-N)*(-M) &:= M*N, \\ (-N)*M := M*(-N) &:= -(M*N). \end{aligned}$$

Moreover, the ordering $\in$ on $\mathbb{N}$ extends to an ordering $\mathbb{Z}$ by the definition

$$\begin{aligned} (-N) &< (-M) \quad \text{if and only } M < N, \\ (-N) &< K \end{aligned}$$

for all $K \in \mathbb{N}$ and $M, N \in \mathbb{N}^*$.

The addition and multiplication thus defined for all integers remain associative and commutative, multiplication distributes over addition, $N + 0 = N$ and $1 * N = N$ for each integer $N \in \mathbb{Z}$. The verifications proceed by cases, depending upon the sign of the operands, and are straightforward but lengthy. See [65, Ch. IV] or the exercises.

Remark 552 Common usage abbreviates each non-negative integer $[(I, 0)]_{\simeq}$ by I. Also, any variable can denote an integer, for instance, $M = [(P, Q)]_{\simeq}$.

3.4.4 Non-Negative Integral Powers of Integers

The J-th power M^J of an integer $M \in \mathbb{Z}$ is the product $M * \cdots * M$ of J factors M. Specifically, from the convention $M^0 := 1$, induction produces higher powers.

Definition 553 (Integral powers) For each integer M, define

$$M^0 := 1.$$

Then for each integer M and for each non-negative integer J define

$$M^{J+1} := (M^J) * M.$$

In M^J, the number M is the **base** while J is the **exponent**.

Remark 554 According to definition 553, the Jth power of M amounts to J multiplications of M, begining with 1. With an alternative but logically equivalent notation, for each $M \in \mathbb{N}$ definition 553 uses a recursive definition with G in theorem 484 replaced by the multiplication function $H^{(M)}$ from remark 492:

$$\begin{aligned} G := H^{(M)} : \mathbb{N} &\to \mathbb{N}, \\ L &\mapsto M * L, \end{aligned}$$

to specify by induction an **exponentiation** function $E^{(M)} : \mathbb{N} \to \mathbb{Z}$, $J \mapsto M^J$, with

$$\begin{aligned} G(L) &:= H^{(M)}(L) = M * L, \\ E^{(M)}(0) &:= 1, \\ E^{(M)}(J+1) &:= G\left[E^{(M)}(J)\right]. \end{aligned}$$

Example 555 Here are the first four non-negative powers of the integer 2:

$$\begin{aligned} 2^0 &= 1; \\ 2^1 &= (2^0) * 2 = 1 * 2 = 2; \\ 2^2 &= (2^1) * 2 = 2 * 2 = 4; \\ 2^3 &= (2^2) * 2 = 4 * 2 = 8. \end{aligned}$$

The following theorem establishes relations between product of bases, sums of exponents, and integral powers.

Theorem 556 *For all $M \in \mathbb{Z}$ and $N \in \mathbb{Z}$, and for all $I \in \mathbb{N}$ and $J \in \mathbb{N}$,*

$$\begin{aligned} (M * N)^J &= (M^J) * (N^J), \\ N^{I+J} &= (N^I) * (N^J). \end{aligned}$$

Proof. This proof proceeds by induction with the exponent J. The first equation, $(M*N)^J = (M^J) * (N^J)$, holds for all integers M and N, and for $J \in \{0, 1\}$:

$$(M * N)^0 = 1 = 1 * 1 = (M^0) * (N^0);$$

$$(M * N)^1 = M * N = (M^1) * (N^1).$$

Hence, to prove that $(M * N)^J = (M^J) * (N^J)$ for every $J \in \mathbb{N}$, let

$$S := \{J \in \mathbb{N} : \forall M \forall N \{[(M \in \mathbb{N}) \wedge (N \in \mathbb{N})] \Rightarrow [(M * N)^J = (M^J) * (N^J)]\}\}.$$

Thus, *assume* that there exists $K \in S$, or, equivalently, $K \in \mathbb{N}$ such that the theorem holds for $J := K$; then

$$\begin{aligned} (M * N)^{K+1} &= [(M * N)^K] * (M * N) && \text{definition 553,} \\ &= [(M^K) * (N^K)] * (M * N) && \text{induction hypothesis,} \\ &= [(M^K) * M] * [(N^K) * N] && \text{associativity and commutativity,} \\ &= (M^{K+1}) * (N^{K+1}) && \text{definition 553 twice,} \end{aligned}$$

whence $K + 1 \in S$, and hence, $S = \mathbb{N}$, which means that the first equation holds.

The second equation, $N^{I+J} = (N^I) * (N^J)$, holds for each integer N, for each nonnegative integer I, and for $J \in \{0, 1\}$:

$$\begin{aligned} N^{I+0} = N^I = (N^I) * 1 = (N^I) * (N^0), \\ N^{I+1} = (N^I) * N = (N^I) * (N^1). \end{aligned}$$

Next, let

$$S := \{J \in \mathbb{N} : \forall I \forall N [N^{I+J} = (N^I) * (N^J)]\}$$

and *assume* that there exists $K \in S$, or, equivalently, $K \in \mathbb{N}$ such that the theorem holds for $J := K$, so that $N^{I+K} = (N^I) * (N^K)$ for all $I \in \mathbb{N}$ and $N \in \mathbb{N}$; then

$$\begin{aligned} N^{I+(K+1)} &= N^{(I+1)+K} && \text{associativity and commutativity of } +, \\ &= (N^{I+1}) * (N^K) && \text{induction hypothesis,} \\ &= [(N^I) * N] * (N^K) && \text{definition 553,} \\ &= (N^I) * [(N^K) * N] && \text{associativity and commutativity of } *, \\ &= (N^I) * (N^{K+1}) && \text{definition 553,} \end{aligned}$$

whence $K + 1 \in S$, and, consequently, $S = \mathbb{N}$, which proves the second equation. □

3.4.5 Exercises

Exercise 661 Calculate $[(2, 4)]_{\simeq} + [(6, 3)]_{\simeq}$.

Exercise 662 Calculate $[(5, 3)]_{\simeq} + [(1, 6)]_{\simeq}$.

Exercise 663 Calculate $[(3,1)]_{\simeq} - [(5,2)]_{\simeq}$.

Exercise 664 Calculate $[(7,3)]_{\simeq} * [(2,5)]_{\simeq}$.

Exercise 665 Prove that Kunen's addition commutes.

Exercise 666 Prove that Landau's addition commutes.

Exercise 667 Prove that Kunen's addition is associative.

Exercise 668 Prove that Landau's addition is associative.

Exercise 669 Prove that Kunen's multiplication commutes.

Exercise 670 Prove that Landau's multiplication commutes.

Exercise 671 Prove that Kunen's multiplication is associative.

Exercise 672 Prove that Landau's multiplication is associative.

Exercise 673 Prove that Kunen's multiplication distributes over addition.

Exercise 674 Prove that Landau's multiplication distributes over addition.

Exercise 675 Prove that subtraction does *not* commute.

Exercise 676 Prove that subtraction is *not* associative.

Exercise 677 Prove that multiplication distributes over subtraction.

Exercise 678 Prove that subtraction does *not* distribute over multiplication.

Exercise 679 Prove that subtraction does *not* distribute over addition.

Exercise 680 Prove that addition does *not* distribute over subtraction.

3.5 RATIONAL NUMBERS

3.5.1 Definition of Rational Numbers

Some practical situations involve comparisons of proportions. Integer arithmetic does not allow for proportions, but a method similar to that for passing from natural numbers to differences also leads from integers to proportions. As a pair of natural numbers (I, J) can represent a difference, a pair of integers (P, Q) can represent a proportion.

For example, the density of the universe at one location involves the mass or number of particles P in a volume Q, which can be summarized by the ordered pair (P, Q). The density at another location involves the mass or number of particles M in a volume N, as summarized by the pair (M, N). A comparison of the densities (P, Q) and (M, N) can proceed through a multiplication by a common factor, for instance, Q or N.

Specifically, in a volume N times larger than Q, the first density of P particles in a volume Q becomes $P * N$ particles in a volume $Q * N$, or $(P * N, Q * N)$.

Likewise in a volume Q times larger than N, the second density of M particles in a volume N becomes $Q * M$ particles in a volume $Q * N$, or $(Q * M, Q * N)$.

Both densities $(P * N, Q * N)$ and $(Q * M, Q * N)$ refer to the same volume $Q * N$; thus, they are identical if and only if their masses equal each other: $P * N = Q * M$.

The foregoing reasoning leads to a relation between pairs of integers.

Definition 557 On the set $\mathbb{Z} \times \mathbb{Z}^*$ of all pairs of integers (P, Q) with $Q \neq 0$, define a relation $\equiv$ by

$$(P, Q) \equiv (M, N)$$
$$\Updownarrow$$
$$P * N = Q * M$$

The relation $\equiv$ on $\mathbb{Z} \times \mathbb{Z}^*$ in definition 557 *differs* from the relation $\triangleq$ on $\mathbb{N} \times \mathbb{N}$ in definition 527. Nevertheless, it is also an equivalence relation (definition 415).

Theorem 558 *The relation $\equiv$ in definition 557 is an equivalence relation on $\mathbb{Z} \times \mathbb{Z}^*$.*

Proof. This proof verifies algebraically that $\equiv$ is reflexive, symmetric, and transitive.

Reflexivity For each pair, $(I, J) \equiv (I, J)$ because $I * J = J * I$.

Symmetry If $(I, J) \equiv (K, L)$, then $I * L = J * K$ by definition of $\equiv$ whence $K * J = L * I$ by commutativity, so that $(K, L) \equiv (I, J)$.

Transitivity If

$$\begin{aligned} (I, J) &\equiv (K, L), \\ (K, L) &\equiv (M, N), \end{aligned}$$

then

$$\begin{aligned} I * L &= J * K, \\ K * N &= L * M, \end{aligned}$$

whence multiplying the left-hand sides and the right-hand sides together gives

$$\begin{aligned} (I * L) * (K * N) &= (J * K) * (L * M), \\ (I * N) * (L * K) &= (J * M) * (K * L), \end{aligned}$$

and hence cancelling the non-zero factor $L * K = K * L$ yields

$$I * N = J * M$$

which means that $(I, J) \equiv (M, N)$. □

Definition 559 The set of **rational** numbers, denoted by $\mathbb{Q}$ (for "quotients"), is the set of all equivalence classes for the relation $\equiv$. A **rational number** is an element of $\mathbb{Q}$.

The equivalence class of a pair (I, J) is called the **ratio** of I to J, and it is denoted by $[(I, J)]$, or also by I/J, or also by $\frac{I}{J}$. A **fraction** is a pair $(I, J) \in (\mathbb{Z} \times \mathbb{Z}^*)$, where I is the **numerator** and J is the **denominator**.

The following theorem provides a means to select a specific fraction from a rational number, or, equivalently, a specific pair (M, N) from an equivalence class P/Q.

Theorem 560 *For each $P/Q \in \mathbb{Q}$, there exists $M/N \equiv P/Q$ such that*

$$M = \min\{I \in \mathbb{N} : \exists J[(J \in \mathbb{N}) \wedge (I/J \equiv P/Q)]\}.$$

Proof. This proof applies the Well-Ordering Principle to the set of all non-negative numerators of a rational number. To this end, for each $P/Q \in \mathbb{Q}$ define the set

$$E := \{I \in \mathbb{N} : \exists J[(J \in \mathbb{N}) \wedge (I/J \equiv P/Q)]\}.$$

Then $E \neq \varnothing$, because if $P \geq 0$ then $P/Q \equiv P/Q$, whence $P \in E$, whereas if $P < 0$ then $(-P)/(-Q) \equiv P/Q$, whence $(-P) \in E$. The Well-Ordering Principle (theorem 509) then guarantees the existence of a first element in E. □

3.5.2 Arithmetic with Rational Numbers

The comparison of two rational numbers P/Q and M/N can proceed through equivalent fractions with a common denominator, for instance, $(P*N)/(Q*N)$ and $(Q*M)/(Q*N)$. Common denominators also lead to an arithmetic with fractions (ordered pairs) and then with rational numbers (equivalence classes).

Definition 561 For all pairs (I,J) and (K,L) in $\mathbb{Z} \times \mathbb{Z}^*$, define functions $+$ and $*$ on $(\mathbb{Z} \times \mathbb{Z}^*) \times (\mathbb{Z} \times \mathbb{Z}^*)$ by their counterparts $+$ and $*$ already defined on $\mathbb{Z} \times \mathbb{Z}^*$:

$$\begin{aligned} (I,J)+(K,L) &:= ([I*L]+[J*K],\ J*L)\,, \\ (I,J)*(K,L) &:= (I*K,\ J*L)\,. \end{aligned}$$

The symbols $+$ and $*$ on the left-hand sides are the functions being defined, whereas the symbols $+$ and $*$ on the right-hand sides are the addition and multiplication of integers. Yet common usage employs $+$ and $*$ for both.

The following theorem shows that equivalent fractions lead to equivalent results.

Theorem 562 *If*

$$\begin{aligned} (I,J) &\equiv (K,L), \\ (M,N) &\equiv (P,Q), \end{aligned}$$

then

$$\begin{aligned} (I,J)+(M,N) &\equiv (K,L)+(P,Q), \\ (I,J)*(M,N) &\equiv (K,L)*(P,Q). \end{aligned}$$

Proof. This proof proceeds through algebraic verifications. By hypotheses, $I*L=J*K$ and $M*Q=N*P$, whence

$$\begin{aligned} (I,J)+(M,N) &= (I*N+J*M, J*N), \\ (K,L)+(P,Q) &= (K*Q+L*P, L*Q), \end{aligned}$$

where

$$\begin{aligned} &(I*N+J*M)*(L*Q)-(J*N)*(K*Q+L*P) \\ &= [I*N*L*Q+J*M*L*Q]-[J*N*K*Q+J*N*L*P] \\ &= [(I*L)*(N*Q)+(J*L)*(M*Q)] \\ &\quad -[(J*K)*(N*Q)+(J*L)*(N*P)] \\ &= [(I*L)*(N*Q)+(J*L)*(M*Q)] \\ &\quad -[(I*L)*(N*Q)+(J*L)*(M*Q)] \\ &= 0. \end{aligned}$$

Thus, $(I,J)+(M,N) \equiv (K,L)+(P,Q)$. Similarly,

$$\begin{aligned} (I,J)*(M,N) &= (I*M, J*N), \\ (K,L)*(P,Q) &= (K*P, L*Q), \end{aligned}$$

where, by hypotheses,

$$\begin{aligned}&(I * M) * (L * Q) - (J * N) * (K * P)\\ &= (I * L) * (M * Q) - (J * K) * (N * P)\\ &= (I * L) * (M * Q) - (I * L) * (M * Q)\\ &= 0.\end{aligned}$$

Thus, $(I, J) * (M, N) \equiv (K, L) * (P, Q)$. □

Definition 563 Define the addition and the multiplication of rational numbers by

$$\begin{aligned}\frac{I}{J} + \frac{K}{L} &:= \frac{I * L + J * K}{J * L},\\ \frac{I}{J} * \frac{K}{L} &:= \frac{I * K}{J * L}.\end{aligned}$$

The following theorems establish algebraic characteristics of rational arithmetic.

Theorem 564 *The addition of rational numbers commutes.*

Proof. For all I/J and K/L in $\mathbb{Q}$,

$$\frac{I}{J} + \frac{K}{L} = \frac{I * L + J * K}{J * L} = \frac{L * I + K * J}{L * J} = \frac{K}{L} + \frac{I}{J}.$$

□

Theorem 565 *The multiplication of rational numbers commutes.*

Proof. For all I/J and K/L in $\mathbb{Q}$,

$$\frac{I}{J} * \frac{K}{L} = \frac{I * K}{J * L} = \frac{K * I}{L * J} = \frac{K}{L} * \frac{I}{J}.$$

□

Theorem 566 *The addition of rational numbers is associative.*

Proof. For all I/J, K/L, and M/N in $\mathbb{Q}$,

$$\begin{aligned}\left(\frac{I}{J} + \frac{K}{L}\right) + \frac{M}{N} &= \frac{I * L + J * K}{J * L} + \frac{M}{N}\\ &= \frac{[(I * L + J * K) * N] + [(J * L) * M]}{(J * L) * N}\\ &= \frac{[I * (L * N) + J * (K * N)] + [J * (L * M)]}{J * (L * N)}\\ &= \frac{[I * (L * N)] + [J * (K * N + L * M)]}{J * (L * N)}\\ &= \frac{I}{J} + \frac{K * N + L * M}{L * N}\\ &= \frac{I}{J} + \left(\frac{K}{L} + \frac{M}{N}\right).\end{aligned}$$

□

Theorem 567 *The multiplication of rational numbers is associative.*

Proof. For all I/J, K/L, and M/N in $\mathbb{Q}$,

$$\begin{aligned}
\left(\frac{I}{J} * \frac{K}{L}\right) * \frac{M}{N} &= \frac{I * K}{J * L} * \frac{M}{N} \\
&= \frac{(I * K) * M}{(J * L) * N} \\
&= \frac{I * (K * M)}{J * (L * N)} \\
&= \frac{I}{J} * \frac{K * M}{L * N} \\
&= \frac{I}{J} * \left(\frac{K}{L} * \frac{M}{N}\right).
\end{aligned}$$

□

Theorem 568 *For each $N \in \mathbb{N}^*$ and each $I/J \in \mathbb{Q}$, multiplications of the numerator and denominator by a non-zero common factor yields the same rational number: $I/J = (I * N)/(J * N)$.*

Proof. Verify the criterion for equivalent fractions: $I * (J * N) = J * (I * N)$. □

Theorem 569 *The multiplication of rational numbers distributes over addition.*

Proof. For all I/J, K/L, and M/N in $\mathbb{Q}$,

$$\begin{aligned}
\left(\frac{I}{J} + \frac{K}{L}\right) * \frac{M}{N} &= \frac{I * L + J * K}{J * L} * \frac{M}{N} \\
&= \frac{(I * L + J * K) * M}{(J * L) * N} \\
&= \frac{(I * L) * M + (J * K) * M)}{J * (L * N)} \\
&= \frac{(I * M) * L + J * (K * M)}{J * (L * N)} \\
&= \frac{(I * M) * (L * N) + (J * N) * (K * M)}{(J * N) * (L * N)} \\
&= \frac{I * M}{J * N} + \frac{K * M}{L * N} \\
&= \left(\frac{I}{J} * \frac{M}{N}\right) + \left(\frac{K}{L} * \frac{M}{N}\right).
\end{aligned}$$

□

The following theorem shows that adding $0/1$ does not produce any change.

Theorem 570 *For each* $K/L \in \mathbb{Q}$, $(K/L) + (0/1) = (K/L)$.

Proof. $(K/L) + (0/1) = ([K * 1 + L * 0]/L * 1) = (K/L)$. □

Thus, the rational number $0/1$ plays the same role as the integer 0 in additions. The following theorem shows that each rational number has an additive inverse.

Theorem 571 *Each* $I/J \in \mathbb{Q}$ *has an additive inverse:* $(I/J) + ([-I]/J) = (0/1)$.

Proof.

$$\begin{aligned}
\frac{I}{J} + \frac{-I}{J} &= \frac{I * J + J * (-I)}{J * J} \\
&= \frac{J * I + J * (-I)}{J * J} \\
&= \frac{J * [I + (-I)]}{J * J} \\
&= \frac{J * 0}{J * J} \\
&= \frac{0}{J * J} \\
&= \frac{0 * (J * J)}{1 * (J * J)} \\
&= \frac{0}{1}.
\end{aligned}$$

□

The following theorem shows that multiplying by $1/1$ does not produce any change.

Theorem 572 *For each* $K/L \in \mathbb{Q}$, $(K/L) * (1/1) = (K/L)$.

Proof. $(K/L) * (1/1) = ([K * 1]/[L * 1]) = (K/L)$. □

Thus, the rational number $1/1$ plays the same role as the integer 1 in multiplications. Similarly, each non-zero rational number has an multiplicative inverse.

Theorem 573 *Each* $I/J \in \mathbb{Q}$ *such that* $I \neq 0$ *has a multiplicative inverse:* $(I/J) * (J/I) = (1/1)$.

Proof.

$$\frac{I}{J} * \frac{J}{I} = \frac{I * J}{J * I} = \frac{I * J}{I * J} = \frac{1 * (I * J)}{1 * (I * J)} = \frac{1}{1}.$$

□

Rational arithmetic thus satisfies the algebraic properties in table 574. Every set with addition and multiplication satisfying the algebraic properties in table 574 is called a **number field.** As a result, the triple $(\mathbb{Q}, +, *)$ is a number field.

The next theorem forms the basis for a concept of division of rational numbers.

Table 574 These properties hold for all $I/J, K/L, P/Q \in \mathbb{Q}$.

(1)	Associativity of +	$[(I/J) + (K/L)] + (P/Q) = (I/J) + [(K/L) + (P/Q)]$
(2)	Commutativity of +	$(I/J) + (K/L) = (K/L) + (I/J)$
(3)	Additive identity	$(K/L) + (0/1) = (K/L) = (0/1) + (K/L)$
(4)	Additive inverse	$(K/L) + ([-K]/L) = (0/1)$
(5)	Associativity of $*$	$[(I/J)(K/L)](P/Q) = (I/J)[(K/L)(P/Q)]$
(6)	Commutativity of $*$	$(I/J)(K/L) = (K/L)(I/J)$
(7)	Multiplicative identity	$(K/L)(1/1) = (K/L) = (1/1)(K/L)$
(8)	Multiplicative inverse	If $(K/L) \neq 0$, then $(K/L)(L/K) = (1/1)$
(9)	Distributivity	$(I/J)[(K/L) + (P/Q)] = [(I/J)(K/L)] + [(I/J)(P/Q)]$

Theorem 575 *For each $K/L \in \mathbb{Q}$, and for each $I/J \in \mathbb{Q}$ such that $I \neq 0$, $[(K/L)*(J/I)]*(I/J) = (K/L)$.*

Proof.

$$\left(\frac{K}{L} * \frac{J}{I}\right) * \frac{I}{J} = \frac{K*J}{L*I} * \frac{I}{J} = \frac{(K*J)*I}{(L*I)*J} = \frac{K*(J*I)}{L*(I*J)} = \frac{K*(I*J)}{L*(I*J)} = \frac{K}{L}.$$

□

Definition 576 For each $K/L \in \mathbb{Q}$, and for each $I/J \in \mathbb{Q}$ such that $I \neq 0$, define

$$\frac{K}{L} \div \frac{I}{J} := \frac{K}{L} * \frac{J}{I}$$

Definition 577 For each $K/L \in \mathbb{Q}$, and for each $J \in \mathbb{N}$, define

$$\left(\frac{K}{L}\right)^0 := \frac{1}{1},$$

$$\left(\frac{K}{L}\right)^{J+1} := \left(\frac{K}{L}\right)^J * \left(\frac{K}{L}\right).$$

Moreover, if $K/L \neq 0/1$, then and for each $J \in \mathbb{Z}$, define

$$\left(\frac{K}{L}\right)^{J-1} := \left(\frac{K}{L}\right)^J * \left(\frac{L}{K}\right).$$

3.5.3 Notation for Sums and Products

The notation introduced here proves convenient to define and investigate sums and products of finite sequences of numbers.

Definition 578 A **finite sequence of numbers** is a function $S : N \to \mathbb{Q}$ defined on some $N \in \mathbb{N}$. The value $S(K)$ is also denoted by S_K; then the function S is also denoted by (S_K).

Example 579 The function $S : 9 \rightarrow \mathbb{Q}$ defined by $S_K := (^2/_3)^K$ is a finite sequence of numbers:

$$\begin{array}{rcccl} S_0 &=& (^2/_3)^0 &=& 1, \\ S_1 &=& (^2/_3)^1 &=& ^2/_3, \\ S_2 &=& (^2/_3)^2 &=& ^4/_9, \\ S_3 &=& (^2/_3)^3 &=& ^8/_{27}, \\ S_4 &=& (^2/_3)^4 &=& ^{16}/_{81}, \\ S_5 &=& (^2/_3)^5 &=& ^{32}/_{243}, \\ S_3 &=& (^2/_3)^6 &=& ^{64}/_{729}, \\ S_4 &=& (^2/_3)^7 &=& ^{128}/_{2187}, \\ S_5 &=& (^2/_3)^8 &=& ^{256}/_{6561}. \end{array}$$

The next definition gives a notation for the product of a finite sequence of numbers.

Definition 580 (Product notation) For each finite sequence of numbers $S : N \rightarrow \mathbb{Q}$, define the "empty product" to be 1:

$$\prod_{K<0} S_K \quad := \quad 1.$$

Then define the product of the first value to be the first value:

$$\prod_{K=0}^{0} S_K \quad := \quad S_0.$$

Hence for each $L \in \mathbb{N}$, such that $0 < L < N$, define the product of the first L values of the sequence S "inductively" [66, p. 5] or "recursively" [35, p. 133] by

$$\prod_{K=0}^{L} S_K \quad := \quad \left(\prod_{K=0}^{L-1} S_K \right) * S_L.$$

Example 581 Consider the finite sequence $S : 9 \rightarrow \mathbb{Q}$ defined by $S_K := (^2/_3)^K$:

$$\begin{array}{rcl} \prod_{K<0} S_K &=& 1, \\ \\ \prod_{K=0}^{0} S_K &=& S_0 \\ &=& 1, \\ \\ \prod_{K=0}^{1} S_K &=& \left(\prod_{K=0}^{0} S_K \right) * S_1 \\ &=& (1) * ^2/_3, \\ \\ \prod_{K=0}^{2} S_K &=& \left(\prod_{K=0}^{1} S_K \right) * S_2 \\ &=& (1 * ^2/_3) * ^4/_9, \\ \\ \prod_{K=0}^{3} S_K &=& \left(\prod_{K=0}^{2} S_K \right) * S_3 \\ &=& (1 * ^2/_3 * ^4/_9) * ^8/_{27}, \\ \\ \prod_{K=0}^{4} S_K &=& \left(\prod_{K=0}^{3} S_K \right) * S_4 \\ &=& (1 * ^2/_3 * ^4/_9 * ^8/_{27}) * ^{16}/_{81}, \\ &\vdots& \\ \prod_{K=0}^{8} S_K &=& \left(\prod_{K=0}^{7} S_K \right) * S_8 \\ &=& (1 * \frac{2}{3} * \frac{4}{9} * \frac{8}{27} * \frac{16}{81} * \frac{32}{243} * \frac{64}{729} * \frac{128}{2187}) * \frac{256}{6561}. \end{array}$$

The next definition gives a notation for the sum of a finite sequence of numbers.

Definition 582 (Sum notation) For each finite sequence of numbers $S : N \to \mathbb{Q}$, define the "empty sum" to be 0:

$$\sum_{K<0} S_K \quad := \quad 0.$$

Then define the sum of the first value to be the first value:

$$\sum_{K=0}^{0} S_K \quad := \quad S_0.$$

Hence for each $L \in \mathbb{N}$, such that $0 < L < N$, define the sum of the first L values of the sequence S inductively by

$$\sum_{K=0}^{L} S_K \quad := \quad \left(\sum_{K=0}^{L-1} S_K\right) + S_L.$$

Example 583 Consider the finite sequence $S : 9 \to \mathbb{Q}$ defined by $S_K := (2/3)^K$:

$$\begin{aligned}
\textstyle\sum_{K<0} S_K &= 0, \\
\textstyle\sum_{K=0}^{0} S_K &= S_0 \\
&= 1, \\
\textstyle\sum_{K=0}^{1} S_K &= \left(\textstyle\sum_{K=0}^{0} S_K\right) + S_1 \\
&= (1) + 2/3, \\
\textstyle\sum_{K=0}^{2} S_K &= \left(\textstyle\sum_{K=0}^{1} S_K\right) + S_2 \\
&= (1 + 2/3) + 4/9, \\
\textstyle\sum_{K=0}^{3} S_K &= \left(\textstyle\sum_{K=0}^{2} S_K\right) + S_3 \\
&= (1 + 2/3 + 4/9) + 8/27, \\
\textstyle\sum_{K=0}^{4} S_K &= \left(\textstyle\sum_{K=0}^{3} S_K\right) + S_4 \\
&= (1 + 2/3 + 4/9 + 8/27) + 16/81, \\
&\;\vdots \\
\textstyle\sum_{K=0}^{8} S_K &= \left(\textstyle\sum_{K=0}^{7} S_K\right) + S_8 \\
&= \left(1 + \tfrac{2}{3} + \tfrac{4}{9} + \tfrac{8}{27} + \tfrac{16}{81} + \tfrac{32}{243} + \tfrac{64}{729} + \tfrac{128}{2187}\right) + \frac{256}{6561}.
\end{aligned}$$

The pattern in the foregoing example, called a **geometric series**, is amenable to an alternative formula, which expresses the entire sum as one ratio.

Theorem 584 (Geometric series) *For every $N \in \mathbb{N}^*$ and every $X \in \mathbb{Q} \setminus \{1\}$,*

$$\sum_{K=0}^{N-1} X^K = \frac{1 - X^N}{1 - X}.$$

Proof. This proof proceeds by induction with N. For $N := 1$, and for every $X \in \mathbb{Q} \setminus \{1\}$,

$$\sum_{K=0}^{1-1} X^K = \sum_{K=0}^{0} X^K = X^0 = 1 = \frac{1 - X^1}{1 - X}.$$

Assume that there exists $I \in \mathbb{N}^*$ such that the theorem holds for $N := I$ and for every $X \in \mathbb{Q} \setminus \{1\}$, so that

$$\sum_{K=0}^{I-1} X^K = \frac{1 - X^I}{1 - X}.$$

Then

$$\begin{aligned}
\sum_{K=0}^{(I+1)-1} X^K &= \left(\sum_{K=0}^{I-1} X^K\right) + X^I \\
&= \frac{1 - X^I}{1 - X} + X^I \\
&= \frac{1 - X^I}{1 - X} + \frac{(1 - X) * X^I}{1 - X} \\
&= \frac{(1 - X^I) + (X^I - X^{I+1})}{1 - X} \\
&= \frac{1 - X^{I+1}}{1 - X}.
\end{aligned}$$

□

An alternative proof of the same formula proceeds along the following outline:

$$\begin{array}{rcccccccccc}
\sum_{K=0}^{N-1} X^K & = & 1 & + & X & + \cdots + & X^{N-2} & + & X^{N-1} \\
X * \sum_{K=0}^{N-1} X^K & = & X & + & X^2 & + \cdots + & X^{N-1} & + & X^N \\
(1 - X) * \sum_{K=0}^{N-1} X^K & = & 1 & + & 0 & + \cdots + & 0 & - & X^N
\end{array}$$

whence dividing both sides by $(1 - X)$ yields

$$\sum_{K=0}^{N-1} X^K = \frac{1 - X^N}{1 - X}.$$

Yet such a proof also requires induction to rearrange the terms in the subtraction.

Example 585 Consider the finite sequence $S : 9 \to \mathbb{Q}$ defined by $S_K := ({}^2\!/_3)^K$:

$$\begin{aligned}
\sum_{K=0}^{8} S_K &= (1 + \tfrac{2}{3} + \tfrac{4}{9} + \tfrac{8}{27} + \tfrac{16}{81} + \tfrac{32}{243} + \tfrac{64}{729} + \tfrac{128}{2187}) + \frac{256}{6561} \\
&= \frac{1 - ({}^2\!/_3)^9}{1 - {}^2\!/_3} \\
&= \frac{1 - {}^{512}\!/_{19,683}}{{}^1\!/_3} \\
&= \frac{{}^{19,171}\!/_{19683}}{{}^1\!/_3} \\
&= {}^3\!/_1 * {}^{19171}\!/_{19683} \\
&= {}^{19171}\!/_{6561}.
\end{aligned}$$

3.5.4 Order on the Rational Numbers

The determination of whether two rational numbers P/Q and M/N coincide can utilize equivalent fractions with a common denominator, for instance, $(P * N)/(Q * N)$ and $(Q * M)/(Q * N)$, and then with the comparison of the numerators $P * N$ and $Q * M$. The same comparison leads to a concept of order on the rational numbers.

Definition 586 Define a relation $<$ on $\mathbb{Q}$ as follows. Firstly, $0 < (P/Q)$ if and only if $0 < P * Q$, so that either P and Q are both positive, or P and Q are both negative:

$$[0 < (P/Q)] \Leftrightarrow [0 < (P * Q)].$$

Secondly, $(I/J) < (P/Q)$ if and only if $0 < [(P/Q) - (I/J)]$:

$$[(I/J) < (P/Q)] \Leftrightarrow \{0 < [(P/Q) - (I/J)]\}.$$

The following theorem shows that the square of any non-zero rational number, and the sum and product of positive rational numbers, are positive rational numbers.

Theorem 587 *If $(M/N) > 0$ and $(I/J) > 0$, then $(M/N) + (I/J) > 0$ and $(M/N) * (I/J) > 0$. Moreover, $(K/L) * (K/L) > 0$ for every $(K/L) \neq (0/1)$.*

Proof. For the square, let $P/Q := (K/L)^2 = (K^2)/(L^2)$. Then $P * Q = (K^2) * (L^2) = (K * L)^2 > 0$ (theorem 549). Thus $(K/L)^2 = P/Q > 0$ (definition 586).

For the product, let $P/Q := (M/N) * (I/J) = (M * I)/(N * J)$. By the hypotheses, $M * N > 0$ and $I * J > 0$. Hence $P * Q = (M * I) * (N * J) = (M * N) * (I * J) > 0$, whence $(M/N) * (I/J) = P/Q > 0$ (definition 586). For the sum, let

$$\frac{P}{Q} := \frac{M}{N} + \frac{I}{J} = \frac{(M * J) + (N * I)}{N * J}.$$

Then

$$P * Q = [(M * J) + (N * I)] * (N * J) = (M * N) * (J * J) + (N * N) * (I * J) > 0.$$

Indeed, $J^2 > 0$ and $N^2 > 0$ (theorem 549), $(M * N) * (J^2) > 0$ and $(N^2) * (I * J) > 0$ by hypothesis, whence $P * Q = (M * N) * (J^2) + (N^2) * (I * J) > 0$. □

The next theorem shows that $<$ is a strict total order (definition 451) on $\mathbb{Q}$.

Theorem 588 *The relation $<$ is a strict total order on the rational numbers.*

Proof. This proof verifies that $<$ is connected, irreflexive, and transitive.

Irreflexivity $(P/Q) \not< (P/Q)$ because $0 \not< 0 = (P/Q) - (P/Q)$.

Connectedness If $(M/N) \neq (P/Q)$, then $P * N \neq Q * M$ whence $(P * N - Q * M) \neq 0$, whence $(Q*N)*(P*N-Q*M) \neq 0$, and then either $(Q*N)*(P*N-Q*M) < 0$, in which case $(P/Q)-(M/N) < (0/1)$, so that $(P/Q) < (M/N)$, or $(Q*N)*(P*N-Q*M) > 0$, in which case $(P/Q) - (M/N) > (0/1)$, so that $(P/Q) > (M/N)$.

Transitivity If $(I/J) < (K/L)$ and $(K/L) < (M/N)$, then $(K/L) - (I/J) > 0$ and $(M/N)-(K/L) > 0$, whence $(M/N)-(I/J) = [(M/N)-(K/L)]+[(K/L)-(I/J)] > 0$ (theorem 587) and then $(M/N) > (I/J)$ (definition 586). □

Definition 589 Define the sets of non-zero rationals ($\mathbb{Q}^*$), negative rationals ($\mathbb{Q}^*_-$), positive rationals ($\mathbb{Q}^*_+$), non-positive rationals ($\mathbb{Q}_-$), and non-negative rationals ($\mathbb{Q}_+$) by

$$\begin{array}{rcl} \mathbb{Q}^* & := & \mathbb{Q} \setminus \{0/1\}, \\ \mathbb{Q}_- & := & \{P/Q \in \mathbb{Q} : P/Q \leq 0/1\}, \\ \mathbb{Q}_+ & := & \{P/Q \in \mathbb{Q} : P/Q \geq 0/1\}, \\ \mathbb{Q}^*_- & := & \{P/Q \in \mathbb{Q} : P/Q < 0/1\}, \\ \mathbb{Q}^*_+ & := & \{P/Q \in \mathbb{Q} : P/Q > 0/1\}. \end{array}$$

Definition 590 The **absolute value** $|P/Q|$ of a rational number P/Q is defined by

$$|P/Q| := \begin{cases} P/Q & \text{if } P/Q \geq 0, \\ -(P/Q) & \text{if } P/Q < 0. \end{cases}$$

Theorem 591 (Triangle Inequality) *For all $K/L \in \mathbb{Q}$ and $M/N \in \mathbb{Q}$,*

$$\left| \frac{K}{L} + \frac{M}{N} \right| \leq \left| \frac{K}{L} \right| + \left| \frac{M}{N} \right|$$

with equality if and only if $(K/L)(M/N) \geq 0$. Also,

$$\left| \left| \frac{K}{L} \right| - \left| \frac{M}{N} \right| \right| \leq \left| \frac{K}{L} - \frac{M}{N} \right|$$

with equality if and only if $(K/L)(M/N) \geq 0$.

Proof. Apply the definition of the absolute value to four cases. □

Theorem 592 (Archimedean Property of the Rationals) *For each rational $P/Q \in \mathbb{Q}$ there exists a natural number $N \in \mathbb{N}$ such that $P/Q < N/1$.*

Proof. If $P/Q \leq 0$, let $N := 1$. If $P/Q > 0$, let $N := |P| + 1$; then

$$\frac{N}{1} - \frac{P}{Q} = \frac{N}{1} - \frac{|P|}{|Q|} = \frac{\{[(|P| + 1) * |Q|] - (1 * |P|)\}}{1 * |Q|} > 0,$$

because $[(|P| + 1) * |Q|] - (1 * |P|) = |P| * (|Q| - 1) + |Q| \geq |Q| > 0$. □

3.5.5 Exercises

Exercise 681 Calculate $(2/3) + (7/5)$.

Exercise 682 Calculate $(5/2) + (1/7)$.

Exercise 683 Calculate $(7/3) - (2/5)$.

Exercise 684 Calculate $(1/2) - (1/3)$.

Exercise 685 Calculate $(2/3) * (7/5)$.

Exercise 686 Calculate $(5/2) * (1/7)$.

Exercise 687 Calculate $(2/3) \div (7/5)$.

Exercise 688 Calculate $(5/2) \div (1/7)$.

Exercise 689 Prove that on $\mathbb{Q}$ the division does *not* commute.

Exercise 690 Prove that on $\mathbb{Q}$ division is *not* associative.

Exercise 691 Prove that on $\mathbb{Q}$ division does *not* distribute over addition.

Exercise 692 Prove that on $\mathbb{Q}$ division does *not* distribute over multiplication.

Exercise 693 Prove that on $\mathbb{Q}$ addition does *not* distribute over division.

Exercise 694 Prove that on $\mathbb{Q}$ multiplication does *not* distribute over division.

Exercise 695 Prove that if $0 < (I/J)$ and $0 < (P/Q)$, then $0 < [(I/J) + (P/Q)]$.

Exercise 696 Prove that if $0 < (I/J)$ and $0 < (P/Q)$, then $0 < [(I/J) * (P/Q)]$.

Exercise 697 Prove that if $Q > 0$ and $R > 0$, then $P/(Q/R) = (P * R)/Q$.

Exercise 698 Prove that for each $P/Q \in \mathbb{Q}$ there exists a *smallest* $N \in \mathbb{N}^*$ such that there exists $M \in \mathbb{Z}$ with $P/Q = M/N$.

Exercise 699 Prove that for each $K/L \in \mathbb{Q}_+$ there exists a *smallest* $N \in \mathbb{N}$ with $K/L \leq N/1$.

Exercise 700 Find rational numbers K/L and M/N such that $(K/L)^2 + (M/N)^2 = 1$.

3.6 FINITE CARDINALITY

3.6.1 Equal Cardinalities

The adjective "cardinal" means "principal" or "of greatest importance" . In the context of sets, the cardinal feature of sets is their size. One way to define the "size" of a set consists in establishing a correspondence with another set of known size, for instance, a natural number, as in figure 3.2. Such a natural number is then the "number" of elements in the set, and the correspondence amounts to an operation of counting. Thus the natural numbers constitute the "standard" sizes with which to "count" sets. More generally, two sets have the same cardinality if and only if there exists a bijection between those two sets.

Figure 3.2: Same cardinality: Earth and Moon, or Pluto and Charon, or $\{\varnothing, \{\varnothing\}\}$.

Definition 593 For all sets A and B, the sets A and B have **the same cardinality** if and only if there exists a bijection $F : A \to B$, a situation denoted by such expressions as

$$\begin{aligned} A &\approx B, \\ \bar{\bar{A}} &= \bar{\bar{B}}, \\ |A| &= |B|, \\ \#(A) &= \#(B), \\ card(A) &= card(B). \end{aligned}$$

Definition 593 does not yet define the concept of cardinality; it only defines the concept of *same cardinality*. Because such a definition leaves the notation $\#(A)$ yet undefined, this exposition adopts the notation $A \approx B$, which merely means that there exists a bijection from A to B.

Example 594 All empty sets have the same cardinality. Indeed, by the axiom of extensionality there exists only one empty set, namely $\varnothing$, and the empty function $\varnothing : \varnothing \to \varnothing$ is a bijection, whence $\varnothing \approx \varnothing$.

Example 595 All singletons have the same cardinality. Indeed, for all sets X and Y and all singletons $\{X\}$ and $\{Y\}$, the function $F : \{X\} \to \{Y\}$ defined by $F := \{(X, Y)\}$ is a bijection. Thus $\{X\} \approx \{Y\}$.

Example 596 The sets $A := \{4, 9\}$ and $B := \{2, 3\}$ have the same cardinality, thanks to the bijection $F : A \to B$ with $F := \{(4, 2), (9, 3)\}$. The other bijection, $G := \{(4, 3), (9, 2)\}$, could also serve to prove that A and B have the same cardinality.

The following theorem forms the basis for the relation between the addition of natural numbers and the union of disjoint sets.

Theorem 597 *For all sets A, B, C, and D, if*

$$\begin{aligned} A &\approx C, \\ B &\approx D, \\ A \cap B &= \varnothing, \\ C \cap D &= \varnothing, \end{aligned}$$

then

$$(A \dot{\cup} B) \approx (C \dot{\cup} D).$$

Proof. The hypotheses $A \approx C$ and $B \approx D$ mean that there exist bijections

$$\begin{array}{rcl} F : A & \to & C, \\ G : B & \to & D. \end{array}$$

Such bijections lead to a bijection

$$H : (A\dot{\cup}B) \to (C\dot{\cup}D)$$

defined by

$$H := (F\dot{\cup}G) \subseteq (A\dot{\cup}B) \times (C\dot{\cup}D)$$

so that

$$[(X,Y) \in H] \Leftrightarrow \begin{cases} Y = F(X) & \text{if } X \in A, \\ Y = G(X) & \text{if } X \in B. \end{cases}$$

The relation H just defined is a function, because for each $X \in (A\dot{\cup}B)$ the relation H contains only one pair (X,Y). Indeed, thanks to $A \cap B = \varnothing$, either $X \in A$ and then $Y = F(X)$, or $X \in B$ and then $Y = G(X)$, but not both (definition 368).

To verify the injectivity of H, assume that $W \in (A\dot{\cup}B)$ and $X \in (A\dot{\cup}B)$ have the same image $H(W) = Y = H(X)$ in $C \cup D$. Because of $C \cap D = \varnothing$, either both images lie in C or both lie in D. In the first case, if both images lie in C, then $W \in A$ and $X \in A$, and then

$$F(W) = H(W) = H(X) = F(X),$$

whence $W = X$ by injectivity of F. In the second case, if both images lie in D, then $W \in B$ and $X \in B$, and then

$$G(W) = H(W) = H(X) = G(X),$$

whence $W = X$ by injectivity of G. To verify the surjectivity of H, assume that $Z \in (C\dot{\cup}D)$. Then either $Z \in C$ or $Z \in D$. In the first case, if $Z \in C$, then the surjectivity of F guarantees the existence of an element $W \in A$ such that

$$Z = F(W) = H(W).$$

In the second case, if $Z \in D$, then the surjectivity of G guarantees the existence of an element $X \in B$ such that

$$Z = G(X) = H(X).$$

Therefore, $H : (A\dot{\cup}B) \to (C\dot{\cup}D)$ is a bijection. □

The following theorem forms the basis for the relation between multiplication of natural numbers and Cartesian products of sets.

Theorem 598 *For all sets A, B, C, and D, if*

$$\begin{array}{rcl} A & \approx & C, \\ B & \approx & D, \end{array}$$

then

$$(A \times B) \approx (C \times D).$$

Proof. The hypotheses $A \approx C$ and $B \approx D$ mean that there exist bijections

$$\begin{array}{rcl} F : A & \to & C, \\ G : B & \to & D. \end{array}$$

Such bijections lead to a bijection H defined by

$$\begin{array}{rcl} H : (A \times B) & \to & (C \times D), \\ (W, X) & \mapsto & (F(W), G(X)). \end{array}$$

The relation H is a function, because F and G are functions, so that for each $W \in A$ and each $X \in B$ there exists at most one $Y \in C$ and at most one $Z \in D$ with $(W, Y) \in F$ and $(X, Z) \in G$, so that there exists at most one $(Y, Z) \in C \times D$ with $\big((W, X),\ (Y, Z)\big) \in H$. To verify the injectivity of H, assume that $(W, X) \in (A \times B)$ and $(U, V) \in (A \times B)$ have the same image $H(W, X) = H(U, V)$ in $C \times D$:

$$\begin{array}{rl} H(W, X) = H(U, V) & \text{hypothesis,} \\ \Updownarrow & \text{definition of } H, \\ (F(W), G(X)) = (F(U), G(V)) & \\ \Updownarrow & \text{equality of pairs,} \\ [F(W) = F(U)] \wedge [G(X) = G(V)] & \\ \Updownarrow & \text{injectivity of } F \text{ and } G, \\ [W = U] \wedge [X = V] & \\ \Updownarrow & \text{equality of pairs.} \\ (W, X) = (U, V) & \end{array}$$

To verify the surjectivity of H, assume that $(R, S) \in (C \times D)$. Then the surjectivity of F guarantees the existence of an element $W \in A$ such that

$$R = F(W),$$

and the surjectivity of G guarantees the existence of an element $X \in B$ such that

$$S = G(X).$$

Consequently,

$$(R, S) = (F(W), G(X)) = H(W, X).$$

Therefore, $H : (A \times B) \to (C \times D)$ is a bijection. □

3.6.2 Finite Sets

The following definition establishes the concept of cardinality for finite sets.

Definition 599 For each set S, the set S is **finite** if and only if there exists $N \in \mathbb{N}$ and a bijection $F : N \to S$. Such a natural number N is then called the **number of elements in** S, or the **cardinality** of S, which is denoted by $\#(S)$, $|S|$, or $\bar{\bar{S}}$.

Example 600 For each natural number $N \in \mathbb{N}$ the set N is finite and has N elements, because the identity function $I_N : N \to N,\ K \mapsto K$ is a bijection.

Remark 601 Because every bijection has an inverse function, a set S is finite if and only if there exist a natural number $N \in \mathbb{N}$ and a bijection $G : S \to N$, for instance, the inverse function $G := F^{\circ -1}$ for any bijection $F : N \to S$.

The following theorem shows that the insertion of a new element into a set corresponds to the arithmetic addition of 1 to its cardinality.

Theorem 602 *The equality* $\#(A \dot{\cup} \{Z\}) = [\#(A)] + 1$ *holds for each finite set* A, *and for each set* $Z \notin A$.

Proof. For each finite set A there exist a natural number N and a bijection $F : N \to A$.

Moreover, for each set Z there exists a bijection of singletons $G : \{N\} \to \{Z\}$.

Consequently, because $A \cap \{Z\} = \varnothing$ by hypothesis, and because $N \cap \{N\} = \varnothing$ by theorem 504, it follows that theorem 597 gives a bijection $N + 1 \to A\dot{\cup}\{Z\}$:

$$H = (F\dot{\cup}G) : N + 1 = (N\dot{\cup}\{N\}) \to (A\dot{\cup}\{Z\}).$$

□

The following theorem shows that the cardinality of the union of two disjoint finite sets equals the arithmetic sum of their two cardinalities.

Theorem 603 *The equality $\#(A\dot{\cup}B) = [\#(A)] + [\#(B)]$ holds for all disjoint finite sets A and B.*

Proof. This proof proceeds by induction with the cardinality of the second set.

If $\#(B) = 0$, then $B = \varnothing$ by definition, whence for each finite set A,

$$\#(A\dot{\cup}B) = \#(A\dot{\cup}\varnothing) = \#(A) = [\#(A)] + 0 = [\#(A)] + [\#(B)].$$

Hence, assume that there exists a natural number $N \in \mathbb{N}$ for which the theorem holds, so that the equality $\#(A\dot{\cup}B) = [\#(A)] + [\#(B)]$ holds for all disjoint finite sets A and B with $\#(B) = N$. For each set C with $N + 1$ elements, there exists a bijection $F : N + 1 \to C$. Consequently, the subset $B := F\text{''}(N) = F\text{''}(\{0, \ldots, N - 1\})$ has N elements, because the restriction $F|_N : N \to B$ is a bijection. Hence, with the element $Z := F(N)$, it follows that $C = B\dot{\cup}\{Z\}$ with $Z \notin B$, whence

$$\begin{aligned}
&\#(A\dot{\cup}C)\\
&= && \text{because } C = B\dot{\cup}\{Z\},\\
&\#(A\dot{\cup}[B\dot{\cup}\{Z\}])\\
&= && \text{associativity of } \cup,\\
&\#[(A\dot{\cup}B)\dot{\cup}\{Z\}]\\
&= && \text{theorem 602},\\
&\#(A\dot{\cup}B) + \#(\{Z\})\\
&= && \text{induction hypothesis},\\
&[\#(A) + \#(B)] + \#(\{Z\})\\
&= && \text{associativity of } +,\\
&\#(A) + [\#(B) + \#(\{Z\})]\\
&= && \text{theorem 602},\\
&\#(A) + \#(C).
\end{aligned}$$

□

The following two theorems confirm that every subset of a finite set is also finite.

Theorem 604 *For each $N \in \mathbb{N}$, every subset $S \subseteq N$ is also a finite set, with at most N elements. Moreover, each proper subset $S \subset N$ has fewer than N elements.*

Proof. This proof proceeds by induction with N.

If $N := 0$, then $N = \varnothing$, and the only subset $S \subseteq N$ is $S = \varnothing$, which is finite.

As an induction hypothesis, assume that there exists a natural number $K \in \mathbb{N}$ such that the theorem holds for $N := K$, so that each subset $S \subseteq K$ is finite with at most K elements, and that each proper subset $S \subset K$ has fewer than K elements. Hence, consider a subset $R \subseteq K + 1$. Two cases arise: either $K \notin R$ or $K \in R$.

If $K \notin R$, then $R \subseteq [(K+1) \setminus \{K\}] = K$ whence R is finite with at most K elements by induction hypothesis.

If $K \in R$, then the set $C := R \setminus \{K\}$ is a subset of $[(K+1) \setminus \{K\}] = K$, whence C is finite with at most K elements by induction hypothesis. Thus, there exists a natural number $L \leq K$ such that $C = R \setminus \{K\}$ has cardinality $L \leq K$, and $L = K$ if and only if $C = K$. Hence, theorem 602 shows that $R = C \dot{\cup} \{K\}$ is finite with cardinality $L + 1 \leq K + 1$, and $L + 1 = K + 1$ if and only if $R = K + 1$. □

Theorem 605 *Every subset of a finite set is also finite, with at most as many elements.*

Proof. For each finite set A there exists a natural number $N \in \mathbb{N}$ and a bijection $F : A \to N$. Hence, for each subset $B \subseteq A$, the restriction $F|_B : B \to N$ is a bijection from B onto a subset $S := F"(B) \subseteq N$. Because every such subset $S \subseteq N$ is also finite with at most N elements, it follows that there exists a natural number $L \leq N$ and a bijection $G : S \to L$. Consequently, the composition $G \circ F|_B$ establishes a bijection from B onto L, which means that B is finite with $L \leq N$ elements. □

Theorem 606 *The equality $\#(A \setminus B) = [\#(A)] - [\#(B)]$ holds for every subset $B \subseteq A$ of every finite set A.*

Proof. The result follows from the disjointness of B and $A \setminus B$, and from theorem 605, which ensures that both B and $A \setminus B$ are finite:

$$\begin{aligned} B \cap (A \setminus B) &= \varnothing, \\ B \cup (A \setminus B) &= A, \\ [\#(B)] + [\#(A \setminus B)] &= \#(A), \end{aligned}$$

whence $\#(A \setminus B) = [\#(A)] - [\#(B)]$ by definition (521) of subtraction. □

Theorem 604 shows that for each subset $S \subseteq N$ there is a bijection $F : S \to L$ with $L \leq N$, but it does not yet prevent the existence of other bijections $G : S \to L$ with $L > N$. The following theorem confirms that there is no such bijection.

Theorem 607 *For all $K, N \in \mathbb{N}$ with $K < N$, there exists no injection $F : N \to K$.*

Proof. This proof uses induction with N. For $N := 1$, the only smaller natural number is $K := 0$, and there exists no function $F : \{1\} \to \varnothing$, whence no injection either.

As an induction hypothesis, assume that there exists a positive integer $L \in \mathbb{N}^*$ such that the theorem holds for $N := L$, so that for every natural number $K < L$ there exists no injection $F : L \to K$. The proof that for every $K < L + 1$ there exists no injection $F : (L+1) \to K$ proceeds by contraposition. Thus, assume that there is such an injection $F : (L+1) \to K$. Let $Z := F(L)$ and $S := K \setminus \{Z\}$. By theorem 606, S is a finite set with $K - 1$ elements, and there is a bijection $G : S \to (K-1)$. Then the restriction $F|_L : L \to S$ is an injection, and the composition $G \circ F|_L : L \to (K-1)$ is an injection. Hence $K - 1 \leq L$ by induction hypotheses, whence $K \leq L + 1$. □

The following theorem shows the equivalence of the concepts of injection, surjection, and bijection between sets with the *same finite* cardinality.

Theorem 608 *For all sets A and B with the* same finite *cardinality, and for each function $F : A \to B$, the following conditions are mutually equivalent:*

(P) *F is injective,*

(Q) *F is surjective,*

(R) *F is bijective.*

Proof. If F is bijective, then F is also injective and surjective, because $(R) \Leftrightarrow [(P) \wedge (Q)]$ by definition of bijectivity (definition 389); therefore both $(R) \Rightarrow (P)$ and $(R) \Rightarrow (Q)$ hold, by theorems 200 and 201.

For the reverse implications, because A and B have the same finite cardinality, there exist a natural number $N \in \mathbb{N}$ and a bijection $G : N \to A$.

If F is injective, then $F"(A) \subseteq B$ is a finite set with cardinality $L \leq N$, so that there exists a bijection $H : F"(A) \to L$. Then the composition

$$H \circ F \circ G : N \xrightarrow{G} A \xrightarrow{F} F"(A) \xrightarrow{H} L$$

is an injection, whence $L \geq N$ by theorem 607. From $L \leq N$ and $N \leq L$ it follows that $L = N$. Hence H, G, and $(H \circ F \circ G) : N \to N$ are bijections, whence F is also surjective, whence bijective, which proves the implication $(P) \Rightarrow (Q)$; therefore $(P) \Rightarrow [(P) \wedge (Q)]$ holds (by theorem 219), whence also $(P) \Rightarrow (R)$.

The proof of the converse, $(Q) \Rightarrow (P)$, uses the contraposition $[\neg(P)] \Rightarrow [\neg(Q)]$. If F is *not* injective, then A contains two distinct elements $X \neq Z$ such that $F(X) = F(Z)$. Let $S := A \setminus \{Z\}$, so that $F"(A) = F"(S)$ and there exists a bijection $J : (N-1) \to S$, by theorem 606. Also, there is a bijection $I : B \to N$, because A and B have the same cardinality, N. Then the composition

$$I \circ F \circ J : (N-1) \xrightarrow{J} S \xrightarrow{F} F"(S) \xrightarrow{I} N$$

cannot be a bijection, because by theorem 607 its inverse could not be an injection from N to $N-1$. Hence F cannot be surjective, which proves $(Q) \Rightarrow (P)$. □

The following theorem shows that the number of elements in a Cartesian product equals the product of the numbers of elements in its factors.

Theorem 609 *For all $K, L \in \mathbb{N}$, the Cartesian product $K \times L$ has $K * L$ elements.*

Proof. This proof uses induction with L. If $L = 0$, then $L = \varnothing$, whence for every $K \in \mathbb{N}$

$$\begin{aligned} K \times 0 = K \times \varnothing &= \varnothing, \\ \#(K \times 0) = \#(\varnothing) &= 0 = K * 0. \end{aligned}$$

As induction hypothesis, assume that there exists a natural number $M \in \mathbb{N}$ such that the theorem holds for $L := M$, so that the equality

$$\#(K \times M) = K * M$$

holds for every $K \in \mathbb{N}$. Hence, from the disjoint union $M + 1 = M \dot{\cup} \{M\}$, and from the distributivity of Cartesian products over unions (theorem 338), it follows that

$$\begin{aligned} K \times (M+1) &= K \times (M \dot{\cup} \{M\}) \\ &= (K \times M) \dot{\cup} (K \times \{M\}), \end{aligned}$$

$$\begin{aligned} \#[K \times (M+1)] &= \#[(K \times M) \dot{\cup} (K \times \{M\})] \\ &= [\#(K \times M)] + [\#(K \times \{M\})] \\ &= (K * M) + K \\ &= K * (M+1), \end{aligned}$$

thanks to the disjoint union $(K \times M) \dot{\cup} (K \times \{M\})$. □

3.6.3 Exercises

Exercise 701 Determine $\#(\varnothing)$.

Exercise 702 Determine $\#[\mathcal{P}(\varnothing)]$.

Exercise 703 Determine $\#(\mathcal{P}[\mathcal{P}(\varnothing)])$.

Exercise 704 Determine $\#\big[\mathcal{P}\big(\mathcal{P}[\mathcal{P}(\varnothing)]\big)\big]$.

Exercise 705 Determine $\#\Big(\mathcal{P}\big[\mathcal{P}\big(\mathcal{P}[\mathcal{P}(\varnothing)]\big)\big]\Big)$.

Exercise 706 Determine $\#\left[\mathcal{P}\left(\left\{\varnothing, \{\varnothing\}, \{\varnothing, \{\varnothing\}\}\right\}\right)\right]$.

Exercise 707 Prove or disprove that all ordered pairs have the same cardinality.

Exercise 708 For all finite sets A and B, prove that $(A \times B) \approx (B \times A)$.

Exercise 709 For all finite sets A, B, C, prove that $[(A \times B) \times C] \approx [A \times (B \times C)]$.

Exercise 710 For all finite sets A and B, prove $[\#(A \Delta B)] = [\#(A \cup B)] - [\#(A \cap B)]$.

3.7 INFINITE CARDINALITY

3.7.1 Infinite Sets

Some sets are *not* finite, for they do *not* admit any bijection onto any natural number.

Definition 610 (Infinite sets) A set Z is **infinite** if and only if Z is **not finite**, which means that there exists no bijection from Z onto any natural number.

For instance, the set $\mathbb{N}$ is infinite.

Theorem 611 *The set* $\mathbb{N}$ *of all natural numbers is* not *finite.*

Proof. For each natural number N and for each function $F : \mathbb{N} \to N$, the restriction $F|_{N+1}\ (N+1) \to N$ cannot be injective, by theorem 607, whence F cannot be injective. Therefore, there exists no such bijection, which means that $\mathbb{N}$ is *not* finite. □

As there exist finite sets with different cardinalities, there also exist infinite sets with different cardinalities. For instance, the following considerations lead to infinite sets with cardinalities different from the cardinality of the natural numbers.

Definition 612 For all sets X, Y, the set Y^X is the set of all functions from X to Y.

Example 613 If $X := \varnothing$, then $2^X = \{\varnothing\}$, because $\varnothing : \varnothing \to \{0,1\}$ is the only function from $X = \varnothing$ to $2 = \{0,1\}$.

Example 614 If $X := \{S\}$ is a singleton, then 2^X consists of two elements, because there are exactly two functions from $X = \{S\}$ to $2 = \{0,1\}$:

$$\begin{aligned} F : \{S\} &\to \{0,1\}, \\ S &\mapsto 0; \end{aligned}$$

$$\begin{aligned} G : \{S\} &\to \{0,1\}, \\ S &\mapsto 1. \end{aligned}$$

Thus, $2^{\{S\}} = \{F, G\}$ has exactly two elements, F and G.

The next theorem shows that each set X is "strictly smaller" than 2^X.

Theorem 615 *For each set X, there exists an injection from X to 2^X. Yet there does* not *exist any surjection from X to 2^X.*

Proof. To establish the existence of an injection $X \hookrightarrow 2^X$, consider the function $J : X \hookrightarrow 2^X$ defined as follows. The function J maps each element $N \in X$ to a function $J(N) : X \to 2$ specified by

$$[J(N)](K) := \begin{cases} 1 & \text{if } K = N, \\ 0 & \text{if } K \neq N. \end{cases}$$

In other words, the function $J(N)$ is the characteristic function $\chi_{\{N\}}$ of the singleton $\{N\}$ (from example 363). Consequently, J is injective; indeed if $M \neq N$, then $[J(M)](M) = 1$ but $[J(N)](M) = 0$, whence $J(M) \neq J(N)$.

To prove the absence of any surjection $X \twoheadrightarrow 2^X$, for each function $J : X \to 2^X$, this proof demonstrates a method known as **Cantor's diagonalization** to show that J is not surjective. Each such function $J : X \to 2^X$ maps each element $N \in X$ to a function $J(N) \in 2^X$, so that $J(N) : X \to 2$. In particular, $J(N)$ maps N to an element $[J(N)](N) \in 2 = \{0, 1\}$. Thus, define a function $F : X \to 2$ by

$$F(N) := \begin{cases} 0 & \text{if } [J(N)](N) = 1, \\ 1 & \text{if } [J(N)](N) = 0. \end{cases}$$

Thus, $F(N) \neq [J(N)](N)$ for every $N \in X$, whence $F \neq [J(N)]$ for every $N \in X$. Consequently $F \notin J"(X)$, whence J is not surjective. □

Example 616 One among several methods to define the set $\mathbb{R}$ of all **real numbers** consists in defining $\mathbb{R}$ as a set of infinite sequences of digits [90, p. 565–566]. Thus the set of all real numbers between 0 and 1 can be defined as the set of all sequences $R \in 3^{\mathbb{N}}$ (subject to the constraint that there does *not* exist any $K \in \mathbb{N}$ such that $R(N) = 2$ for every $N \geq K$). Then every function $J : \mathbb{N} \to \mathbb{R}$ fails to be surjective. Indeed, as in the proof of theorem 615, for each function $J : \mathbb{N} \to \mathbb{R}$ define a function $G : \mathbb{N} \to \mathbb{R}$ by

$$G(N) := \begin{cases} 1 & \text{if } [J(N)](N) = 2, \\ 0 & \text{if } [J(N)](N) = 1. \\ 1 & \text{if } [J(N)](N) = 0. \end{cases}$$

Thus, $G \in \mathbb{R}$ but $G(N) \neq [J(N)](N)$ for every $N \in \mathbb{N}$, whence $G \neq [J(N)]$ for every $N \in \mathbb{N}$. Consequently $G \notin J"(\mathbb{N})$, and therefore J is not surjective. Hence there does not exist any bijection $J : \mathbb{N} \to \mathbb{R}$.

Example 616 reveals that the set $\mathbb{R}$ of all real numbers is "more infinite" than the set $\mathbb{N}$ of all natural numbers. Moreover, applying theorem 615 to $X := \mathbb{R}$ shows that $2^{\mathbb{R}}$ is also "more infinite" than $\mathbb{R}$. Then $2^{(2^{\mathbb{R}})}$ is also "more infinite" than $2^{\mathbb{R}}$. And so forth, thus there exists an "infinite" variety of "infinite" sets.

3.7.2 Denumerable Sets

There exist several infinite sets that have the same cardinality as $\mathbb{N}$ has. For instance, using only addition and multiplication from integer arithmetic, this subsection presents a proof that the set of all non-negative integers $\mathbb{N}$ and the set of all integers $\mathbb{Z}$ have the same cardinality; similarly, $\mathbb{N}$ and the Cartesian product $\mathbb{N} \times \mathbb{N}$ have the same cardinality. The following terminology conforms to [3, p. 152], [19, p. 47], [95, p. 151].

Definition 617 A set is **denumerable** — or has cardinality $\aleph_0$ (read "aleph zero") — if and only if it has the same cardinality as the set $\mathbb{N}$ of all natural numbers.

A set is **countable** if and only if it is either finite or denumerable.

The following theorem shows that the set $\mathbb{Z}$ of *all* the integers has the same cardinality as the set $\mathbb{N}$ of all the non-negative integers.

Theorem 618 *The sets $\mathbb{N}$ and $\mathbb{Z}$ have the same cardinality.*

Proof. Define $F : \mathbb{N} \to \mathbb{Z}$ by

$$F(N) := \begin{cases} N/2 & \text{if there exists } K \in \mathbb{N} \text{ with } N = 2 * K, \\ -(N+1)/2 & \text{if there exists } K \in \mathbb{N} \text{ with } N + 1 = 2 * K. \end{cases}$$

The function F is surjective. Indeed, if $L \in \mathbb{N}$, then $L = F(2 * L)$. Similarly, if $L \in \mathbb{Z} \setminus \mathbb{N}$, then $L = F([2 * L] + 1)$.

The function F is also injective. Indeed, if $F(M) = F(N)$, then either both or neither of $F(M)$ and $F(N)$ are elements of $\mathbb{N}$. If $F(M) \in \mathbb{N}$ and $F(N) \in \mathbb{N}$, then $M/2 = F(M) = F(N) = N/2$, whence $M = N$. If $F(M) \notin \mathbb{N}$ and $F(N) \notin \mathbb{N}$, then $-(M+1)/2 = F(M) = F(N) = -(N+1)/2$, whence $M = N$. □

Theorem 619 *For all disjoint denumerable sets A and B, $A \dot{\cup} B$ is denumerable.*

Proof. By the hypotheses there exist bijections $I : A \to \mathbb{N}$ and $J : B \to \mathbb{N}$. The function $K : \mathbb{N} \to \mathbb{Z}$ with $K(N) := -(N+1)$ is injective, and hence the composition $K \circ I$ is also injective. Therefore, the function $G := (K \circ I) \dot{\cup} J$ is a bijection from $A \dot{\cup} B$ to $\mathbb{Z}$. Hence $F^{\circ -1} \circ G : A \dot{\cup} B \to \mathbb{N}$ is a bijection, with F as in theorem 618. □

To facilitate the proof that $\mathbb{N}$ and the Cartesian product $\mathbb{N} \times \mathbb{N}$ have the same cardinality, the following definition specifies an inductive method to define and compute the sum $1 + 2 + \cdots + (N-1) + N$, known as an **arithmetic series**.

Definition 620 Define a function $T : \mathbb{N} \to \mathbb{N}$ by

$$\begin{aligned} T(0) &:= 0, \\ T(N+1) &:= T(N) + (N+1). \end{aligned}$$

Also, define the notation $0 + 1 + \cdots + N := T(N)$.

Thus,

$$\begin{aligned} T(0) &:= 0, \\ T(1) &:= T(0) + (0+1) = 0 + (0+1) = 1, \\ T(2) &:= T(1) + (1+1) = 1 + (1+1) = 3, \\ T(3) &:= T(2) + (2+1) = 3 + (2+1) = 6, \\ &\vdots \end{aligned}$$

The values of T are called **triangular numbers** because they correspond to the number of elements in the following patterns:

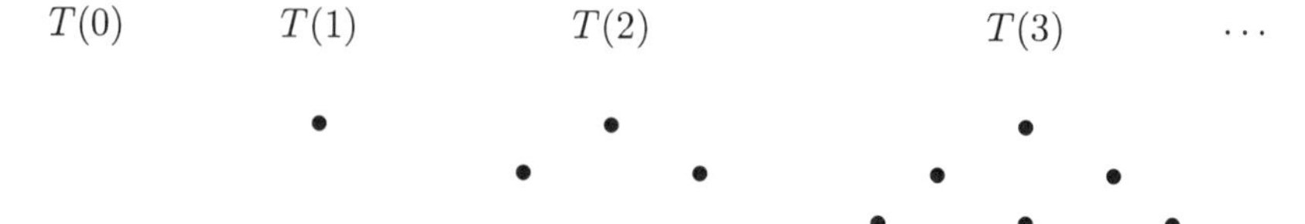

The following theorem provides a different formula to compute the same function T.

Theorem 621 (Arithmetic series) *For each natural number $N \in \mathbb{N}$,*

$$0 + 1 + 2 + \cdots + (N-1) + N = T(N) = \frac{N * (N+1)}{2}.$$

Proof. This proof uses induction with N. If $N := 0$, then $T(0) = 0 = 0 * (0+1)/2$.

Assume that there exists $M \in \mathbb{N}$ such that the theorem holds for $N := M$, so that $T(M) \in \mathbb{N}$ and $2 * T(M) = M * (M+1)$. Then $T(M+1) = T(M) + (M+1) \in \mathbb{N}$, and

$$\begin{aligned} 2 * T(M+1) &= 2 * [T(M) + (M+1)] \\ &= [2 * T(M)] + [2 * (M+1)] \\ &= M * (M+1) + 2 * (M+1) \\ &= (M+2) * (M+1) \\ &= (M+1) * [(M+1) + 1]. \end{aligned}$$

Consequently, $2 * T(M+1) = (M+1) * [(M+1)+1]$, but $T(M+1) \in \mathbb{N}$, whence $(M+1) * [(M+1)+1]/2 \in \mathbb{N}$ and $T(M+1) = (M+1) * [(M+1)+1]/2$. □

An alternative proof of the same formula proceeds along the following outline:

$$\begin{array}{rcccccccccc} T(N) & = & 0 & + & 1 & +\cdots+ & (N-1) & + & N \\ T(N) & = & N & + & (N-1) & +\cdots+ & 1 & + & 0 \\ T(N)+T(N) & = & (N+1) & + & (N+1) & +\cdots+ & (N+1) & + & (N+1) \end{array}$$

whence $2 * T(N) = N * (N+1)$. However, a proof along this outline also requires induction to rearrange the terms of the sum with associativity and commutativity.

The following definition provides a formula for **Cantor's diagonal enumeration** of $\mathbb{N} \times \mathbb{N}$, and the subsequent theorems will verify that indeed it enumerates $\mathbb{N} \times \mathbb{N}$.

Definition 622 Define a function $\mathcal{T} : (\mathbb{N} \times \mathbb{N}) \to \mathbb{N}$ by

$$\begin{aligned} \mathcal{T}(M, N) &:= 1 + M + T(M+N) \\ &= 1 + M + \frac{(M+N) * (M+N+1)}{2}. \end{aligned}$$

The value $\mathcal{T}(M, N)$ corresponds to the sum of the number of elements in the triangular pattern counted by the "triangular number" $T(M+N)$ and a last partial row with $M+1$ elements (instead of a complete last row of $M+N+1$ elements for the next triangular number). For example, with $M := 1$ and $N := 2$,

$\mathcal{T}(1, 2)$

$T(M+N)$ • / • • / • • •

$1 + M$ • •

The following theorem shows that the function $\mathcal{T} : (\mathbb{N} \times \mathbb{N}) \to \mathbb{N}$ is surjective.

Theorem 623 *For each $I \in \mathbb{N}^*$ there exist $M \in \mathbb{N}$ and $N \in \mathbb{N}$ such that*

$$I = 1 + M + \frac{(M+N) * (M+N+1)}{2}.$$

Proof. This proof proceeds by induction with I.

Firstly, if $I := 1$, then $I = 1 + 0 + T(0 + 0)$ with $M := 0$ and $N := 0$.

Secondly, assume that there exists $K \in \mathbb{N}^*$ such that the theorem holds for $I := K$, so that there exist $M, N \in \mathbb{N}$ with $K = 1 + M + T(M + N)$.

If $N > 0$, then $K+1 = 1+[1+M+T(M+N)] = 1+(M+1)+T([M+1]+[N-1])$.

If $N = 0$, then $K + 1 = 1 + [1 + M + T(M + 0)] = 1 + 0 + T(0 + [M + 1])$. □

The following theorem shows that the function $\mathcal{T} : (\mathbb{N} \times \mathbb{N}) \to \mathbb{N}$ is injective.

Theorem 624 *For all $K, L, M, N \in \mathbb{N}$, if $1 + K + T(K + L) = 1 + M + T(M + N)$, then both $K = M$ and $L = N$.*

Proof. If

$$1 + K + T(K + L) = 1 + M + T(M + N),$$

then subtracting M and $T(K + L)$ from both sides gives

$$K - M = T(M + N) - T(K + L).$$

If $K > M$, then $K - M > 0$, so that the left-hand side is positive, but then the right-hand side must also be positive: $T(M + N) - T(K + L) > 0$. Hence $M + N > K + L$, but then

$$T(M + N) - T(K + L) \geq T(M + N) - T(M + N - 1) = M + N$$

by definition of T. Thus

$$K - M = T(M + N) - T(K + L) \geq M + N$$

whence $K \geq (2 * M) + N$. With $M + N > K + L$, this gives

$$M + N > K + L \geq [(2 * M) + N] + L$$

whence subtracting $(2 * M) + N$ from all sides gives

$$-M \geq L,$$

which contradicts the hypothesis that $L \geq 0$. □

The following theorem confirms that $\mathbb{N} \times \mathbb{N}$ is denumerable.

Theorem 625 *The function $\mathcal{T} : (\mathbb{N} \times \mathbb{N}) \to \mathbb{N}$ is bijective.*

Proof. The injectivity and the surjectivity result from theorems 624 and 623. □

The computation of the inverse function $\mathcal{T}^{-1} : \mathbb{N} \to \mathbb{N} \times \mathbb{N}$ can proceed according to the straightforward algorithm provided by the proof of theorem 623. Specifically, for each $I \in \mathbb{N}$, observe that $T(0) = 0 \leq I$, and compute $T(0), T(1), \ldots, T(L)$ until $T(L-1) \leq I < T(L)$. Then let $M := [I - T(L - 1)] - 1$ and $N := (L - 1) - M$.

3.7.3 The Bernstein–Cantor–Schröder Theorem

The following theorem guarantees the existence of a bijection between two sets, provided that there exist injections from one set to the other and vice versa. According to Suppes [95, p. 95], Cantor conjectured the theorem and then Bernstein and Schröder proved it independently of each other in the 1890's. Fraenkel [28, p. 102] credits the following proof to J. M. Whitaker.

Theorem 626 (Bernstein–Cantor–Schröder Theorem) *For all sets A and B, if there exist injections $F : A \to B$ and $G : B \to A$, then there exists a bijection $H : A \to B$.*

Proof. The strategy of this proof consists in producing a subset $E \subseteq A$ such that

$$G"[B \setminus F"(E)] = (A \setminus E)$$

and then in setting

$$H := (F|_E) \dot{\cup} (G^{\circ -1}|_{A \setminus E}).$$

To this end, define

$$\begin{aligned} \mathcal{D} &:= \{C \subseteq A : G"[B \setminus F"(C)] \subseteq (A \setminus C)\} \\ &= \{C \subseteq A : C \subseteq A \setminus G"[B \setminus F"(C)]\}; \end{aligned}$$

the proof will verify that $\bigcup \mathcal{D}$ satisfies the requirements.

Firstly, for all subsets $V, W \subseteq A$, if $V \subseteq W$, then $F"(V) \subseteq F"(W)$, whence $[B \setminus F"(W)] \subseteq [B \setminus F"(V)]$, and hence $G"[B \setminus F"(W)] \subseteq G"[B \setminus F"(V)]$, so that

$$\{A \setminus G"[B \setminus F"(V)]\} \subseteq \{A \setminus G"[B \setminus F"(W)]\}.$$

In particular, if $V \in \mathcal{D}$, then $V \subseteq A \setminus G"[B \setminus F"(V)]$ by definition of $\mathcal{D}$, and $V \subseteq W := \bigcup \mathcal{D}$ by definition of $\bigcup \mathcal{D}$; consequently

$$V \subseteq \{A \setminus G"[B \setminus F"(V)]\} \subseteq \left\{A \setminus G"\left[B \setminus F"\left(\bigcup \mathcal{D}\right)\right]\right\}.$$

Because these inclusions hold for every element $V \in \mathcal{D}$, it follows that they also hold for their union:

$$\bigcup \mathcal{D} \subseteq \left\{A \setminus G"\left[B \setminus F"\left(\bigcup \mathcal{D}\right)\right]\right\}.$$

Let

$$E := A \setminus G"\left[B \setminus F"\left(\bigcup \mathcal{D}\right)\right].$$

From $\bigcup \mathcal{D} \subseteq E$ it follows that

$$A \setminus G"\left[B \setminus F"\left(\bigcup \mathcal{D}\right)\right] \subseteq \{A \setminus G"[B \setminus F"(E)]\},$$

so that

$$E \subseteq \{A \setminus G"[B \setminus F"(E)]\}$$

whence $E \in \mathcal{D}$. Consequently, $E \subseteq \bigcup \mathcal{D}$, whence $E = \bigcup \mathcal{D}$, but then the definition of E gives

$$E = A \setminus G"[B \setminus F"(E)].$$

□

The next theorem demonstrates a use of the Bernstein–Cantor–Schröder Theorem.

Theorem 627 *There exists a bijection $H : \mathbb{Z} \to \mathbb{Q}$.*

Proof. Firstly, there exists an injection $F : \mathbb{Z} \to \mathbb{Q}$ with $F(N) := N/1$.

Secondly, there exists an injection $I : \mathbb{Q} \to (\mathbb{Z} \times \mathbb{Z})$ such that $I(P/Q) := (P, Q)$ with $P \in \mathbb{N}$ minimum (theorem 560).

Also, there exists an injection $J : (\mathbb{Z} \times \mathbb{Z}) \to \mathbb{Z}$, by theorem 625.

Consequently, the composition $G := J \circ I$ is an injection $\mathbb{Q} \to \mathbb{Z}$.

Therefore, the Bernstein–Cantor–Schröder Theorem guarantees the existence of a bijection $H : \mathbb{Z} \to \mathbb{Q}$. □

3.7.4 Other Infinite Sets

The present text has defined a set to be finite if and only if there exists a bijection onto a natural number, and infinite if and only if there does not exist any such bijection. There exists a *different* definition of infinite sets, called "Dedekind-infinite" [95, p. 107], corresponding to Dedekind's definition [17, §V, #64, p. 63].

Definition 628 A set Z is **Dedekind-infinite** if and only if there exists a proper subset $Y \subsetneqq Z$ and a bijection $F : Z \to Y$, or, equivalently, $F^{\circ -1} : Y \to Z$.

Thus, a set is Dedekind-infinite if and only if it has the same cardinality as that of one of its proper subsets.

Example 629 The set $\mathbb{N}$ is Dedekind-infinite because there is a proper subset $\mathbb{N}^* \subset \mathbb{N}$ and a bijection defined by the successor function:

$$\begin{aligned} G : \mathbb{N} &\to \mathbb{N}^*, \\ X &\mapsto X \cup \{X\}. \end{aligned}$$

Theorem 630 *Every Dedekind-infinite set is also infinite.*

Proof. This proof proceeds by contradiction. If a set Z is *not* infinite, in other words, if Z is *finite*, then there exist a natural number $N \in \mathbb{N}$ and a bijection $G : N \to Z$. If Z is also Dedekind-infinite , then there exists a proper subset $Y \subsetneqq Z$ and a bijection $F : Z \to Y$. Then the composition

$$H := G^{\circ -1} \circ F \circ G : N \xrightarrow{G} Z \xrightarrow{F} Y \xrightarrow{G^{\circ -1}} N$$

would be a bijection from N onto a proper subset $G^{\circ -1}"(Y) \subsetneqq N$, which would contradict theorem 607. □

The proof of the converse requires the Axiom of Choice (exercises 729, 730).

Theorem 631 *Every infinite set contains a denumerable subset.*

Proof. Apply recursion (theorem 484). If Z is infinite, then $Z \neq \varnothing$, whence there exists some $X \in Z$. Define $F_0 : \{0\} \to Z$ by $0 \mapsto X$. Assume that there exists an injection $F_N : N \to Z$. Then $Z \neq F_N"(N)$ because Z is infinite, whence $Z \setminus F_N"(N) \neq \varnothing$ and there exists $X \in Z \setminus F_N"(N)$. Let $H_N : \{N\} \to \{X\}$ and let $F_{N+1} := F_N \dot{\cup} H_N$. Finally, let $F := \bigcup_{N \in \mathbb{N}} F_N$. Then $F : \mathbb{N} \to Z$ is an injection. □

Theorem 631 explains the subscript 0 in the notation $\aleph_0$ for the cardinality of $\mathbb{N}$. Because every infinite set Z contains a denumerable subset, there exists an injection $F : \mathbb{N} \to Z$, so that the cardinality of $\mathbb{N}$ cannot exceed that of Z. If there is also an injection $G : Z \to \mathbb{N}$, then the Bernstein–Cantor–Schröder Theorem (theorem 626) guarantees that there is also a bijection $H : \mathbb{N} \to Z$; thus, the cardinality of Z cannot be strictly smaller than that of $\mathbb{N}$. Thus $\aleph_0$ represents the "smallest" infinite cardinality.

Theorem 632 *Every infinite set is also Dedekind-infinite .*

Proof. If W is infinite, then W contains a denumerable subset $Z \subseteq W$, by theorem 631. By example 629, there exists a bijection $F : Z \to Y$ onto a proper subset $Y \subsetneqq Z$. Extend F to all of W by setting $H := F \dot{\cup} \left(I_W|_{W \setminus Z}\right)$. □

Theorem 633 *Every denumerable union of disjoint denumerable sets is denumerable.*

Proof. If $\mathcal{F}$ is denumerable, then there is a bijection $A : \mathbb{N} \to \mathcal{F}$ with $I \mapsto A(I)$. If each $A(I)$ is denumerable, then there is a bijection $F_I : \mathbb{N} \to A$ with $J \mapsto F_I(J) \in A(I)$. Hence the function $G : \mathbb{N} \times \mathbb{N} \to \bigcup \mathcal{F}$ defined by $G(I, J) := F_I(J)$ is a bijection. □

3.7.5 Further Issues in Cardinality

Other Axioms of Infinity

Bolzano [6] and Dedekind [17, §V, Theorem 66, p. 64] argued for the "existence" of an infinite set from practical considerations. Yet the "existence" of an infinite set does *not* follow from the other axioms, but requires an axiom of infinity [3, p. 21]. Zermelo [108] introduced such an axiom with a set such that if X is an element then $\{X\}$ is also an element. The variant adopted here, with $\{X\}$ replaced by $X \cup \{X\}$, is attributed to John von Neumann [102] and has proved more convenient [3, p. 22]. There also exist "inifinitely" many other non-equivalent axioms of infinity [11, §57, p. 342–346].

Peano's Axioms

There is an alternative method to introduce natural numbers, published by Giuseppe Peano, which does not involve set theory. Instead, Peano's system consists of the following axioms, stated here begining with 0, as done in [34, §2, #II], [95, p. 121].

Axiom A0 0 is a natural number.

Axiom A1 $K = L$ means that K and L are the same natural numbers.

Axiom A2 For each natural number N there exists exactly one natural number denoted by N' and called the successor of N.

Axiom A3 $N' \neq 0$ for each natural number N.

Axiom A4 If $K' = L'$, then $K = L$.

Axiom A5 For each set S of natural numbers, if

- 0 is an element of S, and
- if N belongs to S, then K' belongs to S,

then every natural number is an element of S .

From Peano's axioms, and two additional axioms for recursive definitions of addition and multiplication [95, p. 136], the same proofs as in this chapter verify all the algebraic properties of arithmetic and ordering with natural numbers [65, Ch. 1, p. 1–18]. However, in situations that involve other topics, for examples, rational numbers or cardinality of sets, the use of Peano's axioms would require some theoretical link between Peano's natural numbers and other sets, in other words, some means of including Peano's arithmetic and applications within the same framework. This theoretical link can be the development of arithmetic from within set theory, as done here.

Alternate Sequence of Developments

The progession from $\mathbb{N}$ to $\mathbb{Z}$ and then to $\mathbb{Q}$ allows for subtractions with integers without requiring rational numbers, and then allows for divisions and subtractions with rational numbers without requiring the development of "real" numbers. An alternative development proceeds from $\mathbb{N}$ to $\mathbb{Q}_{+}$, then to the non-negative "real" numbers $\mathbb{R}_{+}$ and finally to the "real" numbers $\mathbb{R}$ and the "complex" numbers $\mathbb{C}$, as outlined in [65].

The Generalized Continuum Hypothesis

Because there is *no* surjection from $\mathbb{N}$ to $2^{\mathbb{N}}$ (theorem 615), the question arises whether there is any set S with a cardinality between the cardinalities of $\mathbb{N}$ and $2^{\mathbb{N}}$. In other words, the question pertains to the existence of a set S for which there exist injections

$$\mathbb{N} \stackrel{I}{\hookrightarrow} S \stackrel{J}{\hookrightarrow} 2^{\mathbb{N}}$$

but no injections from $2^{\mathbb{N}}$ back to S and no injections from S back to $\mathbb{N}$. The hypothesis that no such set as S exists is called the **continuum hypothesis**.

More generally, because for every infinite set X there does *not* exist any surjection from X to 2^X, the question arises whether there exists any set S with a cardinality strictly between the cardinalities of X and 2^X. In other words, the question pertains to the existence of a set S for which there exist injections

$$X \stackrel{I}{\hookrightarrow} S \stackrel{J}{\hookrightarrow} 2^X$$

but no injections from 2^X back to S and no injections from S back to X. The hypothesis that no such set as S exists is called the **generalized continuum hypothesis**.

The axioms of set theory (S1–S8) are consistent with both the generalized continuum hypothesis (as proved by K. Gödel [32], [33]) and with the negation of the generalized continuum hypothesis (as proved by P. J. Cohen [14], [15]). Therefore, if the axioms of set theory are consistent, then the generalized continuum hypothesis can be neither proved nor disproved within set theory. The generalized continuum hypothesis is thus logically independent from set theory. Hence there exist two mutually exclusive extensions of set theory, one extension with the generalized continuum hypothesis, the other extension with the negation of the generalized continuum hypothesis.

3.7.6 Exercises

Exercise 711 Prove that if A is infinite and if $A \subseteq B$, then B is also infinite.

Exercise 712 Prove that if A is denumerable and $Z \notin A$, then $A \dot{\cup} \{Z\}$ is denumerable.

Exercise 713 With A denumerable, B finite, disjoint, prove that $A \dot{\cup} B$ is denumerable.

Exercise 714 Prove that if A is denumerable and B finite, then $A \cup B$ is denumerable.

Exercise 715 Prove that if A and B are denumerable, then $A \cup B$ is denumerable.

Exercise 716 Prove that there exists a bijection from $2^{\mathbb{N}}$ to $\mathcal{P}(\mathbb{N})$.

Exercise 717 Prove that $\mathcal{P}(\mathbb{N})$ is *not* countable.

Exercise 718 Prove that if A is denumerable and B is finite, then $A \times B$ is countable.

Exercise 719 For each non-empty set X prove that there is no injection $2^X \hookrightarrow X$.

Exercise 720 Prove that if $[\#(X)] = [\#(Y)]$, then $[\#(2^X)] = [\#(2^Y)]$.

Exercise 721 Prove that if $[\#(X)] < [\#(Y)]$ are both finite, then $[\#(2^X)] < [\#(2^Y)]$.

Exercise 722 Prove that if X is a finite set, then 2^X is also finite.

Exercise 723 Prove that if X is a finite set, then $[\#(X)] < [\#(2^X)]$.

Exercise 724 Prove that every infinite subset $S \subseteq \mathbb{N}$ is denumerable.

Exercise 725 For each denumerable set A prove that every subset $S \subseteq A$ is countable.

Exercise 726 Prove that if A is uncountable and if $A \subseteq B$, then B is also uncountable.

Exercise 727 Prove that if A, B, and C are denumerable, then so is $(A \times B) \times C$.

Exercise 728 Prove that if $A \times B$ is denumerable, then A and B are countable.

Exercise 729 Show where the proof of theorem 633 invokes the Axiom of Choice.

Exercise 730 Show where the proof of theorem 631 invokes the Axiom of Choice.

3.8 ARITHMETIC IN FINANCE

3.8.1 Introduction

This section outlines how the theory proved in this chapter applies to the solutions of financial problems. (Algorithms for digital arithmetics will be proved in chapter 5.)

3.8.2 Percentages and Rates

A rational **percentage** is a fraction with a denominator equal to one hundred. The corresponding **rate** is the rational number represented by that fraction. Thus for all rational numbers $P, Q \in \mathbb{Q}$, define $Q\% := Q/(100)$, and let "Q percent of P" mean the product $[Q/(100)] * P$, or, by commutativity, $P * [Q/(100)]$ (theorems 496, 565).

Example 634 A rate R of six percent can be written $R = 6\%$ or $R = 6/100$.

Example 635 Six percent of fifty equals three: $[6/(100)] * (50) = 300/100 = 3/1$.

Example 636 Two is twenty-five percent of eight. Indeed, solving for the rate R the equation $2 = R * 8$ gives $R = 2/8 = 1/4 = 25/100$ (theorem 568).

3.8.3 Sales and Income Taxes

Some governments levy a sales tax S, equal to the product of a **sales-tax rate** R and the sales price P of goods or services: $S = P * R$. The total cost C to the purchaser is then the sum $C := P + S$ of the price P and the tax S, or, equivalently,

$$\begin{aligned} C &= P + S \\ &= P * 1 + P * R && \text{definition of the sales tax } S, \\ &= P * (1 + R) && \text{theorems 493, 539 565, 569.} \end{aligned}$$

Monetary units are the same for C, P, S; they need not be included in the arithmetic.

Example 637 If $P := 3$ and $R := 7\%$, then $S = R * P = (7/100) * 3 = 21/100$.

Occasionally, the problem arises to determine a price P so that the total cost C equals a specified amount, for instance, an integer, for the buyer's convenience. The solution then consists in solving for the price P the equation $P * (1 + R) = C$:

$$\begin{aligned} P * (1 + R) &= C && \text{definition of the cost } C, \\ P &= \frac{C}{1 + R} && \text{definitions 525, 576.} \end{aligned}$$

Example 638 If a sales tax with rate $R := 8\%$ must yield a cost $C := 27$, then the corresponding price is $P = C/(1 + R) = 27/[(100 + 8)/100] = 2700/108 = 25$.

An **income tax** T is *subtracted* from a taxpayer's taxable income M, which leaves a **net income** N equal to the difference $N := M - T$. The tax T can be a percentage X of the income M, so that $N = M - T = M * 1 - M * X = M * (1 - X)$.

Example 639 With an income-tax rate $X := 25\%$, a taxable income $M := 30\,000$ leaves a net income $N = (30\,000) * (1 - {}^{25}/_{100}) = 22\,500$.

3.8.4 Compounded Interest

An amount P deposited into a savings account — or into any investment — is called a **principal**. If the account earns **interest** at a rate r per period, then at the end of the first period the investment returns an interest $I := P * r$. The total amount A in the account, called the **balance**, is the sum $P + I$ of the principal P and the interest I:

$$\begin{aligned} A &= P + I \\ &= P * 1 + P * r && \text{definition of the interest } I, \\ &= P * (1 + r) && \text{theorems 493, 539 565, 569.} \end{aligned}$$

For accounts with **compounded interest**, the balance $A(N)$ at the end of the Nth period becomes the principal at the beginning of the next period. Consequently,

$$A(N + 1) := A(N) * (1 + r).$$

Hence with $A(0) := P$ mathematical induction reveals the formula

$$A(N) = P * (1 + r)^N. \tag{3.8.4.1}$$

Compounded interest is often specified by an **annual rate** R compounded monthly, which means a rate $r := R/12$ per month.

Example 640 An *annual* rate $R := 12\%$ compounded monthly means a monthly rate $r = R/12 = 1\%$. At this rate, a principal $P := 1000$ produces twenty years hence a balance $1000 * [1 + (1/100)]^{20*12} = 10\,892.55$ (rounded to the nearest hundredth).

If the goal of an investment consists in producing a specified balance $A(N)$ at the end of the Nth period, then the problem arises to compute the necessary principal P. The solution consists in solving for the principal P the equation $P * (1 + r)^N = A(N)$:

$$\begin{aligned} P * (1 + r)^N &= A(N), \\ P &= \frac{A(N)}{(1 + r)^N}. \end{aligned} \tag{3.8.4.2}$$

For this reason, the principal P is also called the **present value** of the balance $A(N)$.

Example 641 To compute a principal P sufficient to yield a $10\,000$ scholarship 18 years hence with an annual rate of 12% compounded monthly, equation 3.8.4.2 gives

$$P = \frac{10\,000}{[1 + (1/100)]^{18*12}} = 1165.70$$

(rounded *up* to the nearest hundredth, because rounding down does not suffice).

3.8.5 Loans, Mortgages, and Savings Plans

Some savings plans include the deposit of a fixed **payment** p at the *end* of *each* period. The payment p is added to the current amount A, producing a balance $B := A + p$ at the beginning of the next period. With interest compounded at a rate r per period, induction gives a formulae for the balance $B(N)$ at the *end* of the Nth period:

$$\begin{aligned}
B(1) &= p, \\
B(2) &= p*(1+r)+p = p*[1+(1+r)], \\
B(3) &= \{p*[1+(1+r)]\}*(1+r)+p = p*[1+(1+r)+(1+r)^2], \\
&\vdots \\
B(N) &= p*[1+(1+r)+\cdots+(1+r)^{N-1}] \\
&= p*\sum_{K=0}^{N-1}(1+r)^K \qquad \text{(definition 582)} \\
&= p*\frac{1-(1+r)^N}{1-(1+r)} \qquad \text{(theorem 584)} \\
&= p*\frac{(1+r)^N-1}{r}. \qquad (3.8.5.3)
\end{aligned}$$

Example 642 To find the savings produced by depositing 100 at the end of each month for 18 years with 12% annual interest compounded monthly, equation 3.8.5.3 gives

$$B(216) = 100*\frac{\{1+[(1/2)/100]\}^{18*12}-1}{(1/2)/100} = 38\,735.32$$

(rounded to the nearest hundredth).

A loan or a mortgage begins by borrowing a principal P. With an interest rate r compounded each period, equation 3.8.4.1 gives the balance due after N periods: $A(N) = P*(1+r)^N$. To pay this balance the borrower can deposit a payment p at the end of each period in an account with the same interest rate r, accumulating a savings specified by equation 3.8.5.3: $B(N) = p*[(1+r)^N - 1]/r$. The savings $B(N)$ must match the balance due $A(N)$, whence solving $B(N) = A(N)$ gives a formula for p:

$$\begin{aligned}
B(N) &= A(N), \\
p*\frac{(1+r)^N-1}{r} &= P*(1+r)^N, \\
p &= P*\frac{r*(1+r)^N}{(1+r)^N-1} \\
&= P*\frac{r}{1-(1+r)^{-N}}. \qquad (3.8.5.4)
\end{aligned}$$

Example 643 To pay a thirty-year mortgage with a principal of 100 000 and an annual interest rate of 6% compounded monthly, equation 3.8.5.4 shows a monthly payment

$$p = (100\,000)*\frac{(6/12)/100}{1-[1+(6/12)/100]^{-30*12}} = 599.56$$

(rounded *up* to the next hundredth).

3.8.6 Perpetuities

A **perpetuity** consists of a principal A earning interest at a rate r compounded each period, and producing a payment p at the end of each period forever. Each payment p withdraws only the interest $r * A$ earned by the principal over the previous period, always leaving the same principal A in the account for the next period. Thus,

$$p = r * A, \tag{3.8.6.5}$$

$$\frac{p}{r} = A. \tag{3.8.6.6}$$

Example 644 To receive payments equal to 1000 at the end of every month forever, and with an annual interest rate of 3% compounded monthly, equation 3.8.6.6 determines that the principal must be $A = p/r = 1000/[({}^3/_{12})/100] = 400\,000$.

The foregoing treatment of perpetuities typifies some of the mathematical methods related to infinities, in the sense that the potentially perpetual stream of income was reduced to a logically equivalent balance of payment and interest within only one finite period. For more detail on the mathematics of finance, see [60].

3.8.7 Exercises

Exercise 731 Calculate the sales price to be charged, so that with a sales-tax rate of 5% the total cost equals 3.

Exercise 732 Calculate the sales price to be charged, so that with a sales-tax rate of 4% the total cost equals 7.

Exercise 733 Calculate the taxable income to be earned, so that with an income-tax rate of 25% the net income equals 36 000.

Exercise 734 Calculate the taxable income to be earned, so that with an income-tax rate of 33% the net income equals 60 000.

Exercise 735 Calculate the taxable income to be earned, subject to an income-tax rate of 33%, to purchase a car with a sales price of 10 000 subject to a sales-tax rate of 8%.

Exercise 736 Find a formula for the taxable income M to be earned, subject to an income-tax rate X, to buy an item with a sales price P subject to a sales-tax rate R.

Exercise 737 A bank offers year-long investments with twelve different monthly rates of interest $R_1, R_2, \ldots, R_{11}, R_{12}$, one rate for each month of the year. The customer decides the order in which to apply each rate, for instance, R_7 in January, R_3 in February, and so forth. Determine the order(s) producing the largest balance at year's end.

Exercise 738 Calculate the maximum cost of a car that can be bought with monthly payments of 275 over 5 years with an annual interest rate of 12% compounded monthly.

Exercise 739 Calculate the monthly payment for a fifteen-year mortgage in the amount of 125 000 with an annual interest rate of 8% compounded monthly.

Exercise 740 Calculate the monthly payment for a twenty-year mortgage in the amount of 250 000 with an annual interest rate of 9% compounded monthly.

3.9 PROJECTS

Project 7 Investigate whether for each rational $P/Q \in \mathbb{Q}_+$ with $0 < P/Q < 1$ there exist $N \in \mathbb{N}$ and a sequence $L : N \to \mathbb{N}^*$ such that if $J > I$, then $L(J) > L(I)$, and

$$\frac{P}{Q} = \sum_{K=0}^{N-1} \frac{1}{L(K)} = \frac{1}{L(0)} + \frac{1}{L(1)} + \cdots + \frac{1}{L(N-1)}.$$

Examples appear in the Rhind Papyrus from Ancient Egypt [100, p. 165]. For instance,

$$\frac{2}{7} = \frac{1}{5} + \frac{1}{12} + \frac{1}{420}.$$

Project 8 Investigate how the concepts and axioms of plane and spatial Euclidean geometry [49] can be derived from logic and set theory.

Chapter 4

Decidability and Completeness

4.0 INTRODUCTION

Figure 4.1: Computers can prove some but not every theorem.

This chapter explains in detail various methods to investigate which of several alternative logics and axioms for mathematics correspond most closely to the common and scientific patterns of reasoning. Some of these logics have sufficient power, while other logical systems lack the power, to decide whether certain logical formulae are theorems, and hence whether they are allowed (within such logics) in common or scientific arguments. Examples demonstrate that all the proofs within the implicational calculus (with implications, but without negation) do not suffice to prove certain implicational tautologies. In other words, the implicational calculus lacks the power to decide the status of certain of its own implicational formulae. The same examples also demonstrate how a computing language based on axioms and rules might fail to generate all sentences in the language.

Because the full propositional calculus is equivalent to the Boolean algebraic logic of Truth tables, it allows for some forms of reasoning by contraposition that are not commonly accepted [73, p. 36]. Therefore, logics between the implicational and full propositional calculi seem to correspond more closely to common patterns of reasoning, for instance, the minimal and intuitionistic logics. However, the first main result presented here reveals that such logics lie outside of the realm of all systems of Truth tables, so that they cannot be described or computed by Truth tables.

The second principal result presented here — called the Completeness Theorem — provides an algorithm to design automatically a proof for each tautology within the full propositional calculus (with implications *and* negations) [11, §15, p. 104–106], [92, p. 191–197]. The Completeness Theorem shows not only that the propositional calculus is amenable to automated proofs by computers (see figure 4.1). The Completeness Theorem also constitutes an example of the type of theory necessary to guarantee the feasibility of any other computing language based on axioms and rules.

The third principal result presented here proves the dependence of one axiom on the others for sets — called well-formed sets — defined in specific ways solely through the axioms of set theory. Such well-formed sets suffice for most of logic, mathematics, computer science, and their applications to the sciences and engineering. Among other features, the result shows that no well-formed set is an element of itself, and that this result is decidable from the other axioms of set theory [64].

4.1 LOGICS FOR SCIENTIFIC REASONING

4.1.1 Scientific Reasoning

The full classical propositional calculus relies on the rules of substitutions and *Modus Ponens*, and any system of axioms equivalent to Łukasiewicz's three axioms (from definition 164 in subsection 1.1.1):

Axiom P1 $(P) \Rightarrow [(Q) \Rightarrow (P)]$.

Axiom P2 $\{(P) \Rightarrow [(Q) \Rightarrow (R)]\} \Rightarrow \{[(P) \Rightarrow (Q)] \Rightarrow [(P) \Rightarrow (R)]\}$.

Axiom P3 $\{[\neg(Q)] \Rightarrow [\neg(P)]\} \Rightarrow [(P) \Rightarrow (Q)]$.

The full propositional calculus forms one of the possible foundations for mathematics and computer science, but it contains principles that might not be necessary in the sciences. This subsection briefly reviews two such principles: the converse law of contraposition and the law of double negation. Subsequent subsections will then investigate logics that do not include these potentially unnecessary principles.

The first such principle pertains to contrapositions. To reject a hypothesis H that is not supported by an experiment that contradicts a conclusion C, a form of reasoning common in the sciences uses the law of contraposition.

Example 645 The following argument is a part of the investigation of Pluto's status:

- A definition of planets: Every planet is larger than every moon.
- If Pluto is a planet, then Pluto must be larger than Ganymede.
- Yet Pluto is smaller than Ganymede.
- Therefore Pluto is not a planet.

(See example 1 in chapter 0.)

Logic captures the reasoning by contraposition in example 645 as follows.

H: "Pluto is a planet."

C: "Pluto is smaller than Ganymede."

$\vdash (H) \Rightarrow (C)$	known by a definition,
$\vdash [(H) \Rightarrow (C)] \Rightarrow \{[\neg(C)] \Rightarrow [\neg(H)]\}$	law of contraposition,
$\vdash [\neg(C)] \Rightarrow [\neg(H)]$	*Modus Ponens*,
$\vdash \neg(C)$	tested by experiments,
$\vdash \neg(H)$	*Modus Ponens*.

Thus observations $\neg(C)$ contrary to an expected conclusion C lead to the rejection $\neg(H)$ of the working hypothesis H. In contrast, the converse law of contraposition

$$\{[\neg(C)] \Rightarrow [\neg(H)]\} \Rightarrow [(H) \Rightarrow (C)],$$

is hardly ever used in a practical reasoning, which would take the following form:

$\vdash [\neg(C)] \Rightarrow [\neg(H)]$	known from definitions,
$\vdash \{[\neg(C)] \Rightarrow [\neg(H)]\} \Rightarrow [(H) \Rightarrow (C)]$	converse law of contraposition,
$\vdash (H) \Rightarrow (C)$	*Modus Ponens*,
$\vdash H$	*verified* by experiments,
$\vdash C$	*Modus Ponens*.

In practice a hypothesis H cannot be *verified* by experiments because it is nearly impossible to verify with total certainty that any hypothesis H is always true (although it might have been thought true for millenia, as in example 160 in chapter 1).

Thus it seems that the converse law of contraposition ought to be omitted from logics that reflect common and scientific patterns of reasoning:

$$\{[\neg(C)] \Rightarrow [\neg(H)]\} \not\Rightarrow [(H) \Rightarrow (C)].$$

Yet without the converse law of contraposition, the next section will establish that the remaining implicational calculus is in some sense incomplete. Hence arises the problem of replacing the converse law of contraposition by other axioms that correspond more closely to common and scientific patterns of reasoning. Each of Brouwer & Heyting's intuitionistic logic or Kolmogorov & Johansson's minimal logic will solve this problem presently. However, the following sections will also reveal that these two logics lie outside of the realm of all the logics defined by Truth tables.

4.1.2 Hypothesis Testing

The second potential unnecessary logical principle pertains to double negations. A common scientific argument relies on the method of **hypothesis testing** to check a conjecture that modifies or contradicts a previous scientific theory. Consequently, the new conjecture is often called the **alternative hypothesis** (denoted by H_1), while the previous theory is called the **null hypothesis** (denoted by H_0). Logically, the null hypothesis is the negation of the alternative hypothesis [52, p. 235]:

(H_0)	the null hypothesis (the previous theory)
$\Updownarrow$	is logically equivalent to
$[\neg(H_1)]$	the negation of the alternative hypothesis.

The corresponding scientific method then includes an experiment to test the alternative hypothesis against the null hypothesis. If the experimental results corroborate the alternative hypothesis H_1, then the method *rejects* the null hypothesis H_0 [52, p. 236]. Rejecting H_0

amounts to supporting its negation $\neg(H_0)$, or, equivalently, $\neg[\neg(H_1)]$ because H_0 was defined as $\neg(H_1)$. Yet rejecting the null hypothesis H_0 and supporting its negation $\neg[\neg(H_1)]$ does not amount to asserting that the alternative hypothesis holds [52, p. 238]. In logical terms, such a reasoning means that the law of double negation $\{\neg[\neg(H_1)]\} \Rightarrow (H_1)$ might not be a scientifically valid argument:

$$\{\neg[\neg(H_1)]\} \not\Rightarrow (H_1).$$

The following subsections present two logics including neither the law of double negation nor the converse law of contraposition.

4.1.3 Brouwer & Heyting's Intuitionistic Logic

Classical logic specifies its concept of negation by the converse law of contraposition. In contrast, **intuitionistic logic** specifies its concept of negation by the special law of reductio ad absurdum and the negation of the antecedent. Moreover, intuitionistic logic includes not only the implication but also the conjunction in its axioms.

Definition 646 (Axioms of intuitionistic logic) A formula is an **axiom** of the intuitionistic logic if and only if it is one of the following formulae [11, p. 141], [79, p. 138]. The first two axioms reproduce the axioms of the implicational calculus.

Axiom I1 $(P) \Rightarrow [(Q) \Rightarrow (P)]$.

Axiom I2 $\{(P) \Rightarrow [(Q) \Rightarrow (R)]\} \Rightarrow \{[(P) \Rightarrow (Q)] \Rightarrow [(P) \Rightarrow (R)]\}$.

The next two axioms define the concept of negation in intuitionistic logic.

Axiom I3 (Special law of reductio ad absurdum) $\{(P) \Rightarrow [\neg(P)]\} \Rightarrow [\neg(P)]$.

Axiom I4 (Law of denial of the antecedent) $[\neg(P)] \Rightarrow [(P) \Rightarrow (Q)]$.

The next three axioms define the concept of conjunction in intuitionistic logic.

Axiom I5 $[(P) \wedge (Q)] \Rightarrow (P)$.

Axiom I6 $[(P) \wedge (Q)] \Rightarrow (Q)$.

Axiom I7 $(P) \Rightarrow \{(Q) \Rightarrow [(P) \wedge (Q)]\}$.

The final three axioms define the concept of disjunction in intuitionistic logic.

Axiom I8 $(P) \Rightarrow [(P) \vee (Q)]$.

Axiom I9 $(Q) \Rightarrow [(P) \vee (Q)]$.

Axiom I10 $[(P) \Rightarrow (R)] \Rightarrow \big([(Q) \Rightarrow (R)] \Rightarrow \{[(P) \vee (Q)] \Rightarrow (R)\}\big)$

Intuitionistic logic also uses the same rules of substitutions and *Modus Ponens* as does classical logic. Because the first two axioms of intuitionistic logic (I1, I2) coincide with the first two axioms of classical logic (P1, P2), it follows that the intuitionistic implicational calculus coincides with the classical implicational calculus. Moreover, because all the axioms of intuitionistic logic are theorems of classical logic (as proved in chapter 1), applications of the rules of inference remain within classical logic, whence it follows that every theorem in intuitionistic logic is also a theorem in classical logic. The converse fails however, because some theorems of classical logic are *not* theorems of intuitionistic logic [11, p. 146, §26.11]. Thus the law of indirect proof

$$\{[\neg(P)] \Rightarrow (Q)\} \Rightarrow \big[\{[\neg(P)] \Rightarrow [\neg(Q)]\} \Rightarrow (P)\big]$$

(exercise 787) and the converse law of contraposition (exercise 785)

$$\{[\neg(Q)] \Rightarrow [\neg(P)]\} \Rightarrow [(P) \Rightarrow (Q)]$$

are *not* theorems of intuitionistic logic.

Moreover, the law of excluded middle $(B) \vee [\neg(B)]$ is not a theorem of intuitionistic logic (example 670). In other words, the pattern of deductive reasoning "either it is, or it is not" does not belong to intuitionistic logic.

Similarly, the law of double negation $\{\neg[\neg(P)]\} \Rightarrow (P)$ is not a theorem of intuitionistic logic (exercise 783). Thus even if the negation of a negation of a proposition is a theorem, $\neg[\neg(P)]$, the proposition P need *not* be a theorem. In other words, if a proposition isn't not P then it may still differ from P.

In contrast, the converse law of double negation $(P) \Rightarrow \{\neg[\neg(P)]\}$ *is* a theorem of intuitionistic logic (theorem 647), so that if a proposition holds, then its negation does not hold. For demonstration purposes, the following three theorems provide examples of proofs within intuitionistic logic. The same theorems also confirm that intuitionistic logic allows for reasoning by the law of double negation, the law of contraposition, and the law of reductio ad absurdum. Thus the first theorem establishes the converse law of double negation within intuitionistic logic.

Theorem 647 *The converse law of double negation is a theorem in intuitionistic logic:*

$$I1, I2, I3, \vdash \quad (P) \Rightarrow \{\neg[\neg(P)]\}.$$

Proof. Apply axioms I4 and I3, and theorems 185 and 175:

$\vdash [\neg(P)] \Rightarrow [(P) \Rightarrow (Q)]$	axiom I4,
$\vdash (P) \Rightarrow \{[\neg(P)] \Rightarrow (Q)\}$	commutation (theorem 185),
$\vdash (P) \Rightarrow \big([\neg(P)] \Rightarrow \{\neg[\neg(P)]\}\big)$	substitution,
$\vdash \big([\neg(P)] \Rightarrow \{\neg[\neg(P)]\}\big) \Rightarrow \{\neg[\neg(P)]\}$	axiom I3,
$\vdash (P) \Rightarrow \{\neg[\neg(P)]\}$	transitivity (theorem 175).

□

The second theorem establishes a precursor of the law of contraposition within intuitionistic logic.

Theorem 648 *The following formula is a theorem in intuitionistic logic:*

$$I1, I2, I4 \vdash \quad [(P) \Rightarrow (Q)] \Rightarrow \big[(P) \Rightarrow \{[\neg(Q)] \Rightarrow [\neg(P)]\}\big].$$

Proof. Apply axiom I4 and theorems 647 and 191:

$\vdash \{\neg[\neg(Q)]\} \Rightarrow \{[\neg(Q)] \Rightarrow [\neg(P)]\}$	axiom I4,
$\vdash (Q) \Rightarrow \{\neg[\neg(Q)]\}$	theorem 647,
$\vdash (Q) \Rightarrow \{[\neg(Q)] \Rightarrow [\neg(P)]\}$	transitivity (theorem 175),
$\vdash [(P) \Rightarrow (Q)] \Rightarrow \big[(P) \Rightarrow \{[\neg(Q)] \Rightarrow [\neg(P)]\}\big]$	theorem 191.

□

The third theorem proves the law of reductio ad absurdum in intuitionistic logic.

Theorem 649 *The law of reductio ad absurdum is a theorem in intuitionistic logic:*

$$I1, I2, I3, \vdash \quad [(P) \Rightarrow (Q)] \Rightarrow \left(\{(P) \Rightarrow [\neg(Q)]\} \Rightarrow [\neg(P)]\right).$$

Proof. Apply axioms I2 and I3, and transitivity (theorem 178):

$$\begin{array}{ll} \vdash [(P) \Rightarrow (Q)] \Rightarrow \left[(P) \Rightarrow \{[\neg(Q)] \Rightarrow [\neg(P)]\}\right] & \text{theorem 648,} \\ \vdash \left[(P) \Rightarrow \{[\neg(Q)] \Rightarrow [\neg(P)]\}\right] & \\ \qquad \Rightarrow \left(\{(P) \Rightarrow [\neg(Q)]\} \Rightarrow \{(P) \Rightarrow [\neg(P)]\}\right) & \text{axiom I2,} \\ \vdash \{(P) \Rightarrow [\neg(P)]\} \Rightarrow [\neg(P)] & \text{axiom I3,} \\ \vdash [(P) \Rightarrow (Q)] \Rightarrow \left(\{(P) \Rightarrow [\neg(Q)]\} \Rightarrow [\neg(P)]\}\right) & \text{theorem 178.} \end{array}$$

□

4.1.4 Kolmogorov & Johansson's Minimal Propositional Calculus

A logical system stricter yet than intuitionistic logic, Kolmogorov and Johansson's **minimal propositional calculus** (henceforth abbreviated as **minimal logic**) specifies its concept of negation solely through the law of reductio ad absurdum (axiom K3) and otherwise retains the implicational axioms from intuitionistic logic.

Definition 650 (Axioms of minimal logic) A formula is an **axiom** of the minimal propositional calculus if and only if it is one of the following formulae [11, p. 142]. The first two axioms are the axioms of the implicational calculus.

Axiom K1 $(P) \Rightarrow [(Q) \Rightarrow (P)]$.

Axiom K2 $\{(P) \Rightarrow [(Q) \Rightarrow (R)]\} \Rightarrow \{[(P) \Rightarrow (Q)] \Rightarrow [(P) \Rightarrow (R)]\}$.

The third axiom defines the concept of negation in minimal logic.

Axiom K3 (Law of reductio ad absurdum)
$[(P) \Rightarrow (Q)] \Rightarrow \left(\{(P) \Rightarrow [\neg(Q)]\} \Rightarrow [\neg(P)]\right)$.

The next three axioms define the concept of conjunction in minimal logic.

Axiom K4 $[(P) \wedge (Q)] \Rightarrow (P)$.

Axiom K5 $[(P) \wedge (Q)] \Rightarrow (Q)$.

Axiom K6 $(P) \Rightarrow \{(Q) \Rightarrow [(P) \wedge (Q)]\}$.

The last three axioms define the concept of disjunction in minimal logic.

Axiom K7 $(P) \Rightarrow [(P) \vee (Q)]$.

Axiom K8 $(Q) \Rightarrow [(P) \vee (Q)]$.

Axiom K9 $[(P) \Rightarrow (R)] \Rightarrow \left([(Q) \Rightarrow (R)] \Rightarrow \{[(P) \vee (Q)] \Rightarrow (R)\}\right)$.

Because axioms K1 and K2 coincide with axioms P1 and P2, the minimal propositional calculus also includes all the classical implicational calculus. The following theorems demonstrate proofs within the minimal propositional calculus. The first theorem also shows that the minimal propositional calculus allows for reasoning by the special law of reductio ad absurdum.

Theorem 651 (Special law of reductio ad absurdum) *The special law of reductio ad absurdum is a theorem of the minimal propositional calculus:*

$$K1, K2, K3 \vdash \quad \{(P) \Rightarrow [\neg(P)]\} \Rightarrow [\neg(P)].$$

Proof. Apply axiom K3 and theorem 173 from the implicational calculus:

$$\begin{array}{ll} \vdash (P) \Rightarrow (P) & \text{theorem 173,} \\ \vdash [(P) \Rightarrow (P)] \Rightarrow \{(P) \Rightarrow [\neg(P)]\} \Rightarrow [\neg(P)] & \text{axiom K3,} \\ \vdash \{(P) \Rightarrow [\neg(P)]\} \Rightarrow [\neg(P)] & \textit{Modus Ponens.} \end{array}$$

□

The second theorem shows that the minimal propositional calculus allows for reasoning by the law of contraposition.

Theorem 652 (Law of contraposition) *The law of contraposition is a theorem of the minimal propositional calculus:*

$$K1, K2, K3 \vdash \quad [(P) \Rightarrow (Q)] \Rightarrow \{[\neg(Q)] \Rightarrow [\neg(P)]\}.$$

Proof. Apply axioms K3 and K1, and transitivity (theorem 183):

$$\begin{array}{ll} \vdash [(P) \Rightarrow (Q)] \Rightarrow (\{(P) \Rightarrow [\neg(Q)]\} \Rightarrow [\neg(P)]) & \text{axiom K3,} \\ \vdash [\neg(Q)] \Rightarrow \{(P) \Rightarrow [\neg(Q)]\} & \text{axiom K1,} \\ \vdash [(P) \Rightarrow (Q)] \Rightarrow \{[\neg(Q)] \Rightarrow [\neg(P)]\} & \text{theorem 183.} \end{array}$$

□

The extent to which the minimal and intuitionistic logics differ from each other and from the full propositional calculus can be described in two ways. The first way consists in appending additional axioms to a smaller logic until it produces all of a larger logic. For instance, the following theorem shows that appending only the law of double negation to the minimal logic yields all of the full propositional calculus.

Theorem 653 *The minimal logic with the law of double negation together form a logic equivalent to the full propositional calculus.*

Proof. The minimal logic and the law of double negation are all theorems from, and hence a part of, the full propositional calculus. For the converse, because the minimal axioms K1 and K2 coincide with the implicational axioms P1 and P2 of the propositional calculus, it suffices to establish that the only remaining classical axiom (P3) is derivable from the minimal logic and the law of double negation:

$$\begin{array}{ll} \vdash \{[\neg(Q)] \Rightarrow [\neg(P)]\} \Rightarrow [\{\neg[\neg(P)]\} \Rightarrow \{\neg[\neg(Q)]\}] & \text{theorem 652,} \\ \vdash (P) \Rightarrow \{\neg[\neg(P)]\} & \text{exercise 741,} \\ \vdash \{[\neg(Q)] \Rightarrow [\neg(P)]\} \Rightarrow [(P) \Rightarrow \{\neg[\neg(Q)]\}] & \text{theorem 183,} \\ \vdash \{\neg[\neg(Q)]\} \Rightarrow (Q) & \text{hypothesis,} \\ \vdash \{[\neg(Q)] \Rightarrow [\neg(P)]\} \Rightarrow [(P) \Rightarrow (Q)] & \text{theorem 184.} \end{array}$$

□

Similarly, using axioms K7, K8, K9, exercise 759 will show that the intuitionistic logic with the law of excluded middle are equivalent to the full propositional calculus.

Exercises will also show that the minimal propositional calculus is strictly smaller than the intuitionistic propositional calculus, by proving that the intuitionistic axiom I4 is *not* provable within the minimal propositional calculus (exercise 788). Moreover, the minimal propositional calculus with the law of excluded middle is also smaller than the intuitionistic propositional calculus (exercise 789).

Remark 654 (Summary) The differences between the minimal, intuitionistic, and classical logics can be described by allowing only one among three additional axioms.

The full classical propositional calculus allows reasoning using the converse law of contraposition, the converse law of double negation, and the law of excluded middle, all of which the minimal and intuitionistic logics reject.

Accepting only the law of double negation in addition to either the minimal or intuitionistic logic reproduces all of the full classical propositional calculus.

Accepting only the law of excluded middle in addition to the intuitionistic logic also reproduces all of the full classical propositional calculus.

Yet accepting only the law of excluded middle with the minimal logic does not yield all of the intuitionistic logic, and hence not the full classical propositional calculus.

The proofs that the minimal and intuitionistic logics reject certain axioms from classical logic appear in section 4.3, using methods presented in section 4.2.

4.1.5 Exercises

Define $(P) \Leftrightarrow (Q)$ by $[(P) \Rightarrow (Q)] \wedge [(Q) \Rightarrow (P)]$. For the following exercises, prove that the stated formula is a theorem of the minimal propositional calculus.

Exercise 741 (Converse law of double negation) $(P) \Rightarrow \{\neg[\neg(P)]\}$

Exercise 742 $[(P) \Leftrightarrow (Q)] \Rightarrow [(P) \Rightarrow (Q)]$

Exercise 743 (Law of triple negation) $(\neg\{\neg[\neg(P)]\}) \Rightarrow [\neg(P)]$

Exercise 744 $[(P) \Leftrightarrow (Q)] \Rightarrow [(Q) \Rightarrow (P)]$

Exercise 745 (Law of contradiction) $\neg\{(H) \wedge [\neg(H)]\}$

Exercise 746 (Converse law of triple negation) $[\neg(P)] \Rightarrow (\neg\{\neg[\neg(P)]\})$

Exercise 747 $[(P) \Rightarrow (Q)] \Rightarrow \{[(Q) \Rightarrow (P)] \Rightarrow [(P) \Leftrightarrow (Q)]\}$

Exercise 748 Explain why every theorem of the minimal propositional calculus is also a theorem of intuitionistic logic.

The next exercises investigate extensions of the minimal and intuitionistic logics.

Exercise 749 Prove that axioms K1, K2, and axiom P3 are equivalent to the axioms of the full propositional calculus.

Exercise 750 Prove that the minimal logic and the converse law of contraposition together form a logic equivalent to the full propositional calculus.

Exercise 751 Prove that axioms I1, I2, and axiom P3 are equivalent to the axioms of the full propositional calculus.

Exercise 752 Prove that the intuitionistic logic with the converse law of contraposition together form a logic equivalent to the full propositional calculus.

Exercise 753 Prove that axioms I1, I2, I4 with $\vdash \{\neg[\neg(Q)]\} \Rightarrow (Q)$ are equivalent to the axioms of the full propositional calculus.

Exercise 754 Prove that the intuitionistic logic with the law of double negation together form a logic equivalent to the full propositional calculus.

Exercise 755 Prove that the minimal logic with the law of denial of the antecedent (axiom I4: $[\neg(P)] \Rightarrow [(P) \Rightarrow (Q)]$) together form a logic equivalent to the intuitionistic logic.

Exercise 756 Prove that axioms K1, K2, with axiom I4 are equivalent to the axioms of the intuitionistic logic.

Exercise 757 Prove that the minimal logic with the law of indirect proof

$$\{[\neg(P)] \Rightarrow (Q)\} \Rightarrow \left[\{[\neg(P)] \Rightarrow [\neg(Q)]\} \Rightarrow (P)\right]$$

together form a logic equivalent to the full propositional calculus.

Exercise 758 Prove that axioms K1, K2 with the law of indirect proof are equivalent to the axioms of the full propositional calculus.

Exercise 759 Prove that the intuitionistic logic with the law of excluded middle $(B) \vee [\neg(B)]$ together form a logic equivalent to the full propositional calculus.

Exercise 760 Prove that the intuitionistic logic with the law of contraposition together form a logic again equivalent to the same intuitionistic logic.

4.2 INCOMPLETENESS

4.2.1 Tautologies and Theorems

This section investigates further the relations between several logics that correspond closely to common and scientific arguments. For instance, tautologies and theorems belong to different types of such logics. Tautologies are the "universal Truths" of logics defined by Truth tables, for instance, Boolean algebraic logic. Theorems are the "universal Truths" of logics defined by axioms and rules, for instance, the implicational, minimal, intuitionistic, or propositional calculi. Nevertheless, two different logics that share the same logical formulae can be compared. Thus the following theorem provides a first comparison between the propositional calculus and Boolean algebraic logic.

Theorem 655 *Every theorem of the propositional calculus is a Boolean tautology.*

Proof. The present proof proceeds by induction on the number of logical steps.

Preliminary step Examinations of their Truth tables confirm that all the axioms of the propositional calculus are tautologies (exercise 191 and examples 74, 80).

Initial step Every proof of every theorem begins with an axiom. Consequently, if the proof consists of exactly one step, then the theorem is an axiom, whence it is a tautology. Thus every theorem provable in a single step is a tautology.

Induction hypothesis Assume that there exists $L \in \mathbb{N}^*$ such that every theorem provable with at most L logical steps is a tautology.

Induction step For every theorem K with a proof consisting of exactly $L+1$ logical steps, the last step is either a substitution in the penultimate step, or an instance of *Modus Ponens*. In either case, the penultimate step is a theorem proved with at most $(L+1)-1 = L$ steps, which is a tautology by the hypothesis of induction.

If the last step is a substitution, then it is a substitution in a tautology, which yields a tautology. If the last step is an instance of *Modus Ponens*, then it produces a theorem K from two preceding steps of the form $\vdash H$ and $\vdash (H) \Rightarrow (K)$. Because $\vdash H$ and $\vdash (H) \Rightarrow (K)$ occur earlier in the proof, their proofs consist of at most L steps. Consequently, by the hypothesis of induction, H and $(H) \Rightarrow (K)$ are tautologies, whence they both have the Truth value True. However, the Boolean Truth table for the connective $\Rightarrow$ contains only one line where H and $(H) \Rightarrow (K)$ both have the Truth value True, and in that line K also has the Truth value True (chapter 0, table 43). Therefore the theorem K is also a tautology. □

In contrast, a tautology cannot be disproved, because its negation is not a theorem.

Theorem 656 *The negation of a Boolean tautology is not a propositional theorem.*

Proof. This proof proceeds by contraposition. For each logical formula P, theorem 655 shows that if $\neg(P)$ is a theorem, then $\neg(P)$ is a tautology. Consequently $\neg(P)$ has the value True regardless of the values of its components. Therefore P cannot be a tautology, for otherwise $\neg(P)$ would have the value False. □

The *converse* of theorem 655 — which states that every Boolean tautology is a propositional theorem — will be proved in section 4.4. However, the converse of theorem 655 fails in other logical systems. In other words, there exist tautologies that are *not* theorems in some logics, for instance, in the implicational calculus.

4.2.2 Incompleteness of the Implicational Calculus

There exist well-formed formulae involving no connective other than implications that can neither be proved nor be disproved within the classical implicational logic. Moreover, some of those undecidable formulae are tautologies, for instance, Peirce's law.

Example 657 (Peirce's law) This example shows that the following formula — called "Peirce's law" after Charles Sanders Peirce (1839–1914) — is a tautology:

$$\{[(P) \Rightarrow (Q)] \Rightarrow (P)\} \Rightarrow (P).$$

Indeed, table 658 confirms that Peirce's law is a tautology.

Because Peirce's law involves only the implicational connective, the question arises, whether Peirce's law is a theorem of the implicational calculus, provable solely from axioms equivalent to Łukasiewicz's implicational axioms P1 and P2 (from definition 164 in subsection 1.1.1):

Axiom P1 $(P) \Rightarrow [(Q) \Rightarrow (P)]$.

Table 658 Implication and Peirce's law in Boolean (two-valued) logic.

P	Q	$(P) \Rightarrow (Q)$	$\{[(P) \Rightarrow (Q)] \Rightarrow (P)\}$	$\{[(P) \Rightarrow (Q)] \Rightarrow (P)\} \Rightarrow (P)$
T	T	T	T	T
T	F	F	T	T
F	T	T	F	T
F	F	T	F	T

Axiom P2 $\{(P) \Rightarrow [(Q) \Rightarrow (R)]\} \Rightarrow \{[(P) \Rightarrow (Q)] \Rightarrow [(P) \Rightarrow (R)]\}$.

A "standard device" [11, p. 112] to investigate whether there is such a proof uses auxiliary tables — different from Boolean Truth tables — for which the axioms and the rules of inference still yield tautologies, but with which Peirce's law is *not* a tautology.

Theorem 659 *There is no proof of Peirce's law in the classical implicational calculus.*

Proof. The three-valued logical implication defined in table 660 redefines the values of the logical implication, with the third value, U, standing for Undefined.

Table 660 A three-valued logical implication.

P	Q	$(P) \Rrightarrow (Q)$
T	T	T
T	F	F
T	U	U
F	T	T
F	F	T
F	U	U
U	T	T
U	F	T
U	U	T

Table 660 will serve merely as an investigative device; it does *not* establish any link between reality and a three-valued logic. (Indeed, other examples will require other three-valued logics. For instance, the three-valued logic used here coincides with one due to Church [11, §19, p. 113] but differs from another one due to J. Donald Monk [79, #8.56, p. 135–136].) In its role as an investigative device, table 660 shows that the axioms of implicational logic remain tautologies (exercise 762 for axiom P1, exercise 761 for axiom P2). Moreover, the first line of table 660 confirms that the rule of *Modus Ponens* remains valid in this three-valued logic, because its first line shows that if P and $(P) \Rrightarrow (Q)$ are both True, then Q is also True. Consequently, the proof of theorem 655 remains valid, so that every theorem of the implicational calculus is a tautology in the three-valued logic of table 660. However, table 661 reveals that Peirce's law is *not* a tautology in this three-valued logic. Therefore, there does not exist a proof of Peirce's law within the classical implicational calculus. □

Besides Peirce's law, there exist other well-formed formulae involving only implications that can neither be proved nor disproved within the classical implicational calculus, in particular, every formula that would imply Peirce's law.

Example 662 The formula $[(P) \Rightarrow (R)] \Rightarrow \left[\{[(P) \Rightarrow (Q)] \Rightarrow (R)\} \Rightarrow (R)\right]$ can be neither proved nor disproved in the classical implicational calculus. For if it were a theorem of the classical implicational calculus, then it would imply Peirce's law:

Table 661 Peirce's law in the three-valued logic of table 660.

P	Q	$(P) \Rightarrow (Q)$	$[(P) \Rightarrow (Q)] \Rightarrow (P)$	$\{[(P) \Rightarrow (Q)] \Rightarrow (P)\} \Rightarrow (P)$
T	T	T	T	T
T	F	F	T	T
T	U	U	T	T
F	T	T	F	T
F	F	T	F	T
F	U	U	T	F
U	T	T	U	T
U	F	T	U	T
U	U	T	U	T

$$\begin{array}{rl}
\vdash [(P) \Rightarrow (R)] \Rightarrow \left[\{[(P) \Rightarrow (Q)] \Rightarrow (R)\} \Rightarrow (R)\right] & \text{hypothesis,} \\
\vdash [(P) \Rightarrow (P)] \Rightarrow \left[\{[(P) \Rightarrow (Q)] \Rightarrow (P)\} \Rightarrow (P)\right] & \text{substitution of } P \text{ for } R, \\
\vdash (P) \Rightarrow (P) & \text{theorem 173,} \\
\vdash \{[(P) \Rightarrow (Q)] \Rightarrow (P)\} \Rightarrow (P) & \textit{Modus Ponens.}
\end{array}$$

Thus if there were a proof of $[(P) \Rightarrow (R)] \Rightarrow \left[\{[(P) \Rightarrow (Q)] \Rightarrow (R)\} \Rightarrow (R)\right]$ in the classical implicational calculus, then appending the foregoing derivation would yield a proof of Peirce's law in the classical implicational calculus, but example 657 has excluded such a proof. Consequently, there does not exist any proof of $[(P) \Rightarrow (R)] \Rightarrow \left[\{[(P) \Rightarrow (Q)] \Rightarrow (R)\} \Rightarrow (R)\right]$ in the classical implicational calculus.

Thus a different implicational logic would result from adding Peirce's law to the implicational axioms P1 and P2. Therefore, Peirce's law is "independent" of the implicational axioms P1 and P2, in the following sense.

Definition 663 In a logical system, an axiom P is **independent** of the other logical axioms if and only if P is *not* a theorem derivable from the other axioms.

Because the classical implicational calculus contains tautologies that cannot be proved within itself, it is called "incomplete" in the following sense.

Definition 664 A logic is **absolutely complete** if and only if every one of its tautologies *is* a theorem. A logic is **absolutely incomplete** if and only if at least one of its tautologies is *not* a theorem.

In contrast, section 4.4 will prove that the propositional calculus is complete. Nevertheless, the implicational calculus is absolutely consistent in the following sense.

Definition 665 A logic is **absolutely consistent** if and only if at least one of its propositional forms is *not* a theorem. A logic is **absolutely inconsistent** if and only if every one of its propositional forms *is* a theorem.

4.2.3 Exercises

The following exercises pertain to the three-valued logic defined in table 660. The first two exercises confirm that Łukasiewicz's implicational axioms P1 and P2 remain tautologies in the three-valued logic defined by table 660.

Exercise 761 Verify that axiom P2 remains a tautology in the three-valued logic:

$$\{(P) \Rightarrow [(Q) \Rightarrow (R)]\} \Rightarrow \{[(P) \Rightarrow (Q)] \Rightarrow [(P) \Rightarrow (R)]\}.$$

Exercise 762 Verify that axiom P1 remains a tautology in the three-valued logic:

$$(P) \Rightarrow [(Q) \Rightarrow (P)].$$

The next two exercises investigate whether Łukasiewicz's other system of implicational axioms Ł1, remains a tautology in the three-valued logic of table 660.

Exercise 763 Use propositional calculus to determine whether axiom Ł1 remains a tautology in the three-valued logic:

$$[(A) \Rightarrow (B)] \Rightarrow \{[(B) \Rightarrow (C)] \Rightarrow [(A) \Rightarrow (C)]\}.$$

Exercise 764 Use three-valued Truth tables to determine whether axiom Ł1 remains a tautology in the three-valued logic:

$$[(A) \Rightarrow (B)] \Rightarrow \{[(B) \Rightarrow (C)] \Rightarrow [(A) \Rightarrow (C)]\}.$$

The next two exercises investigate whether Frege's implicational axioms F1–F3 remain tautologies in the three-valued logic. (Axioms F1 and F2 — from the exercises in subsection 1.3.2 — coincide with axioms P1 and P2.)

Exercise 765 Use propositional calculus to determine whether axiom F3 remains a tautology in the three-valued logic:

$$\{(P) \Rightarrow [(Q) \Rightarrow (R)]\} \Rightarrow \{(Q) \Rightarrow [(P) \Rightarrow (R)]\}.$$

Exercise 766 Use three-valued Truth tables to determine whether axiom F3 remains a tautology in the three-valued logic:

$$\{(P) \Rightarrow [(Q) \Rightarrow (R)]\} \Rightarrow \{(Q) \Rightarrow [(P) \Rightarrow (R)]\}.$$

The next four exercises investigate whether certain propositional forms are theorems within the classical implicational calculus.

Exercise 767 Determine whether the following propositional form can be proved or disproved within the classical implicational calculus.

$$\{[(P) \Rightarrow (Q)] \Rightarrow (R)\} \Rightarrow \{[(R) \Rightarrow (P)] \Rightarrow (P)\}.$$

Exercise 768 Determine whether the following propositional form can be proved or disproved within the classical implicational calculus.

$$\{[(P) \Rightarrow (Q)] \Rightarrow (R)\} \Rightarrow \{[(P) \Rightarrow (R)] \Rightarrow (R)\}.$$

Exercise 769 Determine whether the following propositional form can be proved or disproved within the classical implicational calculus.

$$\{[(P) \Rightarrow (Q)] \Rightarrow (R)\} \Rightarrow \{[(R) \Rightarrow (P)] \Rightarrow [(S) \Rightarrow (P)]\}.$$

Exercise 770 Determine whether the following propositional form can be proved or disproved within the classical implicational calculus.

$$[(R) \Rightarrow (Q)] \Rightarrow (\{[(R) \Rightarrow (Q)] \Rightarrow (P)\} \Rightarrow [(S) \Rightarrow (P)]).$$

4.3 LOGICS NOT AMENABLE TO TRUTH TABLES

4.3.1 Logics with Any Number of Values

One strategy for investigating more realistic logics consists in extending the "standard device" of three-valued logic to any number of values. To this end, for each positive integer $N \in \mathbb{N}^*$, assign the $N + 1$ values

$$T := 0 < 1 < \cdots < N$$

to logical formulae as follows. For each propositional form P define its value $V(P) \in \{T, 1, \ldots, N\}$ inductively by table 666.

Table 666 A logic with $N + 1$ values.

FORMULA	VALUE	CRITERION
$(P) \Rightarrow (Q)$	T	if $V(P) \geq V(Q)$,
	$V(Q)$	if $V(P) < V(Q)$;
$(P) \wedge (Q)$	$\max\{V(P), V(Q)\}$;	
$(P) \vee (Q)$	$\min\{V(P), V(Q)\}$;	
$(P) \Leftrightarrow (Q)$	T	if $V(P) = V(Q)$,
	$\max\{V(P), V(Q)\}$	if $V(P) \neq V(Q)$;
$\neg(P)$	T	if $V(P) = N$,
	N	if $V(P) \neq N$.

Such extended values do *not* establish any link between the natural numbers $1 < \cdots < N$ and the common notions of true and false or the Boolean concepts of True and False; they merely serve as the "standard device" for investigative purposes. As demonstrations, the following theorems provide examples of proofs in the logic of table 666. The first theorem confirms that the rule of *Modus Ponens* holds.

Theorem 667 *The rule of Modus Ponens remains valid in the logic of table 666.*

Proof. If $V(P) = T$ and $V[(P) \Rightarrow (Q)] = T$, then the definition of $V[(P) \Rightarrow (Q)]$ shows that $V(P) \geq V(Q)$, but $T = 0$ is the smallest value, so that $T = V(P) \geq V(Q) \geq T$, whence $V(Q) = T$. □

The second theorem confirms that all the intuitionistic axioms remain tautologies.

Theorem 668 *All the axioms of intuitionistic logic are tautologies with table 666.*

Proof. The following derivation confirms that the law of affirmation of the consequent,

$$(H) \Rightarrow [(C) \Rightarrow (H)],$$

which is axiom I1 (which also coincides with axioms K1 and P1) is a tautology in the logic of table 666. Firstly, by the definition of $V[(P) \Rightarrow (Q)]$ in table 666,

$$V[(C) \Rightarrow (H)] = \begin{cases} T & \text{if } V(C) \geq V(H), \\ V(H) & \text{if } V(C) < V(H). \end{cases}$$

Because $T \leq V(H)$ in either case, it follows that

$$V[(C) \Rightarrow (H)] \leq V(H).$$

Consequently, again by the definition of $V[(P) \Rightarrow (Q)]$ in table 666,

$$V\{(H) \Rightarrow [(C) \Rightarrow (H)]\} = T.$$

Also, the proofs that the other axioms of intuitionistic logic remain tautologies proceed by cases according to the definitions in table 666, as outlined in the exercises. □

The following considerations then show that intuitionistic logic is "smaller" than the classical propositional calculus.

Theorem 669 *Every intuitionistic theorem is a tautology relative to table 666.*

Proof. By theorems 667 and 668, *Modus Ponens* holds and all the intuitionistic axioms are tautologies relative to table 666. Consequently, the proof of theorem 655 also holds for the logic of table 666, so that every intuitionistic theorem remains a tautology. □

Yet there are Boolean tautologies that are *not* tautologies relative to table 666.

Example 670 Table 671 (in Quine's form) reveals that the law of excluded middle

$$(P) \vee [\neg(P)]$$

is *not* a tautology in the logic of table 666.

Table 671 Excluded middle with $N + 1$ values.

(P)	$\vee$	$[\neg$	$(P)]$
T	T	N	T
1	**1**	N	1
⋮	⋮	⋮	⋮
$N-1$	**N-1**	N	$N-1$
N	T	T	N

Consequently, the law of excluded middle is *not* an intuitionistic theorem.

Because every intuitionistic theorem is also a classical theorem, the "standard device" of multiple values thus reveals that intuitionistic logic is only a part, but not all, of the classical propositional calculus. As outlined in the exercises, the minimal logic is only a part, but not all, of intuitionistic logic (and hence not all of classical logic).

4.3.2 Practical Logics Not Amenable to Truth Tables

Multiple values also reveal that the minimal and intuitionistic logics lie beyond all the logics definable by Truth tables of any kind for which all the minimal or intuitionistic axioms remain tautologies, and the rules of substitutions and *Modus Ponens* hold. (Such systems of Truth tables are called **sound** [11, p. 114].)

The strategy consists in selecting for each $N \in \mathbb{N}^*$ *two* sound logics, one with $N + 1$ values and the other one with N values, and a formula R_N, subject to two conditions that must hold for every $N \in \mathbb{N}$.

Firstly, R_N must be a tautology in the logic with N values.

Secondly, R_N may *not* be a tautology in the logic with $N+1$ values.

Hence it will follow that if intuitionistic logic were definable by any system of tables with any number $K+1$ of values, then selecting $N := K+1$ would produce a formula R_N that is a tautology in the alleged system with N values, but is not an intuitionistic theorem because it is not a tautology with $N+1$ values. Consequently for each $K \in \mathbb{N}^*$ the $K+1$ values do not define intuitionistic logic.

To this end, as in classical logic, let $(P) \Leftrightarrow (Q)$ be an abbreviation for $[(P) \Rightarrow (Q)] \wedge [(Q) \Rightarrow (P)]$. Then the following definition provides a formula for R_N.

Definition 672 For each positive integer $N \in \mathbb{N}^*$ define a propositional form R_N by

$$\begin{aligned} R_N &:= \bigvee_{1\leq k<\ell\leq N} [(P_k) \Leftrightarrow (P_\ell)] \\ &:= [(P_1) \Leftrightarrow (P_2)] \vee [(P_1) \Leftrightarrow (P_3)] \vee \cdots \vee [(P_1) \Leftrightarrow (P_N)] \\ &\qquad \vee [(P_2) \Leftrightarrow (P_3)] \vee [(P_2) \Leftrightarrow (P_4)] \vee \cdots \vee [(P_2) \Leftrightarrow (P_N)] \\ &\ \ \vdots \\ &\qquad \vee [(P_{N-1}) \Leftrightarrow (P_N)]\,. \end{aligned}$$

The next theorem confirms that R_N is *not* a tautology for a logic with $N+1$ values.

Theorem 673 *The formula $R_N = \bigvee_{1\leq k<\ell\leq N} [(P_k) \Leftrightarrow (P_\ell)]$ is* not *a tautology relative to table 666 with $N+1$ values, and hence* not *a minimal or intuitionistic theorem.*

Proof. Select the components of R_N so that $V(P_k) = k$ for each $k \in \{1,\ldots,N\}$. (Such components exist by exercise 786.) Then for all $k,\ell \in \{1,\ldots,N\}$ with $k<\ell$, table 666 shows that $(P_k) \Leftrightarrow (P_\ell)$ has the value ℓ. Because $\ell \geq 1 > 0 = T$, it follows that $V[(P_k) \Leftrightarrow (P_\ell)] = \ell \geq 1$. By the definition $V[(P) \vee (Q)] = \min\{V(P), V(Q)\}$ in table 666, it follows that $V(R_N) \geq 1 > T$. Consequently, R_N is *not* a tautology in this logic. Because all the intuitionistic axioms *are* tautologies in this logic, it also follows that R_N is *not* a theorem of the minimal or intuitionistic logic. □

The following theorem confirms that R_N is a tautology for a logic with N values.

Theorem 674 *For every $N > 1$, for every* sound *system of Truth tables with N values, the formula $R_N = \bigvee_{1\leq k<\ell\leq N} [(P_k) \Leftrightarrow (P_\ell)]$ is a tautology.*

Proof. Because $N > N-1$, the following function cannot be injective (theorem 607):

$$\begin{aligned} W : \{1,\ldots,N\} &\to \{1,\ldots,N-1\}, \\ k &\mapsto V(P_k). \end{aligned}$$

Hence there are two indices $i \neq j$ with $V(P_i) = V(P_j)$, whence $V[(P_i) \Leftrightarrow (P_j)] = T$. Because the Truth tables are sound, it follows from axioms I8 and I9, or K7 and K8, that $V\{\cdots \vee [(P_i) \Leftrightarrow (P_j)] \vee \cdots\} = T$. Hence $V(R_N) = T$:

$$\begin{aligned} T &\leq V\left\{ \bigvee_{1\leq k<\ell\leq N} [(P_k) \Leftrightarrow (P_\ell)] \right\} = \min_{1\leq k<\ell\leq N} V[(P_k) \Leftrightarrow (P_\ell)] \\ &\leq V[(P_i) \Leftrightarrow (P_j)] = T. \end{aligned}$$

Consequently, R_N is a tautology in this sound logic with N values. □

4.3.3 Exercises

Odd numbered exercises verify that table 666 is intuitionistically sound.

Exercise 771 Verify that axiom I2,

$$\{(P) \Rightarrow [(Q) \Rightarrow (R)]\} \Rightarrow \{[(P) \Rightarrow (Q)] \Rightarrow [(P) \Rightarrow (R)]\},$$

is a tautology in the logic of table 666, as done in theorem 668.

Exercise 772 Verify that $(P) \Rightarrow (P)$ is a tautology relative to table 666.

Exercise 773 Verify that axiom I7, $(P) \Rightarrow \{(Q) \Rightarrow [(P) \wedge (Q)]\}$, is a tautology in the logic of table 666, as done in theorem 668.

Exercise 774 Verify that $(P) \wedge (P)$ is a tautology relative to table 666.

Exercise 775 Verify that axiom I5, $[(P) \wedge (Q)] \Rightarrow (P)$, is a tautology in the logic of table 666, as done in theorem 668.

Exercise 776 Verify that $(P) \vee (P)$ is a tautology relative to table 666.

Exercise 777 Verify that axiom I6, $[(P) \wedge (Q)] \Rightarrow (Q)$, is a tautology in the logic of table 666, as done in theorem 668.

Exercise 778 Verify that $(P) \Leftrightarrow (P)$ is a tautology relative to table 666.

Exercise 779 Verify that axiom I3, $\{(P) \Rightarrow [\neg(P)]\} \Rightarrow [\neg(P)]$ (the law of special reductio ad absurdum), is a tautology in the logic of table 666, as done in theorem 668.

Exercise 780 Verify that $\neg[(P) \Rightarrow (P)]$ has the value N relative to table 666.

Exercise 781 Verify that axiom I4, $[\neg(P)] \Rightarrow [(P) \Rightarrow (Q)]$ (the law of denial of the antecedent), is a tautology in the logic of table 666, as done in theorem 668.

Exercise 782 Determine $V\left[\{\neg[\neg(P)]\} \Rightarrow (P)\right]$ in the logic of table 666.

Exercise 783 Verify that the law of double negation $\{\neg[\neg(P)]\} \Rightarrow (P)$ is *not* an intuitionistic theorem, as done in example 670.

Exercise 784 Prove that with Truth values as in table 666, $(P) \Leftrightarrow (Q)$ has the same Truth values as $[(P) \Rightarrow (Q)] \wedge [(Q) \Rightarrow (P)]$.

Exercise 785 Prove that $\{[\neg(Q)] \Rightarrow [\neg(P)]\} \Rightarrow [(P) \Rightarrow (Q)]$ (the converse law of contraposition) is *not* an intuitionistic theorem.

Exercise 786 For each $N \in \mathbb{N}^*$ and for each $k \in \{1, \ldots, N\}$ define a connective C_k such that $V[C_k(P)] = k$ for every well-formed logical formula P. As in definition 672, with $R_N = \bigvee_{1 \leq k < \ell \leq N} [(P_k) \Leftrightarrow (P_\ell)]$, identify components $P_1, \ldots, P_N$ such that $V(R_N) \neq T$, so that R_N is *not* a tautology in the logic of table 666.

Exercise 787 Prove that $V\left(\{[\neg(P)] \Rightarrow (Q)\} \Rightarrow [\{[\neg(P)] \Rightarrow [\neg(Q)]\} \Rightarrow (P)]\right)$ (the law of indirect proof) is *not* an intuitionistic theorem.

Exercise 788 Prove that axiom I4 is *not* a theorem of the minimal logic. Hint: with $N := 2$, modify only the values of the negation in table 666, so that axiom I4 is no longer a tautology, but all the minimal axioms remain tautologies.

Exercise 789 Prove that the minimal logic with the law of excluded middle is *not* all of the intuitionistic logic. Hint: with $N := 2$, modify only the values of the negation in table 666, so that axiom I4 is no longer a tautology, but all the minimal axioms *and the law of excluded middle* remain tautologies.

Exercise 790 Let P stand for "Mercury is a planet" and assume that P is True by definition. Also, let Q stand for "Mercury is larger than Ganymede" and assume that measurements show that Q is False. Moreover, let H denote the hypothesis that "if Mercury is a planet, then Mercury is larger than Ganymede" so that $(H) \Leftrightarrow [(P) \Rightarrow (Q)]$. Prove $\neg(H)$ using only the minimal logic.

4.4 AUTOMATED THEOREM PROVING

4.4.1 The Deduction Theorem

The Deduction Theorem presented here is a part of an algorithm to design proofs within the full propositional calculus. A common way of outlining a proof of a logical implication $(H) \Rightarrow (S)$ sketches out a proof that "if H is True, then S is True" without mentioning the situation with H False. Such an outline might suffice in the context of Truth tables, where all the rows with H False show the value True for $(H) \Rightarrow (S)$. Yet Truth tables can become impractical if the hypothesis H contains many, or infinitely many, cases, and then the propositional calculus can become indispensable.

To this end, from a proof of "if H is True, then S is True" ($H \vdash S$) — in other words, from a proof that S is derivable from a True hypothesis H — a theorem known as the Deduction Theorem yields a proof of the logical formula $(H) \Rightarrow (S)$ regardless of the Truth value of H. More generally, from any proof that a logical proposition S is derivable from True hypotheses $H, K, \dots, M, N$ the Deduction Theorem provides a recipe to turn that proof into a proof of

$$(H) \Rightarrow \{(K) \Rightarrow \dots (M) \Rightarrow [(N) \Rightarrow (S)] \dots\}$$

regardless of the Truth values of $H, K, \dots, M, N$.

Theorem 675 (Deduction Theorem) *There is an algorithm to transform any proof of*

$$H, K, \dots, M, N \vdash S$$

within the propositional calculus into a proof of

$$(H) \Rightarrow \{(K) \Rightarrow \dots (M) \Rightarrow [(N) \Rightarrow (S)] \dots\}.$$

Proof. This proof proceeds by induction with the number of hypotheses.

Initial step With zero hypothesis, any proof of $\vdash S$ is a proof that S is a theorem.

Induction hypothesis Assume that there exists a natural number $I \in \mathbb{N}$ such that the theorem holds for every proof of every theorem S with at most I hypotheses $H_1, \dots, H_I \vdash S$.

Induction step The Deduction Theorem removes the hypotheses one at a time, for instance, beginning with the last one listed, here N, from all the steps in the proof. Specifically, from a proof with $I+1$ hypotheses $H_1, \ldots, H_I, H_{I+1} \vdash S$ and logical steps $P, Q, \ldots, R, S$ for $H, K, \ldots, M, N \vdash S$, with H for H_1, ..., with M for H_I, and with N for H_{I+1}, the Deduction Theorem proceeds as follows.

(D1) If a step P of the initial proof is a substitution in one of the axioms, or one of the hypotheses other than the one N being removed here, then in the new proof the Deduction Theorem replaces the old step

$\vdash P$ (axiom or hypothesis)

by a complete proof of $(N) \Rightarrow (P)$, for instance, as in theorem 171:

$\vdash P$ axiom or hypothesis,
$\vdash (P) \Rightarrow [(N) \Rightarrow (P)]$ axiom P1,
$\vdash (N) \Rightarrow (P)$ *Modus Ponens.*

(D2) If the step P in the initial proof is the hypothesis N being removed, then in the new proof the Deduction Theorem replaces the old step

$\vdash N$ (current hypothesis)

by a complete proof of $(N) \Rightarrow (N)$, for instance, that of theorem 173:

$\vdash (N) \Rightarrow \{[(N) \Rightarrow (N)] \Rightarrow (N)\}$ axiom P1,

$\vdash [(N) \Rightarrow \{[(N) \Rightarrow (N)] \Rightarrow (N)\}]$
$\Rightarrow (\{(N) \Rightarrow [(N) \Rightarrow (N)]\} \Rightarrow [(N) \Rightarrow (N)])$ axiom P2,

$\vdash \{(N) \Rightarrow [(N) \Rightarrow (N)]\} \Rightarrow [(N) \Rightarrow (N)]$ *Modus Ponens,*

$\vdash (N) \Rightarrow [(N) \Rightarrow (N)]$ axiom P1,

$\vdash (N) \Rightarrow (N)$ *Modus Ponens.*

(D3) If the step P is derived in the initial proof by *Modus Ponens* from previous True propositions M and $(M) \Rightarrow (P)$, then (D1) and (D2) allow for their replacement by complete proofs of $(N) \Rightarrow (M)$ and $(N) \Rightarrow [(M) \Rightarrow (P)]$ respectively. Specifically, in the new proof, the Deduction Theorem then replaces the old steps

$\vdash (M) \Rightarrow (P)$ (previously proven True),
$\vdash M$ (previously proven True),
$\vdash P$ (*Modus Ponens*),

by a complete proof that $(N) \Rightarrow (P)$, for instance, as in theorem 174:

$\vdash (N) \Rightarrow [(M) \Rightarrow (P)]$ theorem 171,

$\vdash \{(N) \Rightarrow [(M) \Rightarrow (P)]\}$ axiom P2 ...
$\Rightarrow \{[(N) \Rightarrow (M)] \Rightarrow [(N) \Rightarrow (P)]\}$ ... continued,

$\vdash \{[(N) \Rightarrow (M)] \Rightarrow [(N) \Rightarrow (P)]\}$ *Modus Ponens,*
$\vdash (N) \Rightarrow (M)$ theorem 171,
$\vdash (N) \Rightarrow (P)$ *Modus Ponens,*

with the proof of each instance of theorem 171 completely written out.

Still with the hypothesis N, after the completion of any operation (D1)–(D3) on step P, the Deduction Theorem then performs the same operations (D1)–(D3) on each of the following steps, $Q, \ldots, R$. After the completion of operations (D1)–(D3) on all the steps $P, Q, \ldots, R$, for the hypothesis N, the Deduction Theorem gives a proof of

$$H, K, \ldots, M \vdash [(N) \Rightarrow (S)].$$

Then the Deduction Theorem repeats the whole process with the preceding hypotheses, $H, \ldots, M$. The Deduction Theorem terminates with a proof of

$$(H) \Rightarrow \{(K) \Rightarrow \ldots (M) \Rightarrow [(N) \Rightarrow (S)] \ldots\}.$$

□

4.4.2 Example: Law of Assertion from the Deduction Theorem

The following proof shows the use of the Deduction Theorem in designing proofs.

Theorem 676 *The law of assertion* $(A) \Rightarrow \{[(A) \Rightarrow (B)] \Rightarrow (B)\}$ *is a theorem.*

Proof. A finished proof can proceed as follows:

$\vdash [(A) \Rightarrow (B)] \Rightarrow [(A) \Rightarrow (B)]$	theorem 173,
$\vdash (A) \Rightarrow \{[(A) \Rightarrow (B)] \Rightarrow [(A) \Rightarrow (B)]\}$	theorem 171,
$\vdash (A) \Rightarrow \{[(A) \Rightarrow (B)] \Rightarrow (A)\}$	axiom P1,
$\vdash (A) \Rightarrow \{[(A) \Rightarrow (B)] \Rightarrow (B)\}$	theorem 179.

The following considerations explain how to *design* such a proof.

The formula $(A) \Rightarrow \{[(A) \Rightarrow (B)] \Rightarrow (B)\}$ has the pattern $(H) \Rightarrow [(K) \Rightarrow (S)]$ of the Deduction Theorem, with A for H, $(A) \Rightarrow (B)$ for K, and B for S.

Step 1. As in the Deduction Theorem, assume first that the hypotheses H and K are True, and from them derive the conclusion S by proving $H, K \vdash S$. Here, assume that the hypotheses A and $(A) \Rightarrow (B)$ are both True, and prove $A, [(A) \Rightarrow (B)] \vdash B$:

$\vdash A$	first temporary hypothesis,
$\vdash (A) \Rightarrow (B)$	second temporary hypothesis,
$\vdash B$	*Modus Ponens.*

The foregoing derivation shows that if A and $(A) \Rightarrow (B)$ are True, then B is True. Still under the first hypothesis A, the Deduction Theorem allows for the removal of the second hypothesis, $(A) \Rightarrow (B)$, as follows.

Step 2.

Step 2.1 The *first line in step 1* consists of the other hypothesis, A, which is assumed True, whence instructions (D1) in the Deduction Theorem replace A with a complete proof of $[(A) \Rightarrow (B)] \Rightarrow (A)$ as in theorem 171. In other words, replace the first line, $\vdash A$, by the following three lines:

	Proof of theorem 171:
$\vdash A$	temporary hypothesis,
$\vdash (A) \Rightarrow \{\underbrace{[(A) \Rightarrow (B)]}_{Q} \Rightarrow (A)\}$	axiom P1,
$\vdash [(A) \Rightarrow (B)] \Rightarrow (A)$	*Modus Ponens.*
	End of proof of theorem 171.

Step 2.2 Similarly, the *second line in step 1* consists of the hypothesis K being currently removed, here $(A) \Rightarrow (B)$, which instructions (D2) in the Deduction Theorem replace with a complete proof of $(K) \Rightarrow (K)$, here $[(A) \Rightarrow (B)] \Rightarrow [(A) \Rightarrow (B)]$, as in theorem 173. Thus, replace the second line, $\vdash (A) \Rightarrow (B)$, by the following lines:

	Proof of theorem 173 with $[K]$ for $[(A) \Rightarrow (B)]$:
$\vdash [K] \Rightarrow (\{[K] \Rightarrow [K]\} \Rightarrow [K])$	axiom P1,
$\vdash \{[K] \Rightarrow (\{[K] \Rightarrow [K]\} \Rightarrow [K])\}$	axiom P2 ...
$\quad \Rightarrow [([K] \Rightarrow \{[K] \Rightarrow [K]\}) \Rightarrow \{[K] \Rightarrow [K]\}]$	... continued,
$\vdash ([K] \Rightarrow \{[K] \Rightarrow [K]\}) \Rightarrow \{[K] \Rightarrow [K]\}$	*Modus Ponens,*
$\vdash [K] \Rightarrow \{[K] \Rightarrow [K]\}$	axiom P1,
$\vdash [K] \Rightarrow [K]$	*Modus Ponens,*
$\vdash [(A) \Rightarrow (B)] \Rightarrow [(A) \Rightarrow (B)]$	substitution.
	End of proof of theorem 173.

Step 2.3 Finally, the *third line in step 1* invokes *Modus Ponens,* which instructions (D3) replace by an instance of (the proof of) theorem 174:

$\vdash [(A) \Rightarrow (B)] \Rightarrow (A)$	step 2.1,
$\vdash [(A) \Rightarrow (B)] \Rightarrow [(A) \Rightarrow (B)]$	step 2.2,
$\vdash [(A) \Rightarrow (B)] \Rightarrow (B)$	theorem 174.

Hence the proof no longer assumes $(A) \Rightarrow (B)$ as a hypothesis, but it still assumes A as a hypothesis, thus proving that

$$A \vdash \{[(A) \Rightarrow (B)] \Rightarrow (B)\}.$$

Step 3. Finally, the Deduction Theorem allows for the removal of the first hypothesis, A, from step 2. Here step 2 consists of steps 2.1, 2.2, and 2.3.

Step 3.1 In step 2.1 the first line consists of this hypothesis, A, whence instructions (D2) replace A with $(A) \Rightarrow (A)$ by a complete proof of theorem 173. In other words, replace the first line in step 2, $\vdash A$, by the following lines:

	Proof of theorem 173:
$\vdash [A] \Rightarrow (\{[A] \Rightarrow [A]\} \Rightarrow [A])$	axiom P1,
$\vdash \{[A] \Rightarrow (\{[A] \Rightarrow [A]\} \Rightarrow [A])\}$	axiom P1 ...
$\Rightarrow [([A] \Rightarrow \{[A] \Rightarrow [A]\}) \Rightarrow \{[A] \Rightarrow [A]\}]$	... continued,
$\vdash ([A] \Rightarrow \{[A] \Rightarrow [A]\}) \Rightarrow \{[A] \Rightarrow [A]\}$	*Modus Ponens,*
$\vdash [A] \Rightarrow \{[A] \Rightarrow [A]\}$	axiom P1,
$\vdash [A] \Rightarrow [A]$	*Modus Ponens.*
	End of proof of theorem 173.

Step 3.2 The second line in step 2.1 is an instance of axiom P1, which instructions (D1) replace by $(A) \Rightarrow [(A) \Rightarrow \{[(A) \Rightarrow (B)] \Rightarrow (A)\}]$.

Step 3.3 The third line in step 2.1 yields $[(A) \Rightarrow (B)] \Rightarrow (A)$, from *Modus Ponens,* which instructions (D3) replace by a complete proof of $(A) \Rightarrow \{[(A) \Rightarrow (B)] \Rightarrow (A)\}$, as in theorem 174. In this case, however, such a proof would be correct but not necessary, because $(A) \Rightarrow \{[(A) \Rightarrow (B)] \Rightarrow (A)\}$ is merely an instance of axiom P1. Because it is an axiom, all the preceding lines also become superfluous.

Step 3.4 The result of step 2.2, $[(A) \Rightarrow (B)] \Rightarrow [(A) \Rightarrow (B)]$, is True by theorem 173. Hence, instructions (D1) replace it by $(A) \Rightarrow [(A) \Rightarrow (B)] \Rightarrow [(A) \Rightarrow (B)]$.

Step 3.5 Fully written out, the remaining lines in step 2.3 would follow the proof of theorem 174. Removing the hypothesis A then amounts to theorem 179, which forms the last line of the final proof:

$\vdash (A) \Rightarrow \{[(A) \Rightarrow (B)] \Rightarrow (A)\}$	axiom P1 (from 3.3, replacing 2.1),
$\vdash [(A) \Rightarrow (B)] \Rightarrow [(A) \Rightarrow (B)]$	theorem 173 (from step 2.2),
$\vdash (A) \Rightarrow \{[(A) \Rightarrow (B)] \Rightarrow [(A) \Rightarrow (B)]\}$	theorem 171 (from 3.4, replacing 2.2),
$\vdash (A) \Rightarrow \{[(A) \Rightarrow (B)] \Rightarrow (B)\}$	Theorem 179 (from 3.5, replacing 2.3).

Thus the Deduction Theorem has provided some guidance for the construction of a proof of the theorem $(A) \Rightarrow \{[(A) \Rightarrow (B)] \Rightarrow (B)\}$. □

Relying on the same Deduction Theorem, the Completeness Theorem will provide not only guidance but an algorithm to design proofs within the propositional calculus.

4.4.3 The Provability Theorem

The full classical propositional calculus based on axioms P1, P2, P3 is absolutely complete: every tautology is a theorem (definition 664). Moreover, there are algorithms to determine for each propositional form whether it is a theorem, and, if it is, to design a proof of it. The demonstration relies on the following notation.

Definition 677 For each proposition P, define a proposition P' by

$$P' := \begin{cases} P & \text{if } P \text{ is True,} \\ \neg(P) & \text{if } P \text{ is False.} \end{cases}$$

The following theorem constitutes a first step in a proof of completeness.

Theorem 678 (Provability Theorem) *For each propositional form S with variables from a finite list $P, \ldots, R$, there exists a proof of S' from $P', \ldots, R'$:*

$$P', \ldots, R' \vdash S'.$$

Proof. This proof uses induction with the number L of connectives in S.

Initial step A propositional form S with $L := 0$ connective reduces to a single propositional variable, P. Thus, S' is P' and the single line P' is a proof of $P' \vdash P'$.

Induction hypothesis Assume that there exists $I \in \mathbb{N}$ such that the theorem holds for every propositional form with at most $L := I$ instances of logical connectives.

Induction step By the inductive definition of propositional forms, for each propositional form S with exactly $I + 1$ instances of logical connectives, there exist propositional forms V and W, each with at most I instances of logical connectives, and such that S is either $\neg(V)$ or $(V) \Rightarrow (W)$. Each of these two cases splits into several cases.

Negation Suppose that S is $\neg(V)$, and consider two cases.

S **True** If S is True, then S' is S. However, if S is True, V is then False, and V' is $\neg(V)$, which is S and hence also S'. Because V contains at most I instances of logical connectives, it follows from the hypothesis of induction that $P', \ldots, R' \vdash V'$, and substituting S' for V' yields $P', \ldots, R' \vdash S'$.

S **False** In contrast, if S is False, then S' is $\neg(S)$. However, if S is False, V is then True, and V' is V. Because V contains at most I instances of logical connectives, it follows from the hypothesis of induction that $P', \ldots, R' \vdash V'$, and substituting V for V' yields $P', \ldots, R' \vdash V$. Hence, appending a proof of the converse law of double negation (theorem 190, which is also a minimal theorem by exercise 741) produces a proof of $P', \ldots, R' \vdash \{\neg[\neg(V)]\}$, and substituting S for $\neg(V)$ gives $P', \ldots, R' \vdash [\neg(S)]$, whence substituting S' for $\neg(S)$ yields $P', \ldots, R' \vdash S'$.

Implication Suppose that S is $(V) \Rightarrow (W)$, and consider three cases. The first two cases occur if S is True, which occurs if W is True or V is False, or both.

S **True,** W **True** If W is True, then W is W' and by the hypothesis of induction there exists a proof of $P', \ldots, R' \vdash W$. Again because W is True, it follows from theorem 171 that $(V) \Rightarrow (W)$ is also True, and appending a proof of theorem 171 after the proof of $P', \ldots, R' \vdash W$ produces a proof of $P', \ldots, R' \vdash [(V) \Rightarrow (W)]$. However, because $(V) \Rightarrow (W)$ is True and $(V) \Rightarrow (W)$ is S, it also follows that S is S', whence $P', \ldots, R' \vdash [(V) \Rightarrow (W)]$ is $P', \ldots, R' \vdash S'$.

S **True,** W **False** If V is False, then V' is $\neg(V)$ and True. By the hypothesis of induction there exists a proof of $P', \ldots, R' \vdash V'$, which is thus a proof of $P', \ldots, R' \vdash [\neg(V)]$. Hence the law of denial of the antecedent (theorem 187, which is also axiom I4) gives a proof of $[\neg(V)] \Rightarrow [(V) \Rightarrow (W)]$, which is $[\neg(V)] \Rightarrow (S)$, and thence the transitivity of implications (theorem 175) yields a proof of $P', \ldots, R' \vdash S$, which is also a proof of $P', \ldots, R' \vdash S'$.

S **False** The third case occurs if S is False, which occurs if and only if V is True and W is False. Then S' is $\neg(S)$ and W' is $\neg(W)$ but V' is V. By the hypothesis of induction there exist proofs of $P', \ldots, R' \vdash V'$ and $P', \ldots, R' \vdash W'$, which are thus proofs of $P', \ldots, R' \vdash V$ and $P', \ldots, R' \vdash [\neg(W)]$. Appending a proof of theorem 201 (which is also axiom K6) then gives a proof of $P', \ldots, R' \vdash \{(V) \wedge [\neg(W)]\}$, whence the definition (197) of $\wedge$ produces a proof of $P', \ldots, R' \vdash \{\neg[(V) \Rightarrow \{\neg[\neg(W)]\}]\}$. Thence the converse law of double negation, transitivity applied to

$$[(V) \Rightarrow (W)] \Rightarrow \{[(W) \Rightarrow \{\neg[\neg(W)]\}] \Rightarrow [(V) \Rightarrow \{\neg[\neg(W)]\}]\}$$

and contraposition yield a proof of $P', \ldots, R' \vdash \{\neg[(V) \Rightarrow (W)]\}$, which is a proof of $P', \ldots, R' \vdash [\neg(S)]$ and hence also a proof of $P', \ldots, R' \vdash S'$.

□

4.4.4 The Completeness Theorem

The **Completeness Theorem** shows that within the full classical propositional calculus every tautology is a theorem, provable from the axioms and the rules of inference.

Theorem 679 (Completeness Theorem) *Within the full classical propositional calculus, every tautology is a theorem.*

Proof. This proof uses the Deduction Theorem and proceeds by induction with the number of propositional variables that occur in a tautology.

For every tautology S with variables $P, \ldots, Q, R$, theorem 678 produces a proof of $P', \ldots, Q', R' \vdash S$, because S' is S. Two cases arise with the last variable R.

R **True** If R is True, then R' is R, whence from the proof of $P', \ldots, Q', R \vdash S$, the Deduction Theorem gives a proof of $P', \ldots, Q' \vdash [(R) \Rightarrow (S)]$.

R **False** If R is False, then R' is $\neg(R)$, whence from the proof of $P', \ldots, Q', R' \vdash S$, the Deduction Theorem gives a proof of $P', \ldots, Q' \vdash \{[\neg(R)] \Rightarrow (S)\}$.

A proof of $P', \ldots, Q' \vdash S$ follows by the principle of proofs by cases (theorem 221):

$$\vdash ([(R) \Rightarrow (S)] \wedge \{[\neg(R)] \Rightarrow (S)\}) \Rightarrow (S).$$

Thus the Deduction Theorem reduces the number of propositional variables by 1. Therefore, applying the Deduction Theorem as many times as the number of propositional variables in S yields a proof of $\vdash S$. □

4.4.5 Example: Peirce's Law from the Completeness Theorem

The following considerations demonstrate how to plan the design of a proof by the Completeness Theorem (theorem 679), here with the example of Peirce's Law:

$$\{[(P) \Rightarrow (Q)] \Rightarrow (P)\} \Rightarrow (P).$$

To apply the Completeness Theorem, let S designate Peirce's Law. Because S involves only two propositional variables, P and Q, for each of the four combinations of Truth values of P and Q, the Completeness Theorem first invokes the Provability Theorem (theorem 678) for a separate proof of $P', Q' \vdash S'$, here

$$P', Q' \vdash [\{[(P) \Rightarrow (Q)] \Rightarrow (P)\} \Rightarrow (P)]'.$$

In all cases, S has the propositional form $(V) \Rightarrow (W)$, where W is P, and where V is $(H) \Rightarrow (K)$, with $(P) \Rightarrow (Q)$ for H, and P for K:

$$\overbrace{\underbrace{\{\overbrace{[(P) \Rightarrow (Q)]}^{H} \Rightarrow \overbrace{(P)}^{K}\}}_{V} \Rightarrow \underbrace{(P)}_{W}}^{S}.$$

P **True,** *Q* **True** If P is True, then W is also True, because W is P. Hence the Provability Theorem calls for a proof of $P \vdash W$, which is here $P \vdash P$. Thence the Deduction Theorem provides a proof of $(P) \Rightarrow (W)$, in effect here the proof of theorem 173. Because S has the form $(V) \Rightarrow (W)$, the proof just obtained gives the following main steps (the final complete proof replaces every theorem cited by a complete proof of that theorem).

$$\begin{array}{ll} \vdash (P) \Rightarrow (W) & \text{theorem 173,} \\ \vdash (W) \Rightarrow [(V) \Rightarrow (W)] & \text{axiom P1,} \\ & \\ \vdash (P) \Rightarrow [(V) \Rightarrow (W)] & \text{theorem 175,} \\ \vdash (P) \Rightarrow [\underbrace{\{[(P) \Rightarrow (Q)] \Rightarrow (P)\}}_{V} \Rightarrow \underbrace{(P)}_{W}] & \text{substitutions.} \end{array}$$

Alternatively axiom P1 yields the conclusion directly, but the foregoing derivation serves to illustrate the use of the Completeness Theorem.

P **True,** *Q* **False** Because P is again True, the preceding reasoning remains valid because it does not use the Truth value of Q.

P **False,** *Q* **True** If P is False, then so is W. Hence the Provability Theorem calls for a proof of V'. Here V is $[(P) \Rightarrow (Q)] \Rightarrow (P)$, which has the form $(H) \Rightarrow (K)$. With P False, H is True and K is False, whence V is False. Consequently, V' is $\neg(V)$, which has the form $\neg[(H) \Rightarrow (K)]$. Therefore, the Provability Theorem calls for proofs of $P', Q' \vdash H$ and $P', Q' \vdash [\neg(K)]$.

Here $P', Q' \vdash [\neg(K)]$ is $[\neg(P)], Q' \vdash \{\neg[\neg(P)]\}$, which follows from the substitution $[\neg(P)] \Rightarrow [\neg(P)]$ in the proof of theorem 173.

Also, $P', Q' \vdash H$ is $[\neg(P)], Q' \vdash [(P) \Rightarrow (Q)]$, where P is False. Thus the Provability Theorem calls for a proof of $[\neg(P)], Q' \vdash [\neg(P)]$, which again follows from the substitution $[\neg(P)] \Rightarrow [\neg(P)]$ in the proof of theorem 173. Hence $[\neg(P)] \Rightarrow [(P) \Rightarrow (Q)]$ by the law of denial of the antecedent (theorem 187).

These proofs of $[\neg(P)], Q' \vdash H$ and $[\neg(P)], Q' \vdash [\neg(K)]$ complete the proof of $[\neg(P)], Q' \vdash \{\neg[(H) \Rightarrow (K)]\}$, which is $[\neg(P)], Q' \vdash [\neg(V)]$. Again the law of denial of the antecedent gives a proof of $[\neg(P)], Q' \vdash [(V) \Rightarrow (W)]$, which is $[\neg(P)], Q' \vdash S$. The proof just obtained gives the following main steps (the final proof replaces every theorem cited by a complete proof of that theorem).

$$\begin{array}{ll} \vdash [\neg(P)] \Rightarrow [\neg(K)] & \text{theorem 173,} \\ \vdash [\neg(P)] \Rightarrow [\neg(P)] & \text{theorem 173,} \\ \vdash [\neg(P)] \Rightarrow \{[\neg(Q)] \Rightarrow [\neg(P)]\} & \text{axiom P1,} \\ \vdash \{[\neg(Q)] \Rightarrow [\neg(P)]\} \Rightarrow [(P) \Rightarrow (Q)] & \text{axiom P3,} \\ \vdash [\neg(P)] \Rightarrow \underbrace{[(P) \Rightarrow (Q)]}_{H} & \text{theorem 175,} \\ \vdash [\neg(P)] \Rightarrow (H) & \text{substitution;} \end{array}$$

$$
\begin{array}{ll}
\vdash [\neg(P)] \Rightarrow \{(H) \wedge [\neg(K)]\} & \text{theorem 202,} \\
\vdash [\neg(P)] \Rightarrow \{\neg \underbrace{[(H) \Rightarrow (K)]}_{V}\} & \text{definition of } \wedge, \\
\vdash [\neg(P)] \Rightarrow [\neg(V)] & \text{substitution;} \\
\vdash [\neg(V)] \Rightarrow \{[\neg(W)] \Rightarrow [\neg(V)]\} & \text{axiom P1,} \\
\vdash [\neg(P)] \Rightarrow \{[\neg(W)] \Rightarrow [\neg(V)]\} & \text{theorem 175,} \\
\vdash \{[\neg(W)] \Rightarrow [\neg(V)]\} \Rightarrow [(V) \Rightarrow (W)] & \text{axiom P3,} \\
\vdash [\neg(P)] \Rightarrow \underbrace{[(V) \Rightarrow (W)]}_{S} & \text{theorem 175,} \\
\vdash [\neg(P)] \Rightarrow [\underbrace{\{[(P) \Rightarrow (Q)] \Rightarrow (P)\}}_{V} \Rightarrow \underbrace{(P)}_{W}] & \text{substitutions.}
\end{array}
$$

P **False,** Q **False** Because P is again False, the preceding reasoning remains valid because it does not use the Truth value of Q.

From the preceding proofs of $(P) \Rightarrow (S)$ and $[\neg(P)] \Rightarrow (S)$, the principle of proofs by cases (theorem 221) yields a proof of Peirce's Law (S). Subsequent examinations of the proof produced by the Completeness Theorem can yield simplifications.

$$
\begin{array}{ll}
\vdash (P) \Rightarrow [\{[(P) \Rightarrow (Q)] \Rightarrow (P)\} \Rightarrow (P)] & \text{axiom P1,} \\
\vdash (P) \Rightarrow (S) & \text{substitution;} \\
& \\
\vdash [\neg(P)] \Rightarrow \{(H) \wedge [\neg(P)]\} & \text{axiom P1,} \\
\vdash [\neg(P)] \Rightarrow \{\neg \underbrace{[(H) \Rightarrow (K)]}_{V}\} & \text{definition of } \wedge, \\
\vdash [\neg(P)] \Rightarrow [\neg(V)] & \text{substitution;} \\
& \\
\vdash [\neg(V)] \Rightarrow \{[\neg(W)] \Rightarrow [\neg(V)]\} & \text{axiom P1,} \\
\vdash [\neg(P)] \Rightarrow \{[\neg(W)] \Rightarrow [\neg(V)]\} & \text{theorem 175,} \\
\vdash \{[\neg(W)] \Rightarrow [\neg(V)]\} \Rightarrow [(V) \Rightarrow (W)] & \text{axiom P3,} \\
\vdash [\neg(P)] \Rightarrow \underbrace{[(V) \Rightarrow (W)]}_{S} & \text{theorem 175,} \\
\vdash [\neg(P)] \Rightarrow (S) & \text{substitution;} \\
& \\
\vdash (P) \Rightarrow (S) & \text{previous result;} \\
\vdash S & \text{rule of inference;} \\
\vdash \{[(P) \Rightarrow (Q)] \Rightarrow (P)\} \Rightarrow (P) & \text{substitution.}
\end{array}
$$

4.4.6 Exercises

Exercise 791 Assume that $[\neg(P)] \Rightarrow (P)$ holds and prove that P holds. In other words, prove that $\{[\neg(P)] \Rightarrow (P)\} \vdash (P)$.

Exercise 792 Assume that $(P) \Rightarrow [\neg(P)]$ holds and prove that $\neg(P)$ holds. In other words, prove that $\{(P) \Rightarrow [\neg(P)]\} \vdash [\neg(P)]$.

Exercise 793 Apply the Deduction Theorem to prove $\{[\neg(P)] \Rightarrow (P)\} \Rightarrow (P)$.

Exercise 794 Apply the Deduction Theorem to prove $\{(P) \Rightarrow [\neg(P)]\} \Rightarrow [\neg(P)]$.

Exercise 795 Assume $(P) \Rightarrow (Q)$ and prove $\{(P) \Rightarrow [\neg(Q)]\} \Rightarrow [\neg(P)]$. In other words, prove that $[(P) \Rightarrow (Q)] \vdash \left(\{(P) \Rightarrow [\neg(Q)]\} \vdash [\neg(P)]\right)$.

Exercise 796 Apply the Deduction Theorem to prove the law of reductio ad absurdum

$$[(P) \Rightarrow (Q)] \Rightarrow (\{(P) \Rightarrow [\neg(Q)]\} \Rightarrow [\neg(P)]).$$

Exercise 797 Assume $[\neg(P)] \Rightarrow (Q)$ and prove $\{[\neg(P)] \Rightarrow [\neg(Q)]\} \Rightarrow (P)$. In other words, prove that $\{[\neg(P)] \Rightarrow (Q)\} \vdash [\{[\neg(P)] \Rightarrow [\neg(Q)]\} \Rightarrow (P)]$.

Exercise 798 Apply the Deduction Theorem to prove the law of indirect proof

$$\{[\neg(P)] \Rightarrow [\neg(Q)]\} \Rightarrow [\{[\neg(P)] \Rightarrow [\neg(Q)]\} \Rightarrow (P)].$$

Exercise 799 Apply the Deduction Theorem to prove the tautology

$$\{[(P) \Rightarrow (Q)] \Rightarrow [(Q) \Rightarrow (R)]\} \Rightarrow [(Q) \Rightarrow (R)].$$

Exercise 800 Apply the Deduction Theorem to prove the tautology

$$\{[(P) \Rightarrow (Q)] \Rightarrow [(Q) \Rightarrow (R)]\} \Rightarrow \{[(P) \Rightarrow (Q)] \Rightarrow [(P) \Rightarrow (R)]\}.$$

Exercise 801 Apply the Completeness Theorem to prove $\{[\neg(P)] \Rightarrow (P)\} \Rightarrow (P)$.

Exercise 802 Apply the Completeness Theorem to prove the special law of reductio ad absurdum: $\{(P) \Rightarrow [\neg(P)]\} \Rightarrow [\neg(P)]$.

Exercise 803 Apply the Completeness Theorem to prove the law of reductio ad absurdum: $[(P) \Rightarrow (Q)] \Rightarrow (\{(P) \Rightarrow [\neg(Q)]\} \Rightarrow [\neg(P)])$.

Exercise 804 Apply the Completeness Theorem to prove the law of assertion:

$$(P) \Rightarrow \{[(P) \Rightarrow (Q)] \Rightarrow (Q)\}.$$

Exercise 805 Apply the Completeness Theorem to prove

$$\{[(P) \Rightarrow (Q)] \Rightarrow (R)\} \Rightarrow \{[(R) \Rightarrow (P)] \Rightarrow (P)\}.$$

Exercise 806 Apply the Completeness Theorem to prove

$$\{[(P) \Rightarrow (Q)] \Rightarrow (R)\} \Rightarrow \{[(R) \Rightarrow (P)] \Rightarrow [(S) \Rightarrow (P)]\}.$$

Exercise 807 Apply the Completeness Theorem to prove

$$[\{[(P) \Rightarrow (R)] \Rightarrow (Q)\} \Rightarrow (Q)] \Rightarrow \{[(Q) \Rightarrow (R)] \Rightarrow [(P) \Rightarrow (R)]\}.$$

Exercise 808 Apply the Completeness Theorem to prove

$$[(R) \Rightarrow (Q)] \Rightarrow (\{[(R) \Rightarrow (Q)] \Rightarrow (P)\} \Rightarrow [(S) \Rightarrow (P)]).$$

Exercise 809 Apply the Completeness Theorem to prove

$$\{[(R) \Rightarrow (Q)] \Rightarrow [(S) \Rightarrow (P)]\} \Rightarrow \{[(R) \Rightarrow (P)] \Rightarrow [(S) \Rightarrow (P)]\}.$$

Exercise 810 Apply the Completeness Theorem to prove

$$(\{[(R) \Rightarrow (P)] \Rightarrow (P)\} \Rightarrow [(S) \Rightarrow (P)]) \Rightarrow (\{[(P) \Rightarrow (Q)] \Rightarrow (R)\} \Rightarrow [(S) \Rightarrow (P)]).$$

4.5 TRANSFINITE METHODS

4.5.1 Transfinite Induction

Transfinite methods lead to an example of decidability in set theory. On the set $\mathbb{N}$, the Principle of Mathematical Induction (theorem 480) is logically equivalent to the Well-Ordering Principle (theorem 509), which states that every non-empty subset $S \subseteq \mathbb{N}$ has a smallest element. All well-ordered sets also lend themselves to a method of proof known as transfinite *induction*, which relies on "initial intervals" in well-ordered sets.

Definition 680 For each set W well-ordered (definition 458) by a relation $\prec$ and for each $C \in W$, the **initial interval** determined by C is the subset

$$W_C := \{B \in W : (B \prec C) \wedge (B \neq C)\},$$

which consists of all elements preceding C but different from C.

Example 681 If $W := \mathbb{N}$, with $<$ for $\prec$, then $\mathbb{N}_N = N$ for each $N \in \mathbb{N}$.

The Principle of Mathematical Induction extends to all well-ordered sets.

Theorem 682 (Transfinite Induction) *For each set W well-ordered by a relation $\prec$ and for each subset $V \subseteq W$, $V = W$ if and only if the following formula holds:*

$$\forall C\{[(C \in W) \wedge (W_C \subseteq V)] \Rightarrow (C \in V)\}.$$

Proof. If $V = W$, then the formula is a tautology: $[(P) \wedge (Q)] \Rightarrow (P)$. For the converse, let $U := W \setminus V$. If $U \neq \varnothing$, then U has a first element $A \in U$. Thus, if $B \precneqq A$ (so that $B \prec A$ but $B \neq A$), then $B \notin U$, whence $B \in W \setminus U = V$. Hence the initial interval $W_A \subseteq V$, but then $A \in V$ by hypothesis on V, which contradicts $A \in U = W \setminus V$. $\square$

Well-ordered sets also lend themselves to a method of *definition* known as transfinite *construction,* which relies on the concept of "ideal" in a well-ordered set.

Definition 683 For each set W well-ordered by a relation $\prec$ a subset $V \subseteq W$ is an **ideal** of W if and only if V contains every element preceding any of its elements. Thus V is an ideal if and only if $W_C \subseteq V$ for every $C \in V$:

$$\forall B \forall C\{[(B \in W) \wedge (B \prec C) \wedge (C \in V)] \Rightarrow (B \in V)\}.$$

The set of all ideals of W relative to $\prec$ is denoted by $\mathcal{I}_{\prec}(W)$.

Example 684 If $W := \mathbb{N}$, with $<$ for $\prec$, then N is an ideal for each $N \in \mathbb{N}$.

The following two theorems establish relations between ideals and initial intervals.

Theorem 685 *For each set W well-ordered by a relation $\prec$ and for each ideal $V \subseteq W$, if $B \in W \setminus V$, then $V \subseteq W_B$.*

Proof. By definition of an ideal, if $C \in V$ and $B \in W$ with $B \prec C$, then $B \in V$. By contraposition, $C \notin V$ follows from $B \in W \setminus V$, and $C \in W$ with $B \prec C$. Because $\prec$ totally orders W (remark 459), it follows that $B \in W \setminus V$, then $(W \setminus V) \subseteq [W \setminus (W_B \cup \{B\})]$, whence $V \subseteq W_B \cup \{B\}$. Yet $B \notin V$, whence $V \subseteq W_B$. $\square$

Theorem 686 *For each set W well-ordered by a relation $\prec$ and for each ideal $V \subsetneqq W$, there exists a smallest element $A \in W \setminus V$, and for this element $W_A = V$.*

Proof. If $V \subsetneqq W$, then $W \setminus V \neq \varnothing$. Because W is well-ordered by $\prec$ it follows that $W \setminus V$ contains a smallest element $A \in W \setminus V$. Thus for every $B \in W$ such that $B \prec A$ and $B \neq A$, it follows that $B \in V$ by minimality of A in $W \setminus V$. Therefore $W_A \subseteq V$. Moreover, $V \subseteq W_A$ by theorem 685. □

The following theorems show that the set of all ideals is well-ordered by inclusion.

Theorem 687 *For each set W well-ordered by a relation $\prec$ and for each non-empty set $\mathcal{F}$ of ideals of W, the intersection $\bigcap \mathcal{F}$ is also an ideal of W.*

Proof. If $C \in \bigcap \mathcal{F}$, then $C \in V$ for each ideal $V \in \mathcal{F}$. Hence, if $B \in W$ and $B \prec C$, then $B \in V$ because V is an ideal. This conclusion holds for each $V \in \mathcal{F}$, whence $B \in \bigcap \mathcal{F}$. Thus $\bigcap \mathcal{F}$ is an ideal of W. □

Theorem 688 *For each set W well-ordered by a relation $\prec$ and for each non-empty set $\mathcal{F}$ of ideals of W, the smallest element of $\mathcal{F}$ is $\bigcap \mathcal{F}$. Therefore the set $\mathcal{I}_{\prec}(W)$ is well-ordered by inclusion ($\subseteq$).*

Proof. For each non-empty set $\mathcal{F}$ of ideals of W, there exists at least one ideal $U \in \mathcal{F}$, and the intersection $\bigcap \mathcal{F}$ is also an ideal, by theorem 687. Let

$$Z := \left\{C \in W : C \in \left(\bigcup \mathcal{F}\right) \setminus \left(\bigcap \mathcal{F}\right)\right\}.$$

If $Z = \varnothing$, then $\bigcup \mathcal{F} = \bigcap \mathcal{F}$, whence $\mathcal{F}$ contains only one ideal, in effect $U = \bigcap \mathcal{F}$.

If $Z \neq \varnothing$ then it has a smallest element $A \in Z$. Because $\bigcap \mathcal{F}$ is an ideal by theorem 687, and because $A \notin \bigcap \mathcal{F}$, it follows that $W_A = \bigcap \mathcal{F}$ by theorem 686.

Also because $A \notin \bigcap \mathcal{F}$, there exists an ideal $V \in \mathcal{F}$ with $A \notin V$. Hence $V \subseteq W_A$ by theorem 685.

From $V \subseteq W_A$ and $W_A = \bigcap \mathcal{F}$ it follows that $V \subseteq \bigcap \mathcal{F}$, but $\bigcap \mathcal{F} \subseteq V$. Consequently $\bigcap \mathcal{F} = V \in \mathcal{F}$.

Therefore every non-empty set $\mathcal{F} \subseteq \mathcal{I}_{\prec}(W)$ has a smallest element, in effect $\bigcap \mathcal{F}$, so that $\mathcal{I}_{\prec}(W)$ is well-ordered by set inclusions. □

4.5.2 Transfinite Construction

As the Principle of Mathematical Induction leads to a method of definition by induction (theorem 484) — also called a recursive definition or recursion — similarly transfinite induction also yields a method of *definition* by transfinite induction.

Theorem 689 (Transfinite Construction) *For each non-emtpy set W well-ordered by a relation $\prec$ with first element $A \in W$, and for each non-empty set E, let*

$$Y := \bigcup_{B \in W} E^{W_B}$$

denote the set of all functions with domain equal to an initial interval W_B and with range in E (definition 612). For each $Z \in E$, and for each function $P : Y \to E$, there exists exactly one function $F : W \to E$ such that $F(A) = Z$ and $F(B) = P(F|_{W_B})$.

Proof. This proof establishes the uniqueness and existence separately.

Uniqueness There exists at most one such function. Indeed if $F : W \to E$ and $G : W \to E$ are two such functions, with $F(A) = Z = G(A)$ and $F(B) = P(F|_{W_B})$, $G(B) = P(G|_{W_B})$, then let

$$S := \{C \in W : F(C) \neq G(C)\}.$$

If $S \neq \varnothing$, then S has a smallest element, $D \in S$. Hence $F(B) = G(B)$ for every $B \prec D$ in W, which means that $F|_{W_D} = G|_{W_D}$, but then

$$F(D) = P(F|_{W_D}) = P(G|_{W_D}) = G(D)$$

would contradict $D \in S$. Consequently, $S = \varnothing$, so that $F(B) = G(B)$ for every $B \in W$, which means that $F = G$.

Existence Let $\mathcal{F}$ denote the set of all ideals $V \subseteq W$ for which there exists a function $F_V : V \to E$ such that $F_V(A) = Z$ and $F_V(B) = P(F_V|_{W_B})$.

For any other ideal $U \in \mathcal{F}$, applying the uniqueness just proved to the well-ordered set $U \cap V$ instead of W shows that the functions F_U and F_V coincide on the well-ordered subset $U \cap V$ in W.

Hence, define a function $F_{\mathcal{F}} : \bigcup \mathcal{F} \to E$ by setting $F_{\mathcal{F}}(B) := F_U(B)$ for any ideal $U \in \mathcal{F}$ with $B \in U$. In other terms, $F_{\mathcal{F}} = \bigcup_{U \in \mathcal{F}} F_U$. The preceding argument confirms that this definition does not depend on which ideal U contains B, because if $B \in U$ and $B \in V$, then $F_U(B) = F_V(B)$.

Next, if an ideal U is an initial interval, $U = W_B$ for some $B \in W$, and if $W_B \in \mathcal{F}$, then $W_B \cup \{B\} \in \mathcal{F}$. Indeed, a function $F_U : U \to E$ extends to $W_B \cup \{B\}$ by the definition $F_{W_B \cup \{B\}}(B) := P(F_U)$.

Suppose that $W \notin \mathcal{F}$. Then let $\mathcal{G}$ denote the set of all the ideals of W that are not elements of $\mathcal{F}$. In particular, $W \in \mathcal{G}$. Define $V := \bigcap \mathcal{G}$, which is then the smallest ideal of W in $\mathcal{G}$. If V had a *last* element D, then $V = W_D \cup \{D\}$ by definition of W_D and of an ideal; however, $W_D \in \mathcal{F}$, otherwise $V \neq \bigcap \mathcal{G}$, but from $W_D \in \mathcal{F}$ it follows that $W_D \cup \{D\} \in \mathcal{F}$. Thus V cannot have a last element.

If V does not have a last element, then $V = \bigcup_{B \in V} W_B$ by definition of an ideal. Again, it follows that $W_B \in \mathcal{F}$ for each $B \in V$, whence $V = \bigcup_{B \in V} W_B \subseteq \bigcup \mathcal{F}$ and then $V \in \mathcal{F}$ because of the existence of $F_{\mathcal{F}}$, contradicting the definition of V. Therefore, $W \in \mathcal{F}$, which means that F extends to all of W. □

4.5.3 Exercises

Exercise 811 Prove that in every well-ordered set every initial interval is an ideal.

Exercise 812 Provide an example of an ideal that is *not* an initial interval in a well-ordered set.

Exercise 813 Prove that if $\mathbb{Z}$ is well-ordered by $\preccurlyeq$, then $\preccurlyeq$ differs from $\leq$.

Exercise 814 Prove that if $\mathbb{Q}$ is well-ordered by $\preccurlyeq$, then $\preccurlyeq$ differs from $\leq$.

Exercise 815 Provide an example of a well-order $\prec$ on a set of modular integers $\mathbb{Z}_M = \{[0]_M, \ldots, [M-1]_M\}$ and modular integers $[I]_M$, $[K]_M$, $[L]_M$, such that $[K]_M \prec [L]_M$ but $[I]_M + [K]_M \not\prec [I]_M + [L]_M$.

Exercise 816 Provide an example of a well-order $\prec$ on a set of modular integers $\mathbb{Z}_M = \{[0]_M, \ldots, [M-1]_M\}$ and modular integers $[I]_M$, $[K]_M$, $[L]_M$, such that $[0]_M \prec [I]_M$ and $[K]_M \prec [L]_M$ but $[I]_M * [K]_M \not\prec [I]_M * [L]_M$.

Exercise 817 Prove that every subset of a well-ordered set is also well-ordered.

Exercise 818 Determine whether for each set $\mathcal{F}$ of *ideals* in W the union $\bigcup \mathcal{F}$ is also an ideal in W.

Exercise 819 Determine whether for each set $\mathcal{G}$ of *initial intervals* in W the union $\bigcup \mathcal{G}$ is also an initial interval in W.

Exercise 820 Determine whether for each set $\mathcal{G}$ of *initial intervals* in W the intersection $\bigcap \mathcal{G}$ is also an initial interval in W.

4.6 TRANSITIVE SETS AND ORDINALS

4.6.1 Transitive Sets

Sets defined exclusively through the axioms of set theory adopted here are called well-formed sets. They have the advantage of avoiding certain contradictions that would arise from defining sets by means not so strict. The definition of well-formed sets involves the concept of sets that are "transitive" relative to the relation $\in$.

Definition 690 (Transitive Sets) A set A is **transitive** if and only if every element of A is also a subset of A, so that $\forall X[(X \in A) \Rightarrow (X \subseteq A)]$, or, equivalently,

$$\forall Y \forall X\{[(Y \in X) \wedge (X \in A)] \Rightarrow (Y \in A)\}.$$

Example 691 The following sets are transitive:

$$\begin{aligned}
&\varnothing, \\
&\{\varnothing\}, \\
&\{\varnothing, \{\varnothing\}\}, \\
&\Big\{ \varnothing, \{\varnothing\}, \{\varnothing, \{\varnothing\}\} \Big\}, \\
&\Big\{ \varnothing, \{\varnothing\}, \{\{\varnothing\}\}, \{\varnothing, \{\varnothing\}\} \Big\}.
\end{aligned}$$

Counterexample 692 The set $A := \{ \{\varnothing\} \}$ is *not* transitive, because it contains an element $X := \{\varnothing\}$ that is *not* a subset of A: $X \not\subseteq A$, because $\varnothing \in X$ but $\varnothing \notin A$.

Power sets, unions, and intersections of transitive sets are also transitive.

Theorem 693 *If a set A is transitive, then $\mathcal{P}(A)$ is also transitive.*

Proof. If $S \in \mathcal{P}(A)$, then $S \subseteq A$. Thus if $X \in S$, then $X \in A$, and $X \subseteq A$ by transitivity of A. Hence $X \in \mathcal{P}(A)$ for each $X \in S$, whence $S \subseteq \mathcal{P}(A)$. □

Theorem 694 *If a set $\mathcal{F}$ is transitive, then $\bigcup \mathcal{F}$ is also transitive.*

Proof. If $S \in \bigcup \mathcal{F}$, then there exists $A \in \mathcal{F}$ with $S \in A$. Yet $A \subseteq \mathcal{F}$ by transitivity of $\mathcal{F}$. From $S \in A$ and $A \subseteq \mathcal{F}$ follows $S \in \mathcal{F}$, whence $S \subseteq \bigcup \mathcal{F}$. □

Theorem 695 *If a non-empty set $\mathcal{F}$ is transitive, then $\bigcap \mathcal{F}$ is also transitive.*

Proof. If $X \in \bigcap \mathcal{F}$, then $X \in A$ and $X \subseteq A$ for each $A \in \mathcal{F}$, whence $X \subseteq \bigcap \mathcal{F}$. □

4.6.2 Ordinals

Well-formed sets will rely on the concept of ordinals (also called "ordinal numbers" [19, p. 42]). The following definition conforms to Kunen's [64, p. 16].

Definition 696 (Ordinals) A set A is an **ordinal** if and only if it is a transitive set, and the relation $\in$ is a well-ordering and irreflexive on the set A.

Example 697 The following sets are ordinals:

$$\begin{aligned}&\varnothing,\\&\{\varnothing\},\\&\{\varnothing,\{\varnothing\}\},\\&\Big\{\varnothing,\{\varnothing\},\{\varnothing,\{\varnothing\}\}\Big\}.\end{aligned}$$

Counterexample 698 The set

$$A := \Big\{\varnothing,\{\varnothing\},\{\{\varnothing\}\},\{\varnothing,\{\varnothing\}\}\Big\} = \mathcal{P}(\{\varnothing,\{\varnothing\}\})$$

is transitive but *not* an ordinal. Indeed, if

$$\begin{aligned}X &:= \varnothing,\\Y &:= \{\{\varnothing\}\},\end{aligned}$$

then $X \in A$ and $Y \in A$, so that $\{X,Y\} \subseteq A$, but $\{X,Y\}$ does not have any smallest element relative to the relation $\in$, because $X \notin Y$ and $Y \notin X$.

In particular the subset $\{X,Y\}$ is *not* an ordinal, even though A is an ordinal.

Theorem 699 *The empty set is an element of every non-empty ordinal.*

Proof. By definition, every ordinal A is transitive, so that if $X \in A$ then $X \subseteq A$. Consequently, if $Y \in X$ and $X \in A$, then $Y \in A$. Therefore, if $X \in A$ and $X \neq \varnothing$, then X is not the smallest element of A. Yet every non-empty ordinal A has a smallest element. Hence contraposition shows that the smallest element must be $\varnothing$. □

Theorem 700 *If A is an ordinal, then $A \notin A$. Moreover, $A \notin X$ for each $X \in A$. In particular, if A and X are ordinals, then $A \notin X$ or $X \notin A$.*

Proof. If A is an ordinal, then $\in$ is connected (definitions 449, 458): exactly one of $X \in Y$, $X = Y$, or $Y \in X$ holds for all $X, Y \in A$. Because $A = A$, it follows that $A \in A$ cannot hold. Moreover, if A is an ordinal and $X \in A$, then $X \subseteq A$, whence if $Y \in X$ then $Y \in A$. With $Y := A$, it follows by contraposition that $A \notin X$. □

Theorem 701 *Every element of an ordinal is an ordinal.*

Proof. If A is an ordinal and $X \in A$, then $X \subseteq A$ because A is a transitive set. Hence $\in$ well-orders X, because $\in$ well-orders A. If moreover $Z \in Y$ and $Y \in X$, then $Z \in X$ because A is a transitive set, whence $Y \subseteq X$. Thus X is also a transitive set. Furthermore, the relation $\in$ remains irreflexive on the subset $X \subseteq A$. □

Theorem 702 *If A is an ordinal, then either A contains a last element D and $A = D \cup \{D\}$, or $A = \bigcup A$ is the union of all its elements.*

Proof. If A is an ordinal and $B \in A$, then $B \subseteq A$; consequently, $\bigcup A \subseteq A$.

Let $D := \bigcup A$. If $D \neq A$, then there exists $X \in A \setminus D$. Then $X \subseteq D$, because $X \in A$ whence $X \subseteq A$ and $X \subseteq \bigcup A = D$.

Conversely, still with $X \in A \setminus B$, for each $B \in A$ it follows that $X \notin B$, whence $B \in X$, and hence $B \subseteq X$, so that $\bigcup_{B \in A} B \subseteq X$. Thus, $D = \bigcup A = \bigcup_{B \in A} B \subseteq X$.

Therefore, if $D \neq A$, then $A \setminus D = \{D\}$, whence $A = D \cup \{D\}$. □

Theorem 703 *If B is an ordinal, then $B \cup \{B\}$ is also an ordinal.*

Proof. Firstly, $B \cup \{B\}$is transitive. Indeed, if $X \in B \cup \{B\}$, then either $X \in B$, whence $X \subseteq B \subseteq B \cup \{B\}$, or $X \in \{B\}$, whence $X = B \subseteq B \cup \{B\}$.

Secondly, $\in$ well-orders $B \cup \{B\}$. Indeed, for each subset $S \subseteq B \cup \{B\}$, two situations can occur: $S \subseteq B$ or $S \cap \{B\} \neq \varnothing$. If $S \subseteq B$ then S has a smallest element, because B is an ordinal. If $S \cap \{B\} \neq \varnothing$, then either $S = \{B\}$ has the smallest element B, or $S \cap B \neq \varnothing$ and then $S \cap B$ is a subset of B and hence has a smallest element, which is then also a smallest element of $S = (S \cap B) \cup \{B\}$.

Moreover, the relation $\in$ remains irreflexive and transitive on $B \cup \{B\}$. Indeed, $B \notin B$ by theorem 700. Furthermore, if $X \in Y$ and $Y \in Z$ in $B \cup \{B\}$, then $Y \in B$ from either $Z \in B$ or $Z \in \{B\}$. Also, $Y \neq B$ by theorem 700, which forbids $B \in Z$ and $Z \in B \cup \{B\}$, and hence $X \neq B$ also by theorem 700, which forbids $B \in Y$ $Y \in Z$, and $Z \in B \cup \{B\}$. Consequently only two cases can occur: $Z \in B$ or $Z = B$.

If $Z = B$, then $X \neq B$ and $Y \neq B$, whence $X \in Z$.

If $Z \in B$, then X, Y, and Z all three lie in B, whence $X \in Z$ because $\in$ is transitive on the well-ordered set B.

Finally, $\in$ is strict on $B \cup \{B\}$. Indeed, because $\in$ is strict on B, it follows that if $X \in B \cup \{B\}$ and $Y \in B \cup \{B\}$, then two different cases can arise.

If $X \in B$ and $Y \in B$, then $X \in Y$ and $Y \in X$ cannot both hold, for transitivity would yield $X \in X$ which cannot hold by strictness of $\in$ on B.

If $X \in B$ and $Y \in \{B\}$, then $X \in Y$ and $Y \in X$ cannot both hold. Otherwise $Y = B$ and then $X \in B$ and $B \in X$. However, X is also an ordinal by theorem 701, whence $\in$ is also transitive on X, so that $B \in X$ and $X \in B$ yield $B \in B$, which cannot hold by strictness of $\in$ on X. □

4.6.3 Well-Ordered Sets of Ordinals

The following theorems show that every set of ordinals is well-ordered. The first theorem shows that $\subseteq$ is strongly connected (definition 448) on every set of ordinals.

Theorem 704 *For all ordinals A and B, either $A \subset B$, or $A = B$, or $B \subset A$.*

Proof. If $B \not\subseteq A$, then there exists $X \in B \setminus A$, and hence there exists a smallest such element: $X \in B \setminus A$, so that $X \in Y$ for every $Y \in B \setminus A$ with $Y \neq X$. Also, $X \neq \varnothing$, because $X \notin A$ but $\varnothing \in A$. From $X \in B$ it follows that $X \subseteq B$, whence if $Y \in X$, then $Y \in B$, but then $Y \in A$ because of the minimality of X. Thus $X \subseteq A \cap B$.

Conversely, if $Y \in A \cap B$, then $Y \in B$, whence either $Y \in X$ or $X \in Y$, because $X \in B$ also. However, $X \in Y$ cannot occur, because $X \in Y$ and $Y \in A \cap B$ would yield $X \in B$. Thus, if $Y \in A \cap B$, then $Y \in X$, which means that $A \cap B \subseteq X$.

Consequently, $X = A \cap B$.

The foregoing argument with A and B switched shows that if $A \not\subseteq B$, then $Z := B \cap A$ is the smallest element in $A \setminus B$. In particular, $Z = B \cap A = A \cap B = X$.

Consequently, if $A \not\subseteq B$ and $B \not\subseteq A$ both held, then $X := A \cap B =: Z$ would be the smallest element in both $A \setminus B$ and $B \setminus A$.

However, because $(A \setminus B) \cap (B \setminus A) = \varnothing$, it follows that $X \notin (A \setminus B) \cap (B \setminus A)$. Thus at least one of $B \not\subseteq A$ or $A \not\subseteq B$ must fail to hold, which means that $B \subseteq A$ or $A \subseteq B$ or both, so that either $A \subset B$, or $A = B$, or $B \subset A$. □

The second theorem shows that $\in$ is strongly connected on every set of ordinals.

Theorem 705 *For all ordinals A and B, either $A = B$, or $A \in B$, or $B \in A$.*

Proof. Consider the sets $C := A \cup \{A\}$ and $D := B \cup \{B\}$, which are ordinals by theorem 703. Applying theorem 704 to C and D instead of A and B shows that $D \subseteq C$ or $C \subseteq D$. If $C \subseteq D$, then $A \in C \subseteq D = B \cup \{B\}$, so that either $A \in B$ or $A = B$. If $D \subseteq C$, then $B \in D \subseteq C = A \cup \{A\}$, so that either $B \in A$ or $B = A$. □

Theorem 706 *Every set $\mathcal{F}$ of ordinals is well-ordered by $\in$.*

Proof. The relation $\in$ is irreflexive on $\mathcal{F}$ by theorem 700, and it is connected by theorem 705. The relation $\in$ is also transitive on $\mathcal{F}$. Indeed, for all $A, B, C \in \mathcal{F}$, if $A \in B$ and $B \in C$, then $B \subseteq C$ whence $A \in C$.

Moreover, for each non-empty subset $\mathcal{G} \subseteq \mathcal{F}$, there exists some $C \in \mathcal{G}$. For each $B \in \mathcal{G}$, either $B \in C$, or $B = C$, or $C \in B$, by theorem 705. Define

$$E := \{B \in \mathcal{G} : B \in C\} = C \cap \mathcal{G}.$$

If $E = \varnothing$, then C is the smallest element of $\mathcal{G}$, because then $C \in B$ for each $B \in \mathcal{G}$ with $B \neq C$. If $E \neq \varnothing$, then E has a smallest element $A \in E$, because E is a subset of the ordinal C. If $B \in \mathcal{G}$ then $B \notin A$, because $B \in A$ would yield $B \in C$, contradicting the minimality of A. Hence $A \in B$ for every $A \in \mathcal{G}$ with $B \neq A$. □

4.6.4 Unions and Intersections of Sets of Ordinals

The union and the intersection of every non-empty set of ordinals is an ordinal.

Theorem 707 *For each set $\mathcal{F}$ of ordinals, $\bigcup \mathcal{F}$ is also an ordinal.*

Proof. The union $\bigcup \mathcal{F}$ is transitive, by theorem 694.

The union $\bigcup \mathcal{F}$ is well-ordered. Indeed, for each non-empty subset $S \subseteq \bigcup \mathcal{F}$, let

$$E := \{A \in \mathcal{F} : A \cap S \neq \varnothing\}.$$

From $S \neq \varnothing$ it follows that $E \neq \varnothing$. Hence $E \subseteq \mathcal{F}$ has a smallest element A, because $\mathcal{F}$ is an ordinal. Consequently, $S \cap A \neq \varnothing$. Therefore, $S \cap A \subseteq A$ also has a smallest element $B \in S \cap A$. Moreover, for each $C \in S$, there exists $D \in \mathcal{F}$ with $C \in D$. Then either $A = D$, or $A \in D$, or $D \in A$. Yet $D \in A$ cannot occur, by minimality of A. From $B \in A$, with $A \in D$ or $A = D$, it follows that $B \in D$; hence $B \in D$ and $C \in D$. Consequently, either $B \in C$ or $C \in B$ or $B = C$, but $C \in B$ cannot occur by minimality of B and because $B \in A$. Thus, $B \in C$ or $B = C$, which shows that B is the smallest element of S.

By theorem 700 $\in$ is strict on $\bigcup \mathcal{F}$, for every element is an ordinal by theorem 701. □

Theorem 708 *For each non-empty set $\mathcal{F}$ of ordinals, $\bigcap \mathcal{F}$ is also an ordinal.*

Proof. The intersection $\bigcap \mathcal{F}$ is a transitive set by theorem 695. The intersection $\bigcap \mathcal{F}$ is also well-ordered. Indeed, each non-empty subset $S \subseteq \bigcap \mathcal{F}$ is also a subset $S \subseteq A$ of some $A \in \mathcal{F}$ and hence S has a smallest element, because A is an ordinal.

The relation $\in$ is a strict total order on $\bigcap \mathcal{F}$, as it is on every subset of $A \in \mathcal{F}$, where it is strict. Indeed, if $X \in \bigcap \mathcal{F}$ and $Y \in \bigcap \mathcal{F}$, then $X \in A$ and $Y \in A$ whence $X \in Y$ and $Y \in X$ cannot both hold. □

4.6.5 Exercises

Exercise 821 Prove that $\mathbb{N}$ is a transitive set.

Exercise 822 Prove that every natural number $N \in \mathbb{N}$ is a transitive set.

Exercise 823 Prove that $\mathbb{N}$ is an ordinal.

Exercise 824 Prove that every natural number $N \in \mathbb{N}$ is an ordinal.

Exercise 825 Investigate whether $\in$ is a transitive relation on every transitive set.

Exercise 826 Prove that every ordinal is an element of some ordinal.

Exercise 827 Prove that every ordinal is a subset of some ordinal.

Exercise 828 Determine whether every ordinal is a subset of some transitive set.

Exercise 829 Determine whether every transitive set is a subset of some ordinal.

Exercise 830 Verify that it is not necessary to assume that $V \neq \varnothing$ for Transfinite Induction. In other words, prove that for each set W well-ordered by a relation $\prec$ and for each subset $V \subseteq W$, if the following formula is True,

$$\forall C\{[(C \in W) \wedge (W_C \subseteq V)] \Rightarrow (C \in V)\},$$

then either $W = \varnothing$ or $V \neq \varnothing$.

Exercise 831 Determine whether every singleton $\{A\}$ with an ordinal A is an ordinal.

Exercise 832 Determine whether $\{A, B\}$ is an ordinal for all ordinals A and B.

Exercise 833 Determine whether every set of ordinals is an ordinal.

Exercise 834 Determine whether every subset of every ordinal is an ordinal.

Exercise 835 Prove that there exists an ordinal whose power set is not an ordinal.

Exercise 836 Find an ordinal A such that

$$\Big\{\varnothing, \{\varnothing\}, \{\varnothing, \{\varnothing\}\}\Big\} \in R(A).$$

Exercise 837 Find an ordinal C such that

$$\Bigg\{\varnothing, \{\varnothing\}, \{\varnothing, \{\varnothing\}\}, \Big\{\varnothing, \{\varnothing\}, \{\varnothing, \{\varnothing\}\}\Big\}\Bigg\} \in R(C).$$

Exercise 838 Prove that every countable set admits a well-ordering.

Exercise 839 Prove that if $\mathcal{F} \neq \varnothing$, and if there exists a relation $\preccurlyeq$ well-ordering $\bigcup \mathcal{F}$, then there exists a function $F : \mathcal{F} \to \bigcup \mathcal{F}$ with $F(X) \in X$ for every $X \in \mathcal{F}$.

Exercise 840 Prove that the axiom of choice follows from Zermelo's principle: if every set admits a well-ordering, then for every set $\mathcal{F}$ of non-empty sets there exists a function $F : \mathcal{F} \to \bigcup \mathcal{F}$ such that $F(X) \in X$ for every $X \in \mathcal{F}$.

4.7 REGULARITY OF WELL-FORMED SETS

4.7.1 Well-Formed Sets

The following definition establishes sets that contain all the well-formed sets.

Definition 709 For each ordinal A, define a set $R(A)$ by transfinite construction:

- $R(\varnothing) := \varnothing$;
- $R(A \cup \{A\}) := \mathcal{P}[R(A)]$;
- $R(A) := \bigcup_{B \in A} R(B)$ if there does not exist any ordinal B such that $A = B \cup \{B\}$, but if $R(B)$ has been defined for every $B \in A$.

Remark 710 The transfinite construction proceeds as follows. For each ordinal W, let

$$\begin{array}{rcl} E & := & \mathcal{P}[\mathcal{P}(W)], \\ Y & := & \bigcup_{A \in W} E^{W_A}, \\ P\colon Y & \to & E, \\ P(R|_{W_A}) & := & \bigcup_{B \in A} R(B). \end{array}$$

The values of P remain in E, because if $R|_{W_A} : W_A \to E$, then $R(B) \in E = \mathcal{P}[\mathcal{P}(W)]$ for each $B \in W_A = A$, whence $R(B) \subseteq \mathcal{P}(W)$, and hence $\bigcup_{B \in A} R(B) \subseteq \mathcal{P}(W)$, so that $\bigcup_{B \in A} R(B) \in \mathcal{P}[\mathcal{P}(W)]$. Thus, for all ordinals U and V, $U \subseteq V$ or $V \subseteq U$ by theorem 704, so that $\mathcal{P}[\mathcal{P}(U)] \subseteq \mathcal{P}[\mathcal{P}(V)]$ or $\mathcal{P}[\mathcal{P}(V)] \subseteq \mathcal{P}[\mathcal{P}(U)]$. Therefore, the transfinite construction defined with $W := U$ or $W := V$ gives the same definition on $U \cap V$. In other words, for each ordinal A, the definition of $R(A)$ can proceed from any ordinal W containing A, for example, $W := A \cup \{A\}$.

Example 711 The first few sets of the form $R(A)$ are also ordinals:

$$\begin{array}{rcl} R(\varnothing) & = & \varnothing, \\ R(\{\varnothing\}) = R(\varnothing \cup \{\varnothing\}) = \mathcal{P}[R(\varnothing)] = \mathcal{P}[\varnothing] & = & \{\varnothing\}, \\ R(\{\varnothing, \{\varnothing\}\}) = R(\{\varnothing\} \cup \{\{\varnothing\}\}) = \mathcal{P}[R(\{\varnothing\})] = \mathcal{P}[\{\varnothing\}] & = & \{\varnothing, \{\varnothing\}\}. \end{array}$$

The next ordinal,

$$A := \Big\{ \varnothing, \{\varnothing\}, \{\varnothing, \{\varnothing\}\} \Big\},$$

is not of the form $R(B)$ for any set B, but $A = B \cup \{B\}$ for the ordinal $B := \{\varnothing, \{\varnothing\}\}$. Hence, the list continues with

$$\begin{array}{rcl} R\Big(\Big\{ \varnothing, \{\varnothing\}, \{\varnothing, \{\varnothing\}\} \Big\}\Big) & = & R\Big(\{\varnothing, \{\varnothing\}\} \cup \Big\{\{\varnothing, \{\varnothing\}\}\Big\}\Big) \\ & = & \mathcal{P}\left[R\left(\{\varnothing, \{\varnothing\}\}\right)\right] \\ & = & \mathcal{P}\left[\{\varnothing, \{\varnothing\}\}\right] \\ & = & \Big\{ \varnothing, \{\varnothing\}, \{\{\varnothing\}\}, \{\varnothing, \{\varnothing\}\} \Big\}, \end{array}$$

which is of the form $R(A)$, but it is not well-ordered by $\in$ and hence not an ordinal, by counterexample 698. The set $R(A)$ is also different from the ordinal

$$\Big\{ \varnothing, \{\varnothing\}, \{\varnothing, \{\varnothing\}\}, \Big\{\varnothing, \{\varnothing\}, \{\varnothing, \{\varnothing\}\}\Big\} \Big\}.$$

Theorem 712 *For each ordinal Q, the set $R(Q)$ is transitive.*

Proof. This proof proceeds by transfinite induction (theorem 682).

Choose an ordinal W with $Q \in W$, for instance, $W := Q \cup \{Q\}$. Then consider the subset $V \subseteq W$ of all the elements $C \in W$ for which $R(C)$ is transitive.

Consider any element $C \in W$ such that $W_C \subseteq V$. By transfinite induction, it suffices to verify that $C \in V$ to establish that $V = W$. Because

$$W_C = \{B \in W : (B \in C) \wedge (B \neq C)\}$$

by definition 680 for the relation $\in$ instead of $\prec$, it follows that $W_C = C$. From $C = W_C \subseteq V$ it then follows that $R(B)$ is transitive for each $B \in C$ by definition of V. Consequently, $\bigcup_{B \in C} R(B)$ is also transitive, by theorem 694.

Two cases can arise, either C does not have a last element, or C has a last element.

If C does not have a last element, then $R(C) = \bigcup_{B \in C} R(B)$ is transitive.

If C has a last element $Z \in C$, then $C = Z \cup \{Z\}$. From $Z \in C = W_C \subseteq V$, it follows that $R(Z)$ is transitive by definiton of V. Hence $\mathcal{P}[R(Z)]$ is also transitive, by theorem 693. Yet $R(C) = R(Z \cup \{Z\}) = \mathcal{P}[R(Z)]$, whence $R(C)$ is transitive. Thus in either case $R(C)$ is transitive, whence $C \in V$, and thence $V = W$.

Finally, from $Q \in W = V$ it follows that $R(Q)$ is transitive by definition of V. □

4.7.2 Regularity

The following theorems confirm that every element, subset, pairing, power set, union, intersection, and Cartesian product of well-formed sets is again a well-formed set.

Definition 713 A set X is **well-formed** if and only if there exists an ordinal A such that $X \in R(A)$.

Theorem 714 *For each well-formed set X there exists a smallest ordinal A such that $X \in R(A)$.*

Proof. If X is a well-formed set, then there exists an ordinal C such that $X \in R(C) \in \mathcal{P}[R(C)] = R(C \cup \{C\})$. Hence $X \in R(C \cup \{C\})$ by transitivity. Consequently, there exists a smallest ordinal $A \in C \cup \{C\}$ such that $X \in R(A)$.

For every ordinal D, either $D = A$, or $D \in A \in C \cup \{C\}$ and then $X \notin R(D)$, or $A \in D$ and then D is not the smallest such ordinal. □

Theorem 715 *If X is well-formed, then every $Y \in X$ is also well-formed.*

Proof. If X is well-formed, then there exists an ordinal A such that $X \in R(A)$.

If $A = \varnothing$, then $X \in R(A) = R(\varnothing) = \varnothing$, whence $Y \in X$ is also well-formed.

If there exists an ordinal B such that $A = B \cup \{B\}$, then $X \in R(A) = \mathcal{P}[R(B)]$, whence $X \subseteq \mathcal{P}[R(B)]$ by transitivity of $\mathcal{P}[R(B)]$, Consequently, $Y \in R(A) = \mathcal{P}[R(B)]$ is well-formed for each $Y \in X$. If $R(A) = \bigcup_{B \in A} R(B)$ and the theorem holds for each $Z \in R(B)$ for each $B \in A$, then for each $X \in R(A)$, there exists $B \in A$ such that $X \in R(B)$ whence every $Y \in X$ is also well-formed. □

Theorem 716 *If X and Y are well-formed sets, then so are $\{X, Y\}$, $\mathcal{P}(X)$, $\bigcup X$, $X \times Y$, every subset of X, and $\bigcap X$ provided $X \neq \varnothing$.*

Proof. If X and Y are well-formed sets, then there exist ordinals A and B such that $X \in R(A)$ and $Y \in R(B)$. Either $A = B$ (whence $R(A) = R(B)$), or $A \in B$ (whence $R(A) \subseteq R(B)$), or $B \in A$ (whence $R(B) \subseteq R(A)$). For instance, assume that $R(A) \subseteq R(B)$. Thus $X \in R(B)$ and $Y \in R(B)$, so that $\{X, Y\} \in \mathcal{P}[R(B)] = R(B \cup \{B\})$, whence $\{X, Y\}$ is well-formed, because $B \cup \{B\}$ is an ordinal.

Similarly, if $X \in R(B)$ is a well-formed set, then $\subseteq R(B)$ by transitivity, whence $\mathcal{P}(X) \subseteq \mathcal{P}[R(B)] = R(B \cup \{B\})$ and hence $\mathcal{P}(X) \in \mathcal{P}[R(B \cup \{B\})] = R(A \cup \{A\})$ with $A := B \cup \{B\}$. Thus, $\mathcal{P}(X)$ is well-formed.

Consequently, from $\mathcal{P}(X) \in R(A \cup \{A\})$ it follows that $\mathcal{P}(X) \subseteq \mathcal{P}[R(A \cup \{A\})]$, whence if $S \subseteq X$, then $S \in \mathcal{P}[R(A \cup \{A\})]$ is well-formed.

In particular, $\bigcap X$ is well-formed because $\bigcap X \subseteq \bigcup X$ and $\bigcup X$ is well-formed.

Also, if $X \in R(B)$ is a well-formed set, then $X \subseteq R(B)$. Hence, if $Z \in Y$ and $Y \in X$, then $Y \subseteq X$ whence $Z \in X$ and $Z \in R(B)$. This shows that $\bigcup X \subseteq R(B)$. Consequently, $\bigcup X \in \mathcal{P}[R(B)] = R(B \cup \{B\})$ is well-formed.

In particular, if $X \neq \varnothing$, then $\bigcap X$ is well-formed because $\bigcap X \subseteq \bigcup X$ and $\bigcup X$ is well-formed. In particular, because $\{X, Y\}$ is well-formed, it follows that $X \cup Y = \bigcup\{X, Y\}$ is well-formed, whence $\mathcal{P}(X \cup Y)$, $\mathcal{P}[\mathcal{P}(X \cup Y)]$, and $\mathcal{P}\{\mathcal{P}[\mathcal{P}(X \cup Y)]\}$ are also well-formed. Therefore, $X \times Y \in \mathcal{P}\{\mathcal{P}[\mathcal{P}(X \cup Y)]\}$ is also well-formed. □

Theorem 717 *For all well-formed sets X and Y, $\neg[(X \in Y) \wedge (Y \in X)]$. In particular, $X \notin X$ for each well-formed set X.*

Proof. If $X = \varnothing$, then $Y \notin X$ and $X \notin X$.

If $X \neq \varnothing$ and $Y \in X$, then it suffices to verify that $X \notin Y$. To this end, let A be the first ordinal such that $X \in R(A)$; in particular, $X \subseteq R(A)$ by transitivity of $R(A)$.

If $A = Z \cup \{Z\}$, then $X \in R(A) = R(Z \cup \{Z\}) = \mathcal{P}[R(Z)]$, so $X \subseteq R(Z)$, whence $Y \in X \subseteq R(Z)$, and hence $Y \subseteq R(Z)$. However, $X \notin R(Z)$ because A is the first ordinal with $X \in R(A)$. If A does not contain a last element, then $R(A) = \bigcup_{B \in A} R(B)$, whence $X \in R(A)$ means that there exists $B \in A$ with $X \in R(B)$, and then $X \subseteq R(B)$ by transitivity of $R(B)$. Yet $X \notin R(B)$ because $B \in A$ and A is the first ordinal with $X \in R(A)$;

Thus, in either case there exists $C \in A$ such that $X \subseteq R(C)$ and $X \notin R(C)$ and $Y \in R(C)$. (In the first case $C := Z$, while $C := B$ in the second case.)

Therefore $X \notin Y$, because $X \in Y \in R(C)$ and the transitivity of $R(C)$ would yield $X \in R(C)$. In particular, $X \notin X$, because the foregoing argument applied to $Y := X$ shows that if $X \in X$ then $X \notin X$, contrary to the axioms governing $\in$, which state that for all sets X and Y either $X \in Y$ or $X \notin Y$ but not both. □

4.7.3 Exercises

Exercise 841 Prove that the set $\mathbb{N}$ of all natural numbers is a well-formed set.

Exercise 842 Prove that every natural number $N \in \mathbb{N}$ is a well-formed set.

Exercise 843 Prove that the set $\mathbb{Z}$ of all integers is a well-formed set.

Exercise 844 Prove that the set $\mathbb{Q}$ of all rational numbers is a well-formed set.

Exercise 845 Prove that for each well-formed set X there exists $Y \in X$ such that $Y \cap X = \varnothing$.

Exercise 846 Prove that if X is a well-formed set, then so is $\{X\}$.

Exercise 847 For each well-formed set X, prove that if A is the smallest ordinal such that $X \in R(A)$, then there exists an ordinal B such that $A = B \cup \{B\}$.

Exercise 848 Prove that every finite set of well-formed sets is a well-formed set.

Exercise 849 Determine whether every countable set of well-formed sets is a well-formed set.

Exercise 850 Determine whether every set of well-formed sets is a well-formed set.

4.8 FURTHER ISSUES IN DECIDABILITY

Incompleteness of the predicate calculus The example of Peirce's law shows that whereas there does not exist any algorithm to design every proof in the implicational calculus, there exists an algorithm to design every proof in the full propositional calculus. Hence arises the question, whether there exists an algorithm to design every proof in the *predicate* calculus (with logical connectives and quantifiers). The answer is negative. Indeed, Alonzo Church has proved that no algorithm can determine for each formula of the predicate calculus whether the formula is a theorem; in other words, for every algorithm there exists a formula for which the algorithm fails [9], [10]. The absence of any algorithm to design proofs raises the question of whether for each logical formula there exists a proof that the formula is a theorem or a proof that the formula is not a theorem. The answer to this question is also negative. Kurt Gödel has proved a result — known as **Gödel's Incompleteness Theorem** — which states that in every logical system sufficiently large to contain integer arithmetic (for instance, the predicate calculus and the axioms of set theory), there exists a formula for which there exists neither a proof that it is a theorem nor a proof that it is not a theorem [34].

Independence of the axiom of regularity As an extension of theorem 717, every nonempty well-formed set X contains an element Y that does not contain any element of X, so that $Y \cap X = \varnothing$, or, equivalently, so that there does not exist any $Z \in Y$ such that $Z \in X$ (exercise 845). Outside of well-formed sets, however, there exist systems of set theory in which the relation $A \in A$ may hold for some set. One way of preventing the relation $A \in A$ from holding for any set consists in the axiom of regularity,

$$\forall X\{(X \neq \varnothing) \Rightarrow (\exists Y[(Y \in X) \wedge \{\forall Z[(Z \in Y) \Rightarrow (Z \notin X)]\}])\},$$

attributed independently to Zermelo and von Neumann [95, p. 53]. The axiom of regularity has the disadvantage of asserting a condition about sets already defined by previous axioms. Yet within the theory of well-formed sets, exercise 845 confirms that the axiom of regularity is *not* independent but is a theorem derivable from the other axioms. Because well-formed sets suffice for most of logic, mathematics, computer science, and their applications, the foundations of these fields can restrict themselves to well-formed sets [64]. In contrast to the derivability of the axiom of regularity from the other axioms of the theory of well-formed sets, neither the generalized continuum hypothesis nor its negation are derivable from the other axioms of the theory of well-formed sets [32], [33], [14], [15]. Thus, the "axiom of regularity" is an example of an axiom that is "dependent" on the other axioms, whereas the generalized continuum hypothesis is an example of an axiom that is "independent" from the other axioms.

4.9 PROJECTS

Project 9 Investigate whether the intuitionistic logic admits a completeness theorem.

Project 10 Investigate whether the minimal logic admits a completeness theorem.

Part B

Applications

Chapter 5

Number Theory and Codes

5.0 INTRODUCTION

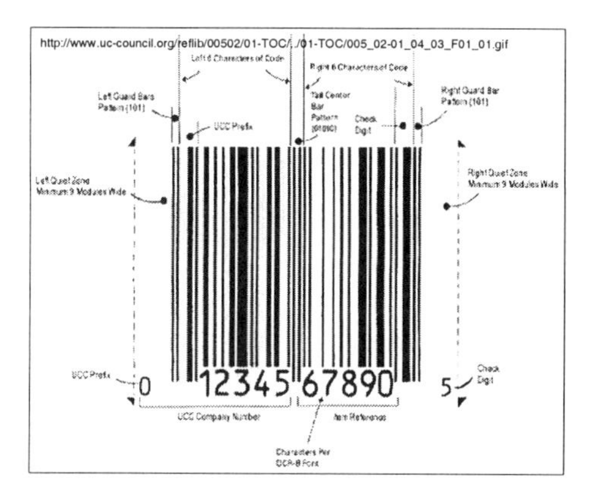

Figure 5.1: The last digit checks a Universal Product Code. Copyright 2001 Uniform Code Council, Inc. Used with Permission.

Various practical situations require codes to encode or decode information. For instance, codes provide some secrecy and security of communications during conflicts. Moreover, through redundancies, codes also provide capabilities for detecting and correcting errors in the storage or transmission of information. The theorems that guarantee such codes, and the algorithms to carry out the codes, rely on the characteristics of integer arithmetic (reviewed in chapter 3), and on the concepts of prime numbers, integer divisions with remainders, and prime factors explained in the present chapter.

5.1 THE EUCLIDEAN ALGORITHM

5.1.1 Division With Integers

Each pair of natural numbers has a sum, a difference, and a product, all among the integers, but some pairs do not have a *quotient* among the integers. Nevertheless, the following theorem produces an integer quotient with an integer remainder.

Theorem 718 (Division with remainder) *For each non-negative integer N and each positive integer M, there exist unique non-negative integers Q and R such that*

$$N = (M * Q) + R,$$

$$M > R.$$

Proof. This proof establishes the existence and the uniqueness separately.

Existence This proof of the existence of Q and R proceeds by induction with N.

Initial step If $N = 0$, then $Q := 0$ and $R := 0 < M$ satisfy $0 = (M * 0) + 0$.

Induction hypothesis Let S denote the set of natural numbers for which the theorem holds:

$$\begin{aligned} S \ := \ & \Big\{ N \in \mathbb{N} : \forall M \Big([(M \in \mathbb{N}) \wedge (0 < M)] \\ & \Rightarrow \big[\exists Q \big(\exists R \{ (Q \in \mathbb{N}) \wedge (R \in \mathbb{N}) \wedge [N = (M * Q) + R] \wedge (R < M) \} \big) \big] \Big) \Big\}. \end{aligned}$$

Assume that there exists $K \in S$, so that the theorem holds for $N := K$. Then for each positive integer M there exist non-negative integers Q and R such that $K = (M * Q) + R$ and $R < M$.

Induction step From $K = (M * Q) + R$ it follows that

$$K + 1 = [(M * Q) + R] + 1 = (M * Q) + (R + 1).$$

From $R < M$ two cases can arise, either $R + 1 < M$ or $R + 1 = M$.

If $R + 1 < M$, then Q and $R + 1$ satisfy the theorem.

If $R + 1 = M$, then $Q + 1$ and 0 satisfy the theorem because

$$K + 1 = (M * Q) + (R + 1) = (M * Q) + M = M * (Q + 1) + 0.$$

In either case, $K + 1 \in S$, whence $S = \mathbb{N}$ (theorem 480).

Uniqueness To verify that there exist only one quotient and only one remainder for the division of N by M, this proof shows that any two quotients Q_1 and Q_2 must coincide with each other, and any two remainders R_1 and R_2 must coincide with each other. With the remainders ranked so that R_1 is the larger of R_1 and R_2, it follows that

$$(M * Q_1) + R_1 = N = (M * Q_2) + R_2,$$

$$M > R_1 \geq R_2 \geq 0.$$

Hence

$$R_1 - R_2 = (M * Q_2) - (M * Q_1) = M * (Q_2 - Q_1)$$

where $R_1 - R_2 \geq 0$ and hence also $Q_2 - Q_1 \geq 0$ because $M > 0$. The proof continues by contradiction. If strict inequalities held, then $R_1 > R_2$ whence $Q_1 > Q_2$ and thence

$$R_1 \geq R_1 - R_2 = M * (Q_2 - Q_1) \geq M$$

would contradict $R_1 < M$. Therefore, $Q_2 = Q_1$ and $R_2 = R_1$. □

Remark 719 The proof of uniqueness in theorem 718 follows the pattern of the law of *reductio ad absurdum* (described in subsection 0.4.2, and also in theorem 196, and theorem 649 or axiom K3 in chapter 4). Define H and K as follows:

H**:** $R_1 > R_2$;

C**:** $R_1 < M$.

Then the proof of uniqueness in theorem 718 proceeds along the following pattern:

$\vdash C$	hypothesis $R_1 < M$,
$\vdash (H) \Rightarrow (C)$	theorem 171,
$\vdash [(H) \Rightarrow (C)] \Rightarrow (\{(H) \Rightarrow [\neg(C)]\} \Rightarrow [\neg(H)])$	*reductio ad absurdum*,
$\vdash \{(H) \Rightarrow [\neg(C)]\} \Rightarrow [\neg(H)]$	*Modus Ponens*,
$\vdash (H) \Rightarrow [\neg(C)]$	proof of theorem 718,
$\vdash \neg(H)$	*Modus Ponens*.

Thus $\neg(H)$ holds, which is $\neg(R_1 > R_2)$, whence $R_1 = R_2$ (because $R_1 \geq R_2$).

Definition 720 If $N = (M * Q) + R$ with $M > R$, then N is the **dividend**, M is the **divisor**, and R is the **remainder**. The quotient Q and the remainder R are the result of **dividing** — or of the **division of** — the dividend N **by** the divisor M.

Example 721 If $N = 9$ and $M = 2$, then $Q = 4$ and $R = 1$, because $9 = (2 * 4) + 1$ with $2 > 1$. The dividend is 9, the divisor 2, the quotient 4, and the remainder 1.

The proof of theorem 718 contains an algorithm to calculate the quotient and the remainder. The algorithm amounts to calculating the products $M * 0, M * 1, \ldots, M * L$ as long as $M * L \leq N$, stopping for the smallest L such that $M * L > N$, and then setting $Q := L - 1$ and $R := N - (M * Q)$. There also exist other algorithms.

5.1.2 Greatest Common Divisors

Any division algorithm can also produce divisors common to two integers.

Definition 722 (Common divisor) An integer M **divides** — or **is a divisor of** — an integer N if and only if there exists an integer L with $M * L = N$.

The notation $M \mid N$ indicates that M divides N.

The notation $M \nmid N$ indicates that M does *not* divide N.

M is a **common divisor** of two integers K and N if and only if M divides K and N.

Example 723 The integer 3 is a common divisor of 6 and 9.

The next theorem shows that two non-zero integers have a greatest common divisor.

Theorem 724 (Greatest common divisor) *For all non-zero integers M and N, there exists a greatest natural number dividing M and N.*

Proof. This proof shows that the set D of all common divisors of M and N is not empty and has an upper bound, and hence also a maximum (by theorem 511).

The first step verifies that $D \neq \varnothing$. Because $M = 1 * M$ and $N = 1 * N$ it follows that 1 divides M and N; thus $1 \in D$.

The second step confirms that D has an upper bound. For each $K \in \mathbb{N}$, if $K > N$ and $Q > 0$, then $K * Q > N * 1 = N$; consequently, K does not divide N. By contraposition, it follows that if $K \in D$ divides both M and N, then $K \leq N$. Thus N is an upper bound for D. Therefore, by the Well-Ordering Principle (theorem 511), D contains a maximum element, which is then the greatest common divisor of M and N. □

The following theorems show that every divisor common to the dividend and the divisor also divides the remainder. The first theorem shows that if an integer divides either factor of a product, then it also divides the product.

Theorem 725 *For all integers $I, J, L \in \mathbb{Z}$, if L divides I, then L divides $I * J$.*

Proof. If L divides I, then there exists $K \in \mathbb{Z}$ such that $I = L * K$. Hence $I * J = (L * K) * J = L * (K * J)$, so that L divides $I * J$. □

The second theorem shows that if an integer divides both summands of a sum, then it also divides the sum.

Theorem 726 *For all $I, J, L \in \mathbb{Z}$, if L divides I and J, then L divides $I + J$ and $I - J$.*

Proof. If L divides both I and J, then there exist $K, P \in \mathbb{Z}$ such that $I = L * K$ and $J = L * P$. Hence $I + J = (L * K) + (L * P) = L * (K + P)$, so that L divides $I + J$. Similarly, $I - J = (L * K) - (L * P) = L * (K - P)$, so that L divides $I - J$. □

The third theorem shows that every divisor common to the dividend and the divisor also divides the remainder.

Theorem 727 *For all integers L, M, N, Q, R with $N = (M * Q) + R$, if L divides M and N, then L divides R. Conversely, if L divides M and R, then L divides N.*

Proof. If L divides M, then L divides $M * Q$ (theorem 725). If moreover L divides N, then L divides $R = N - (M * Q)$ (theorem 726). Conversely, if L divides both M and R, then L divides $M * Q$ and R, whence also $N = (M * Q) + R$ by theorem 726. □

The following theorem provides an algorithm to calculate the greatest common divisor, in a manner amenable to applications, for instance, codes.

Theorem 728 (Euclidean Algorithm) *For all positive integers M and N, there exists a greatest common divisor L, and integers I and J such that*

$$(I * M) + (J * N) = L.$$

Moreover, if an integer H divides M and N, then H divides L.

Proof. This proof proceeds by recursion, keeping the results — quotients and remainders — of all the intermediate divisions.

Specifically, begin a sequence with $R_0 := N$ and $R_1 := M$, and define the next element R_2 as the remainder of the division of R_0 by R_1:

$$\begin{aligned} N &= (M * Q_2) + R_2, \\ R_0 &= (R_1 * Q_2) + R_2. \end{aligned}$$

By theorem 727, if H divides both M and N, then H also divides $R_2 = N - (M * Q_2)$.

Next, define the element R_3 as the remainder of the division of R_1 by R_2:

$$\begin{aligned} M &= (R_2 * Q_3) + R_3, \\ R_1 &= (R_2 * Q_3) + R_3. \end{aligned}$$

Because H divides both R_1 and R_2, theorem 727 shows that H also divides R_3.

With H dividing positive integers $R_{P-1} > R_P > 0$, define the next element R_{P+1} recursively as the remainder of the division of R_{P-1} by R_P:

$$\begin{gathered} R_{P-1} = (R_P * Q_{P+1}) + R_{P+1}, \\ R_P > R_{P+1} \geq 0. \end{gathered}$$

Because H divides R_{P-1} and R_P, theorem 727 shows that H also divides R_{P+1}.

Thus the sequence of positive remainders decreases at each step, which guarantees (by theorem 509) that there exists a smallest index K such that $R_{K+1} = 0$, so that

$$R_{K-1} = (R_K * Q_{K+1}) + 0,$$

and H still divides both R_{K-1} and R_K; in particular, $H \leq R_K$, and also for the greatest common divisor (by setting $H := L$), so that $L \leq R_K$.

Yet R_K divides R_{K-1}, and hence also $R_{K-2}, \ldots, R_1 = M, R_0 = N$. Consequently R_K divides both M and N, whence $R_K \leq L$, and, therefore, $L = R_K$. Because H divides R_K it also follows that H divides L. The penultimate step,

$$R_{K-2} = (R_{K-1} * Q_K) + R_K,$$

yields a first formula for the greatest common divisor L:

$$R_{K-2} - (R_{K-1} * Q_K) = R_K = L.$$

Similarly, the preceding step,

$$R_{K-3} = (R_{K-2} * Q_{K-1}) + R_{K-1},$$

yields

$$R_{K-3} - (R_{K-2} * Q_{K-1}) = R_{K-1}.$$

Hence replacing R_{K-1} by $R_{K-3} - (R_{K-2} * Q_{K-1})$ in

$$R_{K-2} - (R_{K-1} * Q_K) = R_K = L$$

gives

$$\begin{aligned} R_{K-2} - ([R_{K-3} - (R_{K-2} * Q_{K-1})] * Q_K) &= L, \\ R_{K-3} * Q_K + R_{K-2} * (Q_{K-1} * Q_K + 1) &= L. \end{aligned}$$

In general, if

$$L = I_K * R_{K-2} + J_K * R_{K-1},$$

then substituting

$$R_{K-1} = R_{K-3} - (R_{K-2} * Q_{K-1})$$

gives

$$\begin{aligned} L &= I_K * R_{K-2} + J_K * R_{K-1} \\ &= I_K * R_{K-2} + J_K * [R_{K-3} - (R_{K-2} * Q_{K-1})] \\ &= I_{K-1} * R_{K-3} + J_{K-1} * R_{K-2}. \end{aligned}$$

Eventually, $L = I * R_1 + J * R_0 = I * M + J * N$. □

Example 729 Firstly, this example shows how to calculate the greatest common divisors L of $M := 5$ and $N := 7$.

$$\begin{array}{rcrcrcr} N &=& M &*& Q_2 &+& R_2, \\ R_0 &=& R_1 &*& Q_2 &+& R_2, \\ 7 &=& 5 &*& 1 &+& \mathbf{2}; \end{array}$$

$$\begin{array}{rcrcrcr} R_1 &=& R_2 &*& Q_3 &+& R_3, \\ 5 &=& \mathbf{2} &*& 2 &+& \mathbf{1}; \end{array}$$

$$\begin{array}{rcrcrcr} R_2 &=& R_3 &*& Q_4 &+& R_4, \\ \mathbf{2} &=& 1 &*& 2 &+& \mathbf{0}. \end{array}$$

The greatest divisor L is the last non-zero remainder, here $L = R_3 = 1$.

Secondly, this example shows how to calculate integers I and J such that $L = (I * M) + (J * N)$ by solving for the remainders in the foregoing steps.

$$\begin{array}{ccccccccccccccc} 7 & = & 5 & * & 1 & + & \mathbf{2} & \Rightarrow & 7 & - & 5 & * & 1 & = & \mathbf{2}, \\ 5 & = & \mathbf{2} & * & 2 & + & 1 & \Rightarrow & 5 & - & \mathbf{2} & * & 2 & = & \mathbf{1}. \end{array}$$

Substituting the expression $7 - (5 * 1)$ for **2** in the last line then gives

$$\mathbf{1} = 5 - (\mathbf{2} * 2) = 5 - \{[7 - (5 * 1)] * 2\} = (5 * 3) - (7 * 2) = (3 * 5) - (2 * 7).$$

So $I = 3$ and $J = -2$.

Definition 730 (Relative primes) Two integers are **relatively prime** if and only if their greatest common divisor is 1.

Example 731 Neither 8 nor 9 are prime, but they are relatively prime.

5.1.3 Exercises

For the following exercises, calculate the greatest common divisor L of the given integers M and N, and indicate whether M and N are relatively prime. Then calculate integers I and J such that $L = (I * M) + (J * N)$.

Exercise 851 $M := 4$, $N := 6$.

Exercise 852 $M := 6$, $N := 9$.

Exercise 853 $M := 3$, $N := 8$.

Exercise 854 $M := 4$, $N := 9$.

Exercise 855 Prove that if every integer divides an integer N, then $N = 0$.

Exercise 856 Prove that every integer divides 0.

Exercise 857 Prove that if a positive integer M divides a non-negative integer N, then M is the greatest common divisor of M and N.

Exercise 858 Using the division with remainder for *positive* integers (theorem 718), state and prove a similar theorem about the division with remainder for each *integer* $N \in \mathbb{Z}$ and each *non-zero* integer $M \in \mathbb{Z}^*$

Exercise 859 Design an algorithm to calculate the greatest common divisor of *three* positive integers.

Exercise 860 Prove that if P is the greatest common divisor of three positive integers L, M, N, then there exist integers I, J, K such that $P = (I * L) + (J * M) + (K * N)$.

5.2 DIGITAL EXPANSION AND ARITHMETIC

5.2.1 Expansion of Integers in Powers of an Integral Base

While the definition of the natural numbers in terms of the set-theoretic notation

$$\varnothing,\quad \{\varnothing\},\quad \{\varnothing, \{\varnothing\}\},\quad \{\varnothing, \{\varnothing\}, \{\varnothing, \{\varnothing\}\}\},\ldots$$

establishes the connections between numbers and such other mathematical concepts as sets, the set-theoretic notation proves inconvenient for calculations, as would also the adoption of a separate symbol for each separate integer, because it would require infinitely many symbols. Instead, a notation relying upon expansions in powers of a fixed positive integer — called a "base" — greatly facilitates notation and calculations.

Commonly used bases include 60 in the Babylonian civilization and 10 in the Chinese and Greek civilizations several millenia ago, a combination of bases 18 and 20 by the Mayans several centuries ago, and, with digital computers since the 1940s, the bases 2 (Cray-1, DEC VAX G), 3 (SETUN, Moscow State University, 1950s), 8, and 16 (IBM 3090) [48, p. 41 & 51]. Yet other representations of numbers have been considered, for instance, the "level index" of Clenshaw, Olver, and Turner [99].

The following theorem guarantees that every integer admits an expansion in powers of any integral base greater than one.

Theorem 732 (Digital expansion of integers) *For every positive integer $B \geq 2$, and for every positive integer N, there exist a unique non-negative integer K and unique non-negative integers $D_0, \ldots, D_K \in \{0, \ldots, B-1\}$ such that $D_K > 0$ and*

$$N = D_K * B^K + D_{K-1} * B^{K-1} + \cdots + D_1 * B^1 + D_0 * B^0.$$

Proof. This proof uses induction with N, with existence and uniqueness in each step.

Initial step For each non-negative integer $N \in \{0, \ldots, B-1\}$,

$$N = N * 1 = N * B^0.$$

Existence Thus $K := 0$ and $D_0 := N$ confirm the existence of the expansion.

Uniqueness If $K > 0$ and $D_K > 0$, then

$$D_K * B^K + D_{K-1} * B^{K-1} + \cdots + D_1 * B^1 + D_0 * B^0 \geq 1 * B^K \geq B > N,$$

so that the expansion does not equal N. Consequently, $K = 0$ is unique; moreover, if $D_0 \neq N$, then $D_0 * B^0 \neq N$; therefore, D_0 is also unique.

Induction hypothesis Let $S \subseteq \mathbb{N}$ denote the set of all the positive integers for which the theorem holds. Assume that $M \in S$ so that the theorem holds for $N := M$.

Induction step Euclidean division of $N := M + 1$ by B gives a unique quotient $Q \in \mathbb{N}$ and a unique non-negative remainder $R < B$, so that $R \in \{0, \ldots, B-1\}$. Hence, setting $D_0 := R$ yields

$$M + 1 = (B * Q) + R = (B * Q) + (R * B^0) = (B * Q) + (D_0 * B^0).$$

Existence From $B \geq 2$ it follows that

$$Q < B * Q = (M + 1) - R \leq M + 1,$$

whence $Q \leq M$ and $Q \in S$. Consequently, there exist unique non-negative integers K and $C_0, \ldots, C_K \in \{0, \ldots, B - 1\}$ such that

$$Q = C_K * B^K + C_{K-1} * B^{K-1} + \cdots + C_1 * B^1 + C_0 * B^0,$$

and a substitution produces

$$\begin{aligned} N &= (B * Q) + R \\ &= B * (C_K * B^K + C_{K-1} * B^{K-1} + \cdots + C_1 * B^1 + C_0 * B^0) + R * B^0 \\ &= C_K * B^{K+1} + C_{K-1} * B^K + \cdots + C_1 * B^2 + C_0 * B^1 + R * B^0, \end{aligned}$$

which demonstrates the existence of an expansion for $N = M + 1$.

Uniqueness Finally, if $N = M + 1$ admits expansions

$$\begin{aligned} N &= D_K * B^K + D_{K-1} * B^{K-1} + \cdots + D_1 * B^1 + D_0 * B^0 \\ &= C_L * B^L + C_{L-1} * B^{L-1} + \cdots + C_1 * B^1 + C_0 * B^0, \end{aligned}$$

then Euclidean division of N by B guarantees the uniqueness of the remainder, whence $D_0 = R = C_0$, and also the uniqueness of the quotient Q, whence

$$\begin{aligned} Q &= D_K * B^{K-1} + D_{K-1} * B^{K-2} + \cdots + D_1 * B^0 \\ &= C_L * B^{L-1} + C_{L-1} * B^{L-2} + \cdots + C_1 * B^0, \end{aligned}$$

and then the induction hypotheses guarantee the uniqueness of the expansion of the quotient Q, whence $K = L$ and $C_J = D_J$ for every $J \in \{0, \ldots, K\}$. □

Definition 733 (Base, digits) The positive integer $B \geq 2$ is the **base** of the expansion, and the non-negative integers $D_0, \ldots, D_K \in \{0, \ldots, B - 1\}$ are the **digits** of the expansion, often written with the notation

$$N = D_K D_{K-1} \cdots D_1 D_0$$

instead of $N = D_K * B^K + D_{K-1} * B^{K-1} + \cdots + D_1 * B^1 + D_0 * B^0$, and with the base B specified in the context or with a subscript.

The proof of theorem 732 shows how to calculate the expansion of any positive integer through successive divisions by the base.

Example 734 To calculate the expansion of 7 with base 2, divide 7 by 2 to get the remainder D_0:

$$7 = (2 * 3) + 1$$

whence $D_0 = R = 1$ and $Q = 3$. Then divide the quotient 3 by 2 to get the remainder D_1:

$$3 = (2 * 1) + 1$$

whence $D_1 = R = 1$. Next, divide the new quotient 1 by the base 2 to get the remainder D_2:

$$1 = (2 * 0) + 1$$

whence $D_2 = R = 1$. With the quotient 0 the algorithm stops. Indeed all subsequent remainders would also equal 0 because $0 = (2 * 0) + 0$, whence all subsequent digits equal zero. Consequently,

$$7_{\text{ten}} = D_2\, D_1\, D_0 = 1\,1\,1_{\text{two}}$$

with respect to the base $B = 2$. As a verification,

$$\begin{aligned} D_2 * B^2 + D_1 * B^1 + D_0 * B_0 &= 1 * 2^2 + 1 * 2^1 + 1 * 2^0 \\ &= 4 + 2 + 1 \\ &= 7. \end{aligned}$$

Expansions with bases B larger than ten require additional symbols. For instance, hexadecimal expansions (with base sixteen) involve the digits 0 through 9 as well as A for ten, B for eleven, C for twelve, D for thirteen, E for fourteen, and F for fifteen.

Example 735 To calculate the expansion of 3575_{ten} with base sixteen, divide 3575_{ten} by 16_{ten} to get the remainder D_0:

$$3575 = (16 * 223) + 7$$

whence $D_0 = R = 7$ and $Q = 223$. Then divide the quotient 223 by 16 to get the remainder D_1:

$$223 = (16 * 13) + 15$$

whence $D_1 = R = 15 = \text{F}$. Next, divide the new quotient 13 by the base 16 to get the remainder D_2:

$$13 = (16 * 0) + 13$$

whence $D_2 = R = 13 = \text{D}$. With the quotient 0 the algorithm stops. Consequently,

$$3575_{\text{ten}} = D_2\, D_1\, D_0 = \text{D}\,\text{F}\,7_{\text{sixteen}}$$

with respect to the base $B = 16$, so that

$$\begin{aligned} 3575_{\text{ten}} &= \text{D}\,\text{F}\,7 \\ &= \text{D} * (16_{\text{ten}})^2 + \text{F} * (16_{\text{ten}})^1 + 7 * (16_{\text{ten}})^0. \end{aligned}$$

As a verification,

$$\begin{aligned} \text{D} * (16_{\text{ten}})^2 + \text{F} * (16_{\text{ten}})^1 + 7 * (16_{\text{ten}})^0 &= 13 * 16^2 + 15 * 16^1 + 7 * 16^0 \\ &= (13 * 16 + 15) * 16 + 7 \\ &= (223) * 16 + 7 \\ &= 3575. \end{aligned}$$

5.2.2 Digital Integer Arithmetic

The common algorithms to perform arithmetic with digits rely on digitial expansions and on the distributivity of multiplication over addition. Moreover, arithmetic operations produce not only sums and products but digital expansions of sums and products.

Example 736 There is no need to perform any arithmetic to determine the sum of 47 and 35, because the inductive definition of integer addition ensures that $47 + 35$ is a well-defined integer. In other words, the sum of 47 and 35 is $47 + 35$. Arithmetic only becomes necessary to determine the *digital expansion* of $47 + 35$. To this end, the common algorithm makes implicit use of commutativity, associativity, and distributivity:

$$\begin{array}{rr} & 47 \\ + & 35 \\ \hline \end{array} \Leftrightarrow \begin{array}{rr} & 4*10+7 \\ + & 3*10+5 \\ \hline \end{array} \Leftrightarrow \begin{array}{rr} & 4*10 \\ + & 3*10+12 \\ \hline \end{array} \Leftrightarrow \begin{array}{rlr} & 1*10 & \\ & 4*10 & \\ + & 3*10 & +2 \\ \hline = & (1+4+3)*10 & +2 \\ = & (8*10) & +2 \\ = & 82 & \end{array}$$

Thus, $47 + 35 = 82$.

In the foregoing example, the sum of the coefficients $1 + 4 + 3 = 8$ of 10 remains less than 10. In other examples, if this sum equals or exceeds 10, then the same procedure would apply to that sum of coefficients.

Example 737 This example shows the arithmetic for the decimal expansion of $86 + 59$.

$$\begin{array}{rr} & 86 \\ + & 59 \\ \hline \end{array} \Leftrightarrow \begin{array}{rr} & 8*10+6 \\ + & 5*10+9 \\ \hline \end{array} \Leftrightarrow \begin{array}{rr} & 8*10 \\ + & 5*10+15 \\ \hline \end{array} \Leftrightarrow \begin{array}{rlr} & 1*10 & \\ & 8*10 & \\ + & 5*10 & +5 \\ \hline = & 14*10 & +5 \\ = & (10+4)*10 & +5 \\ = & 100+4*10 & +5 \end{array}$$

Hence $86 + 59 = 100 + 4 * 10 + 5 = 145$.

The same algorithm works not only with base ten but also with every integral base.

Example 738 With base two, this example shows how to add five (101_{two}) and seven (111_{two}).

$$\begin{array}{rr} & 101 \\ + & 111 \\ \hline \end{array} \Leftrightarrow \begin{array}{rr} & 10+1 \\ + & 11+1 \\ \hline \end{array} \Leftrightarrow \begin{array}{rr} & 01 \\ & 10 \\ + & 11+0 \\ \hline \end{array} \Leftrightarrow \begin{array}{rlr} & 10 & \\ & 10 & \\ + & 10 & +0 \\ \hline = & (1+1+1)*100 & +0 \\ = & (11)*100 & +0 \\ = & 1100 & \end{array}$$

Thus, $101_{two} + 111_{two} = 1100_{two}$. Indeed, $1100_{two} = 2^3 + 2^2 = 12_{ten}$.

Digital arithmetic with any base also provides means to carry out various algorithms, for instance, the calculation of greatest common divisors.

Example 739 Firstly, this example shows how to calculate the greatest common divisors L of $M := 259$ and $N := 343$.

$$\begin{array}{rcrcrcr} N & = & M & * & Q_2 & + & R_2, \\ R_0 & = & R_1 & * & Q_2 & + & R_2, \\ 343 & = & 259 & * & 1 & + & 84; \\ \\ R_1 & = & R_2 & * & Q_3 & + & R_3, \\ 259 & = & 84 & * & 3 & + & \mathbf{7}; \\ \\ R_2 & = & R_3 & * & Q_4 & + & R_4, \\ 84 & = & 7 & * & 12 & + & \mathbf{0}. \end{array}$$

The greatest divisor L is the last non-zero remainder, here $L = R_3 = 7$.

Secondly, this example shows how to calculate integers I and J such that $L = (I * M) + (J * N)$ by solving for the remainders in the foregoing steps.

$$\begin{array}{rcrcrcrcrcrcrcrcl} 343 & = & 259 & * & 1 & + & 84 & \Rightarrow & 343 & - & 259 & * & 1 & = & 84, \\ 259 & = & 84 & * & 3 & + & 7 & \Rightarrow & 259 & - & 84 & * & 3 & = & 7. \end{array}$$

Substituting the expression $343 - (259 * 1)$ for 84 in the last line then gives

$$7 = 259 - (84 * 3) = 259 - \{[343 - (259 * 1)] * 3\} = (259 * 4) - (343 * 3).$$

So $I = 4$ and $J = -3$. Check: $(259 * 4) - (343 * 3) = 1036 - 1029 = 7$.

5.2.3 Exercises

Exercise 861 Express thirty-seven in base two. Check your answer by converting it back into base ten.

Exercise 862 Express five hundred and eleven in base two. Check your answer by converting it back into base ten.

Exercise 863 Express three hundred and twenty-three in base seven. Check your answer by converting it back into base ten.

Exercise 864 Express five hundred and eleven in base seven. Check your answer by converting it back into base ten.

Exercise 865 Express three hundred and twenty-three in base sixteen. Check your answer by converting it back into base ten.

Exercise 866 Express five hundred and eleven in base sixteen. Check your answer by converting it back into base ten.

Exercise 867 Express three thousand five hundred and ninety-nine in base sixty. Check your answer by converting it back into base ten.

Exercise 868 Express three thousand five hundred and thirty-four in base sixty. Check your answer by converting it back into base ten.

For the following exercises, calculate the greatest common divisor L of the given integers M and N, and indicate whether M and N are relatively prime. Then calculate integers I and J such that $L = (I * M) + (J * N)$.

Exercise 869 $M := 133$, $N := 143$.

Exercise 870 $M := 221$, $N := 323$.

Exercise 871 $M := 361$, $N := 527$.

Exercise 872 $M := 249$, $N := 253$.

Exercise 873 $M := 417$, $N := 961$.

Exercise 874 $M := 207$, $N := 702$.

Exercise 875 $M := 533$, $N := 943$.

Exercise 876 $M := 544$, $N := 459$.

Exercise 877 Explain how the common algorithm for *multiplication* relies on digital expansions, commutativity, associativity, and distributivity [62, p. 577–579].

Exercise 878 Explain how the common algorithm for *square roots* relies on digital expansions, commutativity, associativity, and distributivity.

Exercise 879 Explain how the common algorithm for *division* relies on digital expansions, commutativity, associativity, and distributivity.

Exercise 880 Explain how the common algorithm for *cube roots* relies on digital expansions, commutativity, associativity, and distributivity.

5.3 PRIME NUMBERS

5.3.1 Prime Numbers

The concept of "prime number" proves useful in solving problems involving integers, because every integer is the product of prime numbers.

Definition 740 (Prime or composite number) An integer P is **prime** if and only if $|P| > 1$ and the only divisors of P are 1, -1 and P and $-P$.

An integer is **composite** if and only if it is *not* prime, and neither 1 nor -1.

Though some texts include 1 among the primes, excluding 1 from the primes proves more convenient [37]. The following theorems provide examples of prime numbers.

Theorem 741 *The integer* 2 *is prime.*

Proof. If a positive integer M divides 2, then $M * L = 2$ for some positive integer L.

If both $M > 1$ and $L > 1$, then $M * L \geq 2 * 2 > 1 * 2 = 2$, whence M and L cannot both exceed 1. Consequently, at least one of M or L equals 1.

If $M = 1$, then $L = 1 * L = M * L = 2$; similarly, if $L = 1$, then $M = M * 1 = M * L = 2$. Therefore, 1 and 2 are the only divisors of 2. □

Theorem 742 *The integer* 3 *is prime.*

Proof. If a positive integer M divides 3, then $M * L = 3$ for some positive integer L.

If both $M > 1$ and $L > 1$, then $M * L \geq 2 * 2 = 4 > 3$, whence M and L cannot both exceed 1. Consequently, at least one of M or L equals 1.

If $M = 1$, then $L = 1 * L = M * L = 3$; similarly, if $L = 1$, then $M = M * 1 = M * L = 3$. Therefore, 1 and 3 are the only divisors of 3. □

5.3.2 Prime-Number Factorization

A prime number has no divisors other than 1 and itself (definition 740). This definition admits a logically equivalent statement in terms of which integers a prime number divides. Theorem 743 establishes one implication, while exercise 899 shows the converse.

Theorem 743 *For all positive integers $M, N \in \mathbb{N}^*$ if a prime P divides $M * N$, then P divides M or N.*

Proof. If P divides $M * N$, then there exists $K \in \mathbb{N}$ with $P * K = M * N$. Either P divides M or P does not divide M. If P divides M, then P divides M or N.

If the prime P does not divide M, then the greatest common divisor L of P and M is 1. Consequently, the Euclidean algorithm produces integers $I, J \in \mathbb{Z}$ such that

$$1 = (I * P) + (J * M).$$

A multiplication of both sides by N gives

$$\begin{aligned} N &= (N * I * P) + (N * J * M) \\ &= (P * N * I) + (M * N * J) \\ &= (P * N * I) + (P * K * J) \\ &= P * [(N * I) + (K * J)] \end{aligned}$$

for some positive integer K, because P divides $M * N$. Thus P divides N. □

The next theorem establishes the "unique prime factorization" of positive integers.

Theorem 744 (Fundamental Theorem of Arithmetic) *For every positive integer N, there exists a unique non-negative integer M, unique prime numbers $P_1 < P_2 < \cdots < P_{M-1} < P_M$, and unique positive integers $Q_1, \cdots, Q_M$, such that*

$$N = (P_1^{Q_1}) * (P_2^{Q_2}) * \cdots * (P_{M-1}^{Q_{M-1}}) * (P_M^{Q_M}).$$

Proof. This proof established the existence and the uniqueness separately.

Existence This proof of existence proceeds by induction (theorem 480) with N.

Initial step For the special case $N = 1$, the convention about the "empty product" (definition 580) with $M = 0$ factor gives $N = 1$. For $N = 2$, theorem 741 shows that 2 is prime, whence $N = 2 = 2^1$ with $M = 1$, $P_1 = 2$, and $Q_1 = 1$.

Induction hypothesis Let $S \subseteq \mathbb{N}$ be the set of all positive integers K for which the present theorem holds for every $N \in \{1, \ldots, K\}$; for instance, $K = 2 \in S$, as just proved. Assume that the theorem holds for some $K \in S$.

Induction step For $K+1$ either of two cases can arise: either $K+1$ is prime, or $K+1$ is composite.

If $K + 1$ is prime, then $K + 1 = P_1^{Q_1}$ with $M = 1$, $P_1 = K + 1$, and $Q_1 = 1$.

If $K + 1$ is not prime, then there exist positive integers $J, L \notin \{1, K + 1\}$ such that $J * L = K + 1$. Hence $M < K + 1$ and $J < K + 1$, whence $J, L \in S$, and, consequently, the theorem holds for J and for L:

$$\begin{aligned} J &= (P_1^{Q_1}) * (P_2^{Q_2}) * \cdots * (P_{M-1}^{Q_{M-1}}) * (P_M^{Q_M}), \\ L &= (P_1^{R_1}) * (P_2^{R_2}) * \cdots * (P_{M-1}^{R_{M-1}}) * (P_M^{R_M}), \end{aligned}$$

with the lists of primes for J and L combined into one list, with exponents Q_I set to zero if one of the primes appears in only one list. Hence $K+1 \in S$ because

$$\begin{aligned} K+1 &= J * L \\ &= (P_1^{Q_1}) * (P_2^{Q_2}) * \cdots * (P_{M-1}^{Q_{M-1}}) * (P_M^{Q_M}) \\ &\quad *(P_1^{R_1}) * (P_2^{R_2}) * \cdots * (P_{M-1}^{R_{M-1}}) * (P_M^{R_M}) \\ &= (P_1^{Q_1+R_1}) * (P_2^{Q_2+R_2}) * \cdots * (P_{M-1}^{Q_{M-1}+R_{M-1}}) * (P_M^{Q_M+R_M}), \end{aligned}$$

Uniqueness Suppose that

$$\begin{aligned} N &= (P_1^{Q_1}) * (P_2^{Q_2}) * \cdots * (P_{M-1}^{Q_{M-1}}) * (P_M^{Q_M}) \\ &= (R_1^{J_1}) * (R_2^{J_2}) * \cdots * (R_{L-1}^{J_{L-1}}) * (R_L^{J_L}) \end{aligned}$$

for sequences of primes $P_1 < \cdots < P_M$ and $R_1 < \cdots < R_L$.

For each index $I \in \{1, \ldots, M\}$, the prime P_I divides $N = (R_1^{J_1}) * \cdots * (R_L^{J_L})$. Consequently, P_I divides at least one of the factors R_K, by theorem 743. Because R_K is also prime, it follows that $P_I = R_K$. Therefore, $\{P_1, \ldots, P_M\} \subset \{R_1, \ldots, R_L\}$, and swapping the rôles of the sets yields the reverse inclusion, so that $\{P_1, \ldots, P_M\} = \{R_1, \ldots, R_L\}$. Because all the primes are distinct, it also follows that $M = L$.

Finally, successive divisions by P_I show that $Q_I = J_I$. □

Theorem 745 *There exist infinitely many positive prime integers.*

Proof. This proof proceeds by contradiction Thus, assume that there are only a finite number N of positive prime integers, $P_1, \ldots, P_N$. Then define $P := 1 + (P_1 * P_2 * \cdots * P_{N-1} * P_N)$. In particular, $P > P_J$ for every $J \in \{1, \ldots, N\}$.

Contraposition shows that none of the primes $P_1, \ldots, P_N$ divides P. Indeed, each prime number P_J divides the product $(P_1 * P_2 * \cdots * P_{N-1} * P_N)$; consequently, if P_J divided P, then it would follow that P_J would also divide $1 = P - (P_1 * P_2 * \cdots * P_{N-1} * P_N)$ by theorem 727, which would contradict $1 < P_J$.

Thus there would not exist any prime number that divides P, and consequently P would be prime. However, $P > P_J$ for every $J \in \{1, \ldots, N\}$, whence P would not be any of the finitely many primes $P_1, \ldots, P_N$, thus contradicting that they are the only positive prime integers. □

Remark 746 The proof of theorem 745 follows the pattern of the *special* law of *reductio ad absurdum* (described in exercise 196, theorem 195, theorem 651, and axiom I3). Define H as follows:

H**:** There exist only finitely many prime integers.

Then the proof of theorem 745 proceeds along the following pattern:

$\vdash (H) \Rightarrow [\neg(H)]$	proof of theorem 745,
$\vdash \{(H) \Rightarrow [\neg(H)]\} \Rightarrow [\neg(H)]$	*special law of reductio ad absurdum,*
$\vdash \neg(H)$	*Modus Ponens.*

Thus $\neg(H)$ holds: there exist *infinitely* many prime integers.

5.3.3 Primality Testing

A recursive algorithm to determine whether a positive integer N is prime can consist in testing whether M divides N for every *prime* number M such that $M * M \leq N$. In other words, the algorithm can consist in testing, for every prime $M < N$, whether there exists a positive integer Q such that $N = (M * Q) + 0$.

Algorithm 747 For each $N \in \mathbb{N}$ the algorithm tests whether N is prime as follows.
If $N \in \{0, 1\}$, then N is not prime, and the algorithm stops.
If $N \in \{2, 3\}$, then N is prime, and the algorithm stops.
If $N > 3$, then the algorithm proceeds as follows.
Let $M := 2$.
Divide N by M, for instance, through an integer division with remainder.
If M divides N, then N is *not* prime, and the algorithm stops.
If M does *not* divide N, then the algorithm proceeds to the next prime, increasing M by 1 (replacing M by $M + 1$) until M is prime.
If $M * M > N$, then N is prime, and the algorithm stops.
If $M * M \leq N$, then repeat the algorithm from the division of N by M.

This algorithm need not test any composite integer M. Indeed, if M is composite, then the fundamental theorem of arithmetic shows that there exists a prime P that divides M, so that there exists $L \in \mathbb{N}^*$ with $M = P * L$. If M divides N, then $N = (M * Q) + 0 = (P * L) * Q = P * (L * Q)$, so that every prime divisor P of M also divides N.

This algorithm need not test any prime number M such that $M * M > N$, because if such an integer M divides N, so that there exists a positive integer Q with $N = M * Q$, then $Q * Q < N$ and Q has already been tested by the algorithm.

Example 748 To determine whether $N := 5$ is prime, the algorithm begins with the divisor $M := 2$. Because $5 = (2 * 2) + 1$, it follows that 2 does not divide 5.

Because $3 * 3 > 5$ the algorithm need not test any divisor greater than 2. Consequently, 5 does not admit any divisor; therefore, 5 is prime.

Example 749 To determine whether $N := 23$ is prime, the algorithm tests every prime divisor M such that $M * M \leq N$, beginning with the divisors $M := 2$ and $M := 3$:

$$\begin{aligned} 23 &= (2 * 12) + 1, \\ 23 &= (3 * 7) + 2. \end{aligned}$$

Thus neither 2 nor 3 divides 23. The next prime larger than 2 and 3 is 5. Because $5 * 5 > 23$ the algorithm need not test 5 or any divisor greater than 3. Consequently, 23 does not admit any divisor; therefore, 23 is prime.

There also exist other algorithms to test whether an integer is prime.

5.3.4 Exercises

Exercise 881 Prove that 7 is a prime number.

Exercise 882 Prove that 11 is a prime number.

Exercise 883 Prove that 13 is a prime number.

Exercise 884 Prove that 17 is a prime number.

Exercise 885 Prove that 19 is a prime number.

Exercise 886 Prove that 29 is a prime number.

Exercise 887 Prove that 31 is a prime number.

Exercise 888 Prove that 37 is a prime number.

Exercise 889 Prove that 41 is a prime number.

Exercise 890 Prove that 43 is a prime number.

Exercise 891 Prove that 47 is a prime number.

Exercise 892 Prove that 53 is a prime number.

Exercise 893 Prove that the relation | ("divides") is transitive on $\mathbb{N}^*$.

Exercise 894 Prove that the relation | ("divides") is reflexive on $\mathbb{N}^*$.

Exercise 895 Prove that the relation | ("divides") is anti-symmetric on $\mathbb{N}^*$.

Exercise 896 Prove that the relation | ("divides") is *not* symmetric on $\mathbb{N}^*$.

Exercise 897 Prove that the relation | ("divides") is *not* connected on $\mathbb{N}^*$.

Exercise 898 Indicate which of a preorder, partial order, total order, well-ordering, or equivalence, the relation | ("divides") provides on $\mathbb{N}^*$.

Exercise 899 Prove the following converse of theorem 743. For each positive integer $K > 1$, for all non-negative integers $M \in \mathbb{N}$ and $N \in \mathbb{N}$, if K divides $M * N$ only if K divides M or K divides N, then K is prime.

Exercise 900 For each $K \in \mathbb{N}^*$, let $P_1, \ldots, P_K$ be any sequence of K distinct positive prime integers, and let $\prod_K \mathbb{N}$ be the Cartesian product $\mathbb{N} \times \cdots \times \mathbb{N}$ of K factors $\mathbb{N}$.

Prove that $F : \prod_K \mathbb{N} \to \mathbb{N}$ with $F(N_1, \ldots, N_K) := P_1^{N_1} * \cdots * P_K^{N_K}$ is injective. Conclude that $\prod_K \mathbb{N}$ is denumerable: there exists a bijection $\prod_K \mathbb{N} \to \mathbb{N}$.

5.4 MODULAR ARITHMETIC

5.4.1 Modular Integers

Modular arithmetic pertains to mathematical operations on the remainders of divisions, and it applies to the design and calculation of such codes as universal product bar codes and public cryptography. The name "modular" arises from the existence of only finitely many remainders for each divisor. The following example shows how two integers with the same remainder differ by an integral multiple of the divisor.

Example 750 Consider the positive integer $M := 2$ and the integers $K := 19$ and $L := 11$. The integers 19 and 11 have the same remainder, 1, after division by the divisor 2, and their difference equals an integral multiple of the divisor 2:

$$\begin{aligned} 19 &= 2 * 9 + 1, \\ 11 &= 2 * 5 + 1, \\ 19 - 11 &= 2 * 4. \end{aligned}$$

The following theorem establishes a criterion for two integers to lead to the same remainder after division by any divisor.

Theorem 751 *For all $M \in \mathbb{N}^*$, and $K, L \in \mathbb{Z}$, the divisions of K by M and of L by M produce the same remainder R if and only if M divides $K - L$.*

Proof. For all $M \in \mathbb{N}^*$, and $K, L \in \mathbb{Z}$, the divisions of K by M and of L by M produce quotients $P, Q \in \mathbb{Z}$ and remainders $R, S \in \mathbb{N}$ such that

$$\begin{aligned} K &= (M * P) + R, \\ M &> R \geq 0; \\ L &= (M * Q) + S, \\ M &> S \geq 0. \end{aligned}$$

Hence,

$$\begin{aligned} K - L &= [(M * P) + R] - [(M * Q) + S] \\ &= [M * (P - Q)] + (R - S). \end{aligned}$$

Thus, if $R = S$, then $R - S = 0$ and M divides $K - L = M * (P - Q)$.

Conversely, if M divides $K - L = [M * (P - Q)] + (R - S)$, then M also divides $R - S$, but $|R - S| < M$, whence $R - S = 0$. □

The remainder of a division thus gives a mathematical relation on the integers.

Definition 752 (Congruence modulo divisors) For each positive integer M, the relation of **congruence modulo** M, denoted by $\underset{M}{\equiv}$, relates to one another integers with the same remainder after division by M. Thus, for all integers $K, L \in \mathbb{Z}$,

$$(K \underset{M}{\equiv} L) \Leftrightarrow (M | [K - L]) .$$

The positive integer M is called the **divisor** in Euler's terminology and the **modulus** in Weil's terminology [104, p. 194].

Example 753 For the positive divisor $M := 2$, the integers $K := 19$ and $L := 11$ are congruent modulo 2, because 2 also divides their difference:

$$\begin{aligned} 19 - 11 &= 2 * 4, \\ 19 &\underset{2}{\equiv} 11. \end{aligned}$$

Example 754 For the positive divisor $M := 2$, the integers $K := 18$ and $L := 11$ are *not* congruent modulo 2, because 2 does *not* divide their difference:

$$\begin{aligned} 18 - 11 &= 2 * 3 + 1, \\ 18 &\underset{2}{\not\equiv} 11. \end{aligned}$$

The following theorem shows that congruence modulo an integer is an equivalence relation (as defined in section 2.7).

Theorem 755 *For each $M \in \mathbb{N}^*$, the relation $\underset{M}{\equiv}$ is an equivalence relation on $\mathbb{Z}$.*

Proof. This proof verifies that $\underset{M}{\equiv}$ is reflexive, symmetric, and transitive.

Reflexivity $N \underset{M}{\equiv} N$ because $N - N = 0 * M$ for each integer $N \in \mathbb{Z}$.

Symmetry If $K \underset{M}{\equiv} L$, then $K - L = M * Q$, whence $L - K = M * (-Q)$, and, consequently, $L \underset{M}{\equiv} K$.

Transitivity If $J \underset{M}{\equiv} K$ and $K \underset{M}{\equiv} L$, then $J - K = M * P$ and $K - L = M * Q$, whence

$$J - L = (J - K) - (K - L) = (M * P) - (M * Q) = M * (P - Q),$$

and, consequently, $J \underset{M}{\equiv} L$. □

With $0 \leq R < M$, only M possible values exist for the remainder: $R \in \{0, \ldots, M-1\}$.

Definition 756 (Integers modulo M) The equivalence classes $[0]_M, \ldots, [M-1]_M$ are called the **integers modulo** M, and their set is denoted by the symbol $\mathbb{Z}_M$:

$$\mathbb{Z}_M = \{[0]_M, \ldots, [M-1]_M\}.$$

Example 757

For $M := 2$, $\mathbb{Z}_2 = \{[0]_2, [1]_2\}$.
For $M := 3$, $\mathbb{Z}_3 = \{[0]_3, [1]_3, [2]_3\}$.
For $M := 4$, $\mathbb{Z}_4 = \{[0]_4, [1]_4, [2]_4, [3]_4\}$.
For $M := 5$, $\mathbb{Z}_5 = \{[0]_5, [1]_5, [2]_5, [3]_5, [4]_5\}$.

5.4.2 Modular Addition

This subsection demonstrates the effect of addition of integers on their remainders after division, which amounts to a new type of addition on the remainders. The first theorem shows that any two members of a common equivalence class lead to the same result.

Theorem 758 *For each positive integer M, and for all integers $I, J, K, L \in \mathbb{Z}$, if*

$$\begin{array}{rcl} I & \underset{M}{\equiv} & J, \\ K & \underset{M}{\equiv} & L, \end{array}$$

then

$$(I + K) \underset{M}{\equiv} (J + L).$$

Proof. From the hypotheses

$$\begin{array}{rcl} I & \underset{M}{\equiv} & J, \\ K & \underset{M}{\equiv} & L, \end{array}$$

there exist integers $P, Q \in \mathbb{Z}$ such that

$$\begin{array}{rcl} I - J & = & P * M, \\ K - L & = & Q * M. \end{array}$$

Consequently,

$$\begin{array}{rcl} (I + K) - (J + L) & = & (I - J) + (K - L) \\ & = & (P * M) + (Q * M) \\ & = & (P + Q) * M, \end{array}$$

whence

$$(I + K) \underset{M}{\equiv} (J + L).$$

□

Definition 759 (Addition modulo M) For each positive integer M define the **addition modulo** M in the integers $\mathbb{Z}_M$ modulo M by

$$\oplus : \mathbb{Z}_M \times \mathbb{Z}_M \to \mathbb{Z}_M,$$

$$[I]_M \oplus [K]_M := [I + K]_M.$$

Example 760 Table 761 displays the addition table for the positive divisor $M := 2$.

Table 761 Addition modulo two.

$\oplus$	$[0]_2$	$[1]_2$
$[0]_2$	$[0]_2$	$[1]_2$
$[1]_2$	$[1]_2$	$[0]_2$

Example 762 Table 763 displays the addition table for the positive divisor $M := 3$.

Table 763 Addition modulo three.

$\oplus$	$[0]_3$	$[1]_3$	$[2]_3$
$[0]_3$	$[0]_3$	$[1]_3$	$[2]_3$
$[1]_3$	$[1]_3$	$[2]_3$	$[0]_3$
$[2]_3$	$[2]_3$	$[0]_3$	$[1]_3$

Theorem 764 *For each $M \in \mathbb{N}^*$, the addition modulo M is associative and commutative, with the identity $[0]_M$. Moreover, $[K]_M + [M-K]_M = [0]_M$ for each $K \in \{0, \ldots, M-1\}$.*

Proof. For each positive integer M and for all integers $I, K, P \in \mathbb{Z}$,

$$\begin{aligned}
[0]_M \oplus [I]_M &= [0+I]_M = [I]_M = [I+0]_M = [I]_M \oplus [0]_M; \\
[I]_M \oplus [K]_M &= [I+K]_M = [K+I]_M = [K]_M \oplus [I]_M; \\
([I]_M \oplus [K]_M) \oplus [P]_M &= [I+K]_M \oplus [P]_M \\
&= [(I+K)+P]_M \\
&= [I+(K+P)]_M \\
&= [I]_M \oplus [K+P]_M \\
&= [I]_M \oplus ([K]_M \oplus [P]_M); \\
[K]_M + [M-K]_M &= [K+(M-K)]_M = [M]_M = [0]_M.
\end{aligned}$$

□

Definition 765 A **modular additive opposite** for an integer $[K]_M$ modulo M is an integer $[L]_M$ modulo M such that

$$[K]_M \oplus [L]_M = [0]_M.$$

Such an integer $[L]_M$ is also denoted by $-[K]_M$.

Example 766 With $M := 2$, from $[1]_2 \oplus [1]_2 = [0]_2$ it follows that $-[1]_2 = [1]_2$.

Example 767 With $M := 3$, from $[1]_3 \oplus [2]_3 = [0]_3$ it follows that $-[1]_3 = [2]_3$.

One algorithm to calculate an additive opposite consists of a subtraction from the divisor, followed if necessary by a division with remainder.

Example 768 With $M := 7$ and $[K]_M = [3]_7$, to calculate $-[3]_7$ subtract 3 from 7, which gives $7 - 3 = 4$. Hence $-[3]_7 = [4]_7$. Indeed, $[3]_7 \oplus [4]_7 = [7]_7 = [0]_7$.

5.4.3 Modular Multiplication

This subsection shows the effect of multiplication of integers on their remainders after division, which amounts to a new multiplication on the remainders. The first theorem shows that any two members of a common equivalence class lead to the same result.

Theorem 769 *For each positive integer M, and for all integers $I, J, K, L \in \mathbb{Z}$, if*

$$\begin{array}{rcl} I & \underset{M}{\equiv} & J, \\ K & \underset{M}{\equiv} & L, \end{array}$$

then

$$(I * K) \underset{M}{\equiv} (J * L).$$

Proof. From the hypotheses

$$\begin{array}{rcl} I & \underset{M}{\equiv} & J, \\ K & \underset{M}{\equiv} & L, \end{array}$$

there exist integers $P, Q \in \mathbb{Z}$ such that

$$\begin{array}{rcl} I - J & = & P * M, \\ K - L & = & Q * M. \end{array}$$

Consequently,

$$\begin{array}{rcl} (I * K) - (J * L) & = & (I - J) * K - J * (K - L) \\ & = & (P * M) * K + J * (Q * M) \\ & = & (P * K + J * Q) * M, \end{array}$$

whence

$$(I * K) \underset{M}{\equiv} (J * L).$$

□

Definition 770 (Multiplication modulo M) For each positive integer M define the **multiplication modulo M** in the integers $\mathbb{Z}_M$ modulo M by

$$\begin{array}{c} \circledast : \mathbb{Z}_M \times \mathbb{Z}_M \to \mathbb{Z}_M, \\ [I]_M \circledast [K]_M := [I * K]_M. \end{array}$$

Example 771 Table 772 shows the multiplication table for the divisor $M := 2$.

Table 772 Multiplication modulo two.

$\circledast$	$[0]_2$	$[1]_2$
$[0]_2$	$[0]_2$	$[0]_2$
$[1]_2$	$[0]_2$	$[1]_2$

Example 773 Table 774 shows the multiplication table for the divisor $M := 3$.

Table 774 Multiplication modulo three.

$\circledast$	$[0]_3$	$[1]_3$	$[2]_3$
$[0]_3$	$[0]_3$	$[0]_3$	$[0]_3$
$[1]_3$	$[0]_3$	$[1]_3$	$[2]_3$
$[2]_3$	$[0]_3$	$[2]_3$	$[1]_3$

Theorem 775 *For each $M \in \mathbb{N}^*$, the multiplication modulo M is associative and commutative, with $[1]_M$ as the identity, and it distributes over addition modulo M.*

Proof. For each positive integer M and for all integers $I, K, P \in \mathbb{Z}$,

$$[1]_M \circledast [I]_M = [1 * I]_M = [I]_M = [I * 1]_M = [I]_M \circledast [1]_M,$$

$$[I]_M \circledast [K]_M = [I * K]_M = [K * I]_M = [K]_M \circledast [I]_M,$$

$$\begin{aligned}&([I]_M \circledast [K]_M) \circledast [P]_M = [I * K]_M \circledast [P]_M = [(I * K) * P]_M \\ &= [I * (K * P)]_M = [I]_M \circledast [K * P]_M = [I]_M \circledast ([K]_M \circledast [P]_M),\end{aligned}$$

$$\begin{aligned}&([I]_M \oplus [K]_M) \circledast [P]_M = [I + K]_M \circledast [P]_M = [(I + K) * P]_M \\ &= [I * P + K * P]_M = [I * P]_M \oplus [K * P]_M = ([I]_M \circledast [P]_M) \oplus ([K]_M \circledast [P]_M).\end{aligned}$$

□

5.4.4 Modular Division

This subsection shows the effect of division of integers on their remainders after division, which amounts to a new type of division on the remainders, but which exists only in special circumstances.

Definition 776 A **modular multiplicative inverse**, also called a **modular reciprocal**, for an integer $[K]_M$ modulo M is an integer $[L]_M$ modulo M such that

$$[K]_M \circledast [L]_M = [1]_M.$$

Example 777 With $M := 7$, from $[2]_7 \circledast [4]_7 = [8]_7 = [1]_7$ it follows that $[2]_7^{-1} = [4]_7$.

Every modular integer has at most one modular reciprocal, perhaps none.

Theorem 778 *For each positive integer M, each integer $[K]_M$ modulo M has at most one modular multiplicative inverse.*

Proof. If $[P]_M$ and $[Q]_M$ are multiplicative inverses for an integer $[K]_M$, then

$$[K]_M \circledast [P]_M = [1]_M = [K]_M \circledast [Q]_M.$$

Consequently,

$$\begin{aligned}[P]_M &= [P]_M \circledast [1]_M \\ &= [P]_M \circledast ([K]_M \circledast [Q]_M) \\ &= ([P]_M \circledast [K]_M) \circledast [Q]_M \\ &= [1]_M \circledast [Q]_M \\ &= [Q]_M.\end{aligned}$$

□

The following theorem states a condition for the existence of a modular reciprocal.

Theorem 779 *For each positive integer M, an integer $[K]_M$ modulo M has a multiplicative inverse if and only if the integers K and M are relatively prime.*

Proof. For all positive integers K and M, $[K]_M$ has a multiplicative inverse if and only if there exists an integer P such that $[K]_M * [P]_M = [1]_M$, so that $[K]_M * [P]_M - [1]_M = [0]_M$, or, in yet other words, if and only if there exists an integer N such that

$$(K * P) - 1 = N * M,$$
$$(K * P) - (N * M) = 1.$$

If there exists such an integer N, then every common divisor of K and M also divides $(K * P) - (N * M)$, so that it divides 1, and, consequently, 1 is the only common divisor, which means that K and M are relatively prime.

Conversely, if K and M are relatively prime, then the Euclidean Algorithm (theorem 728) produces integers P and Q such that $(K * P) + (Q * M) = 1$, whence setting $N := -Q$ yields $(K * P) - (N * M) = 1$, so that $[K]_M * [P]_M = [1]_M$. □

Theorem 780 *For each prime integer M, every element in $\mathbb{Z}_M \setminus \{[0]_M\}$ has a multiplicative inverse in $\mathbb{Z}_M$.*

Proof. If M is prime then M and K are relatively prime for each integer K. □

Definition 781 For each $M \in \mathbb{N}^*$, and for each integer K relatively prime to M, the modular reciprocal for $[K]_M$ modulo M is denoted by $[K]_M^{-1}$ or by $1/[K]_M$; thus,

$$[K]_M \circledast ([K]_M^{-1}) = [1]_M.$$

Example 782

- For each positive divisor M and for the integer $K := 1$, which are relatively prime, $[1]_M^{-1} = [1]_M$, because $[1]_M \circledast [1]_M = [1]_M$.
- For the positive divisor $M := 3$ and for the integer $K := 2$, which are relatively prime, $[2]_3^{-1} = [2]_3$, because $[2]_3 \circledast [2]_3 = [4]_3 = [3 + 1]_3 = [1]_3$.
- For the positive divisor $M := 9$ and for the integer $K := 4$, which are relatively prime, $[4]_9^{-1} = [7]_9$, because $[4]_9 \circledast [7]_9 = [28]_9 = [27 + 1]_9 = [1]_9$.

Counterexample 783 For $M := 9$ and $K := 6$, which are *not* relatively prime, $[6]_9^{-1}$ does *not* exist, because $[6]_9 \circledast [N]_9 \neq [1]_9$ for every $N \in \mathbb{Z}$. Indeed,

$$[6]_9 \circledast [N]_9 = ([3]_9 \circledast [2]_9) \circledast [N]_9 = [3]_9 \circledast ([2]_9 \circledast [N]_9)$$

whence a modular multiplication by $[3]_9$ gives

$$\begin{aligned} [3]_9 \circledast \{[6]_9 \circledast [N]_9\} &= [3]_9 \circledast \{[3]_9 \circledast ([2]_9 \circledast [N]_9)\} \\ &= ([3]_9 \circledast [3]_9) \circledast ([2]_9 \circledast [N]_9) \\ &= [9]_9 \circledast ([2]_9 \circledast [N]_9) \\ &= [0]_9 \circledast ([2]_9 \circledast [N]_9) \\ &= [0]_9. \end{aligned}$$

Hence $[6]_9 \circledast [N]_9 \neq [1]_9$ because $[3]_9 \circledast [1]_9 = [3]_9 \neq [0]_9$. Therefore $[6]_9$ does not have any modular reciprocal.

Algorithm 784 One algorithm to calculate modular reciprocals relies on the Euclidean algorithm (theorem 728), which states that for all positive integers M and N, there exists a greatest common divisor L, and integers I and J such that

$$(I * M) + (J * N) = L.$$

If M and N are relatively prime, then $L = 1$ and hence

$$(I * M) + (J * N) = 1.$$

Moreover, $[M]_M = [0]_M$, whence

$$\begin{aligned}(I * M) + (J * N) &= 1, \\ ([I]_M \circledast [M]_M) \oplus ([J]_M \circledast [N]_M) &= [1]_M, \\ ([I]_M \circledast [0]_M) \oplus ([J]_M \circledast [N]_M) &= [1]_M, \\ [J]_M \circledast [N]_M &= [1]_M.\end{aligned}$$

Therefore, $[N]_M^{-1} = [J]_M$.

Example 785 With $M := 5$ and $N := 7$, example 729 showed that

$$(3 * 5) + [(-2) * 7] = 1.$$

Thus $I = 3$ and $J = -2$. Consequently,

$$[7]_5^{-1} = [J]_5 = [-2]_5 = [5 - 2]_5 = [3]_5.$$

Indeed, $[7]_5 \circledast [3]_5 = [2]_5 \circledast [3]_5 = [6]_5 = [5 + 1]_5 = [1]_5$.

Counterexample 786 With $M := 259$ and $N := 343$, example 739 produced the greatest common divisor $7 \neq 1$. Therefore, $[343]_{259}$ does not have any reciprocal.

5.4.5 Divisibility Tests

Modular arithmetic provides algorithms to determine whether an integer M divides an integer N. Some of the resulting algorithms depend on the base of digital expansions. For instance, the following theorem shows that an integer is even — in other words, divisible by two — if and only if the last digit is even, provided that the base is even.

Theorem 787 *If* 2 *divides a base* B, *then* 2 *divides an integer*

$$N = D_K * B^K + \cdots + D_0 * B^0$$

if and only if 2 *divides* D_0.

Proof. If 2 divides B, then $B \underset{2}{\equiv} 0$ modulo 2. Consequently,

$$[B^L]_2 = ([B]_2)^L = ([0]_2)^L = [0]_2$$

for every positive integral power L. Therefore,

$$\begin{aligned}[N]_2 &= [D_K * B^K + D_{K-1} * B^{K-1} + \cdots + D_1 * B^1 + D_0 * B^0]_2 \\ &= [D_K * B^K]_2 + [D_{K-1} * B^{K-1}]_2 + \cdots + [D_1 * B^1]_2 + [D_0 * B^0]_2 \\ &= [D_K]_2 * [B^K]_2 + \cdots + [D_1]_2 * [B^1]_2 + [D_0]_2 * [B^0]_2 \\ &= [D_K]_2 * [0]_2 + [D_{K-1}]_2 * [0]_2 + \cdots + [D_1]_2 * [0]_2 + [D_0]_2 * [1]_2 \\ &= [D_0]_2.\end{aligned}$$

Thus $[N]_2 = [D_0]_2$, so that 2 divides N if and only if 2 divides D_0. □

Example 788 The integer 234_{ten} is even because 4 is even: $234_{\text{ten}} = 2 * 117_{\text{ten}}$

The test does not apply to odd bases.

Counterexample 789 The integer 234_{seven} is *not* even because $234_{\text{seven}} = 2 * 7^2 + 3 * 7^1 + 4 * 7^0 = 123_{\text{ten}}$ is odd.

The next theorem gives a test to determine whether an integer is divisible by seven.

Theorem 790 *With base ten, seven divides an integer*

$$N = D_K * 10^K + D_{K-1} * 10^{K-1} + \cdots + D_1 * 10^1 + D_0 * B^0$$

if and only if

$$\begin{aligned} 0 \underset{7}{\equiv} & \cdots + D_{12} * 5 + D_{11} * 4 + D_{10} * 6 + D_9 * 2 + D_8 * 3 + D_7 * 1 + D_6 * 1 \\ & + D_5 * 5 + D_4 * 4 + D_3 * 6 + D_2 * 2 + D_1 * 3 + D_0 * 1. \end{aligned}$$

Proof. Verify that

$$\begin{aligned} 10^0 &\underset{7}{\equiv} 1, \\ 10^1 &\underset{7}{\equiv} 3, \\ 10^2 &\underset{7}{\equiv} 2, \\ 10^3 &\underset{7}{\equiv} 6, \\ 10^4 &\underset{7}{\equiv} 4, \\ 10^5 &\underset{7}{\equiv} 5, \\ 10^6 &\underset{7}{\equiv} 1, \end{aligned}$$

and hence the sequence has a period of length six, because

$$10^{6*J+I} \underset{7}{\equiv} 10^{6*J} * 10^I \underset{7}{\equiv} (10^6)^J * 10^I \underset{7}{\equiv} (1)^J * 10^I \underset{7}{\equiv} 1 * 10^I \underset{7}{\equiv} 10^I,$$

so that

$$\begin{aligned} 10^{6*J+0} &\underset{7}{\equiv} 1, \\ 10^{6*J+1} &\underset{7}{\equiv} 3, \\ 10^{6*J+2} &\underset{7}{\equiv} 2, \\ 10^{6*J+3} &\underset{7}{\equiv} 6, \\ 10^{6*J+4} &\underset{7}{\equiv} 4, \\ 10^{6*J+5} &\underset{7}{\equiv} 5, \\ 10^{6*J+6} &\underset{7}{\equiv} 1. \end{aligned}$$

Therefore,

$$\begin{aligned} N \underset{7}{\equiv} & \cdots + D_{12} * 5 + D_{11} * 4 + D_{10} * 6 + D_9 * 2 + D_8 * 3 + D_7 * 1 + D_6 * 1 \\ & + D_5 * 5 + D_4 * 4 + D_3 * 6 + D_2 * 2 + D_1 * 3 + D_0 * 1. \end{aligned}$$

□

Example 791 The integer 98_{ten} is divisible by seven because

$$9 * 10^1 + 8 * 10^0 \underset{7}{\equiv} 9 * 3 + 8 * 1 \underset{7}{\equiv} 2 * 3 + 1 * 1 \underset{7}{\equiv} 7 \underset{7}{\equiv} 0.$$

Indeed, $98_{\text{ten}} = 7 * 14_{\text{ten}}$.

The test does not apply to other bases.

Counterexample 792 The integer 98_{sixteen} is *not* divisible by seven because $98_{\text{sixteen}} = 9 * 16^1 + 8 * 16^0 = 152_{\text{ten}} \underset{7}{\equiv} 1 * 2 + 5 * 3 + 2 * 1 \underset{7}{\equiv} = 19 \underset{7}{\equiv} 5$ is *not* divisible by 7.

5.4.6 Exercises

For the following exercises, determine whether the given integer N has a modular reciprocal modulo the given divisor M, and, if so, calculate $[N]_M^{-1}$.

Exercise 901 $M := 3$, $N := 8$.

Exercise 902 $M := 4$, $N := 9$.

Exercise 903 $M := 91$, $N := 51$.

Exercise 904 $M := 16$, $N := 15$.

Exercise 905 $M := 133$, $N := 143$.

Exercise 906 $M := 221$, $N := 323$.

Exercise 907 $M := 361$, $N := 527$.

Exercise 908 $M := 249$, $N := 253$.

Exercise 909 $M := 417$, $N := 961$.

Exercise 910 $M := 207$, $N := 702$.

Exercise 911 $M := 533$, $N := 943$.

Exercise 912 $M := 544$, $N := 459$.

Exercise 913 Design a test to determine whether an integer is divisible by five.

Exercise 914 Prove that $10^p \equiv 1$ $(modulo\, 3)$ for each positive integer p. Recall that the decimal expansion $n = d_k d_{k-1} d_{k-2} \ldots d_2 d_1 d_0$ of a positive integer n means that

$$n = d_k 10^k + d_{k-1} 10^{k-1} + d_{k-2} 10^{k-2} + \cdots + d_2 10^2 + d_1 10 + d_0 \cdot 1.$$

Prove that $3|n$ if and only if $d_k + d_{k-1} + d_{k-2} + \cdots + d_2 + d_1 + d_0 \underset{3}{\equiv} 0$.

Exercise 915 For each positive integer p, prove that if p is odd, then $10^p \underset{11}{\equiv} -1$ $(modulo\, 11)$, whereas if p is even, then $10^p \underset{11}{\equiv} 1$ $(modulo\, 11)$. Design a criterion for divisibility by 11, similar to the one just obtained for divisibility by 3.

Exercise 916 Design a criterion for divisibility by 9.

Exercise 917 Design another algorithm to calculate modular multiplicative inverses. For all relatively prime positive integer M and K, the algorithm shall produce an integer L such that $[K]_M \circledast [L]_M = [1]_M$.

Exercise 918 Consider the sets

$$\begin{aligned} A &:= \mathbb{Z}_4 = \{0/\underset{4}{\equiv}, 1/\underset{4}{\equiv}, 2\underset{4}{\equiv}, 3\underset{4}{\equiv}\} = \{[0]_4, [1]_4, [2]_4, [3]_4\}, \\ B &:= \mathbb{Z}_2 = \{0/\underset{2}{\equiv}, 1/\underset{2}{\equiv}\} = \{[0]_2, [1]_2\}, \end{aligned}$$

and consider the relation $R \subseteq A \times B = \mathbb{Z}_4 \times \mathbb{Z}_2$ defined by

$$\begin{aligned} R &:= \{(n/\underset{4}{\equiv}, n/\underset{2}{\equiv}) : (n/\underset{4}{\equiv} \in \mathbb{Z}_4) \wedge (n/\underset{2}{\equiv} \in \mathbb{Z}_2)\} \\ &= \{([n]_4, [n]_2) : ([n]_4 \in \mathbb{Z}_4) \wedge ([n]_2 \in \mathbb{Z}_2)\}. \end{aligned}$$

Determine whether this relation R is a function.

Exercise 919 Consider the sets

$$\begin{aligned} A &:= \mathbb{Z}_2 = \{0/\underset{2}{\equiv}, 1/\underset{2}{\equiv}\} = \{[0]_2, [1]_2\}, \\ B &:= \mathbb{Z}_4 = \{0/\underset{4}{\equiv}, 1/\underset{4}{\equiv}, 2\underset{4}{\equiv}, 3\underset{4}{\equiv}\} = \{[0]_4, [1]_4, [2]_4, [3]_4\}, \end{aligned}$$

and consider the relation $S \subseteq A \times B = \mathbb{Z}_2 \times \mathbb{Z}_4$ defined by

$$\begin{aligned} S &:= \{(n/\underset{2}{\equiv}, n/\underset{4}{\equiv}) : (n/\underset{2}{\equiv} \in \mathbb{Z}_2) \wedge (n/\underset{4}{\equiv} \in \mathbb{Z}_4)\} \\ &= \{([n]_2, [n]_4) : ([n]_2 \in \mathbb{Z}_2) \wedge ([n]_4 \in \mathbb{Z}_4)\}. \end{aligned}$$

Determine whether this relation S is a function.

Exercise 920 Design a test on P and Q to determine whether the following relation is a function: $\{([N]_P, [N]_Q) : N \in \mathbb{Z}\} \subseteq \mathbb{Z}_P \times \mathbb{Z}_Q$.

5.5 MODULAR CODES

5.5.1 The International Standard Book Number (ISBN) Code

The International Standard Book Number (ISBN) Code identifies each book by a sequence of ten digits

$$d_{10}d_9d_8d_7d_6d_5d_4d_3d_2d_1$$

with decimal digits in the left-most nine positions,

$$d_{10}, d_9, d_8, d_7, d_6, d_5, d_4, d_3, d_2 \in \{0, 1, 2, 3, 4, 5, 6, 7, 8, 9\},$$

and a check digit in base eleven in the right-most position,

$$d_1 \in \{0, 1, 2, 3, 4, 5, 6, 7, 8, 9, X\}.$$

The International Standard Book Number Code defines the check digit d_1 so that

$$10 * d_{10} + 9 * d_9 + 8 * d_8 + 7 * d_7 + 6 * d_6 + 5 * d_5 + 4 * d_4 + 3 * d_3 + 2 * d_2 + d_1 \underset{11}{\equiv} 0.$$

In other words,

$$d_1 \underset{11}{\equiv} -(10d_{10} + 9d_9 + 8d_8 + 7d_7 + 6d_6 + 5d_5 + 4d_4 + 3d_3 + 2d_2).$$

An additional digit becomes necessary to denote ten if the right-hand side gives the equivalence class of ten modulo eleven; any symbol will do, for instance, X.

Example 793 This example demonstrates how to *verify* an International Standard Book Number. Francis Harry Hinsley and Alan Stripp's book *Codebreakers: The inside story of Bletchley Park* [51] has the International Standard Book Number

$$0\,1\,9\,2\,8\,5\,3\,0\,4\,X$$

and indeed

$$\begin{aligned} & 10 * 0 + 9 * 1 + 8 * 9 + 7 * 2 + 6 * 8 + 5 * 5 + 4 * 3 + 3 * 0 + 2 * 4 + 1 * X \\ = \; & 0 + 9 + 72 + 14 + 48 + 25 + 12 + 0 + 8 + 10 \\ \underset{11}{\equiv} \; & 0 + 9 + 6 + 3 + 4 + 3 + 1 + 0 + 8 + 10 \\ \underset{11}{\equiv} \; & (9 + 6) + (3 + 4 + 3 + 1) + (8 + 10) \\ \underset{11}{\equiv} \; & 4 + 0 + 7 \\ \underset{11}{\equiv} \; & 0. \end{aligned}$$

Example 794 This example shows how to *calculate* the check digit for an International Standard Book Number. For Friedrich L. Bauer's book *Decrypted Secrets: methods and maxims of cryptology* [2] the first nine digits of the ISBN are

$$3\,5\,4\,0\,6\,0\,4\,1\,8\,d_1.$$

To calculate the check digit d_1 calculate the additive opposite

$$\begin{aligned}
d_1 &\underset{11}{\equiv} -(10d_{10} + 9d_9 + 8d_8 + 7d_7 + 6d_6 + 5d_5 + 4d_4 + 3d_3 + 2d_2) \\
&\underset{11}{\equiv} -(10*3 + 9*5 + 8*4 + 7*0 + 6*6 + 5*0 + 4*4 + 3*1 + 2*8) \\
&\underset{11}{\equiv} -(30 + 45 + 32 + 0 + 36 + 0 + 16 + 3 + 16) \\
&\underset{11}{\equiv} -(8 + 1 + 10 + 0 + 3 + 0 + 5 + 3 + 5) \\
&\underset{11}{\equiv} -(8 + (1 + 10) + 0 + 3 + 0 + (5 + 3 + 5)) \\
&\underset{11}{\equiv} -(2)\,.
\end{aligned}$$

Thus $[d_1]_{11} = -[2]_{11} = [9]_{11}$ and hence $d_1 = 9$. Indeed the number is

$$3\,5\,4\,0\,6\,0\,4\,1\,8\,9.$$

In the International Standard Book Number (ISBN) Code, the check digit allows for the detection of every single substitution of an erroneous digit, or every transposition (swap) of two digits, in the left-most nine digits.

5.5.2 The Universal Product Code (UPC)

The Universal Code Council, Inc. offers several versions of the Universal Product Code (UPC). One version consists of 12 digits

$$d_1 d_2 d_3 d_4 d_5 d_6 d_7 d_8 d_9 d_{10} d_{11} d_{12}.$$

All digits $d_i \in \{0, \ldots, 9\}$ are decimal digits. The twelfth digit d_{12} plays the role of a check digit, determined so that the digits satisfy the following equation modulo 10:

$$3d_1 + 1d_2 + 3d_3 + 1d_4 + 3d_5 + 1d_6 + 3d_7 + 1d_8 + 3d_9 + 1d_{10} + 3d_{11} + 1d_{12} \underset{10}{\equiv} 0.$$

Example 795 This example demonstrates how to verify a UPC number, for instance,

$$0\,8\,9\,5\,8\,1\,0\,1\,1\,0\,2\,5.$$

Indeed,

$$\begin{aligned}
& 3*\mathbf{0} + 1*\mathbf{8} + 3*\mathbf{9} + 1*\mathbf{5} + 3*\mathbf{8} + 1*\mathbf{1} \\
& \qquad +3*\mathbf{0} + 1*\mathbf{1} + 3*\mathbf{1} + 1*\mathbf{0} + 3*\mathbf{2} + 1*\mathbf{5} \\
\underset{10}{\equiv}\ & 0 + 8 + 27 + 5 + 24 + 1 + 0 + 1 + 3 + 0 + 6 + 5 \\
\underset{10}{\equiv}\ & 0 + 8 + 7 + 5 + 4 + 1 + 0 + 1 + 3 + 0 + 6 + 5 \\
\underset{10}{\equiv}\ & 40 \\
\underset{10}{\equiv}\ & 0.
\end{aligned}$$

5.5.3 The Bank Identification Code

The *Bank Identification Code,* which appears on banks' checks and transaction slips, consists of a nine-digit decimal number, $b_1 b_2 b_3 b_4 b_5 b_6 b_7 b_8 b_9$ with each $b_i \in \{0, \ldots, 9\}$, of which the ninth digit b_9 constitutes a "check digit" so that

$$7b_1 + 3b_2 + 9b_3 + 7b_4 + 3b_5 + 9b_6 + 7b_7 + 3b_8 - b_9 \underset{10}{\equiv} 0.$$

5.5.4 The U.S. postal nine-digit ZIP Code

The U.S. postal nine-digit ZIP code $d_1d_2d_3d_4d_5\ d_6d_7d_8d_9$ may also appear on pieces of mail as a ten-digit bar code $d_1d_2d_3d_4d_5d_6d_7d_8d_9d_{10}$. All digits $d_i \in \{0, \ldots, 9\}$ are decimal digits. The additional tenth digit d_{10} plays the role of a check digit, so that

$$d_1 + d_2 + d_3 + d_4 + d_5 + d_6 + d_7 + d_8 + d_9 + d_{10} \underset{10}{\equiv} 0.$$

For additional references about such modular codes, see [30], [71], [72].

5.5.5 Exercises

Exercise 921 Verify the ISBN 0691029067.

Exercise 922 Verify the ISBN 0486669807.

Exercise 923 Verify the ISBN 3540121595.

Exercise 924 Verify the ISBN 3211828397.

Exercise 925 Calculate the check digit d_1 for the ISBN 038797878d_1.

Exercise 926 Calculate the check digit d_1 for the ISBN 007069180d_1.

Exercise 927 Calculate the check digit d_1 for the ISBN 354013610d_1.

Exercise 928 Calculate the check digit d_1 for the ISBN 038713610d_1.

Exercise 929 Verify the UPC number 040000257318.

Exercise 930 Verify the UPC number 074970051607.

Exercise 931 Verify the UPC number 077652082272.

Exercise 932 Verify the UPC number 080012051818.

Exercise 933 Calculate the missing check digit d_{12} for UPC code 07765208227d_{12}.

Exercise 934 Determine whether the check digit provides a check against a transposition (swap) of *two consecutive digits* in a UPC code. For example, if

$$d_1d_2d_3d_4d_5d_6d_7d_8d_9d_{10}d_{11}d_{12}$$

is a correct UPC code with a correct check digit d_{12}, and if a copying error swaps the first two digits d_1d_2 and produces d_2d_1, determine whether the check digit is wrong in the transposed UPC code

$$d_2d_1d_3d_4d_5d_6d_7d_8d_9d_{10}d_{11}d_{12}.$$

Exercise 935 Determine whether the check digit provides a check against a single substitution of one of the digits in a UPC code. For example, if

$$d_1d_2d_3d_4d_5d_6d_7d_8d_9d_{10}d_{11}d_{12}$$

is a correct UPC code with a correct check digit d_{12}, determine whether the check digit is wrong in the substituted UPC code

$$b_1d_2d_3d_4d_5d_6d_7d_8d_9d_{10}d_{11}d_{12}.$$

Exercise 936 Prove that the formula

$$7b_1 + 3b_2 + 9b_3 + 7b_4 + 3b_5 + 9b_6 + 7b_7 + 3b_8 - b_9 \underset{10}{\equiv} 0$$

is equivalent to

$$7b_1 + 3b_2 + 9b_3 + 7b_4 + 3b_5 + 9b_6 + 7b_7 + 3b_8 + 9b_9 \underset{10}{\equiv} 0.$$

Exercise 937 Calculate the check digit b_9 that corresponds to $b_1b_2b_3b_4b_5b_6b_7b_8 = 21107008$ (First Mutual of Boston, as shown on a check from the Consortium for Mathematics and its Applications [COMAP]).

Exercise 938 Provide an example of a specific number $b_1b_2b_3b_4b_5b_6b_7b_8b_9$ for which a transposition of two *distinct adjacent* digits produces the same check digit b_9, so that the check digit would not detect such a transposition.

Exercise 939 Prove that a single substitution of one digit d_i for another b_i will affect the check digit.

Exercise 940 Calculate the check digit d_{10} for the nine-digit ZIP code of the Mathematical Association of America (MAA): $d_1d_2d_3d_4d_5\ d_6d_7d_8d_9d_{10} = 20077\ 6240d_{10}$.

Exercise 941 Calculate the check digit d_{10} for the nine-digit ZIP code of the Consortium for Undergraduate Mathematics and Its Applications (COMAP):

$$d_1d_2d_3d_4d_5\ d_6d_7d_8d_9d_{10} = 02174\ 4131d_{10}.$$

Exercise 942 Calculate the check digit d_{10} for the nine-digit ZIP code of the Society for Industrial and Applied Mathematics (SIAM):

$$d_1d_2d_3d_4d_5\ d_6d_7d_8d_9d_{10} = 19101\ 7260d_{10}.$$

Exercise 943 Calculate the check digit d_{12} for the *eleven-digit* ZIP code of the Post Office Box of the American Mathematical Society (AMS):

$$d_1d_2d_3d_4d_5\ d_6d_7d_8d_9\ d_{10}d_{11}d_{12} = 02206\ 5904\ 04d_{12}.$$

Its bar code consists of twelve digits $d_1d_2d_3d_4d_5\ d_6d_7d_8d_9\ d_{10}d_{11}d_{12}$ such that $d_1 + d_2 + d_3 + d_4 + d_5 + d_6 + d_7 + d_8 + d_9 + d_{10} + d_{11} + d_{12} \underset{10}{\equiv} 0$.

Exercise 944 Determine whether the check digit provides a check against a swap of two digits in a ZIP code. For example, if $d_1d_2d_3d_4d_5d_6d_7d_8d_9d_{10}$ is a correct ZIP code with a correct check digit d_{10}, determine whether the same check digit is still correct for the swapped ZIP code $d_2d_1d_3d_4d_5d_6d_7d_8d_9d_{10}$.

Exercise 945 Determine whether the check digit checks against a single substitution of one digit in a ZIP code. Thus, if $d_1d_2d_3d_4d_5d_6d_7d_8d_9d_{10}$ is a correct ZIP code with a correct check digit d_{10}, determine whether the same check digit is still correct for the altered ZIP code $b_1d_2d_3d_4d_5d_6d_7d_8d_9d_{10}$ with $b_1 \in \{0, \ldots, 9\}$ but $b_1 \neq d_1$.

The following exercises pertain to Version E of the Universal Product Code, which consists of 8 decimal digits $d_1d_2d_3d_4d_5d_6d_7d_8d$. The eighth digit d_8 plays the role of a check digit, determined so that the digits satisfy the following equation (notice that digits are multiplied by 3, 1 or 0, depending on the situation).

$$1d_1 + 1d_2 + 3d_3 + 3d_4 + 1d_5 + 3d_6 + 1d_7 + 1d_8 \underset{10}{\equiv} 0 \quad \text{if } a_7 \in \{0,1,2\},$$

$$1d_1 + 1d_2 + 3d_3 + 1d_4 + 1d_5 + 3d_6 + 0d_7 + 1d_8 \underset{10}{\equiv} 0 \quad \text{if } a_7 \in \{3\},$$

$$1d_1 + 1d_2 + 3d_3 + 1d_4 + 3d_5 + 3d_6 + 0d_7 + 1d_8 \underset{10}{\equiv} 0 \quad \text{if } a_7 \in \{4\},$$

$$3d_1 + 1d_2 + 3d_3 + 1d_4 + 3d_5 + 1d_6 + 3d_7 + 1d_8 \underset{10}{\equiv} 0 \quad \text{if } a_7 \in \{5,6,7,8,9\}.$$

Exercise 946 Determine whether 12345678 is a correct UPC Version E.

Exercise 947 Determine whether 01216907 is a correct UPC Version E.

Exercise 948 Calculate the missing check digit d_8 for the UPC code $8765432d_8$.

Exercise 949 Calculate the missing check digit d_8 for the UPC code $0419920d_8$.

Exercise 950 Determine whether UPC Version E detects or corrects errors.

5.6 EUCLID, EULER, & FERMAT'S THEOREMS

5.6.1 Fermat's Little Theorem

Several codes rely on theorems about the remainders of integer divisions by powers of integers, for instance, the theorems of Euclid, Euler,and Fermat presented here.

The first theorem shows that if P is a prime number, then for each non-zero integer K the reciprocal of K modulo P is an integral power of K.

Theorem 796 (Fermat's Little Theorem) *For each prime number P, and for each integer K that is* not *divisible by P, there exists a positive integer N for which $K^N \underset{P}{\equiv} 1$ modulo P. Moreover, the smallest such positive integer N divides $P - 1$.*

Proof. This proof establishes the two assertions in the theorem separately.

Existence Because $\mathbb{Z}_P = \{[0]_P, \ldots, [P-1]_P\}$ contains only P elements, the function $\mathbb{N} \to \mathbb{Z}_P$, $N \mapsto [K]_P^N$ is not injective (theorem 607). Hence the sequence

$$[1]_P, [K]_P, [K]_P^2, [K]_P^3, \ldots$$

contains equivalent terms, because there exist distinct integers $I < J$ such that

$$K^I \underset{P}{\equiv} K^J.$$

With P prime, K and K^I have multiplicative inverses, whence

$$1 \underset{P}{\equiv} K^{J-I}.$$

Thus $J - I \in \{N \in \mathbb{N} \setminus \{0\} : K^N \underset{P}{\equiv} 1\}$ whence $\{N \in \mathbb{N} \setminus \{0\} : K^N \underset{P}{\equiv} 1\} \neq \varnothing$. Consequently, $\{N \in \mathbb{N} \setminus \{0\} : K^N \underset{P}{\equiv} 1\}$ has a minimum (theorem 509). Let

$$M := \min\{N \in \mathbb{N} \setminus \{0\} : K^N \underset{P}{\equiv} 1\}.$$

Divisibility For each $Q \in \{1, \ldots, P-1\}$ consider the set S_Q defined by

$$S_Q := \{[Q]_P, [QK]_P, [QK^2]_P, \ldots, [QK^{M-1}]_P\}.$$

The elements of S_Q are distinct because of the minimality of M. The proof proceeds by contradiction. Thus $[Q]_P$ has a multiplicative inverse $[Q]_P^{-1}$ because P is prime, and if there were distinct integers $I, J \in \{0, \ldots, M-1\}$ such that $[QK^I]_P = [QK^J]_P$, then multiplication by $[Q]_P^{-1}$ would give

$$\begin{aligned} [QK^I]_P &= [QK^J]_P, \\ [Q]_P * [K]_P^I &= [Q]_P * [K]_P^J, \\ [K]_P^I &= [K]_P^J, \\ [1]_P &= [K]_P^{J-I}, \\ [K]_P^{I-J} &= [1]_P, \end{aligned}$$

with $0 < |J-I| < M$, contradicting the minimality of M. Thus for each $Q \in \{1, \ldots, P-1\}$ the set S_Q has exactly M elements.

Furthermore, for any two integers Q and R, either $S_Q = S_R$ or $S_Q \cap S_R = \varnothing$. Indeed, if $[QK^I]_P = [RK^J]_P$, then $Q \equiv RK^{J-I}$ modulo P, whence $R \in S_Q$.

Consequently, the $P-1$ sets S_Q partition $\mathbb{Z}_P \setminus \{0\}$, whence M divides $P-1$. □

Example 797 If $P := 5$ and $K := 4$, then $K^2 = 16 = 3 * 5 + 1 \underset{5}{\equiv} 1$ modulo 5. Moreover, $M := 2$ is the smallest positive exponent N such that $K^N \underset{5}{\equiv} 1$ modulo 5, because $[K]_5^1 = [4]_5^1 = [4]_5 \neq [1]_5$. Furthermore, 2 divides $P - 1 = 5 - 1 = 4$.

5.6.2 The Euler–Fermat Theorem

The extension of Fermat's Little Theorem to non-necessarily prime divisors involves the following "totient function" of Euler's.

Definition 798 (Euler's totient function ϕ) For each $H \in \mathbb{N}^*$, let $\phi(H)$ denote the number of positive integers that do not exceed H and are relatively prime to H.

Example 799 There are exactly six positive integers that are less than or equal to 9 and relatively prime to 9, namely 1, 2, 4, 5, 7, 8. Thus $\phi(9) = 6$.

Fermat's Little Theorem generalizes to non-necessarily prime divisors as follows.

Theorem 800 (Euler–Fermat Theorem) *For each positive integer H, and for each non-zero integer K relatively prime to H, $K^{\phi(H)} \underset{H}{\equiv} 1$ modulo H. Moreover, if M is the smallest positive integer such that $K^M \underset{H}{\equiv} 1$ modulo H, then M divides $\phi(H)$.*

Proof. This proof establishes the two assertions in the theorem separately.

Existence Because $\mathbb{Z}_H = \{[0]_H, \ldots, [H-1]_H\}$ contains only H elements, the function $\mathbb{N} \to \mathbb{Z}_H$, $N \mapsto [K]_H^N$, is not injective (theorem 607). Hence the sequence

$$[1]_H, [K]_H, [K]_H^2, [K]_H^3, \ldots$$

contains equivalent terms, because there exist distinct integers $I < J$ such that

$$K^I \underset{H}{\equiv} K^J.$$

With K relatively prime to H, K and K^I have multiplicative inverses; therefore

$$1 \underset{H}{\equiv} K^{J-I}.$$

Thus $J-I \in \{N \in \mathbb{N} \setminus \{0\} : K^N \underset{H}{\equiv} 1\}$ whence $\{N \in \mathbb{N} \setminus \{0\} : K^N \underset{H}{\equiv} 1\} \neq \varnothing$. Consequently, $\{N \in \mathbb{N} \setminus \{0\} : K^N \underset{H}{\equiv} 1\}$ has a minimum (theorem 509). Let

$$M := \min\{N \in \mathbb{N} \setminus \{0\} : K^N \underset{H}{\equiv} 1\}.$$

Divisibility For each positive integer Q relatively prime to H and less than or equal to H, consider the set S_Q defined by

$$S_Q := \{[Q]_H, [QK]_H, [QK^2]_H, \ldots, [QK^{M-1}]_H\}.$$

The elements of S_Q are distinct because of the minimality of M. The proof proceeds by contradiction. Thus $[Q]_H$ has a multiplicative inverse $[Q]_H^{-1}$ because Q and H are relatively prime; if there were distinct integers $I, J \in \{0, \ldots, M-1\}$ such that $[QK^I]_H = [QK^J]_H$, then multiplication by $[Q]_H^{-1}$ would give

$$\begin{aligned} [QK^I]_H &= [QK^J]_H, \\ [Q]_H * [K]_H^I &= [Q]_H * [K]_H^J, \\ [K]_H^I &= [K]_H^J, \\ [1]_H &= [K]_H^{J-I}, \\ [K]_H^{I-J} &= [1]_H, \end{aligned}$$

with $0 < |J - I| < M$, contradicting the minimality of M. Thus for each positive integer Q relatively prime to H and less than or equal to H, the set S_Q has exactly M elements. Furthermore, for any two integers Q and R relatively prime to H, either $S_Q = S_R$ or $S_Q \cap S_R = \varnothing$. Indeed, if $[QK^I]_H = [RK^J]_H$, then $Q \underset{H}{\equiv} RK^{J-I}$ modulo H, whence $R \in S_Q$. Consequently, the sets S_Q partition the set of all the classes $[Q]_H$ for all positive integers Q less than or equal to H and relatively prime to H, of which there are $\phi(H)$, whence M divides $\phi(H)$. Therefore, $\phi(H) - M * L$ for some positive integer L, whence

$$K^{\phi(H)} \underset{H}{\equiv} K^{M*L} \underset{H}{\equiv} (K^M)^L \underset{H}{\equiv} (1^L) \underset{H}{\equiv} 1.$$

□

Example 801 If $H := 9$, then $\phi(H) = 6$. If also $K := 4$, so that K and H are relatively prime, then $K^{\phi(H)} = 4^6 = 4096 = (9 * 455) + 1 \underset{9}{\equiv} 1$. Moreover, $K^3 = 4^3 = 64 = (9 * 7) + 1 \underset{9}{\equiv} 1$ and 3 divides $\phi(9) = 6$.

5.6.3 Exercises

Exercise 951 Calculate $\phi(16)$.

Exercise 952 Calculate $\phi(64)$.

Exercise 953 Calculate $\phi(27)$.

Exercise 954 Calculate $\phi(81)$.

Exercise 955 For each prime P and each $K \in \mathbb{N}^*$, prove that $\phi(P^K) = K + 1$.

Exercise 956 For each $N \in \mathbb{N} \setminus \{0, 1\}$ prove that $\phi(N) = N$ if and only if N is prime.

Exercise 957 For each $N \in \mathbb{N} \setminus \{0, 1\}$ prove that $\phi(N)$ is also the number of positive integers that are relatively prime to N and *less* than N.

Exercise 958 (Preliminary to Euclid's Theorem) For each prime P and for each $N \in \mathbb{N}^*$, prove that the sum of all the positive divisors of P^N equals $P^N + (P^N - 1)/(P - 1)$.

Exercise 959 (Euclid's Theorem) For each $N \in \mathbb{N}^*$, prove that if $P := 2^{N+1} - 1$ is prime, then the sum of all the positive divisors of $M := 2^N * (2^{N+1} - 1)$ equals $2 * M$.

Exercise 960 For each $N \in \mathbb{N} \setminus \{0, 1\}$ prove that $\phi(N) \neq \phi(2)$.

5.7 RIVEST–SHAMIR–ADLEMAN (RSA) CODES

5.7.1 The Rivest–Shamir–Adleman (RSA) Public-Key

The Rivest–Shamir–Adleman (RSA) Public-Key Cryptosystem allows a receiver to broadcast an encoding method that any sender may use but only the receiver may decode. Thus anyone may code and send information through any public channel, but only the initial receiver can easily decode the information. The system relies on the Euler–Fermat Theorem from number theory and it functions as follows.

Algorithm 802 (Rivest–Shamir–Adleman (RSA) Public-Key Cryptosystem)

1. The Receiver Selects and Broadcasts a Code.
   (a) The receiver chooses two positive prime integers P and Q and calculates their product $N := P * Q$.
   (b) The receiver chooses $E \in \mathbb{N}^*$ relatively prime to $H := (P - 1) * (Q - 1)$.
   (c) The receiver broadcasts the pair of integers (N, E).
2. The Sender Codes and Broadcasts a Message.
   (a) The sender writes a message as an integer $M < N$ and encodes it as M^E (modulo N).
   (b) The sender broadcasts M^E (modulo N).
3. The Receiver Decodes the Message.
   (a) The receiver chooses a positive integer D such that $D * E \underset{H}{\equiv} 1$ modulo $H := (P - 1) * (Q - 1)$.
   (b) The receiver decodes the message M^E by computing $(M^E)^D$ (modulo N).

Example 803 (A Rivest–Shamir–Adleman (RSA) Public-Key Cryptosystem) The receiver may select the following code:

$$\begin{aligned} P &:= 13; \\ Q &:= 19; \\ N &:= P * Q = 247; \\ H &:= (P - 1) * (Q - 1) = 216; \\ E &:= 5; \\ D &\underset{H}{\equiv} E^{-1} \underset{H}{\equiv} 173. \end{aligned}$$

Thus, the receiver keeps D secret but broadcasts the pair of integers

$$(N, E) = (216,\ 5).$$

The sender may have a message

$$M = 2,$$

and encodes it as

$$M^E \underset{N}{\equiv} 2^5 \underset{N}{\equiv} 32.$$

The sender then broadcasts

$$32.$$

The receiver decodes the message as

$$\begin{aligned}(M^E)^D &\underset{N}{\equiv} (32)^{173}\\ &\underset{N}{\equiv} 2.\end{aligned}$$

5.7.2 Proof of the Correctness of the RSA Algorithm

The following theorems verify that the Rivest–Shamir–Adelman algorithm decodes the message correctly. The first theorem shows that if $N = P * Q$ with P and Q prime, then $\phi(N) = (P-1)*(Q-1)$.

Theorem 804 *If P and Q are prime, then $\phi(P*Q) = (P-1)*(Q-1)$.*

Proof. If P and Q are prime and $N := P*Q$, then a positive integer M divides $N = P^1 * Q^1$ if and only if M is 1 or a multiple of P or Q, by the Fundamental Theorem of Arithmetic (theorem 744). Thus the only positive divisors of $N = P*Q$ that are less than N are

$$\begin{array}{l}1,\\ P, P*2, \ldots, P*(Q-1),\\ Q, Q*2, \ldots, Q*(P-1),\end{array}$$

of which there are a total of

$$1 + Q + P = (P*Q) - [(P-1)*(Q-1)].$$

Consequently, there remain exactly $(P-1)*(Q-1)$ positive integers less than or equal to N that are relatively prime to $N = P*Q$. Hence $\phi(P*Q) = (P-1)*(Q-1)$. □

The next theorem confirms that the algorithm reproduces the message correctly.

Theorem 805 *If P and Q are prime, if $N = P*Q$ and $H = (P-1)*(Q-1)$, if $E*D \underset{H}{\equiv} 1$, and if $M \in \{1, \ldots, N-1\}$ then $([M]_N^E)^D = [M]_N$.*

Proof. Three cases can arise, according to whether both P and Q divide M, neither P nor Q divides M, or exactly one of P or Q divides M.

If both P and Q divide M, then $N = P*Q$ divides M and $M \underset{N}{\equiv} 0$ whence $M^E \underset{N}{\equiv} 0$ and hence $(M^E)^D \underset{N}{\equiv} 0 \underset{N}{\equiv} M$.

If neither P nor Q divides M, then M and N are relatively prime. The decoded message $(M^E)^D$ then reproduces the initial message M, thanks to the Euler–Fermat Theorem: $M^{\phi(N)} \underset{N}{\equiv} 1$ modulo $P*Q$. Indeed, $\phi(P*Q) = (P-1)*(Q-1)$ whence

$$\begin{aligned}(M^E)^D &= M^{(E*D)}\\ &= M^{(1+L*(P-1)*(Q-1)}\\ &= M * M^{L*(P-1)*(Q-1)}\\ &= M * \left[M^{(P-1)*(Q-1)}\right]^L\\ &\underset{N}{\equiv} M * [1]^L\\ &\underset{N}{\equiv} M.\end{aligned}$$

If exactly one of the factors of N divides M but the other factor does not, for instance, if P divides M but Q does *not* divide M, then Fermat's Little Theorem (theorem 796) shows that

$$M^{Q-1} \underset{Q}{\equiv} 1.$$

Consequently,

$$\begin{aligned} (M^E)^D &= M^{(E*D)} \\ &= M^{(1+L*(P-1)*(Q-1)} \\ &= M * M^{L*(P-1)*(Q-1)} \\ &= M * \left[M^{(Q-1)}\right]^{(P-1)*L} \\ &\underset{Q}{\equiv} M * [1]^{(P-1)*L} \\ &\underset{Q}{\equiv} M. \end{aligned}$$

Thus there exist integers I, J, K such that

$$\begin{aligned} M &= K * P, \\ M^{E*D} &= I * P, \\ M^{E*D} &= (J * Q) + M, \\ I * P &= (J * Q) + M, \\ (I - K) * P &= J * Q. \end{aligned}$$

Because the prime P does not divide the prime Q, it follows that P divides J, so that there exists an integer R such that $J = P * R$. Consequently,

$$M^{E*D} = (J * Q) + M = [(P * R) * Q] + M = [R * (P * Q)] + M = [R * N] + M,$$

so that $M^{E*D} \underset{N}{\equiv} M$. □

Example 806 (A Rivest–Shamir–Adleman (RSA) Public-Key Cryptosystem) The receiver may select the following code:

$$\begin{aligned} P &= 274177; \\ Q &= 616318177; \\ N &= P * Q = 168980268815329; \\ H &= (P-1) * (Q-1) = 168979652222976 \\ &= 2^{13} * 3^3 * 7 * 17 * 37 * 167 * 1039; \\ E &= 82421; \\ D &\underset{H}{\equiv} E^{-1} \underset{H}{\equiv} 32399333229149 \ (\mathit{modulo}\, H). \end{aligned}$$

Thus, the receiver keeps D secret but broadcasts the pair of integers

$$(N, E) = (168980268815329, \ 82421).$$

The sender may have a message

$$M = 4294967297 = 641 * 6700417,$$

and encodes it as

$$M^E \underset{N}{\equiv} 4294967297^{82421} \underset{N}{\equiv} 149216834543948.$$

The sender then broadcasts

$$149216834543948.$$

The receiver decodes the message as

$$\begin{aligned} (M^E)^D &\underset{N}{\equiv} 149216834543948^{32399333229149} \\ &\underset{N}{\equiv} 4294967297 = M. \end{aligned}$$

For additional introductory material about the Rivest–Shamir–Adleman (RSA) Public-Key Cryptosystem, see [77], [78].

5.7.3 Exercises

The following exercises outline a shorter algorithm to calculate powers.

Exercise 961 For all $E, M \in \mathbb{N}$ calculate the number of multiplications involved in the calculation of M^E.

Exercise 962 For all $E, M, N \in \mathbb{N}$ prove that $[M^E]_N$ can be calculated with products of integers not exceeding $N - 1$.

Exercise 963 For all $K, M \in \mathbb{N}$ calculate the number of multiplications involved in the calculation of 1, M, M^2, M^4, M^8, ..., $M^{(2^K)}$.

Exercise 964 For all $K, M, N \in \mathbb{N}$ prove that $[M^K]_N$ can be calculated with products of integers not exceeding $N - 1$.

Exercise 965 For each $E \in \mathbb{N}$ let $E = D_K * 2^K + \cdots + D_1 * 2^1 + D_0 * 2^0$ be its binary representation. Calculate the number of multiplications in the computation of M^E by

$$M^E = M^{D_K * 2^K + \cdots + D_1 * 2^1 + D_0 * 2^0} = M^{D_K * 2^K} * \cdots * M^{D_1 * 2^1} * M^{D_0 * 2^0}.$$

Exercise 966 Calculate the ratio of the numbers of multiplications involved in the calculations of M^E by the algorithm $M * M, \ldots, (M^{E-1}) * M$ and the algorithm $M^{D_K * 2^K} * \cdots * M^{D_1 * 2^1} * M^{D_0 * 2^0}$.

Exercise 967 Calculate $[32^{173}]_{247}$ by iterated squaring and reductions modulo 247.

Exercise 968 Consider the Rivest–Shamir–Adleman (RSA) Public-Key Cryptosystem with the following parameters.

$$\begin{aligned} P &:= 3; \\ Q &:= 5; \\ N &:= P * Q = 15; \\ H &:= (P-1) * (Q-1) = 8; \\ E &:= 7. \end{aligned}$$

Thus, the receiver broadcasts the pair

$$(N, E) = (15,\ 7).$$

Firstly, encode the message $M := 2$.

Secondly, provide some $D \in \mathbb{N}^*$ such that $E * D \underset{H}{\equiv} 1$.

Finally, verify that $(M^E)^D \underset{N}{\equiv} M$.

Exercise 969 Consider the Rivest–Shamir–Adleman (RSA) Public-Key Cryptosystem with the following parameters.

$$\begin{aligned} P &:= 3; \\ Q &:= 7; \\ N &:= P * Q = 21; \\ H &:= (P-1) * (Q-1) = 12; \\ E &:= 11. \end{aligned}$$

Thus, the receiver broadcasts the pair

$$(N, E) = (21,\ 11).$$

Firstly, encode the message $M := 5$.

Secondly, provide some $D \in \mathbb{N}^*$ such that $E * D \underset{H}{\equiv} 1$.

Finally, verify that $(M^E)^D \underset{N}{\equiv} M$.

Exercise 970 Consider the Rivest–Shamir–Adleman (RSA) Public-Key Cryptosystem with the following parameters.

$$\begin{aligned} P &:= 11; \\ Q &:= 19; \\ N &:= P * Q = 209; \\ H &:= (P-1)*(Q-1) = 180; \\ E &:= 97; \end{aligned}$$

Thus, the receiver broadcasts the pair

$$(N, E) = (209,\ 97).$$

Firstly, encode the message $M := 6$.

Secondly, provide some $D \in \mathbb{N}^*$ such that $E * D \underset{H}{\equiv} 1$.

Finally, verify that $(M^E)^D \underset{N}{\equiv} M$.

5.8 FURTHER ISSUES IN NUMBER THEORY

After several millennia of investigations in number theory there still remain simply stated but yet unsolved problems [37].

Open problem 1 (Twin-primes conjecture) Twin primes are pairs (P, Q) of prime numbers P and Q such that $Q = P + 2$. For example, here are pairs of twin primes:

$$\begin{gathered} (3,5),(5,7),(11,13),(17,19),(29,31),(41,43),(59,61),(71,73), \\ (101,103),(107,109),(137,139),(149,151),(179,181),(191,193), \\ \ldots,(99707,99709),(99989,99991),\ldots \end{gathered}$$

The question whether the list of twin primes ends remains unanswered.

Open problem 2 (Perfect-numbers conjecture) For each positive integer $N \in \mathbb{N}^*$, the **aliquot parts** of N are the divisors that are strictly smaller than N. A positive integer is **perfect** if and only if it equals the sum of all its aliquot parts. For example, the aliquot parts of 6 are 1, 2, 3. Hence 6 is perfect because $6 = 1 + 2 + 3$.

More generally, a theorem of Euclid's (exercise 959) states that if $2^{N+1} - 1$ is prime, then $2^N * (2^{N+1} - 1)$ is perfect, and Euclid's theorem generates all *even* perfect numbers [104, p. 54]. For example, the following positive even integers are perfect:

$$\begin{aligned} 2^1 * (2^2 - 1) &= 6 = 1 + 2 + 3, \\ 2^2 * (2^3 - 1) &= 28 = 1 + 2 + 4 + 7 + 14, \\ 2^4 * (2^5 - 1) &= 496 = 1 + 2 + 4 + 8 + 16 + 31 + 2 * 31 + 4 * 31 + 8 * 31, \\ 2^6 * (2^7 - 1) &= 8128 = 1 + 2 + 4 + 8 + 16 + 32 + 64 + 127 + 2 * 127 \\ &\qquad + 4 * 127 + 8 * 127 + 16 * 127 + 32 * 127. \end{aligned}$$

The questions whether the list of perfect numbers ends, whether the list of primes of the type $2^{N+1} - 1$ ends, and whether there is an *odd* perfect number are open.

Open problem 3 (Goldbach's conjecture) Examples suggest that every *even* integer is the sum of two "primes" (in Goldbach's terminology, 1 is prime). For example,

$$\begin{aligned} 2 &= 1+1, \\ 4 &= 2+2, \\ 6 &= 3+3, \\ 8 &= 3+5. \end{aligned}$$

The question whether *every* even integer is the sum of two primes is still open.

Open problem 4 (Carmichael's conjecture) Examples suggest that for each $N \in \mathbb{N}\backslash\{0,1,2\}$ there exists $M \in \mathbb{N} \setminus \{0,1,2\}$ such that $M \neq N$ and $\phi(M) = \phi(N)$:

$$\phi(3) = 2 = \phi(4)$$ (1, 2 are relatively prime to 3, whereas 1, 3 are relatively prime to 4);

$$\phi(5) = 4 = \phi(8)$$ (1, 2, 3, 4 are relatively prime to 5, whereas 1, 3, 5, 7 are relatively prime to 8);

$$\phi(7) = 6 = \phi(9)$$ (1, 2, 3, 4, 5, 6 are relatively prime to 7, whereas 1, 2, 4, 5, 7, 8 are relatively prime to 9).

The question whether there exists some $N > 2$ such that $\phi(N) \neq \phi(M)$ for every $M \neq N$ remains unanswered.

5.9 PROJECTS

Project 11 Design algorithms for arithmetic with Roman or Mayan numerals.

Project 12 Design a working computer program for RSA cryptography.

Chapter 6

Ciphers, Combinatorics, and Probabilities

6.0 INTRODUCTION

Figure 6.1: An ENIGMA ciphering machine.

Cryptography encompasses all **encryption** methods to transform plain text into messages that appear unintelligible, and **decryption** methods to transform seemingly unintelligible messages back into plain text. Among such methods, **codes** substitute words or strings of characters for entire words, phrases, or sentences [55, p. xiv]. In contrast, **ciphers** scramble (permute) the order of an alphabet, substituting single characters for single letters [55, p. xv]. Yet history shows that simple ciphers and codes can simply be deciphered or "broken" by adversaries. Therefore, more secure methods of cryptography require more theory, to design codes and ciphers, and to assess the level of difficulty encountered by foes trying to break the codes. For example, the design of the German `ENIGMA` machine shown in figure 6.1 and the Allies' methods to break the German ciphers relied on the mathematical theory of permutations [93], [105, Ch. 4, p. 59–83]. Secure cryptography must also mask easily identifiable features of a language, because allowing foes to identify such features can enable them to break the code. For example, in English the character "e" occurs more frequently than any other character does [2, p. 270]. A permutation of the alphabet would replace "e" by some other character "x" that would then occur most frequently. Merely by counting repetitions of characters, the foes would then recognize that "x" might stand for "e" and thus start breaking the code. To thwart such attempts, cryptography can also rely on selections — also called combinations — of only some of all possible characters in an alphabet. Such concepts as permutations and combinations form a part of the mathematical field of combinatorics, which leads to the concepts of frequency and probability.

6.1 ALGEBRA

6.1.1 Definitions and Examples of Mathematical Groups

The mathematical concept of "group" constitutes a model for various types of binary operations, including consecutive substitutions (compositions of permutations in ciphers), the addition of integers, and the multiplication of non-zero rationals. As a generalization of the development of arithmetic, the development of "groups" and other related topics forms the subject of the part of mathematics called **algebra**.

Definition 807 (Semi-groups, monoids, groups) A mathematical **semi-group** is a pair $(G, \bullet)$, with a set G and an *associative* binary operation (function)

$$\begin{aligned} \bullet : (G \times G) &\rightarrow G, \\ (p, q) &\mapsto p \bullet q, \end{aligned}$$

such that for all elements $g, p, q \in G$,

$$g \bullet (p \bullet q) = (g \bullet p) \bullet q.$$

An element $e \in G$ is a **left-identity** or a **left-neutral** element, if and only if for each $g \in G$,

$$e \bullet g = g.$$

Similarly, an element $e \in G$ is a **right-identity** or a **right-neutral** element, if and only if for each $g \in G$,

$$g \bullet e = g.$$

An element $e \in G$ is an **identity** or a **neutral** element, if and only if e is both left-neutral and right-neutral, so that $g \bullet e = g = e \bullet g$ for every $g \in G$. A mathematical **monoid** is a semi-group $(G, \bullet)$ for which there exists a neutral element $e \in G$.

In a semi-group with a left-identity e, a **left-inverse** for an element $g \in G$ is an element $p \in G$ such that

$$p \bullet g = e.$$

Similarly, a **right-inverse** for an element $g \in G$ is an element $q \in G$ such that

$$g \bullet q = e.$$

An **inverse** for an element $g \in G$ is an element $q \in G$ that is both a left-inverse and a right-inverse for g, so that $q \bullet g = e = g \bullet q$.

A mathematical **group** is a semi-group $(G, \bullet)$ with a left-inverse e such that each $g \in G$ has a left-inverse $p \in G$, so that $p \bullet g = e$, as summarized in table 808.

A monoid, semi-group, or group $(G, \bullet)$ is **commutative** if and only if $v \bullet w = w \bullet v$ for all $v \in G$ and $w \in G$.

Table 808 These properties hold for all u, v, w in a group $(G, \bullet)$.

(G.1)	Associativity	$[u \bullet v] \bullet w = u \bullet [v \bullet w]$
(G.2)	Left-identity: there exists $e \in G$ with	$e \bullet v = v$
(G.3)	Left-inverse: there exits $r \in G$ with	$r \bullet v = e$

Example 809 As demonstrated in the sequel, enciphering operations form a group, where inverses are the deciphering operations.

Example 810 The integers with addition form a group $(\mathbb{Z}, +)$, with the identity element 0, for which $n + 0 = n = 0 + n$ for every $n \in \mathbb{Z}$. Each integer $n \in \mathbb{Z}$ has an additive inverse $-n \in \mathbb{Z}$ with $n + (-n) = 0 = (-n) + n$. Also, $(\mathbb{Z}, +)$ is commutative because $m + n = n + m$ for all integers m and n.

Example 811 The rationals with addition form a group $(\mathbb{Q}, +)$. The identity element is $0 = {}^0\!/_1$, for which ${}^p\!/_q + 0 = {}^p\!/_q = 0 + {}^p\!/_q$ for every ${}^p\!/_q \in \mathbb{Q}$. Each rational ${}^p\!/_q \in \mathbb{Q}$ has an additive inverse $-{}^p\!/_q \in \mathbb{Q}$ with ${}^p\!/_q + (-{}^p\!/_q) = 0 = (-{}^p\!/_q) + {}^p\!/_q$. Also, $(\mathbb{Q}, +)$ is commutative because ${}^p\!/_q + {}^m\!/_n = {}^m\!/_n + {}^p\!/_q$ for all rationals ${}^p\!/_q$ and ${}^m\!/_n$.

Example 812 The *non-zero* rationals with *multiplication* form a group $(\mathbb{Q}^*, \cdot)$, with the identity element $1 = {}^1\!/_1$, for which ${}^p\!/_q \cdot 1 = {}^p\!/_q = 1 \cdot {}^p\!/_q$ for every ${}^p\!/_q \in \mathbb{Q}^*$. Each *non-zero* rational ${}^p\!/_q \in \mathbb{Q}^*$ has a *multiplicative* inverse $({}^p\!/_q)^{-1} = {}^q\!/_p \in \mathbb{Q}^*$ with ${}^p\!/_q \cdot {}^q\!/_p = 1 = {}^q\!/_p \cdot {}^p\!/_q$. Also, $(\mathbb{Q}^*, \cdot)$ is commutative because ${}^p\!/_q \cdot {}^m\!/_n = {}^m\!/_n \cdot {}^p\!/_q$ for all non-zero rationals ${}^p\!/_q$ and ${}^m\!/_n$.

Example 813 The *positive* rationals with *multiplication* form a group $(\mathbb{Q}^*_+, \cdot)$, with the identity element $1 = {}^1\!/_1$, for which ${}^p\!/_q \cdot 1 = {}^p\!/_q = 1 \cdot {}^p\!/_q$ for every ${}^p\!/_q \in \mathbb{Q}^*_+$. Each *positive* rational ${}^p\!/_q \in \mathbb{Q}^*_+$ has a *multiplicative* inverse $({}^p\!/_q)^{-1} = {}^q\!/_p \in \mathbb{Q}^*_+$ with ${}^p\!/_q \cdot {}^q\!/_p = 1 = {}^q\!/_p \cdot {}^p\!/_q$. Also, $(\mathbb{Q}^*_+, \cdot)$ is commutative because ${}^p\!/_q \cdot {}^m\!/_n = {}^m\!/_n \cdot {}^p\!/_q$ for all positive rationals ${}^p\!/_q$ and ${}^m\!/_n$.

Counterexample 814 The natural numbers with addition, $(\mathbb{N}, +)$, form a semi-group but *not* a group, because positive integers do not have additive inverses in $\mathbb{N}$.

Counterexample 815 The integers with multiplication, $(\mathbb{Z}, *)$, form a semi-group but *not* a group, because integers different from $-1, 0, 1$ do not have reciprocals in $\mathbb{Z}$.

6.1.2 Existence and Uniqueness of Identities and Inverses

The design of successful ciphers, codes, and other applications can depend on judicious uses of theorems — or on the development of appropriate theorems — about mathematical groups. The first theorem presented here shows that in a group every left-identity is also a right-identity.

Theorem 816 *In every group $(G, \bullet)$ each left-identity e is also a right-identity: $g \bullet e = g$ for every $g \in G$.*

Proof. In every group $(G, \bullet)$, each $g \in G$ has a left-inverse $q \in G$ such that $q \bullet g = e$. The same q also has a left-inverse $p \in G$ such that $p \bullet q = e$. Consequently,

$$
\begin{aligned}
g \bullet e &= e \bullet (g \bullet e) && e \text{ is a left-identity,} \\
&= (p \bullet q) \bullet (g \bullet e) && p \bullet q = e, \\
&= p \bullet [q \bullet (g \bullet e)] && \text{associativity,} \\
&= p \bullet [(q \bullet g) \bullet e] && \text{associativity,} \\
&= p \bullet [e \bullet e] && q \bullet g = e, \\
&= p \bullet e && e \text{ is a left-identity,} \\
&= p \bullet (q \bullet g) && q \bullet g = e, \\
&= (p \bullet q) \bullet g && \text{associativity,} \\
&= e \bullet g && p \bullet q = e, \\
&= g && e \text{ is a left-identity.}
\end{aligned}
$$

Thus $g \bullet e = g$ for every $g \in G$, and therefore e is also a right-identity. □

The second theorem shows that in a group every left-inverse is also a right-inverse.

Theorem 817 *In every group $(G, \bullet)$, for each $g \in G$, every left-inverse q with $q \bullet g = e$ is also a right-inverse: $g \bullet q = e$.*

Proof. In every group $(G, \bullet)$, each $g \in G$ has a left-inverse $q \in G$ such that $q \bullet g = e$. The same q also has a left-inverse $p \in G$ such that $p \bullet q = e$. Because in a group a left-identity e is also a right-identity (theorem 816), it follows that

$$
\begin{aligned}
g \bullet q &= e \bullet (g \bullet q) && e \text{ is a left-identity,} \\
&= (p \bullet q) \bullet (g \bullet q) && p \bullet q = e, \\
&= p \bullet [q \bullet (g \bullet q)] && \text{associativity,} \\
&= p \bullet [(q \bullet g) \bullet q] && \text{associativity,} \\
&= p \bullet (e \bullet q) && q \bullet g = e, \\
&= p \bullet q && e \text{ is a left-identity,} \\
&= e && p \bullet q = e.
\end{aligned}
$$

Thus $g \bullet q = e$ and therefore a left-inverse q for g is also a right-inverse for g. □

The next two theorems show the uniqueness of inverses and identities in groups.

Theorem 818 *Each group $(G, \bullet)$ has exactly one identity element.*

Proof. If e_1 and e_2 are two identity elements for $(G, \bullet)$, then $e_1 = e_2$. Indeed,

$$
\begin{aligned}
e_1 &= e_1 \bullet e_2 && e_2 \text{ is a right-identity,} \\
&= e_2 && e_1 \text{ is a left-identity.}
\end{aligned}
$$

Thus each group has at most one identity, but also at least one by definition, hence exactly one identity. □

Theorem 819 *In every group $(G, \bullet)$ each element $g \in G$ has exactly one inverse.*

Proof. If $g \in G$ has two inverses q_1 and q_2, then $q_1 = q_2$:

$$
\begin{aligned}
q_1 &= q_1 \bullet e && e \text{ is a right-identity,} \\
&= q_1 \bullet (g \bullet q_2) && q_2 \text{ is a right-inverse for } g, \\
&= (q_1 \bullet g) \bullet q_2 && \text{associativity of } \bullet, \\
&= e \bullet q_2 && q_1 \text{ is a left-inverse for } g, \\
&= q_2 && e \text{ is a left-identity.}
\end{aligned}
$$

Thus each element has at most one inverse, but also at least one by definition, hence exactly one inverse. □

The uniqueness of a two-sided inverse allows for a specific notation.

Definition 820 In a group $(G, \bullet)$ the inverse of an element g is denoted by g^{-1}, or also by $g^{\bullet -1}$ to specify that the inverse is relative to the operation $\bullet$.

The inverse of a product is the product of the inverses in the reverse order.

Theorem 821 *In every group* $(G, \bullet)$, $(p \bullet q)^{-1} = q^{-1} \bullet p^{-1}$ *for all* $p, q \in G$.

Proof. The associativity of the operation $\bullet$ confirms that

$$
\begin{aligned}
(q^{-1} \bullet p^{-1}) \bullet (p \bullet q) &= q^{-1} \bullet [p^{-1} \bullet (p \bullet q)] && \text{associativity of } \bullet, \\
&= q^{-1} \bullet [(p^{-1} \bullet p) \bullet q] && \text{associativity of } \bullet, \\
&= q^{-1} \bullet (e \bullet q) && \text{left-inverse,} \\
&= q^{-1} \bullet q && \text{left-identity,} \\
&= e && \text{left-inverse.}
\end{aligned}
$$

Thus $q^{-1} \bullet p^{-1}$ is a left-inverse, and hence the inverse, for $p \bullet q$. □

Special circumstances can facilitate the calculation of an inverse. The first theorem to this effect shows that if an element $q \in G$ is the inverse of some element $g \in G$, so that $q = g^{-1}$, then conversely the inverse of q is g.

Theorem 822 *In every group* $(G, \bullet)$, $(g^{-1})^{-1} = g$ *for each* $g \in G$.

Proof. From $g^{-1} \bullet g = e = g \bullet g^{-1}$ (theorem 817) it follows that g is a left-inverse and hence the inverse of g^{-1}. □

The following results establish features of powers of elements in groups, all of which will play a crucial rôle in the analysis of ciphering machines.

Definition 823 In every group $(G, \bullet)$, for each $g \in G$, and for each $n \in \mathbb{N}$, define g^n by induction, with $g^0 := e$ and $g^{n+1} := g^n \bullet g$. The natural number n is called the **exponent** while g^n is the n-th **power** of g.

The first theorem establishes the law of addition for exponent.

Theorem 824 *In every group* $(G, \bullet)$, $g^{m+n} = g^m \bullet g^n$ *for all* $m, n \in \mathbb{N}$ *and* $g \in G$.

Proof. This proof proceeds by induction (theorem 480) with the power n.

If $n = 0$, then $g^{m+0} = g^m = g^m \bullet e = g^m \bullet g^0$. Hence assume that there exists $i \in \mathbb{N}$ such that $g^{m+i} = g^m \bullet g^i$ for every $g \in G$ and every $m \in \mathbb{N}$. Then

$$
\begin{aligned}
g^{m+(i+1)} &= g^{(m+i)+1} && \text{associativity of } + \text{ in } \mathbb{N}, \\
&= g^{m+i} \bullet g && \text{definition of powers,} \\
&= (g^m \bullet g^i) \bullet g && \text{induction hypothesis,} \\
&= g^m \bullet (g^i \bullet g) && \text{associativity of } \bullet, \\
&= g^m \bullet g^{i+1} && \text{definition of powers.}
\end{aligned}
$$

□

The next two theorems establish the commutativity of powers of the same element.

Theorem 825 *In every group* $(G, \bullet)$, $g \bullet g^m = g^m \bullet g$ *for all* $g \in G$ *and* $m \in \mathbb{N}$.

Proof. This proof proceeds by induction (theorem 480) with the power m.

If $m = 0$, then $g^0 \bullet g = e \bullet g = g = g \bullet e = g \bullet g^0$. Hence assume that there exists $i \in \mathbb{N}$ such that $g \bullet g^i = g^i \bullet g$ for every $g \in G$ and every $i \in \mathbb{N}$. Then

$$\begin{array}{rcll} g^{i+1} \bullet g & = & (g^i \bullet g) \bullet g & \text{definition of powers,} \\ & = & (g \bullet g^i) \bullet g & \text{induction hypothesis,} \\ & = & g(\bullet g^i \bullet g) & \text{associativity of } \bullet, \\ & = & g \bullet g^{i+1} & \text{definition of powers.} \end{array}$$

□

Theorem 826 *In every group $(G, \bullet)$, $g^n \bullet g^m = g^m \bullet g^n$ for all $g \in G$ and $m, n \in \mathbb{N}$.*

Proof. This proof proceeds by induction (theorem 480) with the power n.

If $n = 0$, then $g^0 \bullet g^m = e \bullet g^m = g^m = g^m \bullet e = g^m \bullet g^0$. Hence assume that there exists $i \in \mathbb{N}$ such that $g^i \bullet g^m = g^m \bullet g^i$ for every $g \in G$ and every $m \in \mathbb{N}$. Then

$$\begin{array}{rcll} g^{i+1} \bullet g^m & = & (g^i \bullet g) \bullet g^m & \text{definition of powers,} \\ & = & g^i \bullet (g \bullet g^m) & \text{definition of powers,} \\ & = & g^i \bullet (g^m \bullet g) & \text{theorem 825,} \\ & = & (g^i \bullet g^m) \bullet g & \text{associativity of } \bullet, \\ & = & (g^m \bullet g^i) \bullet g & \text{induction hypothesis,} \\ & = & g^m \bullet (g^i \bullet g) & \text{associativity of } \bullet, \\ & = & g^m \bullet g^{i+1} & \text{definition of powers.} \end{array}$$

□

The next theorem shows that the inverse of a power is the power of the inverse.

Theorem 827 *In every group $(G, \bullet)$, $(g^n)^{-1} = (g^{-1})^n$ for all $n \in \mathbb{N}$ and $g \in G$.*

Proof. This proof proceeds by induction (theorem 480) with the power n.

If $n = 0$, then $(g^0)^{-1} = e^{-1} = e = (e^{-1})^0$. Hence assume that there exists $m \in \mathbb{N}$ such that $(g^m)^{-1} = (g^{-1})^m$ for every $g \in G$. Then

$$\begin{array}{rcll} (g^{m+1}) \bullet (g^{-1})^{m+1} & = & [(g^m) \bullet g] \bullet [(g^{-1})^m \bullet g^{-1}] & \text{definition of } {}^{m+1}, \\ & = & [(g^m) \bullet g] \bullet [g^{-1} \bullet (g^{-1})^m] & \text{theorem 826,} \\ & = & (g^m) \bullet \{g \bullet [g^{-1} \bullet (g^{-1})^m]\} & \text{associativity of } \bullet, \\ & = & (g^m) \bullet \{[g \bullet g^{-1}] \bullet (g^{-1})^m\} & \text{associativity of } \bullet, \\ & = & (g^m) \bullet [e \bullet (g^{-1})^m] & \text{definition of } g^{-1}, \\ & = & (g^m) \bullet (g^{-1})^m & e \text{ is the identity,} \\ & = & e & \text{induction hypothesis.} \end{array}$$

□

Definition 828 In every group $(G, \bullet)$, for all $g \in G$ and $n \in \mathbb{N}$, let $g^{-n} := (g^{-1})^n$.

6.1.3 Bijections of Groups

In certain contexts — especially with permutations — the elements of a groups are denoted by Greek letters, for instance, ρ (rho), σ (sigma), η (eta), and ι (iota) for the identity. With such notation, the following theorems establish certain bijections within groups. The first theorem shows that the operation of inversion in a group $(G, \bullet)$ induces a bijection from G onto G. In other words, different elements have different inverses (injectivity), and every element is the inverse of some element (surjectivity).

Theorem 829 *For each group $(G, \bullet)$, the inversion is a bijection from G onto G:*

$$\begin{array}{rcl} \mathcal{I}: G & \to & G, \\ \sigma & \mapsto & \sigma^{-1}. \end{array}$$

Proof. Because $(G, \bullet)$ is a group, $\sigma^{-1} \in G$, whence $\mathcal{I}$ is well defined.

Surjectivity Each $\tau \in G$ lies in the range of $\mathcal{I}$, because by theorem 822

$$\tau = (\tau^{-1})^{-1} = \mathcal{I}(\tau^{-1}) \in \text{Range}\,(\mathcal{I}).$$

Injectivity With ι being the identity of $(G, \bullet)$, and for all $\sigma, \eta \in G$,

$$\begin{array}{rcl} \mathcal{I}(\sigma) & = & \mathcal{I}(\eta) \\ & \Updownarrow & \\ \sigma^{-1} & = & \eta^{-1} \\ & \Updownarrow & \\ \sigma \bullet \sigma^{-1} & = & \sigma \bullet \eta^{-1} \\ & \Updownarrow & \\ \iota & = & \sigma \bullet \eta^{-1} \\ & \Updownarrow & \\ \iota \bullet \eta & = & (\sigma \bullet \eta^{-1}) \bullet \eta \\ & \Updownarrow & \\ \eta & = & \sigma \bullet (\eta^{-1} \bullet \eta) \\ & \Updownarrow & \\ \eta & = & \sigma \bullet \iota \\ & \Updownarrow & \\ \eta & = & \sigma. \end{array}$$

□

Besides inversion, other operations prove useful in ciphers and elsewhere.

Definition 830 For each group $(G, \bullet)$ and each element $\rho \in G$, the function

$$\begin{array}{rcl} L_\rho : G & \to & G, \\ \sigma & \mapsto & \rho \bullet \sigma, \end{array}$$

is the **left-shift** by ρ, whereas the function

$$\begin{array}{rcl} R_\rho : G & \to & G, \\ \sigma & \mapsto & \sigma \bullet \rho, \end{array}$$

is the **right-shift** by ρ.

The next theorem shows that the left-shift and the right-shift induce bijections.

Theorem 831 *For every group $(G, \bullet)$ and for each element $\rho \in G$, the functions L_ρ and R_ρ in definition 830 are bijections from G onto G.*

Proof. This proof establishes the surjectivity and the injectivity separately.

Surjectivity With ι being the identity of $(G, \bullet)$, and for each $\sigma \in G$,

$$L_\rho(\rho^{-1} \bullet \sigma) = \rho \bullet (\rho^{-1} \bullet \sigma) = (\rho \bullet \rho^{-1}) \bullet \sigma = \iota \bullet \sigma = \sigma,$$

$$R_\rho(\sigma \bullet \rho^{-1}) = (\sigma \bullet \rho^{-1}) \bullet \rho = \sigma \bullet (\rho^{-1} \bullet \rho) = \sigma \bullet \iota = \sigma,$$

whence $\sigma \in \text{Range}\,(L_\rho)$ and $\sigma \in \text{Range}\,(R_\rho)$. Thus L_ρ and R_ρ are surjective.

Injectivity For all $\sigma, \eta \in G$,

$$\begin{array}{rcl} L_\rho(\sigma) & = & L_\rho(\eta) \\ & \Updownarrow & \\ \rho \bullet \sigma & = & \rho \bullet \eta \\ & \Updownarrow & \\ \rho^{-1} \bullet (\rho \bullet \sigma) & = & \rho^{-1} \bullet (\rho \bullet \eta) \\ & \Updownarrow & \\ (\rho \bullet \rho^{-1}) \bullet \sigma & = & (\rho \bullet \rho^{-1}) \bullet \eta \\ & \Updownarrow & \\ \iota \bullet \sigma & = & \iota \bullet \eta \\ & \Updownarrow & \\ \sigma & = & \eta \end{array} \qquad \begin{array}{rcl} R_\rho(\sigma) & = & R_\rho(\eta) \\ & \Updownarrow & \\ \sigma \bullet \rho & = & \eta \bullet \rho \\ & \Updownarrow & \\ (\sigma \bullet \rho) \bullet \rho^{-1} & = & (\eta \bullet \rho) \bullet \rho^{-1} \\ & \Updownarrow & \\ \sigma \bullet (\rho \bullet \rho^{-1}) & = & \eta \bullet (\rho \bullet \rho^{-1}) \\ & \Updownarrow & \\ \sigma \bullet \iota & = & \eta \bullet \iota \\ & \Updownarrow & \\ \sigma & = & \eta \end{array}$$

Thus L_ρ and R_ρ are also injective, and hence bijective. □

The injectivity of the left-shift leads to the law of subtraction of exponents.

Theorem 832 *In every group $(G, \bullet)$, $g^{m-n} = g^m \bullet g^{-n}$ for all $g \in G$ and $m, n \in \mathbb{N}$.*

Proof. Because

$$g^{m-n} \bullet g^n = g^{(m-n)+n} = g^m$$

and also

$$(g^m \bullet g^{-n}) \bullet g^n = g^m \bullet (g^{-n} \bullet g^n) = g^m \bullet e = g^m,$$

it follows that $g^{m-n} = g^m \bullet g^{-n}$, from the injectivity of the right-shift by g^n. □

In *finite* groups the inverse of an element is a power of the same element.

Theorem 833 *For every group $(G, \bullet)$ with a finite set G, for each $g \in G$ there exists $\ell \in \mathbb{N}^*$ such that $g^\ell = e$, whence $g^{\ell-1} = g^{-1}$.*

Proof. Because G is finite and $\mathbb{N}$ is denumerable, by theorem 611 the sequence

$$\begin{array}{rcl} s : \mathbb{N} & \to & G, \\ k & \mapsto & g^k, \end{array}$$

cannot be injective. Therefore there exist two distinct integers $m, n \in \mathbb{N}$ such that $m > n$ and $g^m = g^n$. Consequently, from theorems 827 and 832 it follows that

$$g^{m-n} = g^m \bullet (g^{-1})^n = g^n \bullet (g^{-1})^n = e.$$

□

A combination of two operations as described in theorem 831 yields another type of operation, which will also play a crucial rôle in the analysis of ciphering machines.

Definition 834 In each group $(G, \bullet)$ and for all elements $\rho, \sigma \in G$, the **conjugation** by ρ is the function

$$\begin{array}{rcl} C_\rho : G & \to & G, \\ \sigma & \mapsto & \rho \bullet \sigma \bullet \rho^{-1}. \end{array}$$

The **conjugate** of σ by ρ is the element $\rho \bullet \sigma \bullet \rho^{-1}$.

The composition of two conjugations correspond to a conjugation by a product.

Theorem 835 *For each group $(G, \bullet)$ and all elements $\xi, \zeta \in G$,*

$$C_{\xi \bullet \zeta} = C_\xi \circ C_\zeta.$$

Proof. For each $\sigma \in G$, from theorem 821 it follows that

$$\begin{aligned} C_{\xi\bullet\zeta}(\sigma) &= (\xi \bullet \zeta) \bullet \sigma \bullet (\xi \bullet \zeta)^{-1} \\ &= (\xi \bullet \zeta) \bullet \sigma \bullet (\zeta^{-1} \bullet \xi^{-1}) \\ &= \xi \bullet (\zeta \bullet \sigma \bullet \zeta^{-1}) \bullet \xi^{-1} \\ &= C_\xi[C_\zeta(\sigma)] \\ &= (C_\xi \circ C_\zeta)(\sigma). \end{aligned}$$

□

Conjugation also induces a bijection.

Theorem 836 *For each group $(G, \bullet)$ and each $\rho \in G$, the conjugation $C_\rho : G \to G$ is a bijection, with conjugation by ρ^{-1} as the inverse function, so that $C_{\rho^{-1}} = (C_\rho)^{\circ -1}$:*

$$\begin{aligned} C_\rho : G &\to G, \\ \sigma &\mapsto \rho \bullet \sigma \bullet \rho^{-1}. \end{aligned}$$

Proof. From theorem 835 it follows that

$$C_\rho \circ C_{\rho^{-1}} = C_{\rho\bullet\rho^{-1}} = C_\iota = I_G = C_\iota = C_{\rho^{-1}\bullet\rho} = C_{\rho^{-1}} \circ C_\rho.$$

Thus C_ρ and $C_{\rho^{-1}}$ are mutual inverses, whence C_ρ is bijective (theorem 398). □

The following theorem shows that conjugation partitions a group.

Theorem 837 *For each group $(G, \bullet)$ conjugation induces an equivalence relation $\equiv$ defined by $\sigma \equiv \eta$ if and only if there exists $\rho \in G$ such that $\sigma = \rho \bullet \eta \bullet \rho^{-1}$.*

Proof. This proof shows the reflexivity, symmetry, and transitivity separately.

Reflexivity For each $\sigma \in G$, $\sigma = \iota \bullet \sigma \bullet \iota^{-1}$ with the identiy ι, whence $\sigma \equiv \sigma$.

Symmetry If $\sigma \equiv \eta$, then there exists $\rho \in G$ such that $\sigma = C_\rho(\eta)$. Hence $\eta = (C_\rho)^{\circ -1}(\sigma) = C_{\rho^{-1}}(\sigma)$ and thus $\eta \equiv \sigma$.

Transitivity If $\sigma \equiv \eta$ and $\eta \equiv \gamma$, then there exist $\xi \in G$ and $\zeta \in G$ such that $\sigma = C_\xi(\eta)$. and $\eta = C_\zeta(\gamma)$. Consequently,

$$\sigma = C_\xi(\eta) = C_\xi[C_\zeta(\gamma)] = (C_\xi \circ C_\zeta)(\gamma) = C_{\xi\bullet\zeta}(\gamma).$$

Therefore, $\sigma \equiv \gamma$. □

6.1.4 Exercises

Exercise 971 Prove that the set $\{1\}$ with integer multiplication forms a group.

Exercise 972 Prove that the set $\{-1, 1\}$ with integer multiplication forms a group.

Exercise 973 Prove that the set $\{0\}$ with integer addition forms a group.

Exercise 974 Prove that the set $\{0, 1\}$ with integer multiplication forms a monoid but *not* a group.

Exercise 975 Determine whether the set $\{0\}$ with integer multiplication forms a group.

Exercise 976 Determine whether $\{-1, 0, 1\}$ with integer multiplication is a group.

Exercise 977 Prove that the set of all positive powers of 2 with integer multiplication forms a semi-group.

Exercise 978 For each non-zero integer $K \in \mathbb{Z}^*$, prove that the set of all non-negative powers of K with integer multiplication forms a monoid.

Exercise 979 For each $K \in \mathbb{Z}^*$, prove that *if* the set of all non-negative powers of K with integer multiplication forms a group, *then* either $K = -1$ or $K = 1$.

Exercise 980 For every group $(G, \bullet)$, prove that $g^{m*n} = (g^m)^n$ for all $g \in G$, and $m, n \in \mathbb{N}$.

Exercise 981 For all distinct prime integers $P, Q, \in \mathbb{N}^*$, prove that the set

$$G := \{(P/Q)^N \in \mathbb{Z} : N \in \mathbb{Z}\}$$

of all integral powers of P/Q with integer multiplication forms a group.

Exercise 982 For all distinct prime integers $K, L, P, Q, \in \mathbb{N}^*$, prove that the set

$$G := \{(K/L)^M * (P/Q) * N \in \mathbb{Z} : (M \in \mathbb{Z}) \wedge (N \in \mathbb{Z})\}$$

of all products of powers of K/L and P/Q with integer multiplication is a group.

Exercise 983 Prove that each singleton $\{X\}$ with $X \bullet X := X$ forms a group.

Exercise 984 Prove that each pair $\{X, Y\}$ with $X \neq Y$ forms a group with $\bullet$ defined by $X \bullet X := X, X \bullet Y := Y, Y \bullet X := Y, Y \bullet Y := X$.

Exercise 985 For each singleton $\{X\}$ prove that the specification $X \bullet X := X$ is the only definition of $\bullet : \{X\} \times \{X\} \to \{X\}$ such that $(\{X\}, \bullet)$ is a group.

Exercise 986 For each pair $\{X, Y\}$ with $X \neq Y$ determine all the different specifications of $\bullet : \{X, Y\} \times \{X, Y\} \to \{X, Y\}$ such that $(\{X, Y\}, \bullet)$ is a group.

Exercise 987 Prove that the set of all even integers with addition forms a group.

Exercise 988 For each integer $M \geq 2$, prove that $\mathbb{Z}_M = \{[0]_M, \ldots, [M-1]_M\}$ with modular addition forms a group.

Exercise 989 For each prime $P \in \mathbb{N}^*$, prove that $\mathbb{Z}_P^* = \{[1]_P, \ldots, [P-1]_P\}$ with modular multiplication forms a group.

Exercise 990 For each integer $M \geq 2$, prove that the set of all elements $[K]_M \in \{[1]_M, \ldots, [M-1]_M\}$ relatively prime to M is a group with modular multiplication.

6.2 PERMUTATIONS

6.2.1 Permutations and Orbits

The term "permutation" is synonymous with "bijection" but usage commonly employs "permutation" in the context of countable sets, in particular, with finite sets and applications. For instance, the word "swap" results from the word "wasp" by a permutation.

Definition 838 A **permutation** of a set S is a bijection $\sigma : S \to S$, with its domain and range both coinciding with S. A **permutation of** n **elements** is a permutation $\sigma : \{1, \ldots, n\} \to \{1, \ldots, n\}$ of the first n positive integers.

Example 839 For $n := 0$, $S = \varnothing$, and $\varnothing : \varnothing \to \varnothing$ is the only permutation of $\varnothing$.

Example 840 For $n := 1$ and the set $S = \{1\}$ with one element, the identity $\iota : \{1\} \to \{1\}$, $1 \mapsto 1$ is the only permutation of one element.

A common notation for permutations of n elements consists of two rows of integers. The first row lists the domain in increasing order, $1, 2, \ldots, n-1, n$, while the second row lists the range in the order $\sigma(1), \sigma(2), \ldots, \sigma(n-1), \sigma(n)$:

$$\begin{pmatrix} 1 & 2 & 3 & \ldots & k & \ldots & n-2 & n-1 & n \\ \sigma(1) & \sigma(2) & \sigma(3) & \ldots & \sigma(k) & \ldots & \sigma(n-2) & \sigma(n-1) & \sigma(n) \end{pmatrix}.$$

Example 841 A common method to encrypt messages consists of permuting the alphabet (which could represent the first twenty-six positive integers to fit in the framework of definition 838). For instance, the coding machine `ENIGMA D` could encode text through several permutations, including the following permutation [2, p. 110]:

$$\rho_I := \begin{pmatrix} \mathtt{ABCDEFGHIJKLMNOPQRSTUVWXYZ} \\ \mathtt{EKMFLGDQVZNTOWYHXUSPAIBRCJ} \end{pmatrix}.$$

For instance, the first column shows that $\rho_I(\mathtt{A}) = \mathtt{E}$.

Example 842 For $n := 2$ and the set $S = \{1, 2\}$ with two elements, there exist exactly two permutations of two elements. Indeed, for each permutation $\sigma : \{1, 2\} \to \{1, 2\}$, only two alternatives exist for $\sigma(1)$: either $\sigma(1) = 1$ or $\sigma(1) = 2$; in each alternative, only one possibility remains for $\sigma(2)$, because σ must be injective and surjective. Thus, either $\sigma = \iota$, the identity, or $\sigma = \tau$, which swaps 1 and 2:

$$\iota = \begin{pmatrix} 1 & 2 \\ 1 & 2 \end{pmatrix}, \qquad \tau = \begin{pmatrix} 1 & 2 \\ 2 & 1 \end{pmatrix}.$$

For τ, the first column gives $\tau(1) = 2$, and the second column gives $\tau(2) = 1$.

Example 843 With $n := 4$ consider the permutation σ defined by

$$\begin{aligned} \sigma &= \begin{pmatrix} 1 & 2 & 3 & 4 \\ 3 & 4 & 2 & 1 \end{pmatrix} \\ &= \begin{pmatrix} 1 & 2 & 3 & 4 \\ \sigma(1) & \sigma(2) & \sigma(3) & \sigma(4) \end{pmatrix}. \end{aligned}$$

Thus, $\sigma(1) = 3$, $\sigma(2) = 4$, $\sigma(3) = 2$, $\sigma(4) = 1$.

A subsequent discussion will demonstrate how to construct recursively all the permutations of n elements for every n.

Definition 844 The **symmetric group on a set** S is the set $\mathcal{S}(S)$ of all permutations of S with the composition of functions $\circ$. The **symmetric group on n elements** is the set $\mathcal{S}_n$ of all permutations of n elements with the composition of functions $\circ$.

Example 845 The preceding examples have shown that

$$\begin{aligned} \mathcal{S}_0 &= \{\varnothing\}, \\ \mathcal{S}_1 &= \{\iota\} = \left\{\begin{pmatrix} 1 \\ 1 \end{pmatrix}\right\}, \\ \mathcal{S}_2 &= \{\iota, \tau\} = \left\{\begin{pmatrix} 1 & 2 \\ 1 & 2 \end{pmatrix}, \begin{pmatrix} 1 & 2 \\ 2 & 1 \end{pmatrix}\right\}. \end{aligned}$$

Theorem 846 *For each set S, the set $\mathcal{S}(S)$ of all pemutations of S forms a group with the operation of composition of functions $\circ$:*

$$\begin{aligned} \circ : \mathcal{S}(S) \times \mathcal{S}(S) &\rightarrow \mathcal{S}(S), \\ (\sigma, \eta) &\mapsto \sigma \circ \eta. \end{aligned}$$

Proof. This proof verifies the conditions listed in the definition of a group.

Well-defined operation For all permutations $\sigma, \eta \in \mathcal{S}(S)$, the composition $\sigma \circ \eta$ exists and is again a permutation $\sigma \circ \eta \in \mathcal{S}(S)$, because the domain of σ coincides with the domain of η, namely S, and because the composition of two bijections is again a bijection (theorem 394). Thus, the co-domain of the function $\circ : (\mathcal{S}(S) \times \mathcal{S}(S)) \rightarrow \mathcal{S}(S)$ is indeed $\mathcal{S}(S)$.

Associativity The binary operation of composition $\circ$ is associative for all functions (theorem 380), whence, in particular, for permutations.

Identity Moreover, the composition with the identity function $\iota := I_S$ fixes every permutation $\sigma \in \mathcal{S}_n$; indeed, $\iota \circ \sigma = \sigma$ and $\sigma \circ \iota = \sigma$, as it does for every function. Therefore, the identity function $\iota : S \rightarrow S$ is also the identity element in $(\mathcal{S}(S), \circ)$.

Inverses Because every permutation $\sigma \in \mathcal{S}(S)$ is a bijection, it has an inverse $\sigma^{-1} : S \rightarrow S$ (theorem 398). Here $\sigma^{-1} \in \mathcal{S}(S)$, and $\sigma^{-1} \circ \sigma = \iota = \sigma \circ \sigma^{-1}$. □

Example 847 With $n := 4$ consider the permutation σ introduced in example 843:

$$\sigma = \begin{pmatrix} 1 & 2 & 3 & 4 \\ 3 & 4 & 2 & 1 \end{pmatrix}.$$

Swapping both rows and then rearranging the first line in increasing order yields σ^{-1}:

$$\sigma^{-1} = \begin{pmatrix} 3 & 4 & 2 & 1 \\ 1 & 2 & 3 & 4 \end{pmatrix} = \begin{pmatrix} 1 & 2 & 3 & 4 \\ 4 & 3 & 1 & 2 \end{pmatrix}.$$

Thus, $\sigma^{-1}(1) = 4$, $\sigma^{-1}(2) = 3$, $\sigma^{-1}(3) = 1$, $\sigma^{-1}(4) = 2$. As a verification, and as an example of a calculation of a composition,

$$\begin{array}{rclclcl}
(\sigma^{-1} \circ \sigma)(1) & = & \sigma^{-1}[\sigma(1)] & = & \sigma^{-1}(3) & = & 1, \\
(\sigma^{-1} \circ \sigma)(2) & = & \sigma^{-1}[\sigma(2)] & = & \sigma^{-1}(4) & = & 2, \\
(\sigma^{-1} \circ \sigma)(3) & = & \sigma^{-1}[\sigma(3)] & = & \sigma^{-1}(2) & = & 3, \\
(\sigma^{-1} \circ \sigma)(4) & = & \sigma^{-1}[\sigma(4)] & = & \sigma^{-1}(1) & = & 4.
\end{array}$$

Thus $\sigma^{-1} \circ \sigma = \iota$ is the identity, whence σ^{-1} is the inverse of σ.

Each permutation of a set A partitions A into equivalence classes called orbits.

Definition 848 (Orbits of permutations) For each set A, for each $X \in A$, and for each bijection $F : A \to A$, the **orbit** of X by F is the subset $[X]_F \subseteq A$ defined by

$$[X]_F := \{F^{\circ k}(X) \in A : k \in \mathbb{Z}\}.$$

In particular, for each $n \in \mathbb{N}$, for each $i \in \{1, \ldots, n\} =: A$, and for each permutation $F := \sigma \in \mathcal{S}_n$, the **orbit** of i by σ is the subset $[i]_\sigma \subseteq \{1, \ldots, n\}$ defined by

$$[i]_\sigma := \{\sigma^{\circ k}(i) \in \{1, \ldots, n\} : k \in \mathbb{Z}\}.$$

Theorem 849 *For each set A and each bijection $F : A \to A$, the orbits partition A.*

Proof. For each bijection $F : A \to A$, let $\mathcal{F}$ be the set of corresponding orbits:

$$\mathcal{F} := \{[X]_F \in \mathcal{P}(A) : X \in A\}.$$

The first part of the proof shows that the orbits cover the domain of the bijection. Specifically, $X = I_A(X) = F^{\circ 0}(X)$ whence $X \in [X]_F$ for each $X \in A$. Hence

$$A \subseteq \bigcup_{X \in A} [X]_F = \bigcup \mathcal{F}.$$

For the converse inclusion, because $[X]_F \subseteq A$ for every $X \in A$, it follows that

$$\bigcup \mathcal{F} = \bigcup_{X \in A} [X]_F \subseteq A.$$

From $X \in [X]_F$ it also follows that $[X]_F \neq \varnothing$, so that every orbit is non-empty.

The second part of the proof verifies that distinct orbits are disjoint. By contraposition, the proof shows that orbits that have a common element coincide. Indeed, if $X \in [Y]_F \cap [Z]_F$, then there exist $k, \ell \in \mathbb{Z}$ such that

$$F^{\circ k}(Y) = X = F^{\circ \ell}(Z).$$

Consequently,

$$\begin{array}{rcl}
Y & = & F^{\circ -k}(X) = F^{\circ \ell - k}(Z), \\
F^{\circ m}(Y) & = & F^{\circ m + \ell - k}(Z),
\end{array}$$

$$\begin{array}{rcl}
Z & = & F^{\circ -\ell}(X) = F^{\circ k - \ell}(Y), \\
F^{\circ m}(Z) & = & F^{\circ m + k - \ell}(Y).
\end{array}$$

Therefore, $[Y]_F = [Z]_F$. □

Example 850 With $A := \{1, 2\}$, the identity $F := I_A = \iota$ partitions A into two orbits, $[1]_\iota = \{1\}$ and $[2]_\iota = \{2\}$, because $\iota^{\circ k}(1) = 1$ and $\iota^{\circ k}(2) = 2$ for every $k \in \mathbb{Z}$.

In contrast, the permutation

$$F := \tau = \begin{pmatrix} 1 & 2 \\ 2 & 1 \end{pmatrix}$$

partitions A into one orbit, $[1]_\tau = \{1, 2\} = [2]_\tau$. Indeed, $\tau^{\circ 0}(1) = 1$ and $\tau^{\circ 1}(1) = 2$, so that $[1]_\tau = \{1, 2\}$. Similarly, $\tau^{\circ 0}(2) = 2$ and $\tau^{\circ 1}(2) = 1$, so that $[2]_\tau = \{1, 2\}$.

Example 851 With $n := 4$ consider the permutation σ introduced in example 843:

$$\sigma = \begin{pmatrix} 1 & 2 & 3 & 4 \\ 3 & 4 & 2 & 1 \end{pmatrix}.$$

To determine the orbit $[1]_\sigma$ of 1 under σ, calculate all the powers $\sigma^{\circ k}(1)$:

$$\begin{array}{rclclcl} \sigma(1) & = & 3, & & & & \\ \sigma^{\circ 2}(1) & = & \sigma[\sigma(1)] & = & \sigma(3) & = & 2, \\ \sigma^{\circ 3}(1) & = & \sigma[\sigma^{\circ 2}(1)] & = & \sigma(2) & = & 4, \\ \sigma^{\circ 4}(1) & = & \sigma[\sigma^{\circ 3}(1)] & = & \sigma(4) & = & 1. \end{array}$$

Thus the orbit contains all the elements, so that $\{1, 2, 3, 4\} \subseteq [1]_\sigma$. Because the reverse inclusion $[1]_\sigma \subseteq \{1, 2, 3, 4\}$ holds by definition of the orbit, from theorem 290 it follows that for this example $[1]_\sigma = \{1, 2, 3, 4\}$.

6.2.2 Transpositions

Permutations that interchange only two elements correspond to the common transposition of two digits in a number or two letters in a word. For example, the words "net" and "ten" differ from each other by a transposition of their first and last letters.

Definition 852 A **transposition** — also called a **swap** [2, p. 45] — is a permutation $\tau_{r,s} : S \to S$ that swaps two distinct elements $r \neq s$ of S and fixes all the others:

$$\begin{array}{rcll} \tau_{r,s}(r) & = & s, & \\ \tau_{r,s}(s) & = & r, & \\ \tau_{r,s}(x) & = & x & \text{for every } x \in S \setminus \{r, s\}. \end{array}$$

Example 853 The permutation $\tau \in S_2$ defined by

$$\tau := \begin{pmatrix} 1 & 2 \\ 2 & 1 \end{pmatrix}$$

is the transposition $\tau_{1,2}$ that swaps 1 and 2.

The determination of the inverse of a transposition does not require any calculation.

Theorem 854 *Each transposition coincides with its inverse:* $\tau_{r,s} = (\tau_{r,s})^{-1}$.

Proof. Verify that $\tau_{r,s} \circ \tau_{r,s} = \iota_S$, the identity function on S:

$$\begin{array}{rclclcll} (\tau_{r,s} \circ \tau_{r,s})(r) & = & \tau_{r,s}[\tau_{r,s}(r)] & = & \tau_{r,s}(s) & = & r, & \\ (\tau_{r,s} \circ \tau_{r,s})(s) & = & \tau_{r,s}[\tau_{r,s}(s)] & = & \tau_{r,s}(r) & = & s, & \\ (\tau_{r,s} \circ \tau_{r,s})(x) & = & \tau_{r,s}[\tau_{r,s}(x)] & = & \tau_{r,s}(x) & = & x & \text{for every } x \in S \setminus \{r, s\}. \end{array}$$

Thus $(\tau_{r,s} \circ \tau_{r,s})(u) = u$ for every $u \in S$. Therefore, $\tau_{r,s} = (\tau_{r,s})^{-1}$. □

Transpositions facilitate the analysis of permutations by reducing them to permutations of fewer elements.

Theorem 855 *For each positive integer n, each permutation of $n+1$ elements is the composition of a permutation of n elements (extended by fixing $n+1$) preceded by at most one transposition of $n+1$ elements.*

Proof. For each permutation $\sigma \in \mathcal{S}_{n+1}$, either of two cases can arise.

In the first case, if $\sigma(n+1) = n+1$, then σ is the extension to $\{1, \dots, n+1\}$ of its restriction $\sigma|_{\{1,\dots,n\}}$, which is a permutation of n elements.

In the second case, if $\sigma(n+1) \neq n+1$, then consider the transposition $\tau_{n+1,\sigma^{-1}(n+1)} \in \mathcal{S}_{n+1}$, which swaps $n+1$ and its preimage $\sigma^{-1}(n+1)$. Consider the composition

$$\eta := \sigma \circ \tau_{n+1,\sigma^{-1}(n+1)},$$

$$\begin{aligned} \eta(n+1) &= \sigma \circ \tau_{n+1,\sigma^{-1}(n+1)}(n+1) \\ &= \sigma\left[\tau_{n+1,\sigma^{-1}(n+1)}(n+1)\right] \\ &= \sigma\left[\sigma^{-1}(n+1)\right] \\ &= n+1. \end{aligned}$$

Thus, η fixes $n+1$, so that its restriction $\eta|_{\{1,\dots,n\}}$ is a permutation of n elements. Consequently, solving the equation $\eta = \sigma \circ \tau_{n+1,\sigma^{-1}(n+1)}$ for σ yields

$$\begin{aligned} \eta &= \sigma \circ \tau_{n+1,\sigma^{-1}(n+1)}, \\ \eta \circ \left(\tau_{n+1,\sigma^{-1}(n+1)}\right)^{-1} &= \sigma, \\ \eta \circ \tau_{n+1,\sigma^{-1}(n+1)} &= \sigma. \end{aligned}$$

□

Example 856 With $n := 4$ consider the permutation σ introduced in example 843:

$$\sigma = \begin{pmatrix} 1 & 2 & 3 & 4 \\ 3 & 4 & 2 & 1 \end{pmatrix}.$$

Apply theorem 855 with $n+1 = 4$, so that $\sigma^{-1}(n+1) = \sigma^{-1}(4) = 2$ and

$$\tau := \tau_{n+1,\sigma^{-1}(n+1)} = \tau_{4,2} = \begin{pmatrix} 1 & 2 & 3 & 4 \\ 1 & 4 & 3 & 2 \end{pmatrix}.$$

Thus,

$$\eta = \sigma \circ \tau_{4,2} = \begin{pmatrix} 1 & 2 & 3 & 4 \\ 3 & 4 & 2 & 1 \end{pmatrix} \circ \begin{pmatrix} 1 & 2 & 3 & 4 \\ 1 & 4 & 3 & 2 \end{pmatrix} = \begin{pmatrix} 1 & 2 & 3 & 4 \\ 3 & 1 & 2 & 4 \end{pmatrix}.$$

Verification:

$$\eta \circ \tau_{4,2} = \begin{pmatrix} 1 & 2 & 3 & 4 \\ 3 & 1 & 2 & 4 \end{pmatrix} \circ \begin{pmatrix} 1 & 2 & 3 & 4 \\ 1 & 4 & 3 & 2 \end{pmatrix} = \begin{pmatrix} 1 & 2 & 3 & 4 \\ 3 & 4 & 2 & 1 \end{pmatrix} = \sigma.$$

Repeating the procedure specified in the proof of theorem 855 reduces every permutation to a product of transpositions.

Theorem 857 *Each permutation of $n > 0$ elements is the composition of at most $n-1$ transpositions.*

Proof. This proof proceeds by induction (theorem 480) with n.

Initial step Example 845 confirms that for $n = 1$, each permutation is a composition of 0 transposition, and that for $n = 2$, each permutation is the composition of either 0 or 1 transposition.

Induction hypothesis Assume that the theorem holds for $n := k$, so that each permutation of k elements is the composition of at most $k - 1$ transpositions.

Induction step For $n := k + 1$, for each permutation $\sigma \in \mathcal{S}_{k+1}$, theorem 855 shows that two cases arise. In the first case, $\sigma(k + 1) = k + 1$, hence σ permutes only k elements, and the induction hypothesis guarantees that σ is the composition of at most $k - 1 \leq (k+1) - 1$ transpositions. In the second case, $\sigma = \eta \circ \tau$ for some permutation η of k elements and some transposition τ. The induction hypothesis then guarantees that η is the composition of at most $k - 1$ transpositions, whence $\sigma = \eta \circ \tau$ is the composition of at most $(k - 1) + 1 = k = (k + 1) - 1$ transpositions. □

Example 858 With $n := 4$ consider the permutation σ from examples 843 and 856:

$$\sigma = \begin{pmatrix} 1 & 2 & 3 & 4 \\ 3 & 4 & 2 & 1 \end{pmatrix}.$$

Firstly, apply theorem 857 with $n + 1 = 4$, so that $\sigma^{-1}(n + 1) = \sigma^{-1}(4) = 2$ and

$$\tau_1 := \tau_{n+1,\sigma^{-1}(n+1)} = \tau_{4,2} = \begin{pmatrix} 1 & 2 & 3 & 4 \\ 1 & 4 & 3 & 2 \end{pmatrix}.$$

Thus, as in example 856,

$$\eta_1 = \sigma \circ \tau_1 = \begin{pmatrix} 1 & 2 & 3 & 4 \\ 3 & 4 & 2 & 1 \end{pmatrix} \circ \begin{pmatrix} 1 & 2 & 3 & 4 \\ 1 & 4 & 3 & 2 \end{pmatrix} = \begin{pmatrix} 1 & 2 & 3 & 4 \\ 3 & 1 & 2 & 4 \end{pmatrix}.$$

Secondly, apply theorem 857 to η_1 and $n + 1 = 3$, so that $\eta_1^{-1}(n + 1) = \eta_1^{-1}(3) = 1$ and

$$\tau_2 := \tau_{n+1,\eta_1^{-1}(n+1)} = \tau_{3,1} = \begin{pmatrix} 1 & 2 & 3 & 4 \\ 3 & 2 & 1 & 4 \end{pmatrix}.$$

Thus,

$$\eta_2 := \eta_1 \circ \tau_2 = \begin{pmatrix} 1 & 2 & 3 & 4 \\ 3 & 1 & 2 & 4 \end{pmatrix} \circ \begin{pmatrix} 1 & 2 & 3 & 4 \\ 3 & 2 & 1 & 4 \end{pmatrix} = \begin{pmatrix} 1 & 2 & 3 & 4 \\ 2 & 1 & 3 & 4 \end{pmatrix}.$$

Finally, observe that $\eta_2 = \tau_{1,2}$ is a transposition. Consequently, with $\tau_3 := \eta_2$, solving for σ yields

$$\begin{aligned} \tau_3 = \eta_2 = \eta_1 \circ \tau_2 &= (\sigma \circ \tau_1) \circ \tau_2 \\ \tau_3 \circ \tau_2 &= \sigma \circ \tau_1 \\ (\tau_3 \circ \tau_2) \circ \tau_1 &= \sigma. \end{aligned}$$

Verification:

$$\begin{aligned} (\tau_3 \circ \tau_2) \circ \tau_1 &= \left[\begin{pmatrix} 1 & 2 & 3 & 4 \\ 2 & 1 & 3 & 4 \end{pmatrix} \circ \begin{pmatrix} 1 & 2 & 3 & 4 \\ 3 & 2 & 1 & 4 \end{pmatrix}\right] \circ \begin{pmatrix} 1 & 2 & 3 & 4 \\ 1 & 4 & 3 & 2 \end{pmatrix} \\ &= \begin{pmatrix} 1 & 2 & 3 & 4 \\ 3 & 1 & 2 & 4 \end{pmatrix} \circ \begin{pmatrix} 1 & 2 & 3 & 4 \\ 1 & 4 & 3 & 2 \end{pmatrix} \\ &= \begin{pmatrix} 1 & 2 & 3 & 4 \\ 3 & 4 & 2 & 1 \end{pmatrix} \\ &= \sigma. \end{aligned}$$

The next theorem provides the total number of permutations of any finite set.

Theorem 859 *There exist exactly $n!$ permutations of n elements.*

Proof. With $n!$ defined in definition 499, this proof proceeds by induction with n.

Initial step For $n := 0$ and $n := 1$, example 845 confirms that there exist exactly $0! = 1 = 1!$ permutation.

Induction hypothesis Assume that there exists $k \in \mathbb{N}$ such that the theorem holds for some $n := k$, and consider the set $\mathcal{S}_{k+1}$ of all permutations of $k+1$ elements.

Induction step For each permutation $\sigma \in \mathcal{S}_{k+1}$, there exist exactly $k+1$ distinct possibilities for $\sigma(k+1) \in \{1, \ldots, k+1\}$. Consequently, $\mathcal{S}_{k+1}$ splits into $k+1$ disjoint subsets $\mathcal{S}_{k+1}^{(1)}, \ldots, \mathcal{S}_{k+1}^{(k+1)}$ defined by

$$\mathcal{S}_{k+1}^{(m)} := \{\sigma \in \mathcal{S}_{k+1} : \sigma(k+1) = m\}.$$

Thus $\mathcal{S}_{k+1}^{(m)}$ consists of every permutation $\sigma \in \mathcal{S}_{k+1}$ such that $\sigma(k+1) = m$. Within each such subset, consider the composition by the transposition $\tau_{k+1,\sigma^{-1}(k+1)} = \tau_{k+1,m}$, which by theorem 855 yields all $k!$ (induction hypothesis) permutations of k elements:

$$\sigma \circ \tau_{k+1,m} \in \mathcal{S}_k.$$

Thus, $\mathcal{S}_{k+1}$ splits into $k+1$ disjoint subsets with $k!$ distinct permutations each, for a total of $(k+1)k! = (k+1)!$ permutations. □

6.2.3 Exercises

Exercise 991 Write down all the permutations of three elements.

Exercise 992 Write down all the permutations of four elements.

Exercise 993 Determine the number of permutations of $\{0, 1, 2, 3, 4, 5, 6, 7, 8, 9\}$.

Exercise 994 Determine the total number of permutations of the 20 letters of the Roman alphabet, also called the standard alphabet with twenty letters: [2, p. 46–47]:

A B C D D F G H I L M N O P Q R S T V X.

Exercise 995 Determine the total number of permutations of the 26 letters of the English alphabet.

Exercise 996 Determine the total number of permutations of the 29 letters of the German alphabet, which consists of the 26 letters of the English alphabet and Ä, Ö, Ü.

Exercise 997 Determine the total number of permutations of the 30 letters of the Swedish alphabet, which consists of the 29 letters of the German alphabet and Å.

Exercise 998 Determine the total number of permutations of the 31 letters of the Norwegian alphabet, which consists of the 30 letters of the Swedish alphabet and Ø.

Exercise 999 Determine the total number of permutations of the letters of the Russian alphabet, which consists of 33 Cyrillic letters.

Exercise 1000 Determine the number of permutations of the 36 characters consisting of the 26 letters of the English alphabet and the ten digits $\{0,1,2,3,4,5,6,7,8,9\}$.

Exercise 1001 Determine the number of permutations of the 39 characters consisting of the 29 letters of the German alphabet and the ten digits $\{0,1,2,3,4,5,6,7,8,9\}$.

Exercise 1002 Determine the number of permutations of the 40 characters consisting of the 30 letters of the Swedish alphabet and the ten digits $\{0,1,2,3,4,5,6,7,8,9\}$.

Exercise 1003 Determine the number of permutations of the 41 characters consisting of the 31 letters of the Norwegian alphabet and the ten digits $\{0,1,2,3,4,5,6,7,8,9\}$.

Exercise 1004 Determine the number of permutations of the 43 characters consisting of the 33 letters of the Russian alphabet and the ten digits $\{0,1,2,3,4,5,6,7,8,9\}$.

Exercise 1005 Determine whether the set of transpositions of n elements is a group.

Exercise 1006 Factor the following permutation as a product of transpositions:

$$\alpha = \begin{pmatrix} 1 & 2 & 3 & 4 \\ 4 & 3 & 1 & 2 \end{pmatrix}.$$

Exercise 1007 Factor the following permutation as a product of transpositions:

$$\beta = \begin{pmatrix} 1 & 2 & 3 & 4 \\ 3 & 4 & 1 & 2 \end{pmatrix}.$$

Exercise 1008 Determine the orbits of the following permutation:

$$\gamma = \begin{pmatrix} 1 & 2 & 3 & 4 \\ 4 & 3 & 1 & 2 \end{pmatrix}.$$

Exercise 1009 Determine the orbits of the following permutation:

$$\delta = \begin{pmatrix} 1 & 2 & 3 & 4 \\ 3 & 4 & 1 & 2 \end{pmatrix}.$$

Exercise 1010 For each permutation σ, prove that the orbits of σ consist only of singletons if and only if $\sigma = \iota$ (the identity).

6.3 CYCLIC PERMUTATIONS

6.3.1 Cycles

Transpositions consitute a particular type of permutations that shift elements in a cyclic pattern, which also form the basis for the design and deciphering of various ciphers.

Definition 860 A permutation σ of n elements is **cyclic** with **length** ℓ — also abbreviated as an **ℓ-cycle** — if and only if there exists an injection

$$\begin{array}{rcl} s : \mathbb{Z}_\ell = \{[1]_\ell, \ldots, [\ell]_\ell\} & \rightarrow & \{1, \ldots, n\}, \\ {[j]} & \mapsto & n_{[j]}, \end{array}$$

(with the modular integers $\mathbb{Z}_\ell$ defined in chapter 5) such that $\sigma(n_{[j]}) = n_{[j+1]}$:

$$\sigma(n_{[1]}) = n_{[2]}, \ldots, \sigma(n_{[j]}) = n_{[j+1]}, \ldots, \sigma(n_{[\ell-1]}) = n_{[\ell]}, \qquad \sigma(n_{[\ell]}) = n_{[1]},$$

with classes modulo ℓ, and σ fixes every element not in the range of the injection s. The notation

$$\sigma = \begin{pmatrix} n_1 & n_2 & \dots & n_{\ell-1} & n_\ell \end{pmatrix}$$

is common for such cycles [2, p. 44], [4, p. 150], [53, p. 46].

Example 861 According to Suetonius (Gaitus Suetonius Tranquillus, about 69–140 A.D., cited in [2, p. 47]), the Roman emperor Julius Caesar (about 100–44 B.C.) encrypted messages by the following cyclic permutation — in effect, a cyclic shift by three letters to the left — of the twenty letters of the Roman alphabet:

$$\begin{pmatrix} \mathtt{ABCDEFGHILMNOPQRSTVX} \\ \mathtt{DEFGHILMNOPQRSTVXABC} \end{pmatrix}.$$

Example 862 The ENIGMA ciphering machine contains a cyclic shift of the English alphabet one letter to the right [63]:

$$\begin{aligned} \beta &:= \begin{pmatrix} \mathtt{ABCDEFGHIJKLMNOPQRSTUVWXYZ} \\ \mathtt{BCDEFGHIJKLMNOPQRSTUVWXYZA} \end{pmatrix} \\ &= \begin{pmatrix} \mathtt{ABCDEFGHIJKLMNOPQRSTUVWXYZ} \end{pmatrix}. \end{aligned}$$

Example 863 Every transposition $\tau_{r,s} = (r \quad s)$ is a cycle with length $\ell = 2$.

Example 864 The permutation σ from examples 843 and 856,

$$\sigma = \begin{pmatrix} 1 & 2 & 3 & 4 \\ 3 & 4 & 2 & 1 \end{pmatrix} = \begin{pmatrix} 1 & 3 & 2 & 4 \end{pmatrix},$$

is a cycle with length $\ell = 4$, because

$$\sigma(1) = 3, \quad \sigma(3) = 2, \quad \sigma(2) = 4, \quad \sigma(4) = 1.$$

The inverse of a cycle coincides with a power of the same cycle.

Theorem 865 *If σ is a cycle with length ℓ, then $\sigma^{-1} = \sigma^{\ell-1}$.*

Proof. This proof proceeds by induction with powers k of σ. For $k := 0$, $\sigma^k = \sigma^0 = \iota$ is the identity permutation, so that $\sigma^0(m) = m$ for every $m \in \{1, \dots, n\}$.

For $k := 1$, if σ is a cycle with length ℓ, then there exists a sequence such that $\sigma^1 = \sigma = \begin{pmatrix} n_1 & n_2 & \dots & n_{\ell-1} & n_\ell \end{pmatrix}$.

The first part of the proof pertains to every element m *not* in the range of the sequence, so that $\sigma(m) = m$. Assume that there exists $k \in \mathbb{N}$ such that $\sigma^k(m) = m$ for every such m. Then $\sigma^{k+1}(m) = \sigma[\sigma^k(m)] = \sigma(m) = m$, and hence also $\sigma^\ell(m) = m$, for every such m.

The second part of the proof pertains to every element $n_{[j]}$ in the range of the injection, for which $\sigma^1(n_{[j]}) = n_{[j+1]}$ modulo ℓ. Assume that there exists $k \in \mathbb{N}$ such that $\sigma^k(n_{[j]}) = n_{[j+k]}$ for every $j \in \{1, \dots, \ell\}$. Then $\sigma^{k+1}(n_{[j]}) = \sigma[\sigma^k(n_{[j]})] = \sigma(n_{[j+k]}) = \sigma(n_{[j+k+1]})$, and hence also $\sigma^\ell(n_{[j+k+1]}) = n_{[j+\ell]} = n_{[j]}$, for every such $j \in \{1, \dots, \ell\}$. Thus $\iota = \sigma^\ell = \sigma^{\ell-1} \circ \sigma^1$, whence $\sigma^{\ell-1} = \sigma^{-1}$. □

The converse of theorem 865 fails, because for each non-cyclic permutation η of n elements there also exists $\ell \in \mathbb{N}^*$ with $\eta^{\ell-1} = \eta^{-1}$ (theorems 833, 846, 859).

Counterexample 866 The permutation $\eta := \tau_{1,2} \circ \tau_{3,4}$ is *not* cyclic. Yet $\eta^{-1} = \eta^1$.

6.3.2 Factorization of Permutations by Cycles

As every permutation of finitely many elements factors as a product of transpositions (theorem 857), every such permutation also factors as a composition of cycles. Factorizations through cycles will also play a rôle in the analysis of ciphering machines.

Definition 867 Two cycles β and γ are **disjoint** if and only if for each element k either $\beta(k) = k$ or $\gamma(k) = k$.

The first theorem shows that the composition of disjoint cycles commutes.

Theorem 868 *The composition of all disjoint cycles ρ and σ commutes: $\rho \circ \sigma = \sigma \circ \rho$.*

Proof. For each k either $\sigma(k) = k$ or $\rho(k) = k$. If $k = m_{[i]}$ and $\sigma(k) = k$, then

$$\begin{aligned}(\rho \circ \sigma)(k) = \rho[\sigma(m_{[i]})] &= \rho(m_{[i]}) \\ &= m_{[i+1]} \\ &= \sigma(m_{[i+1]}) = \sigma[\rho(m_{[i]})] = (\sigma \circ \rho)(m_{[i]}) = (\sigma \circ \rho)(k).\end{aligned}$$

Similarly, if $k = n_{[j]}$ and $\rho(k) = k$, then

$$\begin{aligned}(\rho \circ \sigma)(k) = \rho[\sigma(n_{[j]})] &= \rho(n_{[j+1]}) \\ &= n_{[j+1]} \\ &= \sigma(n_{[j]}) = \sigma[\rho(n_{[j]})] = (\sigma \circ \rho)(n_{[j]}) = (\sigma \circ \rho)(k).\end{aligned}$$

□

The next theorem shows the effect of non-disjoint transpositions.

Theorem 869 *For all u, v, w, $\tau_{u,v} \circ \tau_{v,w} \circ \tau_{u,v} = \tau_{u,w}$ whereas $\tau_{u,v} \circ \tau_{u,v} \circ \tau_{u,v} = \tau_{u,v}$.*

Proof. For the elements u, v, w,

$$\begin{aligned}(\tau_{u,v} \circ \tau_{v,w} \circ \tau_{u,v})(u) &= (\tau_{u,v} \circ \tau_{v,w})[\tau_{u,v}(u)] \\ &= \tau_{u,v}[\tau_{v,w}(v)] \\ &= \tau_{u,v}(w) \\ &= w \\ &= \tau_{u,w}(u),\end{aligned}$$

$$\begin{aligned}(\tau_{u,v} \circ \tau_{v,w} \circ \tau_{u,v})(w) &= (\tau_{u,v} \circ \tau_{v,w})[\tau_{u,v}(w)] \\ &= \tau_{u,v}[\tau_{v,w}(w)] \\ &= \tau_{u,v}(v) \\ &= u \\ &= \tau_{u,w}(w),\end{aligned}$$

$$\begin{aligned}(\tau_{u,v} \circ \tau_{v,w} \circ \tau_{u,v})(v) &= (\tau_{u,v} \circ \tau_{v,w})[\tau_{u,v}(v)] \\ &= \tau_{u,v}[\tau_{v,w}(u)] \\ &= \tau_{u,v}(u) \\ &= v \\ &= \tau_{u,w}(v).\end{aligned}$$

All the other elements remain fixed. Also, $\tau_{u,v} \circ \tau_{u,v} \circ \tau_{u,v} = \iota \circ \tau_{u,v}$. □

The next theorem shows that every permutation factors as a composition of cycles.

Theorem 870 *For every permutation σ of n elements, there exist disjoint cycles $\gamma_1, \ldots, \gamma_k$ with respective lengths $\ell_1, \ldots, \ell_k$ such that $\sigma = \gamma_k \circ \cdots \circ \gamma_1$.*

Proof. Every permutation $\sigma \in \mathcal{S}_n$ factors as a composition of transpositions $\sigma = \tau_{r_{n-1},s_{n-1}} \circ \cdots \circ \tau_{s_1,s_1}$, by theorem 857.

Define an equivalence relation $\equiv$ on the set of all pairs $\{(r_{n-1}, s_{n-1}), \ldots, (s_1, s_1)\}$ by $(r_i, s_i) \equiv (r_j, s_j)$ if and only if $\{r_i, s_i\} \cap \{r_j, s_j\} \neq \varnothing$. Thus two such pairs are "equivalent" if and only if they contain at least one common index.

If two pairs contain a common index, in the form (r_i, s_i) and (r_j, s_i), then the composition of the transpositions τ_{r_i,s_i} and τ_{r_j,s_i} yields a cycle

$$\tau_{r_i,s_i} \circ \tau_{r_j,s_i} = \tau_{r_i,s_i} \circ \tau_{s_i,r_j} = (r_i, s_i, r_j).$$

In contrast, if two pairs (r_i, s_i) and (r_j, s_j) do *not* contain any common index, so that $\{r_i, s_i\} \cap \{r_j, s_j\} = \varnothing$, then the corresponding transpositions τ_{r_i,s_i} and τ_{r_j,s_j} are disjoint cycles, and hence their composition commutes, by theorem 868.

Consequently, the factorization $\sigma = \tau_{r_{n-1},s_{n-1}} \circ \cdots \circ \tau_{s_1,s_1}$ can be arranged in blocks of mututally equivalent pairs, which form disjoint cycles. □

The number of cycles and their lengths remains invariant under conjugation by transpositions.

Theorem 871 *For each permutation σ factored as a composition $\sigma = \gamma_k \circ \cdots \circ \gamma_1$ of k disjoint cycles $\gamma_1, \ldots, \gamma_k$ with lengths $\ell_1, \ldots, \ell_k$, and for each transposition τ, the conjugation $\tau \circ \sigma \circ \tau$ also factors with k disjoint cycles with lengths $\ell_1, \ldots, \ell_k$.*

Proof. If τ and γ are disjoint cycles, then $\tau \circ \gamma \circ \tau = \tau \circ \tau \circ \gamma = \gamma$ by theorem 868.

If τ and γ are not disjoint, then the factorization of γ as a product of transpositions and theorem 869 show that $\tau \circ \gamma \circ \tau$ merely substitute one index for another. □

6.3.3 Exercises

Exercise 1011 Determine whether all cyclic permutations of n elements form a group.

Exercise 1012 Determine whether all even cyclic permutations of n elements form a group.

Exercise 1013 Prove that every permutation of two elements is a cycle.

Exercise 1014 Factor the following permutation as a product of cycles:

$$\alpha = \begin{pmatrix} 1 & 2 & 3 & 4 \\ 4 & 3 & 1 & 2 \end{pmatrix}.$$

Exercise 1015 Factor the following permutation as a product of cycles:

$$\beta = \begin{pmatrix} 1 & 2 & 3 & 4 \\ 3 & 4 & 1 & 2 \end{pmatrix}.$$

Exercise 1016 Prove that if $\gamma \in \mathcal{S}_n$ is a cycle of length k, then $k \leq n$.

Exercise 1017 Prove that if $\sigma \in \mathcal{S}_n$ is the composition of disjoint cycles $\gamma_1, \ldots, \gamma_h$ of lengths $k_1, \ldots, k_h$, then $\sum_{i=1}^{h} k_i \leq n$.

Exercise 1018 Prove that if $\sigma \in \mathcal{S}_n$ and if ℓ is the smallest natural number such that $\sigma^\ell = \iota$, then ℓ divides $n!$.

Exercise 1019 Provide an example of a permutation $\sigma \in \mathcal{S}_n$ such that $\sigma^\ell = \iota$ with $\ell \leq n$ but ℓ does *not* divide n.

Exercise 1020 Provide an example of a permutation $\sigma \in \mathcal{S}_n$ such that $\sigma^\ell \neq \iota$ for every $\ell \in \{1, \ldots, n\}$.

6.4 SIGNATURES OF PERMUTATIONS

6.4.1 Even and Odd Permutations

As an interlude, this section shows how different permutations can correspond to significantly different physical situations, for instance, in stereochemistry [67, Ch. 3].

Example 872 Each molecule of **lactic acid** consists of three atoms of carbon (C), six atoms of hydrogen (H), and three atoms of oxygen (O). The carbon atom lies near the center of the molecule, while the other atoms form four groups (H, OH, CH_3, COOH) positioned approximately at the vertices of a tetrahedron in space. Each molecule can rotate in space and thus permute the positions of the four groups. Yet not all 24 permutations of the four groups are physically possible. For instance, swapping the H and OH groups produces two different types — **isomers** — of lactic acid molecules, one called **levorotatory** and the other one called **dextrorotatory**, as in figure 6.2.

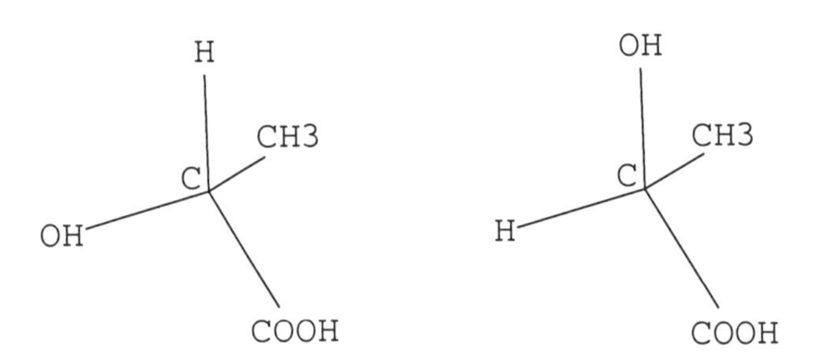

Figure 6.2: Left: levorotatory lactic acid. Right: dextrorotatory lactic acid.

The two isomers of lactic acid exhibit different characteristics, transmitting light in different ways in a phenomenon called **hemihedry**. Thus hemihedry constitutes physical evidence that dextrorotatory and levorotatory molecules are not the same molecules in different positions; they are different kinds of molecules. In other words, the transposition of the hydrogen atom and the group OH cannot occur through a mere rotation in space; it would require a dissociation of one isomer and then a recombination of its individual atoms into the other isomer.

In stereochemistry and in other applications, the permutations that can occur in space differ from those permutations that cannot occur in space by a mathematical feature called their **signature**.

Definition 873 A permutation $\sigma \in \mathcal{S}_n$ of n elements **re-orders** a pair of distinct integers (h, k) if and only if $h < k$ but $\sigma(h) > \sigma(k)$. With $p(\sigma)$ denoting the number of pairs re-ordered by σ, the **signature** [66, p. 53] — also called the **sign** [53, p. 48] or the **signum** [4, p. 319] — of σ is the integer $\text{sgn}\,(\sigma) := (-1)^{p(\sigma)}$. A permutation is **odd** or **even** according to whether its signature is -1 or 1 respectively.

Example 874 The identity permutation ι fixes every integer, hence ι fixes every pair (h, k), and, consequently, ι does not re-order any pair. Therefore, $p(\iota) = 0$ and $\text{sgn}\,(\iota) = (-1)^0 = 1$.

Example 875 The permutation $\tau_{1,2} \in \mathcal{S}_2$ defined by

$$\tau_{1,2} = \begin{pmatrix} 1 & 2 \\ 2 & 1 \end{pmatrix}$$

re-orders the pair $(1, 2)$, because $1 < 2$ but $\tau_{1,2}(1) > \tau_{1,2}(2)$. Also, $\tau_{1,2}$ does not re-order any other pair, because there exists no other pair (h, k) of distinct integers $h < k$ in $\{1, 2\} \times \{1, 2\}$. Consequently, $\tau_{1,2}$ re-orders exactly *one* pair (h, k) with $h < k$. Therefore, $p(\tau_{1,2}) = 1$ and $\operatorname{sgn}(\tau_{1,2}) = (-1)^{p(\tau_{1,2})} = (-1)^1 = -1$.

Example 876 Consider the permutation σ introduced in examples 843 and 856:

$$\sigma = \begin{pmatrix} 1 & 2 & 3 & 4 \\ 3 & 4 & 2 & 1 \end{pmatrix}.$$

To calculate the signature of σ, consider all pairs (h, k) and $(\sigma(h), \sigma(k))$ with $1 \leq h < k \leq 4$:

$$\begin{array}{llcllll} (1,2) & (\sigma(1),\sigma(2)) & = & (3,4) & 3<4 & \text{not re-ordered,} \\ (1,3) & (\sigma(1),\sigma(3)) & = & (3,2) & 3>2 & \text{re-ordered,} \\ (1,4) & (\sigma(1),\sigma(4)) & = & (3,1) & 3>1 & \text{re-ordered,} \\ (2,3) & (\sigma(2),\sigma(3)) & = & (4,2) & 4>2 & \text{re-ordered,} \\ (2,4) & (\sigma(2),\sigma(4)) & = & (4,1) & 4>1 & \text{re-ordered,} \\ (3,4) & (\sigma(3),\sigma(4)) & = & (2,1) & 2>1 & \text{re-ordered.} \end{array}$$

Thus, σ re-orders exactly five pairs, whence $p(\sigma) = 5$ and $\operatorname{sgn}(\sigma) = (-1)^{p(\sigma)} = (-1)^5 = -1$.

Theorem 877 *Every transposition has signature* -1.

Proof. For each transposition that re-orders r and s, with $r < s$,

$$\tau_{r,s} = \begin{pmatrix} 1 & 2 & \ldots & r-1 & r & r+1 & \ldots & s-1 & s & s+1 & \ldots & n-1 & n \\ 1 & 2 & \ldots & r-1 & s & r+1 & \ldots & s-1 & r & s+1 & \ldots & n-1 & n \end{pmatrix},$$

examine all pairs (h, k) such that $h < k$, determine whether $\tau_{r,s}(h) > \tau_{r,s}(k)$, and count all such pairs re-ordered by $\tau_{r,s}$. This proof proceeds according to four cases.

Case 1 If $h \notin \{r, s\}$ and $k \notin \{r, s\}$, then $\tau_{r,s}$ fixes both h and k, so that $h = \tau_{r,s}(h)$ and $k = \tau_{r,s}(k)$. Consequently, $\tau_{r,s}$ does *not* re-order (h, k).

Case 2 If $1 \leq h < r$ and $r \leq k \leq n$, then $\tau_{r,s}(h) = h$ and $\tau_{r,s}(k) \in \{r, \ldots, s, \ldots, n\}$, whence $\tau_{r,s}(h) = h < r \leq \tau_{r,s}(k)$, and $\tau_{r,s}$ does *not* re-order (h, k).

Case 3 If $h = r$ and $r + 1 \leq k \leq s$, then $\tau_{r,s}(h) = s$ but $\tau_{r,s}(k) \in \{r, \ldots, s - 1\}$, so that $\tau_{r,s}$ *re-orders* the pair (h, k). With $h = r$ fixed and with $s - r$ possibilities for $k \in \{r + 1, \ldots, s\}$, the transposition $\tau_{r,s}$ swaps $s - r$ such pairs.

Case 4 If $r + 1 \leq h \leq s - 1$ and $k = s$, then $\tau_{r,s}(h) = h$ but $\tau_{r,s}(k) = \tau_{r,s}(s) = r < h$, so that $\tau_{r,s}$ *re-orders* the pair (h, k). With $k = s$ fixed and with $(s - 1) - r$ possibilities for $h \in \{r + 1, \ldots, s - 1\}$, the transposition $\tau_{r,s}$ re-orders $(s - 1) - r$ such pairs.

Because the preceding four cases exhaust all possibilities for the pairs (h, k) such that $h < k$, and because they produce exactly $[s-r]+[(s-1)-r] = 2(s-r)+1$ such re-orderings, it follows that $p(\sigma) = 2(s - r) + 1$ is an odd integer, whence $\operatorname{sgn}(\sigma) = (-1)^{p(\sigma)} = -1$. □

Example 878 In example 872, the transposition from dextrorotary to levorotary lactic acid has signature -1, because it is a transposition.

The following theorem provides an alternative formula to calculate the signature.

Theorem 879 *For each permutation $\sigma \in \mathcal{S}_n$,*

$$sgn(\sigma) = \frac{\prod_{h<k}[\sigma(k) - \sigma(h)]}{\prod_{h<k}(k-h)}.$$

Proof. For each pair (h, k) with $h < k$, so that $(k-h)$ appears as a factor in the denominator, either $(k-h)$ or $(h-k)$ also appears as a factor in the numerator, depending upon whether σ re-orders $(r, s) := [\sigma^{-1}(h), \sigma^{-1}(k)]$. Indeed, with $r := \sigma^{-1}(h)$ and $s := \sigma^{-1}(k)$, two cases arise. In the first case, $s < r$, and then $(r-s)$ appears in the denominator and $[\sigma(r)-\sigma(s)] = (h-k)$ appears in the numerator, but then $h < k$ and σ re-orders (s, r). In the second case, $r < s$, and then $(s-r)$ appears in the denominator and $[\sigma(s) - \sigma(r)] = (k-h)$ appears in the numerator, but then σ re-orders (r, s). Consequently, the factors of the numerator have the same absolute values as the factors of the denominator, with exactly as many different signs as pairs re-ordered by σ. Therefore, simplifying all the absolute values gives

$$\frac{\prod_{h<k}[\sigma(k) - \sigma(h)]}{\prod_{h<k}(k-h)} = \frac{(-1)^{p(\sigma)}}{1} = \mathrm{sgn}(\sigma).$$

□

Example 880 Consider the permutation σ introduced in examples 843 and 856:

$$\sigma = \begin{pmatrix} 1 & 2 & 3 & 4 \\ 3 & 4 & 2 & 1 \end{pmatrix}.$$

The formula in theorem 879 becomes

$$\begin{aligned}
\mathrm{sgn}(\sigma) &= \frac{\prod_{h<k}[\sigma(k) - \sigma(h)]}{\prod_{h<k}(k-h)} \\
&= \frac{\begin{array}{c}[\sigma(2)-\sigma(1)]*[\sigma(3)-\sigma(2)]*[\sigma(3)-\sigma(1)]*\cdots \\ \cdots*[\sigma(4)-\sigma(3)]*[\sigma(4)-\sigma(2)]*[\sigma(4)-\sigma(1)]\end{array}}{(2-1)(3-2)(3-1)(4-3)(4-2)(4-1)} \\
&= \frac{(4-3)(2-4)(2-3)(1-2)(1-4)(1-3)}{(2-1)(3-2)(3-1)(4-3)(4-2)(4-1)} \\
&= \frac{(4-3)[-(4-2)][-(3-2)][-(2-1)][-(4-1)][-(3-1)]}{(2-1)(3-2)(3-1)(4-3)(4-2)(4-1)} \\
&= \frac{(-1)^5}{1} = (-1)^5 = \mathrm{sgn}(\sigma).
\end{aligned}$$

6.4.2 Signatures of Composite Permutations

The following theorem provides a more general formula to calculate the signature.

Theorem 881 *For each permutation $\sigma \in \mathcal{S}_n$ and each permutation $\eta \in \mathcal{S}_n$,*

$$sgn(\sigma) = \frac{\prod_{h<k}\{\sigma[\eta(k)] - \sigma[\eta(h)]\}}{\prod_{h<k}[\eta(k) - \eta(h)]}.$$

Proof. For each pair (h,k) with $h < k$, so that $[\eta(k) - \eta(h)]$ appears as a factor in the denominator, either $[\eta(k) - \eta(h)]$ or $[\eta(h) - \eta(k)]$ also appears as a factor in the numerator, depending upon whether σ re-orders $(r,s) := (\eta^{-1}[\sigma^{-1}(h)], \eta^{-1}[\sigma^{-1}(k)])$.

Indeed, with $r := \eta^{-1}[\sigma^{-1}(h)])$ and $s := \eta^{-1}[\sigma^{-1}(k)]$, two cases arise.

In the first case, $s < r$, and then $[\eta(r) - \eta(s)]$ appears in the denominator and $\{\sigma[\eta(r)] - \sigma[\eta(s)]\} = (h-k)$ appears in the numerator, but then $h < k$ and σ re-orders (s,r).

In the second case, $r < s$, and then $(s-r)$ appears in the denominator and $\{\sigma[\eta(s)] - \sigma[\eta(r)]\} = (k-h)$ appears in the numerator, but then $h < k$ and σ does not re-order (r,s).

Consequently, the factors of the numerator have the same absolute values as the factors of the denominator, with exactly as many different signs as pairs re-ordered by σ. Therefore, simplifying all the absolute values gives

$$\frac{\prod_{h<k}\{\sigma[\eta(k)] - \sigma[\eta(h)]\}}{\prod_{h<k}[\eta(k) - \eta(h)]} = \frac{(-1)^{p(\sigma)}}{1} = \text{sgn}(\sigma).$$

□

Theorem 882 *The signature of a composition of permutations equals the arithmetic product of their signatures: $sgn(\sigma \circ \eta) = sgn(\sigma) \cdot sgn(\eta)$.*

Proof. Apply theorems 879 and 881:

$$\begin{aligned}
\text{sgn}(\sigma\circ\eta) &= \frac{\prod_{h<k}[(\sigma\circ\eta)(k) - (\sigma\circ\eta)(h)]}{\prod_{h<k}(k-h)} \\
&= \frac{\prod_{h<k}[(\sigma\circ\eta)(k) - (\sigma\circ\eta)(h)]}{\prod_{h<k}[\eta(k) - \eta(h)]} \cdot \frac{\prod_{h<k}[\eta(k) - \eta(h)]}{\prod_{h<k}(k-h)} \\
&= \frac{\prod_{h<k}[(\sigma(\eta(k)) - (\sigma(\eta(h))]}{\prod_{h<k}[\eta(k) - \eta(h)]} \cdot \frac{\prod_{h<k}[\eta(k) - \eta(h)]}{\prod_{h<k}(k-h)} \\
&= \text{sgn}(\sigma)\cdot\text{sgn}(\eta).
\end{aligned}$$

□

Theorem 883 *If σ is the composition of t transpositions, then $sgn(\sigma) = (-1)^t$.*

Proof. This theorem results from theorems 857 and 882. □

Example 884 Consider the permutation σ introduced in examples 843 and 856:

$$\begin{aligned}
\sigma &= \begin{pmatrix} 1 & 2 & 3 & 4 \\ 3 & 4 & 2 & 1 \end{pmatrix} \\
&= \left[\begin{pmatrix} 1 & 2 & 3 & 4 \\ 2 & 1 & 3 & 4 \end{pmatrix} \circ \begin{pmatrix} 1 & 2 & 3 & 4 \\ 3 & 2 & 1 & 4 \end{pmatrix}\right] \circ \begin{pmatrix} 1 & 2 & 3 & 4 \\ 1 & 4 & 3 & 2 \end{pmatrix} \\
&= (\tau_3 \circ \tau_2) \circ \tau_1.
\end{aligned}$$

Because σ is the composition of 3 transpositions, $\text{sgn}(\sigma) = (-1)^3 = -1$.

The following theorem shows every permutation has the same signature as the signature of its inverse permutation.

Theorem 885 *For each permutation, sgn* $(\sigma) =$ *sgn* (σ^{-1}).

Proof. From theorem 882, it follows that

$$1 = \text{sgn}(\iota) = \text{sgn}(\sigma \circ \sigma^{-1}) = \text{sgn}(\sigma) \cdot \text{sgn}(\sigma^{-1}).$$

Hence, $\text{sgn}(\sigma^{-1}) = 1/\text{sgn}(\sigma)$. Because $\text{sgn}(\sigma) \in \{1, -1\}$, however, it also follows that $\text{sgn}(\sigma) = 1/\text{sgn}(\sigma)$. Therefore, $\text{sgn}(\sigma^{-1}) = 1/\text{sgn}(\sigma) = \text{sgn}(\sigma)$. □

Definition 886 The **alternating group of degree** n is the set $\mathcal{A}_n$ consisting of all even permutations of n elements.

6.4.3 Exercises

Exercise 1021 Calculate the signature of each permutation of three elements.

Exercise 1022 Prove that the alternating group $\mathcal{A}_n$ of all even permutations of n elements is indeed a group with the composition of functions, $\circ$.

Exercise 1023 Prove that there exist exactly $(n!)/2$ even permutations of n elements.

Exercise 1024 Prove that the set of all *odd* permutations of n elements is *not* a group with the composition of functions.

Exercise 1025 Prove that every 3-cycle has signature 1.

Exercise 1026 Prove that every k-cycle is the composition of one $(k-1)$-cycle and one transposition.

Exercise 1027 Factor the k-cycle $\begin{pmatrix} 1 & 2 & \ldots & k-1 & k \end{pmatrix}$ as the composition of $k-1$ transpositions.

Exercise 1028 Factor the k-cycle $\begin{pmatrix} n_1 & n_2 & \ldots & n_{k-1} & n_k \end{pmatrix}$ as the composition of $k-1$ transpositions.

Exercise 1029 Prove that every k-cycle has signature $(-1)^{k-1}$.

Exercise 1030 Prove that every $(2k)$-cycle does *not* factor as the composition of two k-cycles.

6.5 ARRANGEMENTS

6.5.1 Arrangements Without Repetition

An arrangement of k elements from among n elements without repetition consists of a selection of k distinct elements in a specific order. For example, the sequence of letters "swap" is an arrangement of four letters without repetition from among the 26 letters of the English alphabet. The sequence of letters "wasp" is also an arrangement of the same four letters, but in a different order. Thus "swap" and "wasp" are different arrangements.

Arrangements with repetitions in which every element appears exactly once are called **perfect pangrams**. There seems to be no widely known and grammatically correct pangrams in English, French, or German [2, p. 238].

Definition 887 An **arrangement without repetition** of k objects selected from among n objects is an injection $\alpha : \{1, \ldots, k\} \to \{1, \ldots, n\}$. The number of arrangements without repetition of k objects selected from among n objects is denoted by A_k^n.

The following theorem provides a formula for the number of arrangements.

Theorem 888 *The number A_k^n of arrangements without repetition of k objects selected from among n objects is*

$$A_k^n = \frac{n!}{(n-k)!} = n * (n-1) * \cdots * [n - (k+1)].$$

Proof. The proof consists of extending each arrangement $\alpha : \{1, \ldots, k\} \to \{1, \ldots, n\}$ to a permutation of n elements. The range $\mathcal{R}(\alpha) = \alpha(\{1, \ldots, k\})$ of α consists of k distinct elements of $\{1, \ldots, n\}$; consequently, the complement $\{1, \ldots, n\} \setminus \mathcal{R}(\alpha)$ consists of $n - k$ elements (theorem 606). For each bijection $\beta : \{k+1, \ldots, n\} \to [\{1, \ldots, n\} \setminus \mathcal{R}(\alpha)]$, the arrangement α extends to a bijection

$$\sigma := \alpha \dot{\cup} \beta : \{1, \ldots, k\} \dot{\cup} \{k+1, \ldots, n\} \to \{1, \ldots, n\},$$

$$\sigma(i) := \begin{cases} \alpha(i) & \text{if } i \in \{1, \ldots, k\}, \\ \beta(i) & \text{if } i \in \{k+1, \ldots, n\}. \end{cases}$$

There exist exactly $(n-k)!$ such bijections β: as many as permutations of $\{1, \ldots, n-k\}$. All arrangements α and all extensions β yield in the manner just described all permutations of n elements, of which there exist exactly $(n!)$. However, each injection α has $(n-k)!$ extensions β. Therefore,

$$A_k^n \cdot (n-k)! = n!$$

□

6.5.2 Arrangements With Repetitions

The mathematical concept of an arrangement with repetition(s) amounts to a sequence of finitely many repeatable elements. In particular, if the elements are letters from an alphabet, then an arrangement with repetition(s) of letters corresponds to a "word" with the alphabet. For example, the sequence of letters "swaps" is an arrangement of five letters with one repetition from among the 26 letters of the English alphabet. The sequence of letters "wasps" is also an arrangement of the same five letters with the same repetition from among the 26 letters of the English alphabet, but in a different order. Thus "swaps" and "wasps" are different arrangements with repetitions.

Arrangements with repetitions where each element appears at least once, such as

"the quick brown fox jumps over a lazy dog"

are called **pangrams**. Pangrams serve to test whether printing and transmission devices process each element correctly.

In certain contexts the "words" need not have any significance and can consist of any sequence of letters from a specified alphabet. For instance, a "password" might consist of any finite arrangement of letters, with repetitions allowed. More generally, an arrangement of k elements from among n elements with repetitions allowed consists of a selection of k non-necessarily distinct elements in a specific order.

Definition 889 An **arrangement with repetition** — also called a **word** — with k objects selected from among n objects is a function $\varphi : \{1, \dots, k\} \to \{1, \dots, n\}$.

The number of arrangements with repetition of k objects selected from among $n > 0$ objects is denoted by R^n_k.

The next theorem gives a formula for the number of arrangements with repetitions.

Theorem 890 *The number R^n_k of arrangements with repetition of k objects selected from among $n > 0$ objects is*

$$R^n_k = n^k.$$

Proof. For each positive integer n, proceed by induction with k. For every positive integer n and for $k := 0$, there exists exactly one function from $\varnothing$ into $\{1, \dots, n\}$, namely the empty function $\varphi = \varnothing \subset \varnothing \times \{1, \dots, n\}$; also, $n^k = n^0 = 1$. Similarly, for $k := 1$, there exist exactly n functions from $\{1\}$ into $\{1, \dots, n\}$, in effect the functions $\varphi_\ell : \{1\} \to \{1, \dots, n\}$, $1 \mapsto \ell$; also, $n^k = n^1 = n$.

For each positive integer n, assume that $R^n_k = n^k$ for some non-negative integer $k \in \mathbb{N}$. For each $\ell \in \{1, \dots, n\}$, consider the function $\gamma_\ell : \{k+1\} \to \{1, \dots, n\}$, $k+1 \mapsto \ell$. Each function $\varphi : \{1, \dots, k\} \to \{1, \dots, n\}$ extends in exactly n ways to a function $\varphi_\ell : \{1, \dots, k+1\} \to \{1, \dots, n\}$ by the definition $\varphi_\ell := \varphi \dot{\cup} \gamma_\ell$. In other words,

$$\varphi_\ell(i) = \begin{cases} \varphi(i) & \text{if } i \in \{1, \dots, k\}, \\ \ell & \text{if } i = k+1. \end{cases}$$

Because there exist exactly n such elements $\ell \in \{1, \dots, n\}$, and because there exist exactly $R^n_k = n^k$ such functions φ, it follows that $R^n_{k+1} = n \cdot n^k = n^{k+1}$. □

6.5.3 Exercises

Exercise 1031 Determine the number of words consisting of two letters from the English alphabet.

Exercise 1032 Determine the number of words consisting of two letters from the German alphabet.

Exercise 1033 Determine the number of words consisting of two letters from the Swedish alphabet.

Exercise 1034 Determine the number of words consisting of two letters from the Norwegian alphabet.

Exercise 1035 Determine the number of words consisting of two letters from the Russian alphabet.

Exercise 1036 Determine the number of words consisting of three letters from the English alphabet.

Exercise 1037 Determine the number of words consisting of three letters from the German alphabet.

Exercise 1038 Determine the number of words consisting of three letters from the Swedish alphabet.

Exercise 1039 Determine the number of words consisting of three letters from the Norwegian alphabet.

Exercise 1040 Determine the number of words consisting of three letters from the Russian alphabet.

Exercise 1041 Determine a necessary and sufficient number k so that there exists at least one million words consisting of k letters from the English alphabet.

Exercise 1042 Determine a necessary and sufficient number k so that there exists at least one million words consisting of k letters from the German alphabet.

Exercise 1043 Determine a necessary and sufficient number k so that there exists at least one million words consisting of k letters from the Swedish alphabet.

Exercise 1044 Determine a necessary and sufficient number k so that there exists at least one million words consisting of k letters from the Norwegian alphabet.

Exercise 1045 Determine a necessary and sufficient number k so that there exists at least one million words consisting of k letters from the Russian alphabet.

Exercise 1046 Determine a necessary and sufficient number k so that there exists at least one billion words consisting of k letters from the English alphabet.

Exercise 1047 Determine a necessary and sufficient number k so that there exists at least one billion words consisting of k letters from the German alphabet.

Exercise 1048 Determine a necessary and sufficient number k so that there exists at least one billion words consisting of k letters from the Swedish alphabet.

Exercise 1049 Determine a necessary and sufficient number k so that there exists at least one billion words consisting of k letters from the Norwegian alphabet.

Exercise 1050 Determine a necessary and sufficient number k so that there exists at least one billion words consisting of k letters from the Russian alphabet.

6.6 COMBINATIONS

6.6.1 Combinations Without Repetition

A mathematical combination without repetition is a selection without repetition and without any ordering. For example, the two words "swap" and "wasp" consist of the same combination of four letters without repetition. The number of combinations without repetition will be related (by theorem 895) to the following binomial coefficients.

Definition 891 For each $n \in \mathbb{N}$ and each $k \in \{0, \ldots, n\}$ define

$$C_k^n := \frac{n!}{(k!) * [(n-k)!]} = \frac{n * (n-1) * \cdots * (n+1-k)}{1 * 2 * \cdots * (k-1) * k}.$$

The numbers C_k^n are called **binomial coefficients**.

The following theorem provides a recurrence relation for the binomial coefficients.

Theorem 892 *For each $n \in \mathbb{N}$ and each $k \in \{1, \ldots, n\}$,*

$$C_k^{n+1} = C_k^n + C_{k-1}^n.$$

Proof. Simplify fractions:

$$\begin{aligned}
C_k^n + C_{k-1}^n &= \frac{n!}{[k!] \cdot [(n-k)!]} + \frac{n!}{[(k-1)!] \cdot [(n-[k-1])!]} \\
&= \frac{(n!) \cdot (n-[k-1]) + (n!) \cdot k}{[k!] \cdot [(n-[k-1])!]} \\
&= \frac{n! \cdot (n+1-k) + n! \cdot k}{[k!] \cdot [([n+1]-k)!]} \\
&= \frac{n! \cdot (n+1)}{[k!] \cdot [([n+1]-k)!]} \\
&= \frac{(n+1)!}{[k!] \cdot [([n+1]-k)!]} \\
&= C_k^{n+1}.
\end{aligned}$$

□

The following theorem shows that the binomial coefficients C_k^n increase as k increases from 0 to $n/2$, and then decrease as k increases from $n/2$ to n.

Theorem 893 *For each $n \in \mathbb{N}$ and all $k \in \{0, \ldots, n\}$ and $\ell \in \{0, \ldots, n\}$, if $k < \ell \leq (n-1)/2$, then $C_k^n < C_\ell^n$. Similarly, if $(n-1)/2 \leq k < \ell \leq n$, then $C_k^n > C_\ell^n$.*

Proof. For each $m \in \{0, \ldots, n\}$,

$$C_m^n = \frac{n * (n-1) * (n-2) * \cdots * (n-[m-2]) * (n-[m-1])}{m * (m-1) * (m-2) * \cdots * 2 * 1}.$$

Consequently, for each $m \in \{0, \ldots, n-1\}$,

$$\begin{aligned}
C_{m+1}^n &= \frac{n * (n-1) * (n-2) * \cdots * (n-[m-2]) * (n-[m-1]) * (n-m)}{(m+1)m * (m-1) * (m-2) * \cdots * 2 * 1} \\
&= \frac{n-m}{m+1} * \frac{n * (n-1) * (n-2) * \cdots * (n-[m+2]) * (n-[m+1])}{m * (m-1) * (m-2) * \cdots * 2 * 1} \\
&= \frac{n-m}{m+1} * C_m^n.
\end{aligned}$$

Therefore, $C_{m+1}^n < C_m^n$ if and only if $n - m < m + 1$, which amounts to $n - 1 < 2 * m$ and thus also to $(n-1)/2 < m$. Similarly, $C_{m+1}^n > C_m^n$ if and only if $n - m > m + 1$, which amounts to $n - 1 < 2 * m$ and thus also to $m < (n-1)/2$. □

The following definition and theorem show that the binomial coefficient C_k^n equals the number of subsets of k elements in a set with n elements.

Definition 894 A **combination** (without repetition) of k elements among n elements is a subset of k elements from $\{1, \ldots, n\}$.

The number of all such combinations is denoted by $\binom{n}{k}$, read "n choose k."

Theorem 895 *For each non-negative integer $n \in \mathbb{N}$, and for each non-negative integer $k \in \{0, \ldots, n\}$, the number $\binom{n}{k}$ of combinations of k elements among n elements, in other words, the number of subsets of k elements from $\{1, \ldots, n\}$, is*

$$\binom{n}{k} = \frac{n!}{[k!] \cdot [(n-k)!]} = C_k^n.$$

Proof. For $n := 0$, and for $k := 0$, there exists exactly *one* subset of $\{1, \ldots, 0\} = \varnothing$ with $k = 0$ element, namely $\varnothing$. Thus, $\binom{0}{0} = 1 = \frac{0!}{[0!] \cdot [(0-0)!]}$.

Proceeding by induction with n, assume that $\binom{n}{\ell} = \frac{n!}{[\ell!] \cdot [(n-\ell)!]}$ for some non-negative integer $n := m \in \mathbb{N}$ and for every $k \in \{0, \ldots, n\}$. Among the subsets of $\{1, 2, \ldots, n, n+1\}$, distinguish between the subsets that contain $n+1$ and the subsets that do *not* contain $n+1$.

Because all the subsets that do *not* contain $n+1$ are also subsets of $\{1, 2, \ldots, n\}$, there exist exactly $\binom{n}{k} = \frac{n!}{[k!] \cdot [(n-k)!]}$ such subsets, by induction hypothesis.

For each subset $S \subseteq \{1, 2, \ldots, n, n+1\}$ that contains $n+1$, $S = \{n+1\} \cup (S \setminus \{n+1\})$, where $S \setminus \{n+1\}$ represents a subset of $\{1, \ldots, n\}$ with $k-1$ elements. Again by induction hypothesis, there exist exactly $\binom{n}{k-1} = \frac{n!}{[(k-1)!] \cdot [(n-[k-1])!]}$ such subsets. Consequently, for each $k \in \{0, \ldots, n\}$,

$$\begin{aligned}
\binom{n+1}{k} &= \binom{n}{k} + \binom{n}{k-1} \\
&= \frac{n!}{[k!] \cdot [(n-k)!]} + \frac{n!}{[(k-1)!] \cdot [(n-[k-1])!]} \\
&= \frac{(n!) \cdot (n-[k-1]) + (n!) \cdot k}{[k!] \cdot [(n-[k-1])!]} \\
&= \frac{n! \cdot (n+1-k) + n! \cdot k}{[k!] \cdot [([n+1]-k)!]} \\
&= \frac{n! \cdot (n+1)}{[k!] \cdot [([n+1]-k)!]} \\
&= \frac{(n+1)!}{[k!] \cdot [([n+1]-k)!]}.
\end{aligned}$$

For the remaining case with $k = n+1$, there exists only one subset of $\{1, 2, \ldots, n, n+1\}$ with $n+1$ elements, and $\binom{n+1}{n+1} = 1 = \frac{(n+1)!}{[(n+1)!] \cdot [([n+1]-[n+1])!]}$. □

The following theorem shows that the binomial coefficients also equal the coefficients of the summands in the expansion of a power of a sum.

Theorem 896 (Binomial formula) *For all $n \in \mathbb{N}$, $p \in \mathbb{Q}$, $q \in \mathbb{Q}$,*

$$(p+q)^n = \sum_{k=0}^{n} C_k^n p^{n-k} q^k.$$

Proof. This proof proceeds by induction with n. For $n := 0$,

$$(p+q)^0 = 1 = 1 * 1 * 1 = C_0^0 p^0 q^{0-0} = \sum_{k=0}^{0} C_k^0 p^{n-k} q^k = \sum_{k=0}^{n} C_k^n p^{n-k} q^k.$$

For $n := 1$,

$$(p+q)^1 = p+q = C_1^1 p^1 q^{1-1} + C_0^1 p^0 q^{1-0} = \sum_{k=0}^{1} C_k^1 p^{n-k} q^k = \sum_{k=0}^{n} C_k^n p^{n-k} q^k.$$

Assume that there exists $\ell \in \mathbb{N}$ such that

$$(p+q)^\ell = \sum_{k=0}^{\ell} C_k^\ell p^{\ell-k} q^k$$

for all $p \in \mathbb{Q}$ and $q \in \mathbb{Q}$. Then

$$\begin{aligned}
(p+q)^{\ell+1} &= (p+q) * (p+q)^\ell \\
&= p * (p+q)^\ell + q * (p+q)^\ell \\
&= p * \sum_{k=0}^{\ell} C_k^\ell p^{\ell-k} q^k + q * \sum_{k=0}^{\ell} C_k^\ell p^{\ell-k} q^k \\
&= \sum_{k=0}^{\ell} C_k^\ell (p * p^{\ell-k} q^k + q * p^{\ell-k} q^k) \\
&= \sum_{k=0}^{\ell} C_k^\ell (p^{\ell+1-k} q^k + p^{\ell+1-(k+1)} q^{k+1}) \\
&= \sum_{k=0}^{\ell} C_k^\ell p^{\ell+1-k} q^k + \sum_{k=0}^{\ell} C_k^\ell p^{\ell+1-(k+1)} q^{k+1} \\
&= \sum_{k=0}^{\ell} C_k^\ell p^{\ell+1-k} q^k + \sum_{j=1}^{\ell+1} C_{j-1}^\ell p^{\ell+1-j} q^j \\
&= C_0^\ell p^{\ell+1-0} q^0 + \sum_{m=1}^{\ell} (C_m^\ell + C_{m-1}^\ell) p^{\ell+1-m} q^m + C_\ell^\ell p^{\ell+1-\ell+1} q^{\ell+1} \\
&= p^{\ell+1} + \sum_{m=1}^{\ell} C_m^{\ell+1} p^{\ell+1-m} q^m + q^{\ell+1} \\
&= \sum_{m=0}^{\ell+1} C_m^{\ell+1} p^{\ell+1-m} q^m.
\end{aligned}$$

□

Theorem 897 *For each non-negative integer $n \in \mathbb{N}$, the set $\{1, \ldots, n\}$ has exactly 2^n subsets. In other words, the power set $\mathcal{P}(\{1, \ldots, n\})$ has exactly 2^n elements.*

Proof. The set $\{1, \ldots, n\}$ has subsets with exactly k elements only for $k \in \{0, \ldots, n\}$, and $\{1, \ldots, n\}$ does not have any subset with more than n elements (theorem 605). Adding the number of subsets with k elements for each $k \in \{0, \ldots, n\}$ gives

$$\sum_{k=0}^{n} \binom{n}{k} = \sum_{k=0}^{n} \binom{n}{k} 1^k 1^{n-k} = (1+1)^n = 2^n.$$

□

6.6.2 Combinations With Repetition

A mathematical combination with repetition is a selection with repetition(s) but without any ordering. Thus a combination with repetition consists of a function that determines the number of repetitions of each element. For example, the two words "swaps" and "wasps" consist of the same combination of five letters with one repetition. In both cases, the function specifies that "s" occurs twice while "a", "p", "w" occur once each (and all the other letters of the alphabet do not occur).

Definition 898 A **combination with repetition** of k elements among n elements is a function $\psi : \{1, \ldots, n\} \to \{0, \ldots, k\}$ such that $\sum_{i=1}^{n} \psi(i) = k$. The number of combinations with repetition of k elements among n elements is denoted by $\left[{n \atop k} \right]$.

In definition 898, the function ψ specifies that i occurs $\psi(i)$ times.

Example 899 Each of the two words "swaps" and "wasps" consist of the same combination with repetition ψ of five elements among 26 elements, with

$$\psi : \{ \texttt{A B C D E F G H I J K L M N O P Q R S T U V W X Y Z} \} \to \{0,1,2,3,4,5\},$$

$$\begin{array}{rcl} \psi(\texttt{A}) & = & 1, \\ \psi(\texttt{P}) & = & 1, \\ \psi(\texttt{S}) & = & 2, \\ \psi(\texttt{W}) & = & 1, \end{array}$$

and with value 0 for all the other letters of the alphabet. Also,

$$\begin{array}{cl} & \psi(\texttt{A}) + \psi(\texttt{B}) + \cdots + \psi(\texttt{P}) + \cdots + \psi(\texttt{S}) + \cdots + \psi(\texttt{W}) + \psi(\texttt{X}) + \psi(\texttt{Y}) + \psi(\texttt{Z}) \\ = & 1 + 0 + \cdots + 1 + \cdots + 2 + \cdots + 1 + 0 + 0 + 0 \\ = & 5. \end{array}$$

The next theorem gives a formula for the number of combinations with repetitions.

Theorem 900 *For every non-negative integer $n \in \mathbb{N}$, and for every non-negative integer $k \in \mathbb{N}$,*

$$\left[{n \atop k} \right] = \binom{n+k-1}{k}.$$

Proof. Define a function $\zeta : \{0, \ldots, n-1\} \to \{0, \ldots, n+k-1\}$ by

$$\zeta(j) := \sum_{i=1}^{j} [\psi(i) + 1] - 1.$$

Thus,

$$\begin{array}{rcl} \zeta(1) & = & [\psi(1) + 1] - 1, \\ \zeta(2) & = & [\psi(1) + 1] + [\psi(2) + 1] - 1, \\ \zeta(3) & = & [\psi(1) + 1] + [\psi(2) + 1] + [\psi(3) + 1] - 1, \\ & \vdots & \\ \zeta(n-1) & = & [\psi(1) + 1] + [\psi(2) + 1] + [\psi(3) + 1] + \cdots + [\psi(n-1) + 1] - 1. \end{array}$$

The auxiliary function ζ is increasing, and hence injective. Indeed, $\psi(i+1) \geq 0$ for each $i \in \{1, \ldots, n-2\}$, whence $[\psi(i+1)+1] > 0$, and thus $\zeta(i+1) = \zeta(i) + [\psi(i+1)+1] > \zeta(i)$. By induction, it then follows that if $i < \ell$ then $\zeta(i) < \zeta(\ell)$. As a verification of the co-domain of ζ,

$$\begin{aligned}
\zeta(n-1) &= [\psi(1)+1] + [\psi(2)+1] + [\psi(3)1] + \cdots + [\psi(n-1)+1] - 1 \\
&= \sum_{i=1}^{n-1} \psi(i) + \sum_{i=1}^{n} 1 - 1 \\
&\leq \sum_{i=1}^{n} \psi(i) - \psi(n) + \sum_{i=1}^{n} 1 - 1 \\
&= k + n - 1.
\end{aligned}$$

Thus, the range of the auxiliary function ζ is a subset of $n-1$ elements from the set $\{1, \ldots, n+k-1\}$, and there exist exactly $\binom{n+k-1}{(n-1)} = \binom{n+k-1}{k}$ such subsets.

Two distinct combinations with repetitions ψ_1 and ψ_2 correspond to two distinct auxiliary functions ζ_1 and ζ_2. Indeed, if ℓ denotes the smallest integer where $\psi_1(\ell) \neq \psi_2(\ell)$, then $\zeta_1(\ell) \neq \zeta_2(\ell)$.

Moreover, to each increasing function $\zeta : \{1, \ldots, n\} \to \{1, \ldots, n+k-1\}$ corresponds a combination with repetitions ψ defined inductively by $\psi(1) := \zeta(1), \ldots, \psi(i+1) := \zeta(i+1) - [\zeta(i)+1]$.

Therefore, there exist exactly as many arrangements with repetitions of k elements among n as there exist subsets with $n-1$ elements from $n+k-1$ elements. □

6.6.3 Permutations With Repetition

Within an arrangement with repetition (word), permuting identical elements does not change the arrangement. For instance, a word that is spelled in the same way forward and backward is a **palindrome**. For example, "dad" is a palindrome. Thus, swapping the two occurrences of "d" reproduces the same word. To produce such a different word as "add" a permutation must permute different elements. Therefore, though the word "dad" has three letters, there exist fewer than 3! permutations that produce different words.

Problems of designing a word or a sentence from a combination with repetition of letters — and also the resulting word or sentence — are called **anagrams**. For instance, the combination with repetition "add" could result in "add" and "dad" as anagrams. As a more elaborate example, Galileo Galilei reportedly communicated his discovery of the phases of the planets to Johannes Kepler through the anagram

"haec immatura a me iam frustra leguntur o. y."

("these unripe things are now read in vain by me"), whence a permutation gives

"cynthia figuras aemulatur mater amorum"

("the mother of love [Venus] imitates Cynthia's [the Moon's] phases") [2, p. 97].

Anagrams correspond to the mathematical concept of permutations with repetition associated with a combination with repetition. Firstly, the set of letters and their number of repetitions corresponds to a combination with repetition

$$\psi : \{A, \ldots, Z\} \to \mathbb{N},$$

so that ψ specifies the number of occurrences of each letter of the alphabet. For example, for the anagram "apssw" $\psi(a) = 1$ because "a" occurs exactly once, but $\psi(s) = 2$ because "s" occurs exactly twice. Secondly, the total number k of letters and their repetition equals the sum of all the values of ψ:

$$k := \psi(A) + \cdots + \psi(Z).$$

Yet in a word the order of different letters matters. Consequently, a way to form a word can proceed by lining up k slots and filling them up with the k letters and repetitions specified by ψ, which corresponds to the mathematical concept of a function

$$\gamma : \{1, \ldots, k\} \to \{A, \ldots, Z\}$$

such that the letter $\gamma(i)$ occurs with multiplicity $\psi[\gamma(i)]$.

Definition 901 A **permutation with repetition** of k elements among n elements is a function $\gamma : \{1, \ldots, k\} \to \{1, \ldots, n\}$ with a combination with repetitions of k elements among n elements, $\psi : \{1, \ldots, n\} \to \{0, \ldots, k\}$ such that for each $i \in \{1, \ldots, n\}$ the preimage $\gamma^{-1}(\{i\})$ contains exactly $\psi(i)$ elements. The number of permutations with repetition of k elements among n elements associated with a combination with repetition ψ is denoted by Q_k^{ψ}, or by $Q_k^{\psi(1),\ldots,\psi(n)}$.

Example 902 The anagram "'apssw" corresponds to permutations of $k := 5$ letters from an alphabet with $n := 26$ letters, and the function ψ from the English alphabet to $\{0, 1, 2, 3, 4, 5\}$ defined with values

$$\begin{aligned} \psi : \{ \texttt{ABCDEFGHIJKLMNOPQRSTUVWXYZ} \} &\to \{0, 1, 2, 3, 4, 5\}, \\ \psi(\texttt{A}) &= 1, \\ \psi(\texttt{P}) &= 1, \\ \psi(\texttt{S}) &= 2, \\ \psi(\texttt{W}) &= 1, \end{aligned}$$

and with value 0 for all the other letters of the alphabet. The function

$$\gamma : \{1, 2, 3, 4, 5\} \to \{ \texttt{ABCDEFGHIJKLMNOPQRSTUVWXYZ} \}$$

is a permutation with repetition associated with ψ:

$$\begin{aligned} \gamma(1) &:= \texttt{W} \text{ occurs } \psi(\texttt{W}) = 1 \text{ time}, \\ \gamma(2) &:= \texttt{A} \text{ occurs } \psi(\texttt{A}) = 1 \text{ time}, \\ \gamma(3) &:= \texttt{S} \text{ occurs } \psi(\texttt{S}) = 2 \text{ times}, \\ \gamma(4) &:= \texttt{P} \text{ occurs } \psi(\texttt{P}) = 1 \text{ time}, \\ \gamma(5) &:= \texttt{S} \text{ occurs } \psi(\texttt{S}) = 2 \text{ times}. \end{aligned}$$

The next theorem gives a formula for the number of permutations with repetition.

Theorem 903 *For each non-negative integer $n \in \mathbb{N}$ and for each combination with repetition of k elements among n elements ψ,*

$$Q_k^{\psi} = \frac{k!}{\prod_{i=1}^{n} [\psi(i)]!}.$$

Proof. Proceed by induction with i. For $i := 1$, the preimage $\gamma^{-1}(\{1\})$ is a subset of $\{1, \ldots, k\}$ with $\psi(1)$ elements, and there exist exactly $\binom{k}{\psi(1)}$ such subsets.

For $i := 2$, the preimage $\gamma^{-1}(\{2\})$ is then a subset with $\psi(2)$ elements of $\{1, \ldots, k\} \setminus \gamma^{-1}(\{1\})$, which contains only $k - \psi(1)$ elements, and there exist exactly $\binom{k-\psi(1)}{\psi(2)}$ such subsets. Consequently, the number of preimages $\gamma^{-1}(\{1, 2\})$ is

$$\binom{k}{\psi(1)} \cdot \binom{k - \psi(1)}{\psi(2)}.$$

For $i := 3$, the preimage $\gamma^{-1}(\{3\})$ is then a subset with $\psi(3)$ elements of $\{1, \ldots, k\} \setminus [\gamma^{-1}(\{1\}) \cup \gamma^{-1}(\{2\})]$, which contains only $k - [\psi(1) + \psi(2)]$ elements, and there exist exactly $\binom{k-[\psi(1)+\psi(2)]}{\psi(3)}$ such subsets. Consequently, the number of preimages $\gamma^{-1}(\{1, 2, 3\})$ is

$$\binom{k}{\psi(1)} \cdot \binom{k - \psi(1)}{\psi(2)} \cdot \binom{k - [\psi(1) + \psi(2)]}{\psi(3)}.$$

Hence assume that for some $i := \ell \in \{1, \ldots, k - 1\}$ the number of distinct subsets that may be the preimage $\gamma^{-1}(\{1, \ldots, \ell\})$ equals

$$\prod_{j=1}^{\ell} \binom{k - \sum_{h=1}^{j-1} \psi(h)}{\psi(j)}.$$

Then the preimage $\gamma^{-1}(\{\ell + 1\})$ is a subset with $\psi(\ell + 1)$ elements of $\{1, \ldots, k\} \setminus \gamma^{-1}(\{1, \ldots, \ell\})$, which contains only $k - \sum_{j=1}^{\ell} \psi(j)$ elements, and there exist exactly $\binom{k-\sum_{j=1}^{\ell} \psi(j)}{\psi(\ell+1)}$ such subsets. Hence the number of preimages $\gamma^{-1}(\{1, \ldots, \ell + 1\})$ is

$$\left[\prod_{j=1}^{\ell} \binom{k - \sum_{h=1}^{j-1} \psi(h)}{\psi(j)}\right] \cdot \binom{k - \sum_{j=1}^{\ell} \psi(h)}{\psi(\ell + 1)} = \prod_{j=1}^{\ell+1} \binom{k - \sum_{h=1}^{j-1} \psi(h)}{\psi(j)}.$$

Finally,

$$\begin{aligned}
Q_k^{\psi} &= \prod_{j=1}^{n} \binom{k - \sum_{h=1}^{j-1} \psi(h)}{\psi(j)} \\
&= \prod_{j=1}^{n} \frac{\left[k - \sum_{h=1}^{j-1} \psi(h)\right]!}{[\psi(j)!] \cdot \left\{\left[k - \sum_{h=1}^{j-1} \psi(h) - \psi(j)\right]!\right\}} \\
&= \prod_{j=1}^{n} \frac{\left[k - \sum_{h=1}^{j-1} \psi(h)\right]!}{[\psi(j)!] \cdot \left\{\left[k - \sum_{h=1}^{j} \psi(h)\right]!\right\}} \\
&= \frac{1}{\prod_{j=1}^{n} [\psi(j)!]} \cdot \frac{k!}{[k - \psi(1)]!} \cdot \frac{[k - \psi(1)]!}{[k - [\psi(1) + \psi(2)]]!} \cdots \frac{\left[k - \sum_{h=1}^{n-1} \psi(h)\right]!}{[k - \sum_{h=1}^{n} \psi(h)]!} \\
&= \frac{k!}{\prod_{j=1}^{n} [\psi(j)!]},
\end{aligned}$$

because the denominator of each fraction cancels the numerator of the next fraction, except for the first, $k!$, and the last, $[k - \sum_{h=1}^{n} \psi(h)]! = [k - k]! = 1$. □

6.6.4 Exercises

Exercise 1051 For each set with exactly n elements, determine the number P_1^n of partitions consisting only of singletons.

Exercise 1052 For each set with exactly $n = 2\ell$ elements, determine the number $P_2^{2\ell}$ of partitions consisting only of subsets with exactly two elements.

Exercise 1053 For each set with exactly $n = 3\ell$ elements, determine the number $P_3^{3\ell}$ of partitions consisting only of subsets with exactly three elements.

Exercise 1054 For each set with exactly $n = 4\ell$ elements, determine the number $P_4^{4\ell}$ of partitions consisting only of subsets with exactly three elements.

Exercise 1055 For each set with exactly $n = k * \ell$ elements, determine the number $P_k^{k*\ell}$ of partitions consisting only of subsets with exactly k elements.

Exercise 1056 Find the number of 2-cycles among the permutations of 2 elements.

Exercise 1057 Find the number of 3-cycles among the permutations of 3 elements.

Exercise 1058 Find the number of 4-cycles among the permutations of 4 elements.

Exercise 1059 Find the number of n-cycles among the permutations of n elements.

Exercise 1060 Find the number of k-cycles among the permutations of n elements.

6.7 PROBABILITY

6.7.1 Probability Spaces

The mathematical concept of "probability" consists of a set $\mathcal{A}$ (modeling "events") and a function P (for "probabilities") subject to certain conditions [27, p. 286], [52, p. 12]. In applications, the elements of $\mathcal{A}$ correspond to all the events under consideration. Thus the union (defined by axiom S6) of all the elements of $\mathcal{A}$ exhausts all events.

Definition 904 The **space** $\mathcal{S}_{\mathcal{A}}$ of a set of sets $\mathcal{A}$ is their union $\mathcal{S}_{\mathcal{A}} := \bigcup \mathcal{A}$.

To allow for the negation of an event and for the disjunction of events, the set $\mathcal{A}$ must contain complements and unions of events.

Definition 905 An **algebra** of sets is a *non-empty* set $\mathcal{A}$ with space $\mathcal{S}_{\mathcal{A}} := \bigcup \mathcal{A}$ satisfying the following two conditions.

($\mathcal{A}$.1) For each set B, if $B \in \mathcal{A}$, then also its complement $\mathcal{S}_{\mathcal{A}} \setminus B \in \mathcal{A}$.

($\mathcal{A}$.2) For all sets B and C, if $B \in \mathcal{A}$ and $C \in \mathcal{A}$, then $B \cup C \in \mathcal{A}$.

Moreover, an algebra of sets $\mathcal{A}$ is a **σ-algebra** if and only if

($\mathcal{A}$.3) For each countable (definition 617) set $\mathcal{F} \subseteq \mathcal{A}$ of subsets, $(\bigcup \mathcal{F}) \in \mathcal{A}$.

Example 906 For each set S, the set $\mathcal{A} := \{\varnothing, S\}$ is an algebra of sets. In particular, for this algebra of sets, the space is

$$\mathcal{S}_{\mathcal{A}} := \left(\bigcup \mathcal{A}\right) = \left(\bigcup \{\varnothing, S\}\right) = \varnothing \cup S = S.$$

Example 907 For any set $S := \{\downarrow, \uparrow\}$ with two distinct elements (with $\downarrow$ called "spin down" and $\uparrow$ called "spin up") the power set $\mathcal{P}(S)$ is an algebra of sets:

$$\mathcal{A} := \mathcal{P}(S) = \mathcal{P}(\{\downarrow, \uparrow\}) = \{\varnothing,\ \{\downarrow\},\ \{\uparrow\},\ S\}.$$

Example 908 More generally, for each set S, the power set $\mathcal{A} := \mathcal{P}(S)$ is an algebra and a σ-algebra, because the complements in S and every union of subsets of S are also subsets of S. For this algebra of sets, the space is

$$\mathcal{S}_{\mathcal{A}} := \left(\bigcup \mathcal{A}\right) = \left(\bigcup \mathcal{P}(S)\right) = S.$$

Example 909 In particular, for any alphabet S, for instance, for the English alphabet

$$S := \{\ \mathtt{A\,B\,C\,D\,E\,F\,G\,H\,I\,J\,K\,L\,M\,N\,O\,P\,Q\,R\,S\,T\,U\,V\,W\,X\,Y\,Z}\ \}$$

the power set $\mathcal{A} := \mathcal{P}(S)$ is an algebra of sets by example 908.

In applications, each singleton from $\mathcal{A}$ can represent the event of the selection of a single letter. For instance, $\{\mathtt{L}\} \in \mathcal{A}$ corresponds to the selection of the letter L.

Similarly, each set with exactly two letters represents the event of the selection of either letter. For instance, $\{\mathtt{L}, \mathtt{M}\} \in \mathcal{A}$ corresponds to the selection of a single letter that can be either L or M.

More generally, each element of $\mathcal{A} = \mathcal{P}(S)$ that is a finite subset of the alphabet S represents the event of the selection of any one letter from that subset. For instance, $\{\mathtt{K}, \mathtt{L}, \mathtt{M}\} \in \mathcal{A}$ corresponds to the selection of a single letter: either K or L or M.

Every algebra of sets also contains differences and intersections.

Theorem 910 *For each algebra of sets $\mathcal{A}$, if $B, C \in \mathcal{A}$ then $B \cap C \in \mathcal{A}$ and $B \setminus C \in \mathcal{A}$.*

Proof. Condition $(\mathcal{A}.1)$ states that $\mathcal{S}_{\mathcal{A}} \setminus B \in \mathcal{A}$ and $\mathcal{S}_{\mathcal{A}} \setminus C \in \mathcal{A}$. Consequently, condition $(\mathcal{A}.2)$ gives $(\mathcal{S}_{\mathcal{A}} \setminus B) \cup (\mathcal{S}_{\mathcal{A}} \setminus C) \in \mathcal{A}$ whence de Morgan's first law gives

$$\mathcal{S}_{\mathcal{A}} \setminus (B \cap C) = (\mathcal{S}_{\mathcal{A}} \setminus B) \cup (\mathcal{S}_{\mathcal{A}} \setminus C) \in \mathcal{A}.$$

Hence, condition $(\mathcal{A}.1)$ yields

$$B \cap C = \mathcal{S}_{\mathcal{A}} \setminus [\mathcal{S}_{\mathcal{A}} \setminus (B \cap C)] \in \mathcal{A}.$$

Complements and intersections then yield differences: $B \setminus C = B \cap (\mathcal{S}_{\mathcal{A}} \setminus C) \in \mathcal{A}$. □

Every algebra of sets also contains the empty set and the whole space.

Theorem 911 *Every algebra of sets $\mathcal{A}$ contains the elements $\varnothing$ and $\mathcal{S}_{\mathcal{A}}$.*

Proof. Because $\mathcal{A} \neq \varnothing$ by definition, $\mathcal{A}$ contains at least one element $B \in \mathcal{A}$.
Hence condition $(\mathcal{A}.1)$ guarantees that $(\mathcal{S}_{\mathcal{A}} \setminus B) \in \mathcal{A}$.
Consequently, $\varnothing = B \cap (\mathcal{S}_{\mathcal{A}} \setminus B) \in \mathcal{A}$ by theorem 910.
Finally, condition $(\mathcal{A}.1)$ confirms that $\mathcal{S}_{\mathcal{A}} = \mathcal{S}_{\mathcal{A}} \setminus \varnothing \in \mathcal{A}$. □

With events represented by elements of an algebra of sets $\mathcal{A}$, the function P assigns a number — called a "likelihood" or a "probability" — to each event.

Definition 912 A **probability function** is a function P defined on an algebra of sets $\mathcal{A}$ and whose range consists only of non-negative numbers, with the following conditions.

(P.1) $P(\mathcal{S}_{\mathcal{A}}) = 1$.

(P.2) If $B \in \mathcal{A}$ and $C \in \mathcal{A}$ with $B \cap C = \varnothing$, then $P(B \dot{\cup} C) = P(B) + P(C)$.

Moreover, if $\mathcal{A}$ is a σ-algebra, then

(P.3) for each pairwise disjoint countable set $\mathcal{F} \subseteq \mathcal{A}$,

$$P\left(\dot{\bigcup}\mathcal{F}\right) = \sum_{C\in\mathcal{F}} P(C),$$

where $\sum_{C\in\mathcal{F}} P(C)$ denotes the sum of all the values $P(C)$ for every $C \in \mathcal{F}$.

Definition 913 A **probability space** is a triple $(\mathcal{S}_{\mathcal{A}}, \mathcal{A}, P)$ consisting of an algebra (or a σ-algebra) $\mathcal{A}$ with space $\mathcal{S}_{\mathcal{A}}$, and a probability function P with domain $\mathcal{A}$.

The following example corresponds to a situation with only one likely event.

Example 914 For each set S, the set $\mathcal{A} := \{\varnothing, S\}$ is an algebra of sets (example 906), and the function P defined by

$$\begin{aligned} P(\varnothing) &:= 0, \\ P(S) &:= 1, \end{aligned}$$

is a probability function. Indeed, verifying S and all disjoint unions confirms that

$$\begin{aligned} P(S) &= 1, \\ P(\varnothing\dot{\cup}\varnothing) &= P(\varnothing) = 0 = 0 + 0 = P(\varnothing) + P(\varnothing), \\ P(\varnothing\dot{\cup}S) &= P(S) = 1 = 0 + 1 = P(\varnothing) + P(S). \end{aligned}$$

This example corresponds to a situation in which only one event, S, is probable. The probability that S occurs is 1, and the probability that S does not occur is 0.

The next example corresponds to a situation with only two equally likely events.

Example 915 For any set with two elements, for instance, $S := \{\downarrow, \uparrow\}$, and the corresponding algebra of sets (example 907)

$$\mathcal{A} := \mathcal{P}(S) = \mathcal{P}(\{\downarrow, \uparrow\}) = \{\varnothing,\ \{\downarrow\},\ \{\uparrow\},\ S\},$$

the function $P : \mathcal{A} \to \mathbb{Q}_+$ defined for each $C \in \mathcal{A}$ by

$$\begin{aligned} P(\varnothing) &:= 0, \\ P(\{\downarrow\}) &:= 1/2, \\ P(\{\uparrow\}) &:= 1/2, \\ P(S) &:= 1, \end{aligned}$$

is a probability function. Indeed, verifying S and all disjoint unions confirms that

$$\begin{aligned} P(S) &= 1, \\ P(\varnothing\dot{\cup}\varnothing) &= P(\varnothing) = 0 = 0 + 0 = P(\varnothing) + P(\varnothing), \\ P(\varnothing\dot{\cup}S) &= P(S) = 1 = 0 + 1 = P(\varnothing) + P(S). \\ P(\varnothing\dot{\cup}\{\uparrow\}) &= P(\{\uparrow\}) = 1/2 = 0 + 1/2 = P(\varnothing) + P(\{\uparrow\}), \\ P(\varnothing\dot{\cup}\{\downarrow\}) &= P(\{\downarrow\}) = 1/2 = 0 + 1/2 = P(\varnothing) + P(\{\downarrow\}), \\ P(\{\downarrow\}\dot{\cup}\{\uparrow\}) &= P(S) = 1 = 1/2 + 1/2 = P(\{\downarrow\}) + P(\{\uparrow\}). \end{aligned}$$

This example corresponds to the situation in which exactly one among two possibilities can occur (either $\downarrow$ or $\uparrow$) and both possibilities are equally likely.

More generally, the following examples correspond to N equally likely events.

Example 916 For each positive integer $N \in \mathbb{N}^*$, define an algebra of sets $\mathcal{A}$ by

$$\begin{aligned} S &:= \{0, 1, \ldots, N-1\} = \{K \in \mathbb{N} : K < N\}, \\ \mathcal{A} &:= \mathcal{P}(S). \end{aligned}$$

Then define a probability function $P : \mathcal{P}(S) \to \mathbb{Q}_+$ by

$$P(\{K\}) = 1/N$$

for every $K \in S$, so that with $\#(E)$ denoting the cardinality of E for each $E \in \mathcal{P}(S)$,

$$P(E) = [\#(E)]/N.$$

The probability of a disjoint union of events equals the sum of their probabilities, as the cardinality of their union equals the sum of their cardinalities (theorem 603):

$$P(\{0\} \dot{\cup} \{1\}) = P(\{0, 1\}) = 2/N = (1+1)/N = (1/N) + (1/N) = P(\{0\}) + P(\{1\}).$$

Such a situation occurs if each event E consists of a selection of any number K of possibilities from N mutually exclusive but equally likely possibilities.

Example 917 For the English alphabet with $N := 26$ letters,

$$S := \{ \texttt{A B C D E F G H I J K L M N O P Q R S T U V W X Y Z} \}$$

the power set $\mathcal{A} := \mathcal{P}(S)$ is an algebra of sets by example 908, and the probability function from example 916 corresponds to equally likely selections of letters. For instance, $\{\texttt{L}\} \in \mathcal{A}$ corresponds to the selection of the letter L, and $\{\texttt{M}\} \in \mathcal{A}$ corresponds to the selection of the letter M, and either selection occurs with the same probability:

$$P(\{\texttt{L}\}) = \frac{1}{26} = P(\{\texttt{M}\}).$$

Similarly, the selection of any two different letters occurs with the same probability:

$$P(\{\texttt{K}, \texttt{L}\}) = P(\{\texttt{K}\}) + P(\{\texttt{L}\}) = \frac{1}{13} = P(\{\texttt{L}\}) + P(\{\texttt{M}\}) = P(\{\texttt{L}, \texttt{M}\}).$$

The next theorem gives a formula for the probability of complementary events.

Theorem 918 *For each probability function P with domain $\mathcal{A}$, if $E \in \mathcal{A}$ then*

$$P(\mathcal{S}_{\mathcal{A}} \setminus E) = 1 - P(E).$$

In particular, $P(\varnothing) = 0$.

Proof. From conditions (P.1) and (P.2) it follows that $1 = P(\mathcal{S}_{\mathcal{A}}) = P(\mathcal{S}_{\mathcal{A}}) + P(E)$. □

Example 919 For the English alphabet with $N := 26$ letters,

$$S := \{ \texttt{A B C D E F G H I J K L M N O P Q R S T U V W X Y Z} \}$$

and the probability function from example 917 with equally likely selections of individual letters, the selection of any letter *except* the letter L corresponds to the complement of the selection of L. Consequently,

$$P(S \setminus \{\texttt{L}\}) = 1 - P(\{\texttt{L}\}) = 1 - \frac{1}{26} = \frac{25}{26}.$$

Similarly, the selection of any one letter *except* K and L corresponds to the complement of the selection of either K or L

$$P(S \setminus \{\texttt{K}, \texttt{L}\}) = 1 - P(S \setminus \{\texttt{K}, \texttt{L}\}) = 1 - \frac{1}{13} = \frac{11}{13}.$$

The following theorem shows that the range of a probability function lies between 0 and 1.

Theorem 920 *For each probability function P with domain $\mathcal{A}$, if $B \in \mathcal{A}$ and $C \in \mathcal{A}$, with $B \subseteq C$, then $P(B) \leq P(C)$. In particular, for each $E \in \mathcal{A}$,*

$$0 \leq P(E) \leq 1.$$

Proof. Because $P(C \setminus B) \geq 0$ it follows that $P(C) = P(C \setminus B) + P(B) \geq P(B)$. In particular, $0 \leq P(E) \leq P(\mathcal{S}_{\mathcal{A}}) = 1$. □

6.7.2 The Exclusion-Inclusion Principle

The definition of a probability function states that the probability of the union of two disjoint events equals the sum of their probabilities. The following considerations provide formulae for the probability of the union of events that need not be disjoint. The first theorem proceeds by adding the probabilities of two events, which counts their intersection twice, and then subtracting the probability of their intersection once.

Theorem 921 *For each probability space $(\mathcal{S}_{\mathcal{A}}, \mathcal{A}, P)$,*

$$P(E_1 \cup E_2) = P(E_1) + P(E_2) - P(E_1 \cap E_2)$$

for all $E_1 \in \mathcal{A}$ and $E_2 \in \mathcal{A}$.

Proof. From the disjoint unions

$$\begin{aligned} E_1 &= (E_1 \setminus E_2)\dot{\cup}(E_1 \cap E_2), \\ E_2 &= (E_2 \setminus E_1)\dot{\cup}(E_1 \cap E_2), \\ E_1 \cup E_2 &= (E_1 \setminus E_2)\dot{\cup}(E_1 \cap E_2)\dot{\cup}(E_2 \setminus E_1), \end{aligned}$$

it follows that

$$\begin{aligned} &P(E_1 \cup E_2) \\ &= P[(E_1 \setminus E_2)\dot{\cup}(E_1 \cap E_2)\dot{\cup}(E_2 \setminus E_1)] \\ &= P(E_1 \setminus E_2) + P(E_1 \cap E_2) + P(E_2 \setminus E_1) \\ &= P(E_1 \setminus E_2) + P(E_1 \cap E_2) + P(E_2 \setminus E_1) + [P(E_1 \cap E_2) - P(E_1 \cap E_2)] \\ &= [P(E_1 \setminus E_2) + P(E_1 \cap E_2)] + [P(E_1 \cap E_2) + P(E_2 \setminus E_1)] - P(E_1 \cap E_2) \\ &= P[(E_1 \setminus E_2)\dot{\cup}(E_1 \cap E_2)] + P[(E_1 \cap E_2)\dot{\cup}(E_2 \setminus E_1)] - P(E_1 \cap E_2) \\ &= P(E_1) + P(E_2) - P(E_1 \cap E_2). \end{aligned}$$

□

Example 922 For the English alphabet with $N := 26$ letters,

$$S := \{ \texttt{A B C D E F G H I J K L M N O P Q R S T U V W X Y Z} \}$$

and the probability function from example 917 with equally likely selections of individual letters, the selection of any one letter from either $\{\texttt{K}, \texttt{L}\}$ or $\{\texttt{L}, \texttt{M}\}$, in other words, from $\{\texttt{K}, \texttt{L}, \texttt{M}\}$ occurs with probability

$$\begin{aligned} P(\{\texttt{K}, \texttt{L}, \texttt{M}\}) &= P(\{\texttt{K}, \texttt{L}\} \cup \{\texttt{L}, \texttt{M}\}) \\ &= P(\{\texttt{K}, \texttt{L}\}) + P(\{\texttt{L}, \texttt{M}\}) - P(\{\texttt{K}, \texttt{L}\} \cap \{\texttt{L}, \texttt{M}\}) \\ &= P(\{\texttt{K}, \texttt{L}\}) + P(\{\texttt{L}, \texttt{M}\}) - P(\{\texttt{L}\}) \\ &= \tfrac{1}{13} + \tfrac{1}{13} - \tfrac{1}{26} \\ &= \tfrac{3}{26}. \end{aligned}$$

As a verification,

$$\begin{aligned} P(\{\mathtt{K},\mathtt{L},\mathtt{M}\}) &= P(\{\mathtt{K}\}\dot{\cup}\{\mathtt{L}\}\dot{\cup}\{\mathtt{M}\}) \\ &= P(\{\mathtt{K}\}) + P(\{\mathtt{L}\}) + P(\{\mathtt{M}\}) \\ &= \tfrac{1}{26} + \tfrac{1}{26} + \tfrac{1}{26} \\ &= \tfrac{3}{26}. \end{aligned}$$

Induction leads to similar formulae for the probability of any finite union of events.

6.7.3 Conditional Probabilities

The restriction of considerations from a space to a smaller space specified by conditions corresponds to the mathematical concept of "conditional" probabilities.

Definition 923 For each probability space $(\mathcal{S}_{\mathcal{A}}, \mathcal{A}, P)$, for each $B \in \mathcal{A}$ with $P(B) > 0$, and for each $E \in \mathcal{A}$, define the **conditional probability of E given B** by

$$P(E|B) := \frac{P(E \cap B)}{P(B)}.$$

Also, let $\mathcal{A}_B := \{E \cap B \in \mathcal{P}(\mathcal{S}_{\mathcal{A}}) : E \in \mathcal{A}\}$.

The following theorem verifies that the conditional probability defines a probability function with a smaller space.

Theorem 924 *For each probability space $(\mathcal{S}_{\mathcal{A}}, \mathcal{A}, P)$, for each $B \in \mathcal{A}$ such that $P(B) > 0$, and for each $E \in \mathcal{A}$, the set $\mathcal{A}_B$ is an algebra of sets, and the function $P(\ \ |B)$ defined on $\mathcal{A}_B$ by $P(E|B)$ is a probability function.*

Proof. Firstly, $B \in \mathcal{A}_B$, because $B = B \cap B$ with $B \in \mathcal{A}$. Consequently, $\mathcal{S}_{\mathcal{A}_B} = B$, because from the definition of $\mathcal{A}_B$ and $B \in \mathcal{A}_B$ it follows that $B \subseteq \bigcup \mathcal{A}_B \subseteq B$.

If $C \in \mathcal{A}_B$, then there exists $E \in \mathcal{A}$ with $C = E \cap B$. Hence $\mathcal{S}_{\mathcal{A}} \setminus E \in \mathcal{A}$ and

$$\mathcal{S}_{\mathcal{A}_B} \setminus C = B \setminus C = B \setminus (E \cap B) = B \setminus E = B \cap (\mathcal{S}_{\mathcal{A}} \setminus E) \in \mathcal{A}_B.$$

Finally, if $C \in \mathcal{A}_B$ and $D \in \mathcal{A}_B$, then there exist $E \in \mathcal{A}$ and $F \in \mathcal{A}$ such that $C = E \cap B$ and $D = F \cap B$. Hence $E \cup F \in \mathcal{A}$, whence

$$C \cup D = (E \cap B) \cup (F \cap B) = (E \cup F) \cap B \in \mathcal{A}_B.$$

Similarly, if $\mathcal{F} \subseteq \mathcal{A}_B$, then for each $C \in \mathcal{F}$ there exists $E_C \in \mathcal{A}$ such that $C = E_C \cap B$. Consequently, $\bigcup_{C \in \mathcal{F}} E_C \in \mathcal{A}$, whence

$$\bigcup \mathcal{F} = \bigcup_{C \in \mathcal{F}} C = \bigcup_{C \in \mathcal{F}} (E_C \cap B) = \left(\bigcup_{C \in \mathcal{F}} E_C\right) \cap B \in \mathcal{A}_B.$$

Moreover,

$$P(B|B) = \frac{P(B \cap B)}{P(B)} = \frac{P(B)}{P(B)} = 1.$$

Finally, if $C \in \mathcal{A}_B$ and $D \in \mathcal{A}_B$ and $C \cap D = \varnothing$, then there exist $E \in \mathcal{A}$ and $F \in \mathcal{A}$ such that $C = E \cap B$ and $D = F \cap B$, with

$$\varnothing = C \cap D = (E \cap B) \cap (F \cap B) = (E \cap F) \cap B.$$

Therefore,

$$\begin{aligned} P(C \cup D|B) &= \frac{P([C \cup D] \cap B)}{P(B)} \\ &= \frac{P([(C \cap B) \cup (D \cap B)])}{P(B)} \\ &= \frac{P(C \cap B) + P(D \cap B)}{P(B)} \\ &= \frac{P(C|B)P(B) + P(D|B)P(B)}{P(B)} \\ &= P(C|B) + P(D|B). \end{aligned}$$

□

Definition 925 For each probability space $(\mathcal{S}_{\mathcal{A}}, \mathcal{A}, P)$, for each $D \in \mathcal{A}$ and each $E \in \mathcal{A}$, the sets D and E are **independent** of each other if and only if $P(D \cap E) = P(D) \cdot P(E)$. More generally, for each non-empty set of subsets $\mathcal{F} \subseteq \mathcal{A}$, the elements of $\mathcal{F}$ are mutually independent if and only if $P\left(\bigcap \mathcal{F}\right) = \prod_{E \in \mathcal{F}} P(E)$.

Theorem 926 *For each probability space $(\mathcal{S}_{\mathcal{A}}, \mathcal{A}, P)$, for each event $B \in \mathcal{A}$ such that $P(B) > 0$, and for each event $E \in \mathcal{A}$, the sets B and E are independent of each other if and only if $P(E|B) = P(E)$.*

Proof. From the definitions of conditional probabilities and mutual independence,

$$\begin{aligned} P(E|B) &= P(E) \\ &\Updownarrow \\ P(E|B)P(B) &= P(E)P(B) \\ &\Updownarrow \\ P(E \cap B) &= P(E)P(B). \end{aligned}$$

□

Example 927 For the English alphabet with $N := 26$ letters,

$$E := \{ \texttt{A B C D E F G H I J K L M N O P Q R S T U V W X Y Z} \},$$

a two-letter word is a sequence w of two non-necessarily distinct letters,

$$w : \{0, 1\} \to \{ \texttt{A B C D E F G H I J K L M N O P Q R S T U V W X Y Z} \},$$

or, equivalently, an arrangement with repetition(s) of two letters. As a probability space, let $S := E^2$ be the set of all sequences of two letters in the alphabet E, of which there exist exactly $R_2^N = N^2 = 26^2 = 676$. For instance,

$$\begin{aligned} w_{\text{is}} : \{0, 1\} &\to E, \\ 0 &\mapsto \text{i}, \\ 1 &\mapsto \text{s}; \end{aligned}$$

$$\begin{aligned} w_{\text{it}} : \{0, 1\} &\to E, \\ 0 &\mapsto \text{i}, \\ 1 &\mapsto \text{t}, \end{aligned}$$

are such two-letter sequences. Moreover, consider the algebra of all sets of such sequences, $\mathcal{A} := \mathcal{P}(S)$. Furthermore, consider the equal likelihood probability function $P : \mathcal{P}(S) \to \mathbb{Q}_{+}$, defined by $P(C) = [\#(C)]/(N^2)$. Thus, the two words are equally likely, so that

$$P(w_{\text{is}}) = \frac{1}{676} = P(w_{\text{it}}).$$

Next consider the set $C \in \mathcal{A}$ consisting of all sequences of two letters beginning with "i" and with any second letter; there exist exactly $N = 26$ such sequences, from w_{ia} through w_{iz}. Because all such sequences are equally likely and disjoint,

$$\begin{aligned}
P(C) &= P(\{w_{\text{ia}}, w_{\text{ib}}, \ldots, w_{\text{iy}}, w_{\text{iz}}\}) \\
&= P(\{w_{\text{ia}}\} \dot{\cup} \{w_{\text{ib}}\} \dot{\cup} \cdots \dot{\cup} \{w_{\text{iy}}\} \dot{\cup} \{w_{\text{iz}}\}) \\
&= P(\{w_{\text{ia}}\}) + P(\{w_{\text{ib}}\}) + \cdots + P(\{w_{\text{iy}}\}) + P(\{w_{\text{iz}}\}) \\
&= \tfrac{1}{676} + \tfrac{1}{676} + \cdots + \tfrac{1}{676} + \tfrac{1}{676} \\
&= \tfrac{26}{676} \\
&= \tfrac{1}{26}.
\end{aligned}$$

Similarly consider the set $D \in \mathcal{A}$ consisting of all sequences of two letters ending with "s" and with any first letter; there exist exactly $N = 26$ such sequences, from w_{as} through w_{zs}, and $P(D) = 1/26$.

Then the two "events" C and D are mutually independent, because

$$P(C \cap D) = P(\{w_{\text{is}}\}) = \frac{1}{676} = \frac{1}{26} * \frac{1}{26} = P(C) * P(D).$$

Thus, the selection of the second letter is independent of the selection of the first letter. Equivalently, the probability of selecting the second letter given the selection of the first letter equals the product of the probability of selecting the first letter and the probability of selecting the second letter.

Induction leads to a similar result for any finite number of events.

Theorem 928 (Bayes's Formula) *For each probability space $(\mathcal{S}_{\mathcal{A}}, \mathcal{A}, P)$, for each partition $\mathcal{F} \subseteq \mathcal{A}$ of $\mathcal{S}_{\mathcal{A}}$, and for each event $A \in \mathcal{A}$,*

$$P(A) = \sum_{E \in \mathcal{F}} P(A|E) \cdot P(E)$$

and

$$P(E|A) = \frac{P(A|E) \cdot P(E)}{\sum_{S \in \mathcal{F}} P(A|S) \cdot P(S)}.$$

Proof. For each $E \in \mathcal{A}$, the intersection distributes over the union (theorem 328):

$$E = E \cap \mathcal{S}_{\mathcal{A}} = E \cap \left(\bigcup \mathcal{A}\right) = E \cap \left(\dot{\bigcup} \mathcal{F}\right) = \dot{\bigcup}_{C \in \mathcal{F}} (E \cap C),$$

with disjoint unions because $\mathcal{F}$ is a partition (definition 419). Hence

$$P(E) = P\left(\dot{\bigcup}_{C \in \mathcal{F}} (E \cap C)\right) = \sum_{C \in \mathcal{F}} P(E \cap C) = \sum_{C \in \mathcal{F}} [P(E|C) P(C)].$$

Consequently, for each $D \in \mathcal{F}$,

$$P(D|E) = \frac{P(D \cap E)}{P(E)} = \frac{P(E|D)P(D)}{P(E)} = \frac{P(E|D)P(D)}{\sum_{C \in \mathcal{F}}[P(E|C)P(C)]}.$$

□

6.7.4 Applications of Binomial Probabilities

Such physical particles as electrons, protons, and neutrons have a feature, called a **spin,** which characterizes their rest (motionless) states and their interactions with magnetic fields that might surround them [26, Ch. 35], [25, Ch. 6 & Ch. 12]. Each measurement of such interactions yields either one of only two values, either $+1/2$ or $-1/2$. Depending on the magnetic field, each of the two values can occur with different probabilities, p and q, with $0 < p < 1$ and $q := 1 - p$. In the context of spins, the values $+1/2$ and $-1/2$ are abbreviated by $\uparrow$ and $\downarrow$ (read "spin up" and "spin down") respectively.

With a single particle, $S := \{\downarrow, \uparrow\}$ denotes the set of measurement values. With the corresponding algebra of sets (example 907)

$$\mathcal{A} := \mathcal{P}(S) = \mathcal{P}(\{\downarrow, \uparrow\}) = \{\varnothing,\ \{\downarrow\},\ \{\uparrow\},\ S\},$$

the function $P : \mathcal{A} \to \mathbb{Q}_+$ defined for each $C \in \mathcal{A}$ by

$$\begin{aligned} P(\varnothing) &:= 0, \\ P(\{\downarrow\}) &:= p, \\ P(\{\uparrow\}) &:= q = 1 - p, \\ P(S) &:= 1, \end{aligned}$$

is a probability function. Indeed, as in example 915,

$$\begin{aligned} P(S) &= 1, \\ P(\varnothing \dot\cup \varnothing) &= P(\varnothing) = 0 = 0 + 0 = P(\varnothing) + P(\varnothing), \\ P(\varnothing \dot\cup S) &= P(S) = 1 = 0 + 1 = P(\varnothing) + P(S), \\ P(\varnothing \dot\cup \{\uparrow\}) &= P(\{\uparrow\}) = p = 0 + p = P(\varnothing) + P(\{\uparrow\}), \\ P(\varnothing \dot\cup \{\downarrow\}) &= P(\{\downarrow\}) = q = 0 + q = P(\varnothing) + P(\{\downarrow\}), \\ P(\{\downarrow\} \dot\cup \{\uparrow\}) &= P(S) = 1 = p + (1 - p) = p + q = P(\{\downarrow\}) + P(\{\uparrow\}). \end{aligned}$$

In a system with N particles, each particle still has a spin, with value $\uparrow$ with probability p, and value $\downarrow$ with probability q. Each **state** of the system is then a function

$$s : \{1, \ldots, N\} \to \{\downarrow, \uparrow\},$$

also denoted by

$$(s_1, s_2, s_3, \ldots, s_{N-2}, s_{N-1}, s_N),$$

where $s_k \in \{\downarrow, \uparrow\}$ for each $k \in \{1, \ldots, N\}$. For instance,

$$(\uparrow, \uparrow, \uparrow, \ldots, \uparrow, \uparrow, \uparrow)$$

represents one state, whereas

$$(\downarrow, \downarrow, \uparrow, \ldots, \downarrow, \uparrow, \downarrow)$$

represents another state. Thus, the set S of all states of the system is the set

$$S := \{\downarrow, \uparrow\}^{\{1, \ldots, N\}}$$

of all functions $s : \{1, \ldots, N\} \to \{\downarrow, \uparrow\}$. In particular, the set S has exactly 2^N elements. The corresponding algebra of sets $\mathcal{A} = \mathcal{P}(S)$ has $2^{(2^N)}$ elements.

For some physical systems, the numbers p and q are the same for all particles, and the probabilities that the spins have the value $\uparrow$ or $\downarrow$ are independent of one another. By the definition of independent probabilities, this means that the probability $\widetilde{P}$ of a state

$$(s_1, s_2, s_3, \ldots, s_{N-2}, s_{N-1}, s_N)$$

equals the product of the probabilities of the values $s_1, \ldots, s_N$ of the spins of all the particles:

$$\begin{aligned}&\widetilde{P}(s_1, s_2, s_3, \ldots, s_{N-2}, s_{N-1}, s_N)\\ &= P(\{s_1\}) \cdot P(\{s_2\}) \cdot P(\{s_3\}) \cdots P(\{s_{N-2}\}) \cdot P(\{s_{N-1}\}) \cdot P(\{s_N\}).\end{aligned}$$

For instance,

$$\begin{aligned}&\widetilde{P}(\uparrow, \uparrow, \uparrow, \ldots, \uparrow, \uparrow, \uparrow)\\ &= P(\{\uparrow\}) \cdot P(\{\uparrow\}) \cdot P(\{\uparrow\}) \cdots P(\{\uparrow\}) \cdot P(\{\uparrow\}) \cdot P(\{\uparrow\})\\ &= p \cdot p \cdot p \cdots p \cdot p \cdot p,\end{aligned}$$

whereas

$$\begin{aligned}&\widetilde{P}(\downarrow, \downarrow, \uparrow, \ldots, \downarrow, \uparrow, \downarrow)\\ &= P(\{\downarrow\}) \cdot P(\{\downarrow\}) \cdot P(\{\uparrow\}) \cdots P(\{\downarrow\}) \cdot P(\{\uparrow\}) \cdot P(\{\downarrow\})\\ &= q \cdot q \cdot p \cdots q \cdot p \cdot q.\end{aligned}$$

Thus, if M particles have a spin with value $\uparrow$ and the remaining $N - M$ particles have a spin with value $\downarrow$, then the state of the system occurs with probability $p^M \cdot q^{N-M}$.

Example 929 A system of $N = 4$ particles has $2^N = 2^4 = 16$ states:

$$\begin{array}{rclcrcl}
\widetilde{P}(\uparrow,\uparrow,\uparrow,\uparrow) &=& p^4, & & \widetilde{P}(\downarrow,\uparrow,\uparrow,\uparrow) &=& p^3 \cdot q,\\
\widetilde{P}(\uparrow,\uparrow,\uparrow,\downarrow) &=& p^3 \cdot q, & & \widetilde{P}(\downarrow,\uparrow,\uparrow,\downarrow) &=& p^2 \cdot q^2,\\
\widetilde{P}(\uparrow,\uparrow,\downarrow,\uparrow) &=& p^3 \cdot q, & & \widetilde{P}(\downarrow,\uparrow,\downarrow,\uparrow) &=& p^2 \cdot q^2,\\
\widetilde{P}(\uparrow,\uparrow,\downarrow,\downarrow) &=& p^2 \cdot q^2, & & \widetilde{P}(\downarrow,\uparrow,\downarrow,\downarrow) &=& p \cdot q^3,\\
\widetilde{P}(\uparrow,\downarrow,\uparrow,\uparrow) &=& p^3 \cdot q, & & \widetilde{P}(\downarrow,\downarrow,\uparrow,\uparrow) &=& p^2 \cdot q^2,\\
\widetilde{P}(\uparrow,\downarrow,\uparrow,\downarrow) &=& p^2 \cdot q^2, & & \widetilde{P}(\downarrow,\downarrow,\uparrow,\downarrow) &=& p \cdot q^3,\\
\widetilde{P}(\uparrow,\downarrow,\downarrow,\uparrow) &=& p^2 \cdot q^2, & & \widetilde{P}(\downarrow,\downarrow,\downarrow,\uparrow) &=& p \cdot q^3,\\
\widetilde{P}(\uparrow,\downarrow,\downarrow,\downarrow) &=& p \cdot q^3, & & \widetilde{P}(\downarrow,\downarrow,\downarrow,\downarrow) &=& q^4.
\end{array}$$

Example 930 In example 929, there are exactly six states where any two particles have spins with the value $\uparrow$ and the other two particles have spins with the value $\downarrow$:

$$\begin{array}{rclcrcl}
\widetilde{P}(\uparrow,\uparrow,\downarrow,\downarrow) &=& p^2 \cdot q^2, & & \widetilde{P}(\downarrow,\uparrow,\uparrow,\downarrow) &=& p^2 \cdot q^2,\\
\widetilde{P}(\uparrow,\downarrow,\uparrow,\downarrow) &=& p^2 \cdot q^2, & & \widetilde{P}(\downarrow,\uparrow,\downarrow,\uparrow) &=& p^2 \cdot q^2,\\
\widetilde{P}(\uparrow,\downarrow,\downarrow,\uparrow) &=& p^2 \cdot q^2, & & \widetilde{P}(\downarrow,\downarrow,\uparrow,\uparrow) &=& p^2 \cdot q^2.
\end{array}$$

Because the singletons containing these states are pairwise disjoint, the probability that exactly two particles have spins with the value $\uparrow$ equals the sum of their probabilities:

$$\widetilde{P}\Big(\{(\uparrow,\uparrow,\downarrow,\downarrow), (\uparrow,\downarrow,\uparrow,\downarrow), (\uparrow,\downarrow,\downarrow,\uparrow), (\downarrow,\uparrow,\uparrow,\downarrow), (\downarrow,\uparrow,\downarrow,\uparrow), (\downarrow,\downarrow,\uparrow,\uparrow)\}\Big) =$$

$$\widetilde{P}(\uparrow,\uparrow,\downarrow,\downarrow) + \widetilde{P}(\uparrow,\downarrow,\uparrow,\downarrow) + \widetilde{P}(\uparrow,\downarrow,\downarrow,\uparrow) + \widetilde{P}(\downarrow,\uparrow,\uparrow,\downarrow) + \widetilde{P}(\downarrow,\uparrow,\downarrow,\uparrow) + \widetilde{P}(\downarrow,\downarrow,\uparrow,\uparrow)$$

$$= p^2 \cdot q^2 + p^2 \cdot q^2 + p^2 \cdot q^2 + p^2 \cdot q^2 + p^2 \cdot q^2 + p^2 \cdot q^2$$

$$= 6 \cdot p^2 \cdot q^2.$$

For instance, if both spins are equally likely, so that $p = 1/2 = q$, then the probability that for a state to occur with equally many spins up and down is

$$6 \cdot p^2 \cdot q^2 = 6 \cdot (1/2)^2 \cdot (1/2)^2 = 6/2^4 = 6/16 = 3/8,$$

which is also the ratio of the number 6 of states with equally many spins up and down to the total number 16 of possible states for the system, because $p = q$.

For physical systems, listing all possible states of a system can be impossible. For instance, a laboratory system containing $6.23 * 10^{23}$ particles has $2^{(6.23*10^{23})}$ states, which are impossible to list, because they outnumber the total number of particles in the known universe. Therefore, instead of a list, a *calculation* becomes necessary.

The first step toward such a calculation consists in deriving a formula to count the number of states

$$(\uparrow, \downarrow, \downarrow, \ldots, \uparrow, \downarrow, \uparrow)$$

with a specified number K of occurrences of $\uparrow$ anywhere and a specified number $L = N - K$ of occurrences of $\downarrow$ anywhere else. Specifying one such state amounts to specifying the K coordinates containing $\uparrow$ among all N coordinates of the state, which also amounts to selecting a combination of K coordinates from among N. Because there are exactly C_K^N such combinations, it follows that there are exactly C_K^N states with exactly K of occurrences of $\uparrow$ anywhere and exactly $L = N - K$ of occurrences of $\downarrow$ anywhere else. Because the probability of each such state occurring equals $p^K \cdot q^L$, the probability of any such state occurring equals the sum of their probabilities:

$$C_K^N \cdot p^K \cdot q^L.$$

Example 931 In example 930, where $p = 1/2 = q$ with $N = 4$ particles, the probability of a state occurring with $K = 2$ spins $\uparrow$ and $L = 2$ spins $\downarrow$ anywhere equals

$$C_K^N \cdot p^K \cdot q^L = C_2^4 \cdot (1/2)^2 \cdot (1/2)^2 = \frac{4 \cdot 3}{1 \cdot 2} \cdot \frac{1}{2^2} \cdot \frac{1}{2^2} = \frac{3}{1 \cdot 2 \cdot 2^2} = \frac{3}{8}.$$

The following example demonstrates how to calculate the probability that the number of spins in a specified orientation lies within a range.

Example 932 A system of $N := 10$ particles having spins up or down has exactly $2^N = 2^{10} = 1024$ possible states. With spins up or down with probabilities $p = 1/2 = q$, consider the problem of calculating the probability that the number of spins up lies between 4 and 6 inclusive. The number of states occurring with exactly $K = 5$ spins $\uparrow$ and exactly $L = 5$ spins $\downarrow$ anywhere equals

$$C_K^N = C_5^{10} = \frac{10 \cdot 9 \cdot 8 \cdot 7 \cdot 6}{1 \cdot 2 \cdot 3 \cdot 4 \cdot 5} = 252.$$

Similarly, the number of states occurring with $K = 4$ spins $\uparrow$ and $L = 6$ spins $\downarrow$ is

$$C_K^N = C_4^{10} = \frac{10 \cdot 9 \cdot 8 \cdot 7}{1 \cdot 2 \cdot 3 \cdot 4} = 210.$$

Because $C_K^N = C_{N-K}^N$ it follows that $C_6^{10} = C_4^{10} = 210$. Hence the number of states with a number of spins up between 4 and 6 inclusive equals

$$C_4^{10} + C_5^{10} + C_6^{10} = 210 + 252 + 210 = 672.$$

Thus the probability that the number of spins up lies between 4 and 6 inclusive is

$$\frac{C_4^{10} + C_5^{10} + C_6^{10}}{2^{10}} = \frac{672}{1024} = \frac{21}{32}.$$

Therefore, the number of spins up lies from 4 through 6 with a probability near $2/3$. In other words, the system has a larger majority of spins up (more than 6) or a smaller minority of spins up (less than 4) only with a probability near $1/3$.

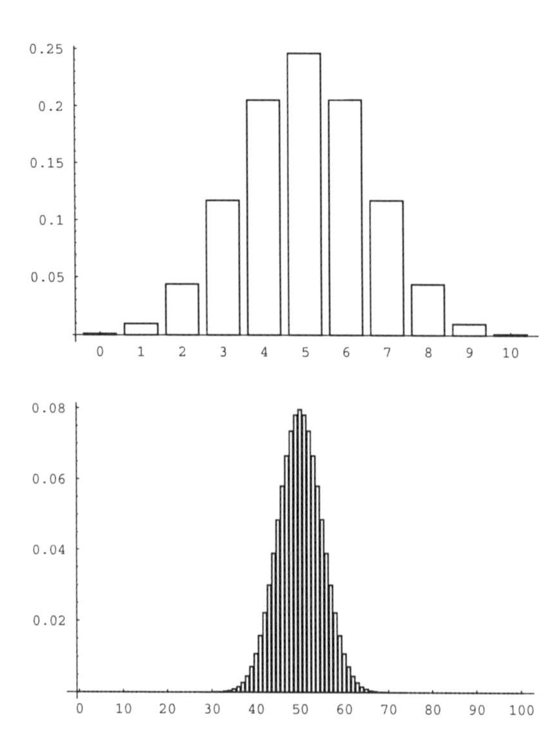

Figure 6.3: Plots of C_K^N versus K. Top: $N = 10$. Bottom: $N := 100$.

Figure 6.3 shows how as the total number of particles increases, the probability increases that the number of particles with spins up lies within a narrow range about one half of the total number of particles, a phenomenon called the Weak Law of Large Numbers [27, p. 294]. For instance, in a laboratory system of $N = 6.23 * 10^{23}$ particles with $p = 1/2 = q$, similar calculations show that the odds that states occur with a number K of spins $\uparrow$ deviating by more than 1% away from $N/2$ are less than 99 to $6.23 * 10^{21}$, which amounts to less than 100 to 10^{21}, or, equivalently, less than 1 to 10^{19}. The spins of a real system are thus unlikely to align spontaneously.

6.7.5 Exercises

Exercise 1061 Calculate the probability of selecting "swap" from all the four-letter words from twenty-six letters.

Exercise 1062 Calculate the probability of selecting "wasp" from all the four-letter words from twenty-six letters.

Exercise 1063 Calculate the probability of selecting either "swap" or "wasp" from all the four-letter words from twenty-six letters.

Exercise 1064 Calculate the probability of selecting either "paws" or "swap" or "wasp" from all the four-letter words from twenty-six letters.

Exercise 1065 Calculate the probability of selecting either "passwords" or "sapswords" or "sodswraps" or "swordpass" from all the nine-letter words from twenty-six letters.

Exercise 1066 Calculate the probability of selecting "ÖLS" from all the three-letter words from twenty-nine letters.

Exercise 1067 Calculate the probability of selecting either "BÄR" or "FÜR" or "ÖLS" or "ÖSE" or "RÄT" or "ÜBE" from all the three-letter words from twenty-nine letters.

Exercise 1068 For each set $\mathcal{A}$ with space $\mathcal{S}_{\mathcal{A}}$, prove that $\mathcal{A} \subseteq \mathcal{P}(\mathcal{S}_{\mathcal{A}})$.

Exercise 1069 For a system as in example 932 but with 20 particles, calculate the probability that the number of particles with spins up lies in the range from 8 inclusive through 12 inclusive.

Exercise 1070 For a system as in example 932 but with 40 particles, calculate the probability that the number of particles with spins up lies in the range from 16 inclusive through 24 inclusive.

6.8 THE ENIGMA MACHINES

6.8.1 Design and Operation of the ENIGMA Machines

ENIGMA machines are electro-mechanical devices, as shown in figure 6.1, to encipher plaintext and to decipher encrypted messages. The following description of the machines and their enciphering schemes follows accounts from Marian Rejewski [82], [83], [84], and Alan Stripp [93].

To accept plain or enciphered text, the ENIGMA machine contains a keyboard, which consists of $n := 26$ keys for the twenty-six letters of the English alphabet, or for the numbers $\{01, \ldots, 26\}$. Some versions of the ENIGMA machines allow for $n := 28$ characters, including either Ä and Ü, or Ä Ö, Ü, but not X [2, p. 107]. Through the keyboard, an operator can type individual characters one at a time.

To show plain or enciphered text, the ENIGMA machine also contains a display, which consists of n characters arranged in the same pattern as the keyboard, but on a panel separate from the keyboard. To display a character, the machine illuminates the character by a lamp behind it.

As the operator types in text through the keyboard, the same operator or a second operator can see the enciphered character illuminated on the display.

If the first operator types in plaintext, then the second operator reads out an enciphered message, which can be given to a radio operator for transmission. Conversely, and by design, if the first operator types in an enciphered message, then the second operator reads out the corresponding plaintext.

Between the keyboard and the display, the ENIGMA machine enciphers or deciphers messages through electro-mechanical devices, which perform ever changing compositions of permutations of the alphabet.

By design, an ENIGMA machine carries out the same operations for deciphering and enciphering. For either task, the machine contains a series of typically three but possibly more electro-mechanical rotors, numbered by Roman numerals, for example, I, II, III .... The rotors differ from one another by performing different permutations of the alphabet, so that the rotor numbered J performs a permutation ρ_J:

$$\begin{array}{rcl} \rho_I : \{1,\dots,n\} & \to & \{1,\dots,n\}, \\ \rho_{II} : \{1,\dots,n\} & \to & \{1,\dots,n\}, \\ \rho_{III} : \{1,\dots,n\} & \to & \{1,\dots,n\}, \\ & \vdots & \end{array}$$

Example 933 Rotors I, II, III of the ENIGMA used by the German Army ("Wehrmacht") perform the following permutations [2, p. 110]:

$$\rho_I := \begin{pmatrix} \texttt{ABCDEFGHIJKLMNOPQRSTUVWXYZ} \\ \texttt{EKMFLGDQVZNTOWYHXUSPAIBRCJ} \end{pmatrix},$$

$$\rho_{II} := \begin{pmatrix} \texttt{ABCDEFGHIJKLMNOPQRSTUVWXYZ} \\ \texttt{AJDKSIRUXBLHWTMCQGZNPYFVOE} \end{pmatrix},$$

$$\rho_{III} := \begin{pmatrix} \texttt{ABCDEFGHIJKLMNOPQRSTUVWXYZ} \\ \texttt{BDFHJLCPRTXVZNYEIWGAKMUSQO} \end{pmatrix}.$$

Also, the rotors can be installed in any order, for instance, III-I-II, which composes their premutations in any order, for instance,

$$\rho_{III} \circ \rho_I \circ \rho_{II}.$$

Moreover, every rotor carries the letters of the alphabet printed on a movable ring around its periphery, and each rotor can be installed with its ring rotated to any one of twenty-six positions, called **ring-settings**. Each turn of the ring by one position corresponds to a conjugation by the permutation β representing the cyclic shift of the alphabet by one letter to the right:

$$\begin{array}{rcl} \beta & := & \begin{pmatrix} \texttt{ABCDEFGHIJKLMNOPQRSTUVWXYZ} \\ \texttt{BCDEFGHIJKLMNOPQRSTUVWXYZA} \end{pmatrix} \\ & = & \begin{pmatrix} \texttt{ABCDEFGHIJKLMNOPQRSTUVWXYZ} \end{pmatrix}. \end{array}$$

Each turn of the ring by one letter on rotor numbered J precedes the rotor permutation ρ_J by β^1 relative to the preceding rotor, and follows ρ_J by β^{-1} relative to the following rotor [2, p. 110].

Thus, rotating the ring to bring B into position A amounts to conjugating ρ_J by β^1. For instance, with ρ_I as in example 933,

$$\begin{array}{rcl} (\beta^{-1} \circ \rho_I \circ \beta)(\texttt{A}) & = & \beta^{-1}\{\rho[\beta(\texttt{A})]\} \\ & = & \beta^{-1}\{\rho[\texttt{B}]\} \\ & = & \beta^{-1}\{\texttt{K}\} \\ & = & \texttt{J}. \end{array}$$

More generally, rotating the ring on rotor J by any number $i \in \{0,\dots,n-1\}$ of positions in one direction conjugates the permutation ρ_J by β^i, so that rotor J performs the permutation

$$\omega_{J,i} := \beta^{-i} \circ \rho_J \circ \beta^i.$$

Furthermore, before every message the operator can turn each rotor installed in the machine to any of twenty-six starting positions, called **indicator-settings** or **message-settings**, and identified by which letter on the rotor shows up through an opening on the machine. As with ring-settings, rotating rotor J by k positions conjugates the permutation $\omega_{J,i}$ by β^k, which produces

$$\beta^{-k} \circ \omega_{J,i} \circ \beta^k = \beta^{-(k+i)} \circ \rho_J \circ \beta^{i+k} = \omega_{J,i+k}.$$

For example, installing rotor III in position `A` gives the permutation ρ_{III}, but installing rotor I in position `E` yields the permutation $\beta^{-4} \circ \rho_I \circ \beta^4$, and installing rotor II in position `D` yields the permutation $\beta^{-3} \circ \rho_{II} \circ \beta^3$. Thus, the starting position `AED` and the rotor-order III-I-II mean that the series of rotors begins with `A` showing on rotor III, followed by `E` showing on rotor I, followed by `D` showing on rotor II, which yields the composite permutation

$$\rho_{III} \circ (\beta^{-4} \circ \rho_I \circ \beta^4) \circ (\beta^{-3} \circ \rho_{II} \circ \beta^3) = \omega_{III,0} \circ \omega_{I,4} \circ \omega_{II,3}.$$

In general, selecting the rotor-order III-I-II with ring-settings i_{III}, i_I, i_{II} and rotating the rotors to the indicator-setting k_{III}, k_I, k_{II} sets up the machine for its first permutation

$$\omega_{III,i_{III}+k_{III}} \circ \omega_{I,i_I+k_I} \circ \omega_{II,i_{II}+k_{II}}.$$

Before and after the rotors, a plugboard allows for an optional permutation, through a set of cable-pairs and twenty-six pairs of sockets arranged in the same pattern as the keyboard. Each pair of sockets corresponds to one letter of the alphabet. Each cable-pair connecting two pairs of sockets to each other performs a transposition, swapping the letter for the first pair of sockets and the letter for the second pair of sockets. Any pair of sockets not connected to a cable-pair fixes its corresponding letter. Thus the plugboard performs yet another permutation of the alphabet,

$$\sigma : \{1, \ldots, n\} \to \{1, \ldots, n\}.$$

To prepare the machine for enciphering, the operator turns the ring on each rotor to the prescribed ring-setting, for example, `OPQ`; the operator then installs the rotors in the prescribed, rotor-order, for instance, III-I-II. The operator then connects any number of pairs of sockets with cable-pairs as prescribed.

While enciphering plaintext, after the operator types a letter by pressing a key on the keyboard, an electric current travels from the key and passes first through the plugboard, then through the rotors in one direction, then back through the same rotors but in the reverse direction, and finally again through the plugboard to a lamp on the display. Then the rotors turn to prepare another sequence of permutations for the next letter. Specifically, the first rotor turns by one position, which amounts to a cyclic shift of the alphabet by one letter to the right; moreover, for each complete revolution of one rotor by twenty-six positions, the next rotor turns by one position. Thus the manner in which the rotors turn does *not* depend on which letter has just been typed. This feature allows for the matching of the enciphering machine with the deciphering machines.

To communicate the machine settings to the operators of other machines about to receive the ciphered message, the enciphering operator selects any starting positions for the rotors, for instance, `OPQ`, which forms the **indicator-setting**, and selects any sequence of three letters and repeats it once for redundancy, for instance, `FGHFGH`, which forms the **message-setting**. The operator then enciphers the message-setting, which might give any sequence of six letters, for instance, `UVWXYZ`, which forms the **message-indicator**. Because the machine changes the permutation after each letter, the repeating pattern of the message-setting `FGHFGH` need not cause a repeating pattern in the enciphered indicator `UVWXYZ`. Finally the operator transmits the indicator-setting `OPQ` and the indicator `UVWXYZ`, in plaintext. Thereafter the operator resets the starting position to the message-setting, here `FGH`, and begins typing the plaintext, which the machine enciphers.

To prepare a machine for deciphering, the operator must adjust each rotor to the same ring-setting, set the rotors in the same order, here III-I-II, and connect the same pairs of sockets as done by the enciphering operator. Therefore the use of such machines for communication requires a prior coordination of the enciphering and deciphering operators, either by the operators or by a higher authority. The deciphering operator then sets the same indicator-setting, `OPQ`, and types the indicator `UVWXYZ`, which the machine deciphers to `FGHFGH`. The deciphering operator next resets the indicator-setting to `FGH`, and then types the ciphered message, which the machine restores into plaintext.

The total number m of rotors depends upon the particular version of the machine [2, p. 107–108]. For instance, in 1923 the first commercial versions — `ENIGMA A` and `ENIGMA B` — contained four rotors, whereas in 1926 the next commercial versions — `ENIGMA C` and `ENIGMA D` — contained three rotors. The last rotor — called the reflecting rotor — reflected the current back through the other rotors; on some versions it remained fixed but on the `ENIGMA D` it could turn and be set by the operator as any of the other rotors. Thus the `ENIGMA` machine enciphers each letter through a permutation of the form

$$\begin{aligned}&\sigma^{-1} \circ (\beta^{i_1+k_1} \circ \rho_1^{-1} \circ \beta^{-(i_1+k_1)}) \circ \cdots \circ (\beta^{i_{m-1}+k_{m-1}} \circ \rho_{m-1}^{-1} \circ \beta^{-(i_{m-1}+k_{m-1})})\\ &\quad \circ \rho_m \circ\\ &\quad (\beta^{-(i_{m-1}+k_{m-1})} \circ \rho_{m-1} \circ \beta^{i_{m-1}+k_{m-1}}) \circ \cdots \circ (\beta^{-(i_1+k_1)} \circ \rho_1 \circ \beta^{i_1+k_1}) \circ \sigma.\end{aligned}$$

Different versions of the `ENIGMA` also includes other variations. For instance, if the first rotor completes one revolution and moves the second rotor by one position, then on some versions the third rotor also moves by one position. On other versions, the third rotor moves only as the second rotor completes one revolution, which occurs only once in twenty-six revolutions of the first rotor.

6.8.2 Breaking the ENIGMA Ciphers

`ENIGMA` machines became widely used by the German armed forces in the late 1920s and through the 1930s [2, p. 107], and hence arose the need for the Allies to break the `ENIGMA` ciphers. Yet reconstructing the internal wiring of every rotor and then "breaking the code" through purely logical and mathematical analyses of enciphered messages intercepted by radio seemed impossible [105, p. 12]. In other words, breaking the code seemed to require access to a military version of an `ENIGMA` machine. As a first step to this end, the Polish cipher bureau obtained a commercial version of the `ENIGMA`, and three Polish mathematicians (Marian Rejewski, Jerzy Rozycki, and Henryk Zygalski) started to develop a mathematical method to reconstruct the inner wiring of a military `ENIGMA` machine from intercepted messages. They successfully tested their method with intercepted messages and machine settings given to them by a German. A few days before Germany's invasion of Poland and the beginning of the Second World War, the Poles delivered their deciphering methods and two replicas of the military `ENIGMA` to the Allies [105, p. 15].

Thus the Allies had the following information about the `ENIGMA` ciphers.

- The alphabet in use consists of n characters; usually $n = 26$ or $n = 28$.
- The plugboard performs a permutation σ equal to the product of an unknown number k (usually ten [51, plate 4]) of *disjoint* 2-cycles (swaps):

$$\sigma = \tau_{r_1,s_1} \circ \cdots \circ \tau_{r_k,s_k}.$$

- The machine contains a sequence of ℓ rotors chosen from a set of m different interchangeable rotors numbered I, II, III .... Usually $\ell \in \{3,4\}$, $m \in \{3,8\}$, and the values of ℓ and m in use can be inferred from the origin of the message (army, navy, or airforce radio networks).

- Each rotor permutes the alphabet of n letters through a permutation ω_{J,i_J+k_J} that depends on the type of the rotor $J \in \{I, II, \ldots\}$, on any of its n ring-settings, and on any of the n positions of the rotor during the encryption of the ith letter of a message, so that $\omega_{J,i_J+k_J} = \beta^{-(i_J+k_J)} \circ \rho_J \circ \beta^{i_J+k_J}$.
- The position of each rotor begins with the message-setting FGH, transmitted as a message-indicator UVWXYZ which is enciphered from FGHFGH through the known initial indicator-setting OPQ. Moreover, the rotors turn by one position after each letter.
- The last (mth) rotor sends each partially enciphered letter back through the other rotors and the plugboard in the reverse order.

Consequently, the ENIGMA machine transforms the ith plaintext letter p_i into the ith ciphered letter q_i through the permutation α_i, so that

$$q_i = \alpha_i(p),$$

with

$$\alpha_i = \sigma^{-1} \circ \left(\beta^{i_1+k_1} \circ \rho_1^{-1} \circ \beta^{-(i_1+k_1)}\right) \circ \cdots \circ \left(\beta^{i_{m-1}+k_{m-1}} \circ \rho_{m-1}^{-1} \circ \beta^{-(i_{m-1}+k_{m-1})}\right) \circ \rho_m \circ \left(\beta^{-(i_{m-1}+k_{m-1})} \circ \rho_{m-1} \circ \beta^{i_{m-1}+k_{m-1}}\right) \circ \cdots \circ \left(\beta^{-(i_1+k_1)} \circ \rho_1 \circ \beta^{i_1+k_1}\right) \circ \sigma.$$

This formula is known as the **ENIGMA equation**.

One of the significant methods used in breaking the ENIGMA ciphers utilized those instances with repetitions in the message-indicator at the same locations as repetitions in the repeated message-setting [105, p. 63]. For example, with a plaintext indicator setting LTS, an unknown repeated message-setting FGHFGH might get enciphered into a message-indicator VBYQGY, with a repetition of Y at the same location as the repetition of H. In contrast, the enciphered message-indicator usually does not contain any repetition because of the rotation of the rotors between letters. Consequently, the presence of any such repetition in the enciphered message indicator reveals the existence of 3-cycles. Therefore, the interception of any such repetition can exclude from the search all the permutations that do not admit any 3-cycle as a factor.

6.8.3 Reconstructing the ENIGMA Wiring

The determination of the ENIGMA's internal wiring and its enciphering algorithm solely through logical and mathematical analyses of intercepted messages seemed impossible to the Allies with the resources available during the Second World War [105, p. 12]. Nevertheless, on several occasions the Germans would introduce a new rotor with a different wiring [2, p. 108], which the Allies were able to determine in large part (but not always [105, p. 113–115]) through logical and mathematical analyses. For instance, starting from a commercial version of the ENIGMA rotors, the Poles reconstructed the inner wiring of the military version of the ENIGMA rotors largely through two of Marian Rejewski's theorems [63] (proofs of which form the object of exercises).

Theorem 934 (Rejewski's First Theorem) *If two permutations $\xi, \zeta \in \mathcal{S}_n$ are each a composition of $n/2$ disjoint 2-cycles (transpositions), so that*

$$\begin{aligned} \xi &= \tau_{p_{n/2},q_{n/2}} \circ \cdots \circ \tau_{p_1,q_1}, \\ \zeta &= \tau_{r_{n/2},s_{n/2}} \circ \cdots \circ \tau_{r_1,s_1}, \end{aligned}$$

then $\zeta \circ \xi$ is a composition of pairs of disjoint cycles of equal lengths.

Theorem 935 (Rejewski's Second Theorem) *If a permutation $\omega \in \mathcal{S}_n$ is a composition of pairs of disjoint cycles of equal lengths, then there exist two permutations $\xi, \zeta \in \mathcal{S}_n$ that are each a composition of $n/2$ disjoint 2-cycles (transpositions), so that $\omega = \zeta \circ \xi$.*

The following considerations outline Rejewski's method to reconstruct the ENIGMA's wiring from observations of the effects of the yet unknown permutations α_i performed by the machine to encipher the i-th letter of a message. Though α_i remained unknown until the end of the analysis, the Allies knew from the design of the ENIGMA that $\alpha_i = \alpha_i^{-1}$, and they knew from the operative procedures for the ENIGMA that the message indicator UVWXYZ encrypted a yet unknown repetition FGHFGH of the yet unknown message indicator FGH.

The first considerations pertain to the deciphering of the message-indicator UVWXYZ from the repeated unknown message-setting FGHFGH through six yet unknown permutations $\alpha_1, \ldots, \alpha_6$, defined to parallel the motions of the rotors so that

$$\begin{aligned} U &= \alpha_1(F), \\ V &= \alpha_2(G), \\ W &= \alpha_3(H), \\ X &= \alpha_4(F) = (\alpha_4 \circ \alpha_1^{-1})(U) = (\alpha_4 \circ \alpha_1)(U), \\ Y &= \alpha_5(G) = (\alpha_5 \circ \alpha_2^{-1})(V) = (\alpha_5 \circ \alpha_2)(V), \\ Z &= \alpha_6(H) = (\alpha_6 \circ \alpha_3^{-1})(W) = (\alpha_6 \circ \alpha_3)(W). \end{aligned}$$

From the same daily ring-setting, if the daily messages contain sufficiently many messages so that every letter of the alphabet appears in every position, then observations of the message-indicators yield formulae for the permutations $\alpha_4 \circ \alpha_1$, $\alpha_5 \circ \alpha_2$, $\alpha_6 \circ \alpha_3$.

After collecting enough data to specify the three compositions $\alpha_4 \circ \alpha_1$, $\alpha_5 \circ \alpha_2$, $\alpha_6 \circ \alpha_3$, the first mathematical task consists in factoring each of them as the product of disjoint cycles (theorem 870).

The second mathematical task then consists in applying Rejewski's theorems (theorem 934 and 935) to factor each permutation as compositions of disjoint transpositions.

For an ENIGMA machine with three turning rotors numbered L, M, N and one fixed reflective rotor numbered R,

$$\begin{aligned} \alpha_1 &= \sigma^{-1} \circ \beta^{-1} \circ \rho_N^{-1} \circ \beta^1 \circ \rho_M^{-1} \circ \rho_L^{-1} \circ \rho_R \circ \rho_L \circ \rho_M \circ \beta^{-1} \circ \rho_N \circ \beta^1 \circ \sigma, \\ \alpha_2 &= \sigma^{-1} \circ \beta^{-2} \circ \rho_N^{-1} \circ \beta^2 \circ \rho_M^{-1} \circ \rho_L^{-1} \circ \rho_R \circ \rho_L \circ \rho_M \circ \beta^{-2} \circ \rho_N \circ \beta^2 \circ \sigma, \\ \alpha_3 &= \sigma^{-1} \circ \beta^{-3} \circ \rho_N^{-1} \circ \beta^3 \circ \rho_M^{-1} \circ \rho_L^{-1} \circ \rho_R \circ \rho_L \circ \rho_M \circ \beta^{-3} \circ \rho_N \circ \beta^3 \circ \sigma, \\ \alpha_4 &= \sigma^{-1} \circ \beta^{-4} \circ \rho_N^{-1} \circ \beta^4 \circ \rho_M^{-1} \circ \rho_L^{-1} \circ \rho_R \circ \rho_L \circ \rho_M \circ \beta^{-4} \circ \rho_N \circ \beta^4 \circ \sigma, \\ \alpha_5 &= \sigma^{-1} \circ \beta^{-5} \circ \rho_N^{-1} \circ \beta^5 \circ \rho_M^{-1} \circ \rho_L^{-1} \circ \rho_R \circ \rho_L \circ \rho_M \circ \beta^{-5} \circ \rho_N \circ \beta^5 \circ \sigma, \\ \alpha_6 &= \sigma^{-1} \circ \beta^{-6} \circ \rho_N^{-1} \circ \beta^6 \circ \rho_M^{-1} \circ \rho_L^{-1} \circ \rho_R \circ \rho_L \circ \rho_M \circ \beta^{-6} \circ \rho_N \circ \beta^6 \circ \sigma. \end{aligned}$$

Setting

$$\gamma := \rho_M^{-1} \circ \rho_L^{-1} \circ \rho_R \circ \rho_L \circ \rho_M$$

leads to the equations for the yet unknown permutations σ, ρ_N, and γ in terms of the known permutations $\alpha_4 \circ \alpha_1$, $\alpha_5 \circ \alpha_2$, $\alpha_6 \circ \alpha_3$:

$$\alpha_4 \circ \alpha_1 =$$
$$\sigma^{-1} \circ \beta^{-4} \circ \rho_N^{-1} \circ \beta^4 \circ \gamma \circ \beta^{-4} \circ \rho_N \circ \beta^3 \circ \rho_N^{-1} \circ \beta^1 \circ \gamma \circ \beta^{-1} \circ \rho_N \circ \beta^1 \circ \sigma,$$

$$\alpha_5 \circ \alpha_2 =$$
$$\sigma^{-1} \circ \beta^{-5} \circ \rho_N^{-1} \circ \beta^5 \circ \gamma \circ \beta^{-5} \circ \rho_N \circ \beta^3 \circ \rho_N^{-1} \circ \beta^2 \circ \gamma \circ \beta^{-2} \circ \rho_N \circ \beta^2 \circ \sigma,$$

$$\alpha_6 \circ \alpha_3 =$$
$$\sigma^{-1} \circ \beta^{-6} \circ \rho_N^{-1} \circ \beta^6 \circ \gamma \circ \beta^{-6} \circ \rho_N \circ \beta^3 \circ \rho_N^{-1} \circ \beta^3 \circ \gamma \circ \beta^{-3} \circ \rho_N \circ \beta^3 \circ \sigma.$$

In December of 1932, the Polish cipher bureau received from the French cipher bureau information containing a plugboard permutation σ to be used by the Germans for some time (presumably three months [2, p. 108]). The knowledge of σ then yielded six equations for the remaining two unknown permutations ρ_N and γ:

$$\sigma \circ \alpha_1 \circ \sigma^{-1} = \beta^{-1} \circ \rho_N^{-1} \circ \beta^1 \circ \rho_M^{-1} \circ \rho_L^{-1} \circ \rho_R \circ \rho_L \circ \rho_M \circ \beta^{-1} \circ \rho_N \circ \beta^1,$$

$$\sigma \circ \alpha_2 \circ \sigma^{-1} = \beta^{-2} \circ \rho_N^{-1} \circ \beta^2 \circ \rho_M^{-1} \circ \rho_L^{-1} \circ \rho_R \circ \rho_L \circ \rho_M \circ \beta^{-2} \circ \rho_N \circ \beta^2,$$

$$\sigma \circ \alpha_3 \circ \sigma^{-1} = \beta^{-3} \circ \rho_N^{-1} \circ \beta^3 \circ \rho_M^{-1} \circ \rho_L^{-1} \circ \rho_R \circ \rho_L \circ \rho_M \circ \beta^{-3} \circ \rho_N \circ \beta^3,$$

$$\sigma \circ \alpha_4 \circ \sigma^{-1} = \beta^{-4} \circ \rho_N^{-1} \circ \beta^4 \circ \rho_M^{-1} \circ \rho_L^{-1} \circ \rho_R \circ \rho_L \circ \rho_M \circ \beta^{-4} \circ \rho_N \circ \beta^4,$$

$$\sigma \circ \alpha_5 \circ \sigma^{-1} = \beta^{-5} \circ \rho_N^{-1} \circ \beta^5 \circ \rho_M^{-1} \circ \rho_L^{-1} \circ \rho_R \circ \rho_L \circ \rho_M \circ \beta^{-5} \circ \rho_N \circ \beta^5,$$

$$\sigma \circ \alpha_6 \circ \sigma^{-1} = \beta^{-6} \circ \rho_N^{-1} \circ \beta^6 \circ \rho_M^{-1} \circ \rho_L^{-1} \circ \rho_R \circ \rho_L \circ \rho_M \circ \beta^{-6} \circ \rho_N \circ \beta^6.$$

Solving this system yields the permutation ρ_N, which gives the wiring of rotor N.

6.8.4 Exercises

The following exercises demonstrate that for the ENIGMA machine to decipher and encipher messages through the same process, without enciphering any character by the same character, it is necessary and sufficient that the reflecting rotor perform $n/2$ disjoint transpositions.

Exercise 1071 For all permutations $\alpha \in \mathcal{S}_n$ and $\omega \in \mathcal{S}_n$, prove that if $\alpha(i) \neq i$ for every $i \in \{1, \ldots, n\}$, then $(\omega^{-1} \circ \alpha \circ \omega)(j) \neq j$ for every $j \in \{1, \ldots, n\}$.

Exercise 1072 For all permutations $\alpha \in \mathcal{S}_n$ and $\omega \in \mathcal{S}_n$, prove that if $\alpha(i) = i$ for some $i \in \{1, \ldots, n\}$, then $(\omega^{-1} \circ \alpha \circ \omega)(j) = j$ for some $j \in \{1, \ldots, n\}$.

Exercise 1073 Prove that if the reflecting rotor does not fix any character, then neither does the ENIGMA machine. In other words, prove that if $\rho_R(i) \neq i$ for every $i \in \{1, \ldots, n\}$, then

$$[\sigma^{-1} \circ \cdots \circ \beta^{i_J} \rho_J^{-1} \circ \beta^{-i_J} \circ \cdots \circ \beta^{-i_J} \rho_J \circ \beta^{i_J} \circ \cdots \circ \sigma](j) \neq j$$

for every $j \in \{1, \ldots, n\}$.

Exercise 1074 Prove that if the reflecting rotor fixes a character, then the ENIGMA machine also fixes a character. In other words, prove that if $\rho_R(i) = i$ for some $i \in \{1, \ldots, n\}$, then

$$[\sigma^{-1} \circ \cdots \circ \beta^{i_J} \rho_J^{-1} \circ \beta^{-i_J} \circ \cdots \circ \beta^{-i_J} \rho_J \circ \beta^{i_J} \circ \cdots \circ \sigma](j) = j$$

for some $j \in \{1, \ldots, n\}$.

Exercise 1075 Prove that if the reflecting rotor performs a self-inverse permutation, then the ENIGMA machine also performs a self-inverse permutation. In other words, prove that if $\rho_R^{-1} = \rho_R$, then $\alpha^{-1} = \alpha$, with α as in the ENIGMA equation.

Exercise 1076 Prove that if the ENIGMA machine also performs a self-inverse permutation, then the reflecting rotor performs a self-inverse permutation, In other words, prove that if $\alpha^{-1} = \alpha$, then $\rho_R^{-1} = \rho_R$ with α as in the ENIGMA equation.

Exercise 1077 Prove that if a permutation $\eta \in \mathcal{S}_n$ is self-inverse, so that $\eta^{-1} = \eta$, then η is the composition of disjoint transpositions.

Exercise 1078 Prove that if a permutation $\eta \in \mathcal{S}_n$ is the composition of disjoint transpositions, then η is self-inverse, so that $\eta^{-1} = \eta$.

Exercise 1079 Prove that if there exists a self-inverse permutation $\rho_R \in \mathcal{S}_n$ that does not fix any character, then n is even.

Exercise 1080 Prove that if n is even, then there exists a self-inverse permutation $\rho_R \in \mathcal{S}_n$ that does not fix any character.

The following exercises demonstrate the impracticality of breaking the ENIGMA ciphers by trying out all possible permutations.

Exercise 1081 Determine the total number of self-inverse permutations that do not fix any character of an alphabet with $n := 28$ letters.

Exercise 1082 Determine the total number of self-inverse permutations of an alphabet with $n := 28$ letters.

Exercise 1083 Determine the total number of permutations of an alphabet with $n := 28$ letters, which is also the total number of possible wirings of each rotor.

Exercise 1084 Determine the total number of permutations of an alphabet with $n := 28$ letters, which is also the total number of possible wirings of each rotor.

Exercise 1085 Determine the total number of arrangements of three rotors from a selection of eight rotors numbered I, II, III, IV, V, VI, VII, VIII.

Exercise 1086 Determine the total number of message-settings, each message-setting consisting of a sequence of three letters, with repetitions allowed, from the alphabet with $n = 28$ letters.

Exercise 1087 Determine the total number of plugboard-settings, each plugboard-setting consisting of a selection of fourteen pairs of letters, without repetitions, from the alphabet with $n = 28$ letters.

The following exercises demonstrate that the plugboard allows for the largest number of permutations provided that it does *not* swap all characters.

Exercise 1088 Determine the total number of selections of k pairs of letters, without repetitions, from an alphabet with $n = 2\ell$ letters.

Exercise 1089 Determine the total number of self-reflecting permutations of $2k$ letters that do not fix any letter.

Exercise 1090 Determine the total number of plugboard-settings, each plugboard-setting consisting of a selection of k pairs of letters, without repetitions, from an alphabet with $n = 2\ell$ letters.

Exercise 1091 For $\ell := 13$ and for $\ell := 14$, determine the number k of pairs of letters that maximizes the total number of plugboard-settings, each plugboard-setting consisting of a selection of k pairs of letters, without repetitions, from an alphabet with $n = 2\ell$ letters.

The following exercises explain why the same ENIGMA machine can decipher and encipher messages.

Exercise 1092 For each group $(G, \bullet)$ and all elements $p, q \in G$ such that $q = q^{-1}$, prove that $p \bullet q \bullet p^{-1}$ is its own inverse: $p \bullet q \bullet p^{-1} = (p \bullet q \bullet p^{-1})^{-1}$.

Exercise 1093 For all permutations σ and $\rho_{J_j,i}$ such that $\rho_{J_m,i} = \rho_{J_m,i}^{-1}$, prove that the permutation η defined by

$$\eta := \sigma^{-1} \circ (\rho_{J_1,i}^{-1} \circ \cdots \circ \rho_{J_{m-1},i}^{-1}) \circ \rho_{J_m,i} \circ (\rho_{J_{m-1},i} \circ \cdots \circ \rho_{J_1,i}) \circ \sigma$$

is its own inverse: $\eta = \eta^{-1}$.

The following exercises explain why only with an alphabet consisting of an even number of characters can the same ENIGMA machine decipher and encipher messages without ever enciphering any letter by itself.

Exercise 1094 For each permutation $\eta \in \mathcal{S}_n$, prove that $\eta = \eta^{-1}$ if and only if η is the composition of disjoint 2-cycles (swaps).

Exercise 1095 For each permutation $\eta \in \mathcal{S}_n$, prove that if $\eta = \eta^{-1}$, and if η does not fix any element, so that $\eta(i) \neq i$ for every $i \in \{1, \ldots, n\}$, then n is even.

Exercise 1096 Prove Rejewski's First Theorem (theorem 934).

Exercise 1097 Prove Rejewski's Second Theorem (theorem 935).

Exercise 1098 Find the probability of *guessing* the correct message-setting.

Exercise 1099 Find the probability of *guessing* the correct rotor selection and order.

Exercise 1100 Find the probability of *guessing all* the correct message, rotor, and ring settings.

6.9 PROJECTS

Project 13 Investigate whether there exist perfect pangrams.

Project 14 Design a working (real or virtual) version of an ENIGMA machine.

Chapter 7

Graph Theory

7.0 INTRODUCTION

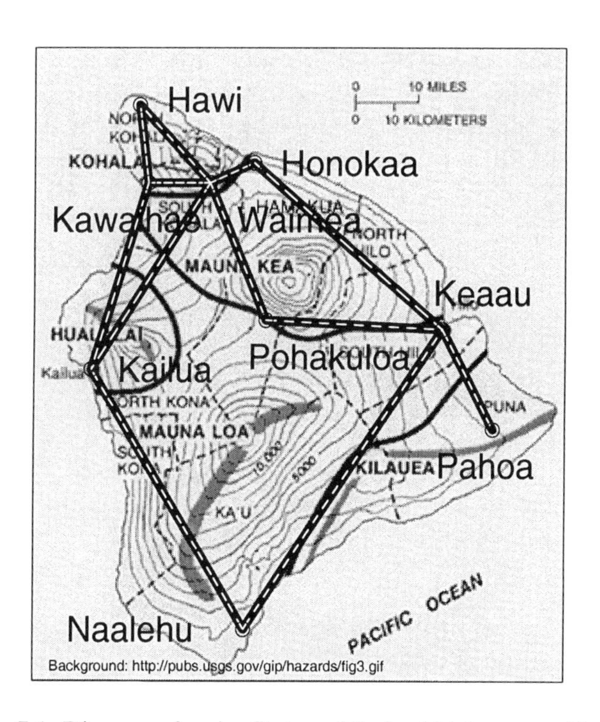

Figure 7.1: Diagram of major State and Federal highways on Hawaii.

Several situations include networks, for instance, cities connected by roads, as in figure 7.1, or generators and consumers connected by power lines, or computers connected by data chanels, where many types of problems can arise [13], [16], [81].

- In designing printed circuit boards, one problem consists in connecting components with lines lying within a plane without crossing one another.

- In delivering data, goods, or power, one problem consists in determining the maximum flow between two points through a network of transmission lines.

- In planning emergency or transportation services, one problem consists in locating the shortest route between providers and consumers.

- In maintaining roads, one problem consists in planning an itinerary that follows each stretch of road exactly once in each direction.

- The "traveling salesman's problem" consists in tracing an itinerary starting and returning to the same city, and passing through every city exactly once.

Such practical problems can be stated and to some extent solved by the mathematical "graph theory" presented here. (Algorithms suited for the solution of large problems in economics or industry, however, rely on the more advanced mathematical theory of "linear programming" [13, Part III], [16, Ch. 14–21].)

7.1 MATHEMATICAL GRAPHS

7.1.1 Directed Graphs

Graph theory requires no new axiom, because it relies on set theory and induction. Corresponding to nodes connected by links, the mathematical concept of a "directed graph" thus consists of a set for the nodes and a relation (definition 341) for the edges.

Definition 936 A **directed graph** is a pair $G = (V, D)$ consisting of a set V and a relation $D \subseteq V \times V$. In this exposition, all graphs have a *finite* set V of vertices.

The **vertices** — or **nodes** — of the graph G are the elements of the set V.

The **directed edges** of the graph G are the elements of the set D. A pair of vertices (X, Y) is a **directed edge** of G if and only if $(X, Y) \in D$.

A directed edge (X, Y) **starts** — or **originates** — at its first coordinate, X, and **ends** — or **terminates** — at its second coordinate, Y.

A **loop** is an edge (X, X) starting and ending at the same vertex.

A graph is **loop-free** if and only if $(X, X) \notin D$ for every $X \in V$.

An edge $(X, Y) \in D$ is **incident** on a vertex $Z \in V$ if and only if $Z \in \{X, Y\}$.

A **subgraph** of a directed graph $G = (V, D)$ is a pair $F := (U, C)$ consisting of a subset of vertices $U \subseteq V$ and a subset of directed edges $C \subseteq D$.

A common graphical representation of a directed graph marks each vertex by a circle, and each directed edge by an arrow (or by a circular arrow for a loop).

Example 937 For the set V and the relation D specified by

$$V := \{0, 1, 2, 3, 4, 5, 6\},$$

$$D := \{(2,2), (2,4), (2,6), (3,3), (3,6), (5,5)\},$$

the directed graph $G = (V, D)$ admits the following representation.

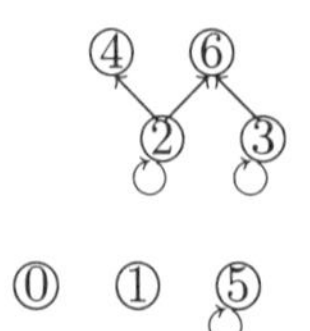

The subsets U and C specified by

$$U := \{2,3,4,6\},$$

$$C := \{(2,2),(2,4),(2,6),(3,3),(3,6)\},$$

form a subgraph $F := (U, C)$, which admits the following representation.

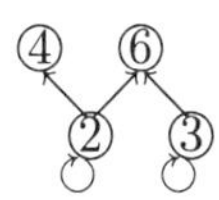

As example 937 shows, a graph need not have an edge between every pair of vertices. Only those graphs called "complete" do. While some texts require complete graphs to have *exactly* one edge between every pair of distinct vertices, the following terminology — requiring only *at least* one edge — conforms to [70, p. 179].

Definition 938 A directed graph is **complete** if and only if it contains *at least* one edge between each pair of *distinct* vertices.

Example 939 The following graphs are complete directed graphs with two vertices.

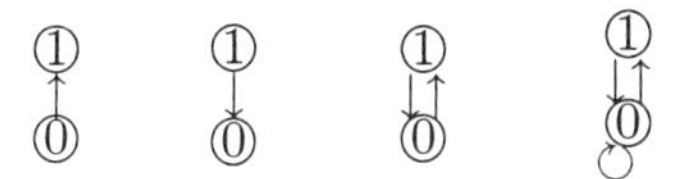

Example 940 The following graphs are complete directed graphs with three vertices.

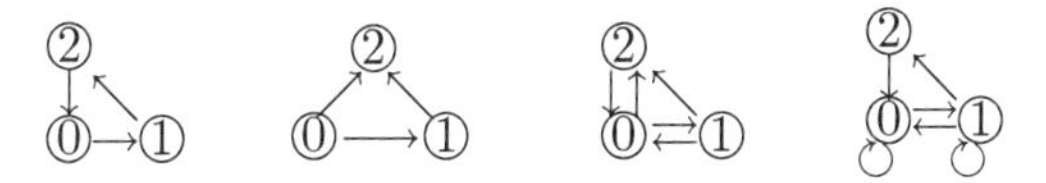

Example 941 The following graphs are complete directed graphs with four vertices.

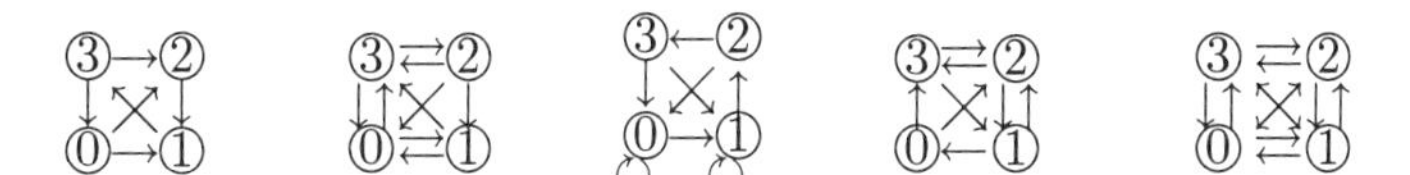

7.1.2 Undirected Graphs

An "undirected graph" corresponds to a mathematical graph where edges do not have directions or where the directions of edges does not matter. There exist several definitions, for instance, in terms of equivalence classes (definition 425) of directed edges.

Definition 942 For each set V define a relation $\rightleftharpoons$ on $V \times V$ by relating each pair $(X, Y) \in V \times V$ only to itself, (X, Y), and to (Y, X):

$$\forall X \Big(\forall Y \{ \forall U [\forall W ($$
$$[(U, W) \rightleftharpoons (X, Y)] \Leftrightarrow \{[(U, W) = (X, Y)] \vee [(U, W) = (Y, X)]\})] \} \Big).$$

Theorem 943 *The relation $\rightleftharpoons$ in definition 942 is an equivalence relation on $V \times V$.*

Proof. This proof verifies the reflexivity, symmetry, and transitivity separately.

Reflexivity For each $(X, Y) \in V \times V$, set $(U, W) := (X, Y)$ in definition 942:

$$\begin{gathered}[(U, W) \rightleftharpoons (X, Y)] \\ \Updownarrow \\ \{[(X, Y) = (X, Y)] \vee [(X, Y) = (Y, X)]\} \\ \Uparrow \\ [(X, Y) = (X, Y)].\end{gathered}$$

Hence $(X, Y) \rightleftharpoons (X, Y)$ follows from $(X, Y) = (X, Y)$.

Symmetry For all $(U, W), (X, Y) \in V \times V$, the definition of ordered pairs shows that $(U, W) = (Y, X)$ if and only if $(W, U) = (X, Y)$:

$$\begin{gathered}(U, W) = (Y, X) \\ \Updownarrow \\ (U = Y) \wedge (W = X) \\ \Updownarrow \\ (W = X) \wedge (U = Y) \\ \Updownarrow \\ (W, U) = (X, Y).\end{gathered}$$

Hence, the symmetry of equality $(=)$ confirms that

$$\begin{gathered}(U, W) \rightleftharpoons (X, Y) \\ \Updownarrow \\ [(U, W) = (X, Y)] \vee [(U, W) = (Y, X)] \\ \Updownarrow \\ [(X, Y) = (U, W)] \vee [(X, Y) = (W, U)] \\ \Updownarrow \\ (X, Y) \rightleftharpoons (U, W).\end{gathered}$$

Thus, $(U, W) \rightleftharpoons (X, Y)$ if and only if $(X, Y) \rightleftharpoons (U, W)$.

Transitivity For all $(P, Q), (U, W), (X, Y) \in V \times V$, from

$$[(P, Q) = (U, W)] \Leftrightarrow [(Q, P) = (W, U)]$$

it follows that $(P, Q) \rightleftharpoons (U, W)$ if and only if $(Q, P) \rightleftharpoons (W, U)$:

$$\begin{gathered}(P, Q) \rightleftharpoons (U, W) \\ \Updownarrow \\ [(P, Q) = (U, W)] \vee [(P, Q) = (W, U)] \\ \Updownarrow \\ [(Q, P) = (W, U)] \vee [(Q, P) = (U, W)] \\ \Updownarrow \\ (Q, P) \rightleftharpoons (W, U).\end{gathered}$$

Hence, if $(P,Q) \rightleftharpoons (U,W)$ and $(U,W) \rightleftharpoons (X,Y)$, then $(P,Q) \rightleftharpoons (X,Y)$:

$$\begin{array}{c}
[(P,Q) \rightleftharpoons (U,W)] \wedge [(U,W) \rightleftharpoons (X,Y)] \\
\Updownarrow \\
\{[(P,Q) = (U,W)] \vee [(P,Q) = (W,U)]\} \wedge \{[(U,W) = (X,Y)] \vee [(U,W) = (Y,X)]\} \\
\Updownarrow \\
\{[(P,Q) = (U,W)] \vee [(Q,P) = (U,W)]\} \wedge \{[(U,W) = (X,Y)] \vee [(U,W) = (Y,X)]\} \\
\Updownarrow \\
\{[(P,Q) = (U,W)] \wedge [(U,W) = (X,Y)]\} \vee \{[(P,Q) = (U,W)] \wedge [(U,W) = (Y,X)]\} \\
\vee\{[(Q,P) = (U,W)] \wedge [(U,W) = (X,Y)]\} \vee \{[(Q,P) = (U,W)] \wedge [(U,W) = (Y,X)]\} \\
\Downarrow \\
[(P,Q) = (X,Y)] \vee [(P,Q) = (Y,X)] \vee [(Q,P) = (X,Y)] \vee [(Q,P) = (Y,X)] \\
\Updownarrow \\
[(P,Q) = (X,Y)] \vee [(P,Q) = (Y,X)] \vee [(P,Q) = (Y,X)] \vee [(P,Q) = (X,Y)] \\
\Updownarrow \\
[(P,Q) = (X,Y)] \vee [(P,Q) = (Y,X)] \\
\Updownarrow \\
(P,Q) \rightleftharpoons (X,Y).
\end{array}$$

□

The following definition translates concepts from relations into terms with graphs.

Definition 944 In a directed graph $G = (V, D)$, the equivalence class (relative to $\rightleftharpoons$) of a directed edge (X, Y) is the set $[(X, Y)]_{\rightleftharpoons}$ defined by

$$[(X,Y)]_{\rightleftharpoons} := \{(X,Y),(Y,X)\}.$$

For each subset $D \subseteq V \times V$, the set of all equivalence classes (relative to $\rightleftharpoons$) of edges in D is the set $D/\rightleftharpoons$ defined by

$$(D/\rightleftharpoons) := \{[(X,Y)]_{\rightleftharpoons} \in \mathcal{P}(V \times V) : (X,Y) \in D\}.$$

The **quotient map** (example 429) relative to $\rightleftharpoons$ is the function denoted by π (the Greek letter "pi") or $\pi_{\rightleftharpoons}$ and defined by

$$\begin{array}{rcl}
\pi_{\rightleftharpoons} : (V \times V) & \to & (V \times V)/\rightleftharpoons, \\
(X,Y) & \mapsto & [(X,Y)]_{\rightleftharpoons}.
\end{array}$$

In particular, $\pi_{\rightleftharpoons}(X,Y) = [(X,Y)]_{\rightleftharpoons}$ and $\pi_{\rightleftharpoons}(D) = D/\rightleftharpoons$.

The following definition shows one way of specifying undirected graphs.

Definition 945 For each directed graph $G = (V, D)$, the associated **undirected graph** is the pair $(V, D/\rightleftharpoons)$. The **undirected edge between** two vertices $X, Y \in V$ is the equivalence class $[(X,Y)]_{\rightleftharpoons}$.

A common graphical representation of an undirected graph marks each vertex by a circle, and each undirected edge by a line or a curve, for instance, a circular arc.

Example 946 For the set V and the relation D specified by

$$\begin{array}{rcl}
V & := & \{0,1,2,3,4,5,6\}, \\
D & := & \{(2,2),(2,4),(2,6),(3,3),(3,6),(5,5)\},
\end{array}$$

the undirected graph $G := (V, D/ \rightleftharpoons)$ admits the following representation.

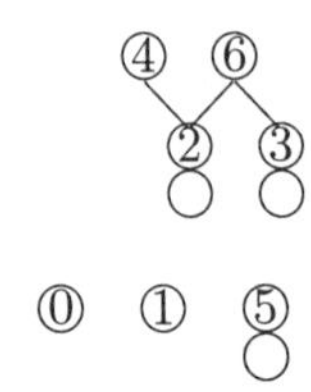

Complete undirected graphs connect every pair of distinct vertices with an edge.

Definition 947 An undirected graph is **complete** if and only if it contains *at least* one edge between each pair of *distinct* vertices.

The loop-free complete undirected graph with n vertices is denoted by K_n.

Example 948 Here are the first four loop-free complete undirected graphs:

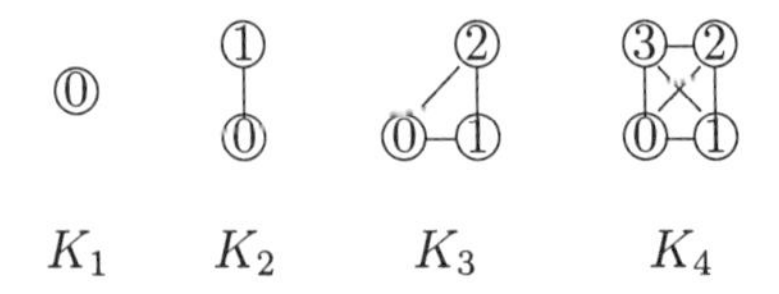

K_1 K_2 K_3 K_4

7.1.3 Degrees of Vertices

The number of edges that start or end at each vertex provides a criterion to determine features of a graph, for instance, the existence of certain connections between vertices.

Definition 949 The **inward degree** of a vertex X in a directed graph $G = (V, D)$ is the number $\deg_{G,\text{in}}(X)$ of edges *ending* at X.

The **outward degree** of a vertex X in a directed graph $G = (V, D)$ is the number $\deg_{G,\text{out}}(X)$ of edges *starting* at X.

The **degree** of a vertex X in a graph $G = (V, D)$ is the number $\deg_G(X)$ of edges *incident* on X. In a directed graph, $\deg_G(X) = \deg_{G,\text{in}}(X) + \deg_{G,\text{out}}(X)$.

Example 950 For the directed graph $G = (V, D)$ in example 937,

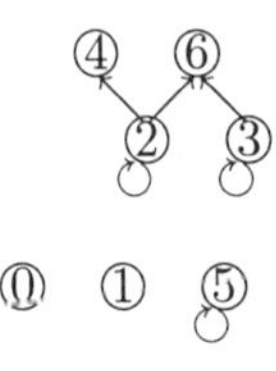

the vertices have the following degrees:

$$\begin{array}{lllllllll}
\deg_{G,\text{in}}(0) & = & 0, & \deg_{G,\text{out}}(0) & = & 0, & \deg_G(0) & = & 0; \\
\deg_{G,\text{in}}(1) & = & 0, & \deg_{G,\text{out}}(1) & = & 0, & \deg_G(1) & = & 0; \\
\deg_{G,\text{in}}(2) & = & 1, & \deg_{G,\text{out}}(3) & = & 0, & \deg_G(2) & = & 4; \\
\deg_{G,\text{in}}(3) & = & 1, & \deg_{G,\text{out}}(3) & = & 2, & \deg_G(3) & = & 3; \\
\deg_{G,\text{in}}(4) & = & 1, & \deg_{G,\text{out}}(4) & = & 0, & \deg_G(4) & = & 1; \\
\deg_{G,\text{in}}(5) & = & 1, & \deg_{G,\text{out}}(5) & = & 1, & \deg_G(5) & = & 2; \\
\deg_{G,\text{in}}(6) & = & 2, & \deg_{G,\text{out}}(6) & = & 0, & \deg_G(6) & = & 2.
\end{array}$$

The following considerations demonstrate that the sum of all the degrees equals twice the number of edges.

Definition 951 In every directed graph $G = (V, D)$, for each vertex $X \in V$ let $O(X)$, $I(X)$, $E(X)$ be the sets of edges starting from X, ending at X, and incident on X:

$$\begin{aligned} O(X) &:= \{(Z,Y) \in D : Z = X\}, \\ I(X) &:= \{(W,Z) \in D : Z = X\}, \\ E(X) &:= I(X) \cup O(X). \end{aligned}$$

The non-empty sets $O(X)$ or $I(X)$ partition the set D of edges. (See definition 419 for "partitions" and definition 325 for the notation $\dot{\cup}$ of pairwise disjoint unions.)

Theorem 952 *In each graph $G = (V, D)$,*

$$\dot{\bigcup}_{X \in V} O(X) = D = \dot{\bigcup}_{X \in V} I(X).$$

Proof. By definition 951, $O(X) \subseteq D$ and $I(X) \subseteq D$ for every $X \in V$. Consequently,

$$\begin{aligned} \bigcup_{X \in V} O(X) &\subseteq D, \\ \bigcup_{X \in V} I(X) &\subseteq D. \end{aligned}$$

Conversely, $(X,Y) \in O(X)$ and $(X,Y) \in I(Y)$ for each $(X,Y) \in D$. Therefore,

$$\begin{aligned} D &\subseteq \bigcup_{X \in V} O(X), \\ D &\subseteq \bigcup_{X \in V} I(X). \end{aligned}$$

Finally, if $Y \neq Z$, then $I(Y) \cap I(Z) = \varnothing$, because if $(X,Y) \in I(Y)$ and $(X,Y) \in I(Z)$, then $Y = Z$. Similarly $O(Y) \cap O(Z) = \varnothing$. Thus each union is a disjoint union. □

Theorem 953 *In each graph $G = (V, D)$, the sum of the degrees of all the vertices equals twice the number of edges:*

$$\sum_{X \in V} \deg_G(X) = 2 * [\#(D)].$$

Proof. Each edge contributes a summand 2 to the sum of all degrees:

$$\begin{aligned} \sum_{X \in V} \deg_G(X) &= \sum_{X \in V} \deg_{G,\text{in}}(X) + \sum_{X \in V} \deg_{G,\text{out}}(X) \\ &= \sum_{X \in V} \#[I(X)] + \sum_{X \in V} \#[O(X)] \\ &= \# \left[\dot{\bigcup}_{X \in V} I(X) \right] + \# \left[\dot{\bigcup}_{X \in V} O(X) \right] \\ &= \#(D) + \#(D) \\ &= 2 * [\#(D)] \end{aligned}$$

by induction (theorem 480) and theorem 603 on the cardinality of disjoint unions. □

Theorem 954 *In each graph the number of vertices with an odd degree is even.*

Proof. For each graph $G = (V, D)$, let $W \subseteq V$ be the subset of all vertices with even degrees, and let $U \subseteq V$ be the subset of all vertices with odd degrees. Then

$$\begin{aligned} V &= U \dot{\cup} W, \\ \varnothing &= U \cap W. \end{aligned}$$

By theorem 953 ,

$$2 * [\#(D)] = \sum_{X \in V} \deg_G(X) = \sum_{X \in U} \deg_G(X) + \sum_{X \in W} \deg_G(X)$$

whence

$$\sum_{X \in U} \deg_G(X) = 2 * [\#(D)] - \sum_{X \in W} \deg_G(X)$$

is the difference of two even integers and hence also even, by theorem 726. □

7.1.4 Exercises

Exercise 1101 Draw K_4 so that no edge crosses any other edge.

Exercise 1102 Draw a complete undirected graph K_5 with 5 vertices.

Exercise 1103 Determine the degree of each vertex in the following graphs.

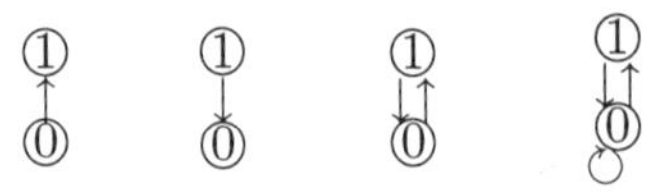

Exercise 1104 Determine the degree of each vertex in the following graphs.

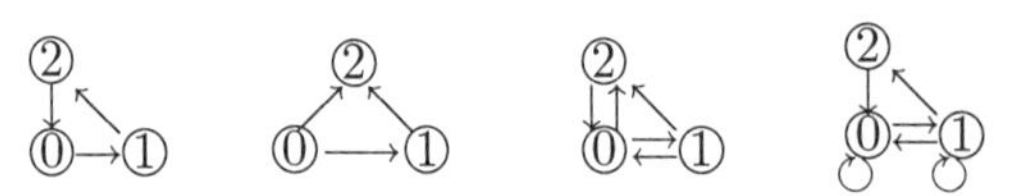

Exercise 1105 Determine the degree of each vertex in the following graphs.

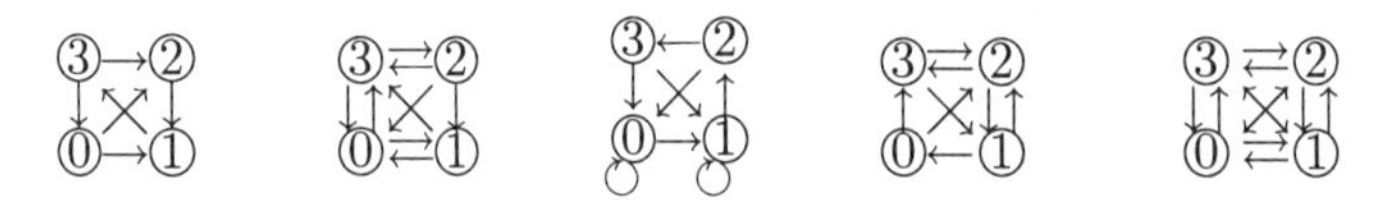

Exercise 1106 Determine the degree of each vertex in the following graphs.

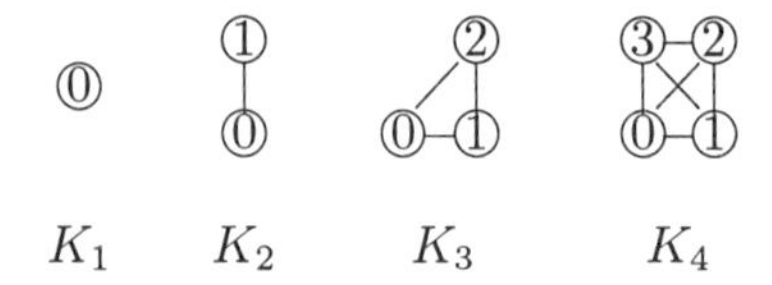

K_1 K_2 K_3 K_4

Exercise 1107 Find the degree of each vertex of the graph in figure 7.1.

Exercise 1108 Find the degree of each vertex of the graph in figure 7.4.

Exercise 1109 Verify the formula in theorem 953 for the graph in figure 7.1.

Exercise 1110 Verify the formula in theorem 953 for the graph in figure 7.4.

Exercise 1111 Determine whether there exists a graph with 5 vertices with respective degrees $(1, 2, 3, 4, 5)$.

Exercise 1112 Determine whether there exists a graph with 6 vertices with respective degrees $(1, 2, 3, 4, 5, 6)$.

Exercise 1113 Determine whether there exists a graph with 5 vertices with respective degrees $(1, 2, 3, 4, 4)$.

Exercise 1114 Determine whether there exists a graph with 6 vertices with respective degrees $(2, 3, 4, 5, 6, 7)$.

Exercise 1115 Prove theorem 953 by induction.

Exercise 1116 Determine whether there exists a graph with N vertices $(X_1, \ldots, X_N)$ with respective degrees $(J_1, \ldots, J_N)$ for each finite sequence $(J_1, \ldots, J_N)$ of natural numbers with an even number of odd values, and with all the other values even.

Exercise 1117 Determine the number of undirected edges in a complete undirected graph with N vertices.

Exercise 1118 Determine the minimum and maximum numbers of directed edges in a complete directed graph with N vertices.

Exercise 1119 Determine the maximum number of directed edges in a graph with N vertices.

Exercise 1120 Determine the minimum number of vertices in a graph with M directed edges.

7.2 PATH-CONNECTED GRAPHS

7.2.1 Walks in Graphs

Whereas directed graphs represent relations, the common terminology for graphs differs from the common terminology for relations. For instance, a *connected relation* relates any two different elements by *one* pair, in either order (definition 449). In contrast, a *connected graph* relates any two different vertices by a *sequence* of pairs, called a "walk" or a "curve" between the two vertices.

Definition 955 (Walks in graphs) A **walk** — also called an **undirected walk** — in a graph $G = (V, D)$ is a finite sequence g of N pairs of vertices,

$$\begin{aligned} g : N &\to V \times V, \\ i &\mapsto g(i) = (X_i, Y_i), \end{aligned}$$

with $Y_{i-1} = X_i$, and $(X_i, Y_i) \in D$ or $(Y_i, X_i) \in D$, for each $i \in \{1, \ldots, N-1\}$.

The walk g is **directed** if and only if $(X_i, Y_i) \in D$ for each $i \in \{0, \ldots, N-1\}$.

A walk **from** X **to** Y is a walk g in G such that $X = X_0$, and $Y = Y_{N-1}$.

A directed walk g **starts**, or **originates**, at X_0 and **ends**, or **terminates**, at Y_{N-1}.

An undirected walk g **traverses** an edge (X, Y) if and only if $(X, Y) \in \text{Range}(g)$ or $(Y, X) \in \text{Range}(g)$ (the range of the function g, as in definition 355).

A directed walk g **traverses** an edge (X, Y) if and only if $(X, Y) \in \text{Range}(g)$.

A directed or undirected walk $g : N \to V \times V$ **passes through** a vertex $Y \in V$ if and only if there exist at least two non-necessarily distinct indices $i, j \in \{0, \ldots, N-1\}$ such that $Y = Y_i$ and $Y = X_j$. The walk **contains** a vertex $Y \in V$ if and only if either it originates from Y, or it passes through Y, or it terminates at Y.

Traversing two consecutive sequences of edges yields a composite walk.

Definition 956 In a graph $G = (V, D)$, two directed walks $p : M \to V \times V$ and $q : N \to V \times V$ are **consecutive** if and only if one ends where the other starts:

$$p(M-1) = (X_{M-1}, Y_{M-1}),$$

$$q(0) = (Y_{M-1}, Z_M).$$

Their **composition** is the **composite walk** $q \bullet p$ defined by

$$\begin{array}{rcll} q \bullet p : M + N & \to & V \times V, & \\ (q \bullet p)(i) & := & p(i) & \text{for every } i \in M, \\ (q \bullet p)(i) & := & q(i - M) & \text{for every } i \in \{M, \ldots, M + N - 1\}. \end{array}$$

Thus p coincides with the restriction (definition 365) $(q \bullet p)|_M$ of $q \bullet p$ to M.

The composition of walks *differs* from the composition of functions. However, the composition of consecutive walks is also associative.

Theorem 957 *If g ends where p starts, and if p ends where q starts, then*

$$q \bullet (p \bullet g) = (q \bullet p) \bullet g.$$

Proof. In a directed graph $G = (V, D)$, if a walk $g : K \to V \times V$ ends where a walk $p : L \to V \times V$ starts, so that

$$g(K-1) = (X_{K-1}, Y_{K-1}),$$
$$p(0) = (Y_{K-1}, Z_K),$$

then the composite walk $p \bullet g$ is defined by

$$\begin{array}{rcll} p \bullet g : K + L & \to & V \times V, & \\ (p \bullet g)(i) & := & g(i) & \text{for every } i \in K, \\ (p \bullet g)(i) & := & p(i - K) & \text{for every } i \in \{K, \ldots, K + L - 1\}. \end{array}$$

If the walk $p : L \to V \times V$ ends where the walk $q : M \to V \times V$ starts, so that

$$p(L-1) = (X_{L-1}, Y_{L-1}),$$
$$q(0) = (Y_{L-1}, Z_L),$$

then the composite walk $q \bullet p$ is defined by

$$\begin{array}{rcll} q \bullet p : L + M & \to & V \times V, & \\ (q \bullet p)(i) & := & p(i) & \text{for every } i \in L, \\ (q \bullet p)(i) & := & q(i - L) & \text{for every } i \in \{L, \ldots, L + M - 1\}. \end{array}$$

Consequently, $q \bullet p$ starts where g ends, with

$$[(q \bullet p) \bullet g] : K + L + M \to V \times V.$$

Moreover, $[(q \bullet p) \bullet g](i) = [q \bullet (p \bullet g)](i)$ for every i; firstly, for every $i \in K$,

$$\begin{array}{rcll} [(q \bullet p) \bullet g](i) & = & g(i) & \text{for every } i \in K, \\ & = & (p \bullet g)(i) & \text{for every } i \in K, \\ & = & [q \bullet (p \bullet g)](i) & \text{for every } i \in K. \end{array}$$

Secondly, for every $i \in \{K, \dots, K+L-1\}$,

$$\begin{aligned}
&[(q \bullet p) \bullet g](i) \\
&= (q \bullet p)(i-K) \quad \text{for every } i \in \{K, \dots, K+L-1\} \\
&= p(i-K) \quad \text{for every } i \in \{K, \dots, K+L-1\} \\
&= (p \bullet g)(i) \quad \text{for every } i \in \{K, \dots, K+L-1\} \\
&= [q \bullet (p \bullet g)](i) \quad \text{for every } i \in \{K, \dots, K+L-1\}.
\end{aligned}$$

Thirdly, for every $i \in \{K+L, \dots, K+L+M-1\}$,

$$\begin{aligned}
&[(q \bullet p) \bullet g](i) \\
&= (q \bullet p)(i-K) \quad \text{for every } i \in \{K+L, \dots, K+L+M-1\} \\
&= q(i-[K+L]) \quad \text{for every } i \in \{K+L, \dots, K+L+M-1\} \\
&= [q \bullet (p \bullet g)](i) \quad \text{for every } i \in \{K+L, \dots, K+L+M-1\}.
\end{aligned}$$

□

Traversing a sequence of edges in reverse yields a reverse walk, denoted here by $g^{\bullet -1}$ to distinguish it from an inverse function and from a reciprocal.

Definition 958 (Reverse walks) For each walk

$$\begin{aligned}
g : M &\rightarrow V \times V, \\
g(i) &:= (X_i, Y_i),
\end{aligned}$$

in a graph $G = (V, D)$, the **reverse walk** $g^{\bullet -1}$ is defined by

$$\begin{aligned}
g^{\bullet -1} : M &\rightarrow V \times V, \\
g^{\bullet -1}(i) &:= (Y_i, X_i).
\end{aligned}$$

7.2.2 Path-Connected Components of Graphs

Graphs need not connect all their vertices to one another by walks. For instance, the Hawaiian road network connects cities within each island, but there are no roads between cities on different islands. Only the graphs called "path-connected" connect all their vertices to one another through walks.

Definition 959 (Path-connected graphs) A directed or undirected graph $G = (V, D)$ is **path-connected** if and only if for each pair of distinct vertices $(X, Y) \in V \times V$ there exists an *undirected* walk from X to Y.

Example 960 The directed graph $G = (V, D)$ in example 937 is *not* path-connected; for instance, it does not contain any edge from 0 to 1 or from 1 to 0.

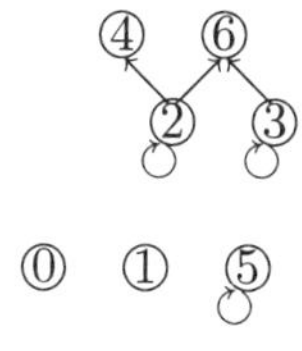

However, the subgraph $F = (U, C)$ in example 937 *is* path-connected, because it contains a walk between any two vertices. For instance, ($(3,6), (6,2), (2,4)$) is an undirected walk between 3 and 4, because $(3,6) \in D$, $(2,6) \in D$, $(2,4) \in D$.

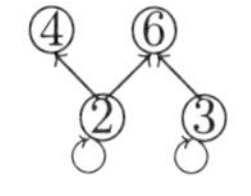

As the graphs in example 960 suggest, a graph splits into mutually disjoint path-connected subgraphs. Such disjoint path-connected subgraphs constitute equivalence classes (definition 425) for the following relation.

Definition 961 For each undirected graph $G = (V, D)$, define a relation $\leftrightarrow$ on V so that $\leftrightarrow$ relates two vertices X and Y if and only if $X = Y$ or there exists an undirected walk in G from X to Y.

Theorem 962 *The relation $\leftrightarrow$ in definition 961 is an equivalence relation.*

Proof. This proof verifies that $\leftrightarrow$ is reflexive, symmetric, and transitive.

Reflexivity For each $X \in V$, $X = X$ whence $X \leftrightarrow X$ by definition 961.

Symmetry For all $X, Y \in V$, two cases can arise: $X = Y$ or $X \neq Y$.

If $X = Y$, then $Y \leftrightarrow X$ by definition 961,

If $X \neq Y$ and $X \leftrightarrow Y$, then there exists an undirected walk g in G from X to Y. Consequently, the reverse walk $g^{\bullet-1}$ in definition 958 is an undirected walk in G from Y to X. Therefore $Y \leftrightarrow X$.

Transitivity For all $X, Y, Z \in V$, if $X \leftrightarrow Y$, and $Y \leftrightarrow Z$, then three cases can arise.

If $X = Y$, then $X \leftrightarrow Z$ follows from the hypothesis $Y \leftrightarrow Z$.

If $Y = Z$, then $X \leftrightarrow Z$ follows from the hypothesis $X \leftrightarrow Y$.

If $X \neq Y$ and $Y \neq Z$, then there exist undirected walks $p : M \to V \times V$ from X to Y and $q : N \to V \times V$ from Y to Z. Then the composite walk $q \bullet p : M + N \to V \times V$ in definition 956 is an undirected walk in G from X to Z. Thus $X \leftrightarrow Z$.

Therefore $\leftrightarrow$ is an equivalence relation on the set of vertices V. □

Definition 963 (Path-connected components) A **path-connected component** of a graph $G = (V, D)$ is a path-connected subgraph $F := (U, C)$ whose set of vertices U forms an equivalence class for the relation $\leftrightarrow$ in definition 961, and such that C contains all the edges from D incident on any vertex in U.

Example 964 The directed graph $G = (V, D)$ in example 937,

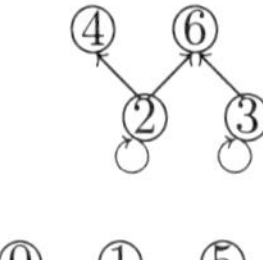

⓪ ① ⑤

has four path-connected components: ⓪, ①, ⑤, and the subgraph $F = (U, C)$:

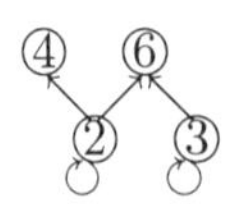

Example 965 Roads within each island form a path-connected component of the road network. (See figure 7.2.)

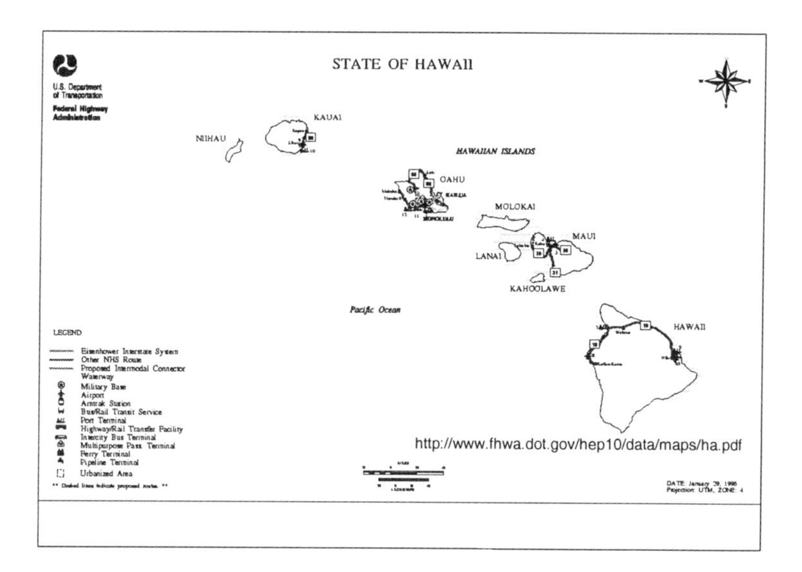

Figure 7.2: Path-connected components of the Federal highway network in Hawaii.

7.2.3 Longer Walks

The composition of walks yields a partial order (definition 443) on walks.

Definition 966 For each undirected graph $G = (V, D)$, define a relation $\preccurlyeq$ on walks such that $\preccurlyeq$ relates two walks g and q,

$$g \preccurlyeq q,$$

if and only if there is a walk p that starts where g ends and ends where q starts, so that

$$q = p \bullet g.$$

The walk q then **extends** the walk g. In the particular case where p is not the empty walk, so that $q \neq g$, then q is a **proper extension** of g, a relation denoted by

$$g \prec q.$$

Theorem 967 *The relations $\preccurlyeq$ and $\prec$ in definition 966 are partial orders on walks.*

Proof. This proof verifies the transitivity and the anti-symmetry separately.

Transitivity If $u \prec v$ and $v \prec w$, then there exist walks p and q such that $v = p \bullet u$ and $w = q \bullet v$. Consequently, $u \prec w$ because

$$(q \bullet p) \bullet u = q \bullet (p \bullet u) = q \bullet v = w$$

by associativity of the composition of walks (theorem 957). The same proof of transitivity also applies to $\preccurlyeq$.

Anti-symmetry If $u \prec v$ and $v \prec u$, then there exist walks p and q such that $v = p \bullet u$ and $u = q \bullet v$. From $v = p \bullet u$ it follows that v first traverses all the edges traversed by u. From $u = q \bullet v$ it follows that u first traverses all the edges traversed by v. Hence $u = v$. The same proof of anti-symmetry also applies to $\preccurlyeq$.

Irreflexivity If $u \prec v$, then there exists a non-empty walk p such that $v = p \bullet u$. Thus v first traverses all the edges traversed by u and then the edges traversed by p. Hence $u \neq v$. By contraposition, it follows that $u \not\prec u$.

Reflexivity With p denoting the empty walk, it follows that $u \preccurlyeq u$ for each walk u.

Thus $\preccurlyeq$ is a weak partial order and $\prec$ is a strict partial order. □

7.2.4 Longest Paths

A graph need not have any maximal walk, because walks can traverse the same edges repeatedly. In contrast, there is an upper bound (definition 461) on the lengths of walks — called "paths" — that traverse every edge at most once.

Definition 968 A **path** is an injective walk (which traverses each edge at most once).

A **simple path** is a path passing through each vertex at most once.

Theorem 969 *Every sequence of paths that is non-decreasing relative to $\preccurlyeq$ has a maximal element.*

Proof. If (p_k) is a *non-decreasing* sequence of paths relative to $\preccurlyeq$, then $p_k \preccurlyeq p_{k+1}$ for each $k \in \mathbb{N}$, whence there exists a sequence of paths (q_k) such that $q_k \bullet p_k = p_{k+1}$.

By injectivity of paths, if q_k is not the empty path. then p_{k+1} traverses edges not traversed by p_k. Because the graph is finite, there exist at most finitely many values of k for which q_k is not empty. In other words, there exists $\ell \in \mathbb{N}$ such that $q_k = \varnothing$ for every $k > \ell$. Therefore p_ℓ is a maximal path in the sequence (p_k). □

Theorem 970 *Every edge is contained in a maximal path. Every edge is also contained in a maximal simple path.*

Proof. This proof builds a maximal path with a definition by induction (theorem 484).

For each finite directed graph $G = (V, D)$ there exists $N \in \mathbb{N}$ and a bijection $F : N \to D, I \mapsto (X_I, Y_I)$. For each edge $(X_0, Y_0) \in D$, let p_1 be the path consisting of this edge, so that $p_1 : \{0\} \to D$ with $p_1(0) = (X_0, Y_0)$.

Assume that there exists $\ell \in \mathbb{N}$ such that there exists a path (or a simple path) p_ℓ in G with $p_k \preccurlyeq p_\ell$ for every $k \in \{1, \ldots, \ell\}$, and $p_\ell(\ell - 1) = (X_{\ell-1}, Y_{\ell-1})$.

If there exists an edge $F(I) = (X, Y) \in D$ with $X = Y_{\ell-1}$ that p_ℓ has not traversed (and such that p_ℓ has not passed through Y for simple paths), then select such an edge $F(I) = (X_\ell, X_\ell)$ with the smallest value of I and define $p_{\ell+1}$ by appending (X_ℓ, X_ℓ) to p_ℓ. If there exists no such edge, then p_ℓ is maximal. Because D is finite, it follows that the sequence of paths (p_k) has a maximal element (theorem 969). □

Every walk also contains a simple path between the same endpoints.

Theorem 971 *For each walk there exists a simple path with the same initial and terminal vertices.*

Proof. For each directed walk $g : N \to D$, two cases can arise.

If g traverses each vertex at most once, then the directed walk is a simple path.

If g traverses a vertex $X \in V$ more than once, then there exist two distinct indices $k, \ell \in N$, for instance, with $0 < k < \ell < N$, such that $g(k) = (X_k, Y_k)$ and $g(\ell) = (X_\ell, Y_\ell)$

with $X_k = X = X_\ell = Y_{\ell-1}$. Hence define a directed walk $g^{(1)} : [N - (\ell + 1 - k)] \to D$ by deleting from g all the edges from (X_k, Y_k) through $(X_{\ell-1}, Y_{\ell-1})$:

$$g^{(1)}(i) := \begin{cases} g(i) & \text{if } 0 \leq i < k, \\ g(i + \ell - k) & \text{if } k \leq i < N - (\ell - k). \end{cases}$$

Thus $g^{(1)}$ is a directed walk with the same start and end as for g, but traversing the vertex X_k one fewer times than g does. Repeating the foregoing process at most N times yields a simple path, from the same start to the same end.

The proof for undirected walks is exactly the same but with undirected edges. □

7.2.5 Exercises

Exercise 1121 Design a graph with exactly 4 vertices and exactly 6 non-empty undirected paths.

Exercise 1122 Design a graph with exactly N vertices and exactly $N*(N-1)/2$ non-empty undirected paths.

Exercise 1123 Design a graph with exactly 3 vertices and exactly 7 non-empty undirected paths.

Exercise 1124 Design a graph with exactly N vertices and exactly $N * (N - 1) + 1$ non-empty undirected paths.

Exercise 1125 Determine a lower bound on the number of paths in a path-connected graph with N vertices.

Exercise 1126 Determine a lower bound on the number of paths in a path-connected graph with M edges.

Exercise 1127 Determine an upper bound on the number of paths in a path-connected graph with N vertices.

Exercise 1128 Determine a lower bound on the number of paths in a path-connected graph with M edges.

Exercise 1129 Determine an upper bound on the number of paths in a path-connected graph with M edges.

Exercise 1130 Design an algorithm to identify all the path-connected subgraphs of a graph from its lists of edges and vertices.

7.3 EULER AND HAMILTONIAN CIRCUITS

7.3.1 Circuits

Some situations require paths that originate and terminate at the same point.

Definition 972 A walk is **closed** if and only if it starts and ends at the same vertex.

A **circuit** is a closed path. A **simple circuit** is a simple closed path.

Remark 973 An alternative way to specify a closed walk

$$\begin{aligned} g: N &\rightarrow D, \\ i &\mapsto (X_i, Y_i) \end{aligned}$$

consists in replacing the domain $N = \{0, \ldots, N-1\}$ by the set $\mathbb{Z}_N = \{[0]_N, \ldots, [N-1]_N\}$ of integers modulo N (definition 756), and in replacing g by the function

$$\begin{aligned} c: \mathbb{Z}_N &\rightarrow D, \\ [i]_N &\mapsto (X_i, Y_i). \end{aligned}$$

The edges $c([N-1]_N)$ and $C([0]_N)$ are consecutive, because $c([N-1]_N) = (X_{N-1}, Y_{N-1})$ and $c([N]_N) = c([0]_N) = (X_0, Y_0)$ with $Y_{N-1} = X_0$.

7.3.2 Hamiltonian Paths and Circuits

The traveling salesman's problem consists in tracing an itinerary — called a Hamiltonian circuit — that starts from one "home" city, passes through every other city exactly once, and returns to the "home" city. The name of "traveling salesman's problem" serves only as a quick way to refer to the problem, because in practice such other considerations as costs of travel between certain cities might raise the cost of a Hamiltonian circuit above the cost of other "walks" that might pass through some cities more than once. Moreover, the traveling salesman's problem also has other applications.

Definition 974 A **Hamiltonian path** is a simple path passing through every vertex.
A **Hamiltonian circuit** is a simple circuit passing through every vertex.

Example 975 The graph ⓪⇄① has a Hamiltonian circuit: $((0, 1), (1, 0))$.

Example 976 The graph ⓪→①→② has a Hamiltonian path: $((0, 1), (1, 2))$.

Theorem 977 *In every Hamiltonian circuit, the set of vertices $U \subseteq V$ and the set of edges $C \subseteq D$ have the same number of elements,*

$$[\#(U)] = [\#(C)].$$

Moreover, for every vertex $X \in U$ in the subgraph $H := (U, C)$,

$$\deg_H(X) = 2.$$

Proof. A Hamiltonian circuit in a graph $G = (V, D)$ consists of an injective function

$$\begin{aligned} g: \mathbb{Z}_N &\rightarrow V \times V, \\ [i]_N &\mapsto (X_i, Y_i), \end{aligned}$$

such that $X_i = Y_{i-1}$, and for each vertex $X \in V$ there exists $[i]_N \in \mathbb{Z}_N$ with $X \in \{X_i, Y_i\}$. Hence the first projection

$$\begin{aligned} p_1: U \times U &\rightarrow V, \\ (X, Y) &\mapsto X \end{aligned}$$

is surjective. By injectivity of g, the first projection is also injective. Thus p_1 establishes a bijection from the vertices to the edges. Also, each vertex has exactly two edges incident on it, by definition of a circuit. □

Theorem 978 *In every Hamiltonian path that is not a circuit, the number of vertices exceeds the number of edges by* 1*:*

$$[\#(U)] = [\#(C)] + 1.$$

Moreover, every Hamiltonian path that is not a circuit contains exactly two terminal vertices, while every other vertex has degree 2*.*

Proof. Completing a Hamiltonian path by inserting one edge from the end to the start produces a Hamiltonian circuit. □

Counterexample 979 The following directed graph has no Hamiltonian paths:

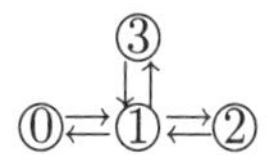

Indeed, any Hamiltonian circuit would have exactly four edges, but removing any two edges from the graph leaves at least one vertex out of any circuit. Similarly, any Hamiltonian path would have exactly three edges, but any path-connected subgraph with exactly three edges has at least three terminal vertices instead of the required two.

Besides exhaustive searches, which would try every circuit, there exists no known efficient algorithm to determine whether a graph has a Hamiltonian circuit. Nevertheless, the following theorem describes one class of graphs that contain Hamiltonian circuits.

Theorem 980 *Every complete directed graph contains a Hamiltonian path.*

Proof. This proof proceeds by induction with the number N of vertices.

Initial step Every graph with $N \in \{0, 1\}$ vertex has a Hamiltonian empty circuit.

Induction hypothesis Assume that for some $K \in \mathbb{N}^*$ the theorem holds for $N := K$, so that every complete directed graph with K vertices has a Hamiltonian circuit.

Induction step For each complete directed graph $G = (V, D)$ with $K + 1$ vertices, select any vertex $X_0 \in V$. Define the subgraph $F := (U, C)$ by $U := V \setminus \{X_0\}$, and with C obtained by removing from D all the edges incident on X_0. Then $F = (U, C)$ is a complete directed graph with K vertices.

By the hypothesis of induction, F admits a Hamiltonian path p. Number the vertices of F by $(X_1, \ldots, X_K)$, so that p traverses the edges in the order (X_1, X_2), ..., (X_{K-1}, X_K). Several cases can arise.

If $(X_0, X_1) \in D$, then preceding p by (X_0, X_1) gives a Hamiltonian path in G.

If $(X_0, X_1) \notin D$, then $(X_1, X_0) \in D$ by completeness. If also $(X_0, X_2) \in D$, splicing $(X_1, X_0), (X_0, X_2)$ between X_1 and X_2 in p gives a Hamiltonian path in G.

If $(X_0, X_2) \notin D$, then $(X_2, X_0) \in D$ by completeness. If also $(X_0, X_3) \in D$, splicing $(X_2, X_0), (X_0, X_3)$ between X_2 and X_3 in p gives a Hamiltonian path in G.

By induction, assume that there exists $J \in \{1, \ldots, K - 1\}$ such that $(X_0, X_I) \notin D$ but $(X_I, X_0) \in D$ for every $I \in \{1, \ldots, J\}$.

If $(X_0, X_{J+1}) \in D$, then splicing $(X_J, X_0), (X_0, X_{J+1})$ between X_J and X_{J+1} in p gives a Hamiltonian path in G.

In contrast, if $(X_0, X_I) \notin D$ but $(X_I, X_0) \in D$ for every $I \in \{1, \ldots, K - 1\}$, then for $I := K - 1$ two cases can arise.

Either $(X_0, X_K) \in D$, and then splicing $(X_{K-1}, X_0), (X_0, X_K)$ between X_{K-1} and X_K in p gives a Hamiltonian path in G.

Or $(X_0, X_K) \notin D$, whence $(X_K, X_0) \in D$ by completeness, and following p by (X_K, X_0) also gives a Hamiltonian path in G. □

7.3.3 Eulerian Paths and Circuits

The maintenance of roads, for instance, snow removal or weed control, can save time and energy if it proceeds along itineraries that follow each stretch of road exactly once in each direction. There exist algorithms to determine the feasibility of such an itinerary — called an Eulerian circuit — and then to trace it.

Definition 981 An **Eulerian path** is a path that traverses *every edge exactly once*.
An **Eulerian circuit** is a closed Eulerian path.

Example 982 The graph ⓪⇄① has an Eulerian circuit: $(\,(0,1),(1,0)\,)$.

The following theorems reveal that a graph has an Eulerian circuit if and only if every vertex has an even dgree.

Theorem 983 *In every graph with an Eulerian circuit, every vertex has an even degree.*

Proof. Every vertex without any incident edge has the even degree 0. Every vertex Z with at least one incident edge lies on an edge of the Eulerian circuit,

$$\begin{array}{rcl} c:\mathbb{Z}_N & \to & D, \\ c([i]_N) & = & (X_i,Y_i), \\ X_0 & = & Y_{N-1}, \end{array}$$

because such a circuit traverses every edge. Consequently, every edge incident on Z has the form (X_i,Z), or (Z,Y_i). for some $[i]_N \in \mathbb{Z}_N$.

If $(X_i,Z) \in D$, then $(Z,Y_{i+1}) \in D$. If $(Z,Y_i) \in D$, then $(X_{i-1},Z) \in D$.

In either case both edges add an even integer to the degree of Z. □

Counterexample 984 With odd degrees, the graph ⓪–① has *no* Eulerian circuit.

Theorem 985 *In every path-connected graph $G = (V,D)$ where every vertex has an even degree, for all vertices X and Y in V there exists an Eulerian path from X to Y. In particular, if $X = Y$, then there exists an Eulerian circuit.*

Proof. This proof proceeds by induction with the number M of edges.

Initial step If $M := 0$, then the empty function $g : \varnothing \to \varnothing$ is an Euler circuit.

If $M := 1$ and every vertex has an even degree, then the single edge starts and ends at the same vertex, which is thus an Euler circuit.

Induction hypothesis Assume that there exists $K \in \mathbb{N}$ such that the theorem holds for every graph with $M := K$ edges.

Induction step If $G = (V,D)$ is a graph with $M := K+1$ edges, then it contains at least one edge $(X_0,Y_0) \in D$ because $K+1 \geq 1$. By theorem 970, the edge (X_0,Y_0) is contained in a maximal path $p : N \to D$.

If p passes through a vertex X_i, then it traverses an even number of edges incident on X_i. Because the last vertex Y_{N-1} has an even degree, it follows from the maximality of p that p can be extended, unless $Y_{N-1} = X_0$. Thus, $Y_{N-1} = X_0$ and p is a circuit.

Let F be the subgraph of G with the same vertices but with all the edges traversed by p removed. Because each connected component of F has fewer edges, it follows by induction hypothesis that each connected component of F has an Euler circuit.

Because the initial graph G is path-connected, it follows that each connected component of F has a vertex on the path p. Splicing the Euler circuit for the connected component into p then gives an Euler circuit for G. □

Example 986 The following directed graph has an Eulerian circuit,

③
⇅
⓪⇄①⇄②

for instance, $(\ (0,1),(1,2),(2,1),(1,3),(3,1),(1,0)\ )$.

Counterexample 987 With $V := \{0,1,2,3\}$ and $E := \{(0,1),(1,2),(1,3)\}$,

③
|
⓪–①–②

the graph $G := (V, E)$ has *no* Eulerian circuit, because ① has an odd degree.

7.3.4 Exercises

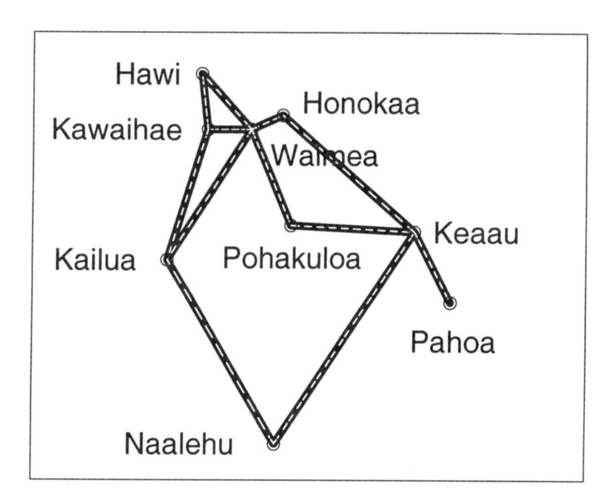

Figure 7.3: Diagram of major State and Federal highways on Hawaii.

For the following exercises, each undirected edge in the graph in figure 7.3 consists of two directed edges, one in each direction.

Exercise 1131 Determine whether the graph in figure 7.3 has an Eulerian circuit.

Exercise 1132 Determine whether the graph in figure 7.3 has a Hamiltonian circuit.

Exercise 1133 Prove that every path-connected graph with exactly two vertices has a Hamiltonian path.

Exercise 1134 Prove that every path-connected graph with exactly three vertices has a Hamiltonian path.

Exercise 1135 Prove that every path-connected graph with exactly two vertices has an Eulerian path.

Exercise 1136 Prove that every path-connected graph with exactly three vertices has an Eulerian path.

Exercise 1137 Determine whether K_4 (example 948) has an Eulerian circuit.

Exercise 1138 Determine whether K_5 (definition 947) has an Eulerian circuit.

Exercise 1139 Determine whether K_4 (example 948) has a Hamiltonian circuit.

Exercise 1140 Determine whether K_5 (definition 947) has a Hamiltonian circuit.

7.4 MATHEMATICAL TREES

7.4.1 Trees

Some situations require graphs that do not contain any closed walk.

Definition 988 (Trees) An **undirected tree** — also called a **free tree** or simply a **tree** — is an undirected path-connected graph without any circuit.

A vertex is **terminal** if and only if it has degree 1.

Theorem 989 *Every tree with at least one edge has at least two terminal vertices.*

Proof. For each edge (X, Y) in a tree, $X \neq Y$ because (X, Y) is not a loop. Consequently, there exists at least one maximal path $p : N \to D$ starting at X with $p(0) = (X, Y)$, and ending with some edge $p(N-1) = (X_{N-1}, Y_{N-1})$.

Because the tree has no circuit, and because the simple path p is maximal, it follows that there does not exist any edge other than (X_{N-1}, Y_{N-1}) incident on Y_{N-1}. Consequently, Y_{N-1} has degree 1. The same argument yields a vertex $Z \in V$ of degree 1 at the end of a maximal simple path q starting at Y with $q(0) = (Y, X)$. Then $Z \neq Y_{N-1}$ because the graph has no circuit. Therefore the graph has at least two terminal vertices. □

Theorem 990 *Every tree with exactly $N \in \mathbb{N}^*$ edges has exactly $N + 1$ vertices.*

Proof. This proof proceeds by induction with the number of edges.

If a tree $G = (V, D)$ has exactly 1 edge, then it has exactly 2 vertices because it is path-connected and circuit-free, by definition.

Assume that there exists $K \in \mathbb{N}^*$ such that the theorem holds for $N := K$, so that every tree with exactly K edges has exactly $K + 1$ vertices.

If a tree has exactly $K + 1$ edges, then it has at least one terminal vertex. Removing this terminal vertex and the only edge incident on this terminal vertex yields a tree with exactly K edges and hence exactly $K + 1$ vertices. Consequently the initial tree has $(K + 1) + 1$ vertices. □

7.4.2 Spanning Trees

For situations that require graphs without any closed walk, each graph has a subgraph without any closed walk.

Definition 991 A **spanning tree** is a subgraph that is a tree with the same vertices.

Example 992 Figure 7.4 shows a spanning tree for the major highways on Hawaii.

Theorem 993 *Every path-connected graph has a spanning tree.*

Proof. This proof proceeds by induction with the number M of edges.

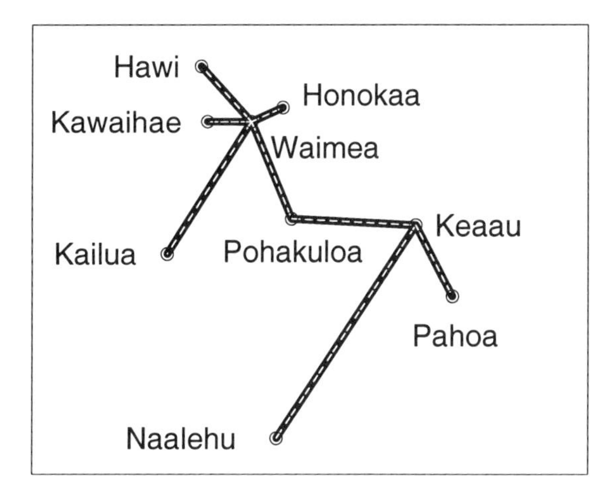

Figure 7.4: Spanning tree for State and Federal highways on Hawaii.

Initial step Every empty path-connected graph with $M := 0$ edge has exactly 0 or 1 vertex and in either case it is its own spanning tree.

Induction hypothesis Assume that for some $K \in \mathbb{N}$ the theorem holds for $M := K$, so that every path-connected graph with at most K edges has a spanning tree.

Induction step For every path-connected graph $G = (V, D)$ with exactly $K + 1$ edges, if G does not contain any closed walk, then G is its own spanning tree. If G contains a closed walk, then removing any one edge from the loop gives a path-connected subgraph $F = (V, C)$ with K edges but with the same vertices as in G. By the hypothesis of induction, the subgraph F admits a spanning tree, which is then also a spanning tree for G, because F and G have the same set of vertices. □

7.4.3 Directed Trees and Star-Shaped Graphs

Some situations, for instance, compiler design, require trees with directed edges [81].

Definition 994 A directed graph $G = (V, D)$ is **star-shaped** relative to a vertex $X \in V$ if and only if for each vertex $Y \in V$ there is a directed walk from X to Y in G.

Example 995 The graph ⓪⇄① is star-shaped relative to each vertex.

Example 996 The graph ⓪→①→② is star-shaped relative to ⓪

Counterexample 997 The graph ⓪→①←② is not star-shaped.

Definition 998 A **directed tree** is a star-shaped directed graph without any circuit.

7.4.4 Exercises

Exercise 1141 Determine a spanning tree for the complete graph K_3.

Exercise 1142 Determine a spanning tree for the complete graph K_4.

Exercise 1143 Determine a spanning tree for the complete graph K_5.

Exercise 1144 Determine a spanning tree for the complete graph K_6.

Exercise 1145 Determine a spanning tree for the complete graph K_N.

Exercise 1146 Determine a spanning tree different from the one in figure 7.4 for the road network in figure 7.3.

Exercise 1147 Prove that a directed tree is star-shaped relative to only one vertex. (Such a vertex is called the **root** of the directed tree.)

Exercise 1148 Prove that every directed tree is path-connected.

Exercise 1149 Prove that in a directed tree the root has an inward degree equal to 0.

Exercise 1150 Prove that in a directed tree every vertex other than the root has an inward degree equal to 1.

7.5 WEIGHTED GRAPHS AND MULTIGRAPHS

7.5.1 Weighted Graphs

Some situations involve additional information about each link in a graph. For instance, graphs of roads can include the length of each stretch of road. Similarly, graphs of power grids can include the maximum flow allowed by each stretch of power line.

Definition 999 A **weighted directed graph** is a pair $G = ((V, D), W)$ consisting of a directed graph (V, D) and a non-negative function $W : V \times V \to \mathbb{Q}_+$ such that $W(X, Y) = 0$ for every edge $(X, Y) \notin D$. The value $W(X, Y)$ is called the **weight** of the edge (X, Y). The **weight** of a walk $p : N \to D$ is the sum

$$\sum_{i=0}^{N-1} W[p(i)] = \sum_{i=0}^{N-1} W[(X_i, Y_i)]$$

of the weights of all the edges traversed by the walk.

A common graphical representation of weighted graphs marks each edge (X, Y) with its weight $W(X, Y) \in \mathbb{Q}$.

Example 1000 In the weighted directed graph $⓪\xrightarrow{3/2}①$ the edge has weight $3/2$.

Similar definitions apply to undirected graphs.

Example 1001 In the weighted undirected graph $⓪\overset{5/7}{-}①$ the edge has weight $5/7$.

Definition 1002 A **multiple graph** is a weighted graph $G = ((V, D), W)$ with a non-negative *integer* weight function $W : V \times V \to \mathbb{N}$.

A common graphical representation of multigraphs marks each edge (X, Y) with weight $N \in \mathbb{N}$ by N arrows from X to Y.

Example 1003 The graph ⓪⇉① is a multiple directed graph with a double edge.

Similar definitions apply to undirected graphs.

Example 1004 The graph ⓪=① is a multiple undirected graph with a double edge.

Example 1005 The graph ⓪≡① is a multiple undirected graph with a triple edge.

7.5.2 Shortest Path and Minimum-Weight Paths or Trees

Some situations require a shortest path between two points, or a spanning tree entailing the smallest cost and hence the minimum weight.

Theorem 1006 *In a path-connected weighted graph with at least one edge, there exists a path with minimum weight between any two vertices.*

Proof. Each path-connected graph has only finitely many edges, whence also finitely many paths between any two vertices. Hence the set of the weights for all paths between the two vertices is finite and non-empty, and, consequently, it has a minimum. Therefore there exists at least one path between the two vertices with the smallest weight.

This proof also outlines an inductive algorithm attributed to Edsger W. Dijkstra to construct such a minimum path [35, p. 403]. Let P be the set of all paths in a weighted directed graph $G = ((V, D), W)$. For each pair of vertices $(X_0, Z_0) \in V \times V$, define a "starting" subset $A_0 \subseteq V$, and a "terminating" subset $B_0 \subseteq V$ by

$$\begin{aligned} A_0 &:= \varnothing, \\ B_0 &:= V. \end{aligned}$$

Moreover, select any strict upper bound U for the weight of every path in the graph, for instance,

$$U := 1 + \sum_{(X,Y) \in D} W(X,Y).$$

Furthermore, define a "routing" function C_0 by

$$\begin{aligned} C_0 : V &\to P \times \mathbb{Q}, \\ Y &\mapsto (P_0(Y), W_0(Y)) := \begin{cases} (\varnothing, U) & \text{for every } Y \neq X_0, \\ (\varnothing, 0) & \text{for } Y := X_0. \end{cases} \end{aligned}$$

Thus for each vertex in the graph the initial routing function C_0 gives an empty path $\varnothing$, corresponding to the initial stage of the algorithm, which has not yet determined any path to any vertex.

At stage K the algorithm has determined the minimum-weight path, with weight $W_K(X_0, X)$, from X_0 to every vertex X in a subset $A_K \subseteq V$, but not for any vertex $Y \in B_K = V \setminus A_K$.

Hence for each $K \in \mathbb{N}$ define subsets $A_{K+1} \subseteq V$ and $B_{K+1} := V \setminus A_{K+1} \subseteq V$, and a routing function $C_{K+1} : V \to P \times \mathbb{Q}$, as follows.

For each vertex $Y \in B_K$, with minimum weight among all the vertices in B_K, examine every vertex $Z \in V$ such that $(Y, Z) \in D$, and set

$$\begin{aligned} A_{K+1} &:= A_K \cup \{Y\}, \\ B_{K+1} &:= V \setminus A_{K+1}, \\ C_{K+1} : V &\to P \times \mathbb{Q}, \\ S &\mapsto (P_{K+1}(S), W_{K+1}(S)), \\ W_{K+1}(S) &:= \begin{cases} W_K(S) & \text{if } (Y,S) \notin D, \\ \min\{W_K(S), W_K(Y) + W(Y,S)\} & \text{if } (Y,S) \in D, \end{cases} \\ P_{K+1}(S) &:= \begin{cases} P_K(S) & \text{if } (Y,S) \notin D, \\ \min\{P_K(S), (Y,S) \bullet P_K(Y)\} & \text{if } (Y,S) \in D. \end{cases} \end{aligned}$$

Each step of the algorithm removes one vertex from B_K, and the graph is finite; hence the algorithm stops at some step $K := L$ with $A_L = V$ and $B_L = \varnothing$. In particular, $Z_0 \in A_L$

and $C_L(Z_0)$ contains a path $P_L(Z_0)$ from X_0 to Z_0. Moreover, the weight $W_L(Z_0)$ of the path $P_L(Z_0)$ is minimum among the weights of all the paths from X_0 to Z_0, because the algorithm selects edges with minimum weights at each step. □

Theorem 1007 *In a path-connected weighted graph, there exists a spanning tree with minimum weights.*

Proof. Each path-connected graph has only finitely many spanning trees, whence at least one of them has the smallest weight. This proof also outlines an inductive algorithm attributed to Prim to construct such a minimum spanning tree from any vertex $X_0 \in V$ [35, p. 418].

Let P be the set of all paths in a weighted directed graph $G = ((V,D),W)$. For each vertex $X_0 \in V$, define a "starting" subset $A_0 \subseteq V$, a "terminating" subset $B_0 \subseteq V$, a set of edges D_0, and an initial weight W_0 by

$$\begin{aligned} A_0 &:= \{X_0\}, \\ B_0 &:= V \setminus \{X_0\}, \\ D_0 &:= \varnothing, \\ W_0 &:= 0. \end{aligned}$$

Thus A_0 consists of the initial point X_0, B_0 consists of all the other vertices, and the initial set of edges D_0 is empty. Hence for each $K \in \mathbb{N}$ define subsets $A_{K+1} \subseteq V$, $B_{K+1} := V \setminus A_{K+1} \subseteq V$, $D_{K+1} \subseteq D$, and a weight W_{K+1} as follows.

For every vertex $X \in A_K$ and connected by at least one edge in D to some vertex in B_K, select any edge $(X,Y) \in D$ with $Y \in B_K$ and with minimum weight among all the edges from X to any vertex in B_K. Then insert the edge (X,Y) to D_K and move Y from B_K to A_K to define new sets by

$$\begin{aligned} A_{K+1} &:= A_K \cup \{Y\}, \\ B_{K+1} &:= V \setminus A_{K+1}, \\ D_{K+1} &:= D_K \cup \{(X,Y)\}, \\ W_{K+1} &:= W_K + W(X,Y). \end{aligned}$$

Because each step of the algorithm removes one vertex, and because the graph is finite, it follows that the algorithm terminates at some step $K := L$ with $A_L = V$ and $B_L = \varnothing$. Then the subgraph $T := (A_L, D_L)$ is a minimum spanning tree for G. □

7.5.3 Maximal Flows in Transportation Graphs

Some situtations, for instance, the transportation of a maximum flow of goods or power from one point (the "source" or "producer" of goods or power) to another point (the "sink" or "consumer" of goods or power).

Definition 1008 A **transportation graph** is a path-connected weighted directed graph $G = ((V,D),W)$ with at least two vertices $S_+, S_- \in V$ subject to the following conditions. The vertex, $S_+ \in V$, which is called a **source**, has an inward degree equal to zero, so that no edge terminates at the source. The vertex, $S_- \in V$, which is called a **sink** has an outward degree equal to zero, so that no edge originates at the sink.

Networks subject flows of goods or services to several constraints. Firstly, if the weight of each edge represents its capacity to sustain traffic, then along each edge the flow cannot exceed the weight of the edge. Secondly, at every vertex other than the source or the sink, the total flow into the vertex equals the total flow out of the vertex.

Definition 1009 A **flow** in a transportation graph $G = ((V, D), W)$ is a function

$$F : V \times V \to \mathbb{Q}$$

with two constraints. Firstly, the flow through an edge does not exceed its capacity:

$$0 \leq F(X,Y) \leq W(X,Y) \qquad \text{for every edge } (X,Y) \in D.$$

Secondly, at each vertex other than the source or the sink, the total inward flow equals the total outward fow:

$$\sum_{(X,Y)\in D} F(X,Y) = \sum_{(Y,Z)\in D} F(Y,Z) \qquad \text{for every vertex } Y \in V \setminus \{S_+, S_-\}.$$

The sum

$$T(F) := \sum_{(S_+,Y)\in D} F(S_+,Y)$$

is the **total** of the flow F.

The determination of the maximum flow that a weighted graph can sustain relies on the following concepts.

Definition 1010 A **cut** in a graph $G = (V, D)$ is an ordered partition (A, B) of the set V of vertices. (See definition 419 for partitions.) Thus

$$\begin{aligned} V &= A \dot{\cup} B, \\ \varnothing &= A \cap B. \end{aligned}$$

Definition 1011 The **capacity** of a cut (A, B) in a weighted graph $G = ((V, D), W)$ is the sum

$$C(A,B) := \sum_{(X\in A)\wedge(Y\in B)} W(X,Y).$$

The following two theorems reveal that the maximum flow sustainable by a network equals the minimum capacity among all cuts in the network [16, Ch. 19], [70, Ch. 10].

Theorem 1012 *The total of each flow does not exceed the capacity of any cut.*

Proof. At the source, the inward flow equals zero, so that

$$T(F) = \sum_{(S_+,Y)\in D} F(S_+,Y) = \sum_{(S_+,Y)\in D} F(S_+,Y) - \sum_{(Y,S_+)\in D} F(Y,S_+).$$

At every vertex $Y \in A \setminus \{S_+\}$ the inward flow equals the outward flow, so that the net flow equals zero:

$$\sum_{(Y,Z)\in D} F(Y,Z) - \sum_{(X,Y)\in D} F(X,Y) = 0.$$

Consequently, only $S_+ \in A$ contributes a non-zero summand to the total flow, which thus equals the sum of all the net flows through all the vertices in A:

$$T(F) = \sum_{(S_+,Y)\in D} F(S_+,Y) = \sum_{Y\in A} \left[\sum_{(Y,Z)\in D} F(Y,Z) - \sum_{(X,Y)\in D} F(X,Y) \right].$$

Moreover, the change of notation from Y and Z to X and Y shows that

$$\sum_{Y\in A}\sum_{Z\in A} F(Y,Z) = \sum_{Y\in A}\sum_{X\in A} F(X,Y).$$

Therefore,

$$\begin{aligned}
\sum_{(S_+,Y)\in D} F(S_+,Y) &= \sum_{Y\in A}\left[\sum_{(Y,Z)\in D} F(Y,Z) - \sum_{(X,Y)\in D} F(X,Y)\right] \\
&= \sum_{Y\in A}\sum_{(Y,Z)\in D} F(Y,Z) - \sum_{Y\in A}\sum_{(X,Y)\in D} F(X,Y) \\
&= \left[\sum_{Y\in A}\sum_{Z\in A} F(Y,Z) + \sum_{Y\in A}\sum_{Z\in B} F(Y,Z)\right] \\
&\quad - \left[\sum_{Y\in A}\sum_{X\in A} F(X,Y) + \sum_{Y\in A}\sum_{X\in B} F(X,Y)\right] \\
&= \left[\sum_{Y\in A}\sum_{Z\in A} F(Y,Z) - \sum_{Y\in A}\sum_{X\in A} F(X,Y)\right] \\
&\quad + \left[\sum_{Y\in A}\sum_{Z\in B} F(Y,Z) - \sum_{Y\in A}\sum_{X\in B} F(X,Y)\right] \\
&= 0 + \left[\sum_{Y\in A}\sum_{Z\in B} F(Y,Z) - \sum_{Y\in A}\sum_{X\in B} F(X,Y)\right] \\
&\leq \sum_{Y\in A}\sum_{Z\in B} F(Y,Z) \\
&\leq \sum_{Y\in A}\sum_{Z\in B} W(Y,Z) \\
&= C(A,B).
\end{aligned}$$

□

Theorem 1013 (Max-flow min-cut theorem) *There exists a total flow equal to the capacity of some cut.*

Proof. This proof proceeds by induction with the number of directed edges.

Initial step With exactly one edge, the total flow $W(S_+, S_-)$ equals the capacity of the cut $\{\,\{S_+\}, \{S_-\}\,\}$.

Induction hypothesis Assume that the theorem holds for all graphs with K edges.

Induction step If $G = ((V,D),W)$ is a transportation graph with $K+1$ edges, it contains at least one path $p : N \to D$ from the source to the sink. Let W_0 be the smallest of the weights of the edges traversed by p:

$$W_0 := \min_{0\leq i<N} W[p(i)],$$

and select any edge (X_0, Y_0) along p with weight W_0. Define a flow F_0 by

$$\begin{array}{rcll} F_0 : V \times V & \to & \mathbb{Q}_+ & \\ (X,Y) & \mapsto & W_0 & \text{if } p \text{ traverses } (X,Y), \\ (X,Y) & \mapsto & 0 & \text{if } p \text{ does not traverse } (X,Y). \end{array}$$

Remove the edge (X_0, Y_0) from the graph, and subtract $W[p(i)]$ from the weight of the edge $p(i)$, which gives a graph G_1. The graph G_1 has one edge fewer than G has. In contrast, if G_1 does *not* contain any path from the source to the sink, then every flow in G must pass entirely through (X_0, Y_0) and hence has a total W_0. Then F_0 is a maximal flow, equal to the capacity of the cut (A_0, B_0) defined by the path-connected components A_0 and B_0 of G_1.

If G_1 contains a path from the source to the sink, then by the hypothesis of induction G_1 admits a flow F_1 with the capacity of some cut (A, B), which is also a cut in the initial graph.

Restoring the edge (X_0, Y_0) and adding both flows yields a flow $F := F_0 + F_1$ with a total flow equal to the capacity of the same cut (A, B) in the inital graph. □

7.5.4 Exercises

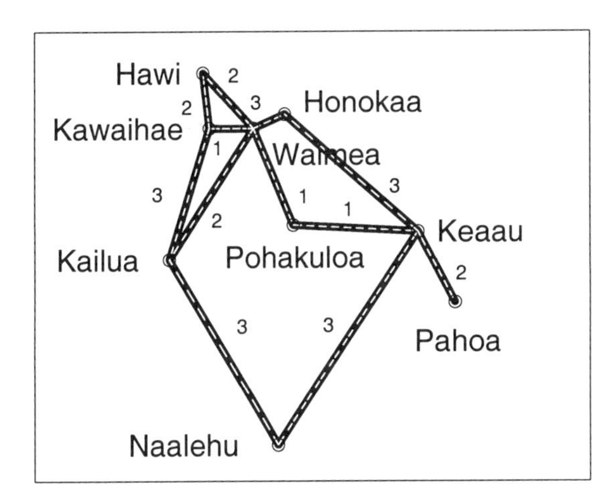

Figure 7.5: Diagram of traffic capacities of major highways on Hawaii.

Exercise 1151 Determine a maximum traffic flow from Hawi to Naalehu in the road network in figure 7.5.

Exercise 1152 Determine a maximum traffic flow from Kailua to Keaau in the road network in figure 7.5.

Exercise 1153 Determine a shortest route from Honokaa to Naalehu in the road network in figure 7.6.

Exercise 1154 Determine a shortest route from Kailua to Keaau in the road network in figure 7.6.

Exercise 1155 Determine a shortest route from Kawaihae to Pahoa in the road network in figure 7.6.

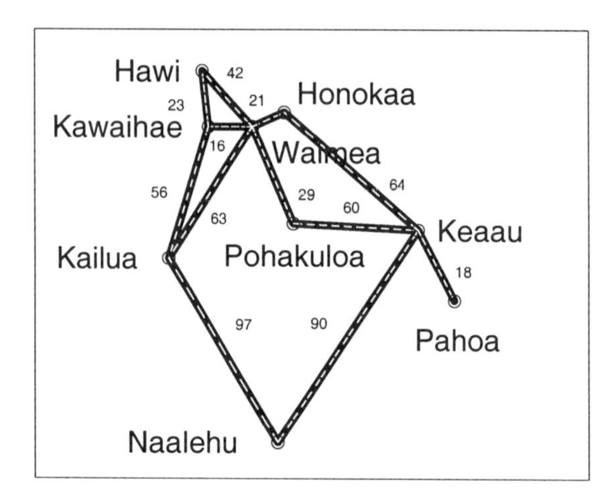

Figure 7.6: Distances (in kilometers) along major highways on Hawaii.

Exercise 1156 Determine a shortest route from Naalehu to Pohakuloa in the road network in figure 7.6.

Exercise 1157 Determine a minimum spanning tree in the road network in figure 7.6.

Exercise 1158 Determine a maximum spanning tree in the road network in figure 7.5.

Exercise 1159 Prove that if a graph has several sources, then there exists an equivalent graph with only one source.

Exercise 1160 Prove that if a graph has several sinks, then there exists an equivalent graph with only one sink.

7.6 BIPARTITE GRAPHS AND MATCHINGS

7.6.1 Bipartite Graphs

The problem of assigning available resources in a set A to various uses in a set B amounts to the problem of selecting one use $Y \in B$ for each resource $X \in A$, subject to constraints in the form of a set of pairs (X, Y) specifying which resource X the use Y requires. In such problems, resources are not connected to one another, and uses are not connected to one another. The corresponding graph is called a "bipartite" graph.

Definition 1014 A **bipartite graph** is a graph $G = (V, D)$ with an ordered partition (A, B) of V such that no edge connects any two vertices from the same subset of the partition:

$$[(A \times A) \cap D] = \varnothing = [(B \times B) \cap D].$$

The bipartite graph is **complete** if and only if it contains an edge between each vertex of A and each vertex of B. If $\#(A) = M$ and $\#(B) = N$, then the corresponding complete bipartite graph is denoted by $K_{M,N}$.

Example 1015 With $V := \{2,3,4,6,8,9\}$ and $D := \{(2,4),(2,6),(2,8),(3,6),(3,9)\}$,

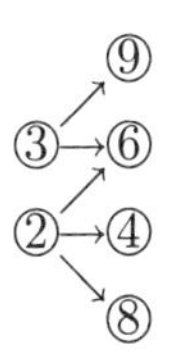

the graph $G = (V, D)$ is bipartite, with partition $A := \{2,3\}$ and $B := \{4,6,8,9\}$.

Example 1016 Here are some complete bipartite graphs.

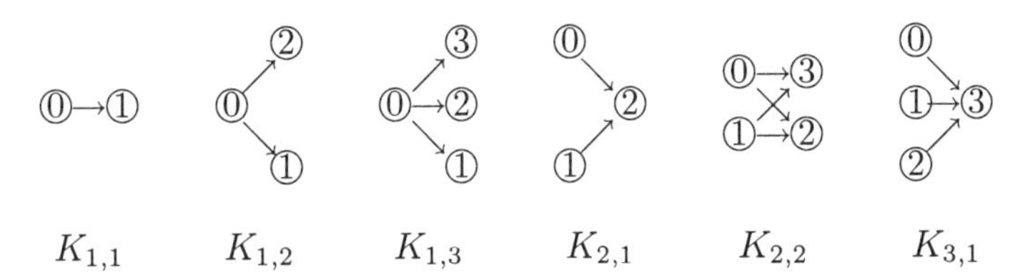

$K_{1,1}$ $K_{1,2}$ $K_{1,3}$ $K_{2,1}$ $K_{2,2}$ $K_{3,1}$

Definition 1017 A **matching** in a bipartite graph $G = (V, D)$ with partition (A, B) of V is an injection $M : R \to B$ of a subset $R \subseteq A$ such that $M \subseteq D$. In other words, a matching is a set of edges M each of which is incident to exactly one vertex in R and exactly one vertex in B. A **complete matching** is a matching $M : A \to B$.

Example 1018 The set $\{(2,4),(3,9)\}$ is a matching for the graph in example 1015.

As with any relation $D \subseteq V \times V$, for each subset $S \subseteq A$ in a bipartite graph $G = (V, D)$ with a partition (A, B) of V, the set (see definition 370 for D")

$$D"(S) = \{Y \in B : \exists X\{(X \in A) \wedge [(X, Y) \in D]\}\}$$

consists of all the vertices in B connected to any vertices in S.

Definition 1019 For each subset $S \subseteq A$ in a bipartite graph $G = (V, D)$ with a partition (A, B) of V, define the **excess supply** $E(S)$ of S by the difference

$$E(S) := \#(S) - \#[D"(S)].$$

Also, define the excess supply of the graph G by

$$\max_{S \subseteq A} E(S),$$

which is the maximum excess supply for all subsets of A.

Because $E(\varnothing) = \#(\varnothing) - \#[D(\varnothing)] = \#(\varnothing) - \#(\varnothing) = 0$, it follows that the excess supply of every graph is non-negative.

Example 1020 In example 1015, $E(\{2\}) = 1 - 3 = -2$ and $E(\{3\}) = 1 - 2 = -1$, whence the graph has excess supply $E(\varnothing) = 0$.

Definition 1021 For each subset $S \subseteq A$ in a bipartite graph $G = (V, D)$ with a partition (A, B) of V, and for each matching $M : R \to B$, define the **shortage of supply** of M by $\#(A) - \#(R)$.

7.6.2 Maximum Matching

The largest subset of supply $R \subseteq A$ that can be matched to a demand $D"(R) \subseteq B$ corresponds to the maximum flow of resources to uses. Hence maximum matchings are amenable to methods for maximum flows [16, Ch. 15], [70, Ch. 11].

Theorem 1022 *The excess supply of the graph cannot exceed the shortage of supply of any matching.*

There is a matching with shortage of supply equal to the excess supply of the graph.

Proof. For each bipartite graph $G = (V, D)$ with a partition (A, B) of V, define a transportation graph $T := ((U, H), W)$ as follows.

For any two vertices (sets) S_+ and S_- not in V, let $U := V \dot{\cup} \{S_+, S_-\}$. Then let

$$\begin{array}{rcll} P & := & \{(S_+, X) : X \in A\} & \text{(edges from the source } S_+ \text{ to } A\text{)}, \\ Q & := & \{(Y, S_-) : Y \in B\} & \text{(edges from } B \text{ to the sink } S_-\text{)}, \\ H & := & D \cup P \cup Q & \text{(all edges from } G\text{, } P\text{, and } Q\text{)}. \end{array}$$

Then define weights by

$$\begin{array}{rcll} W(S_+, X) & := & 1 & \text{for each } X \in A, \\ W(Y, S_-) & := & 1 & \text{for each } Y \in B, \\ W(X, Y) & := & \#(A) + 1 & \text{for each } (X, Y) \in D. \end{array}$$

Select any subset $S \subseteq A$ with maximal excess supply, so that $E(S)$ equals the excess supply of G. Then define a cut (I, J) in T so that I consists of the source S_+, all the vertices in S, and their images in B:

$$I := \{S_+\} \cup S \cup D"(S).$$

Then $J = H \setminus I$ because the cut (I, J) partitions the set H of vertices of T.

By definition of $D"(S)$, there does not exist any edge connecting any vertex from S to any vertex in $B \setminus D"(S)$. Consequently, the cut (I, J) does not separate any vertex in S from any vertex in $B \setminus D"(S)$. By definition of the set I, it follows that $S \subseteq I$ and $D"(S) \subseteq I$ whence $A \setminus S \subseteq J$ and $B \setminus D"(S) \subseteq J$. Consequently, the cut (I, J) does not separate any vertex in $A \setminus S$ from any vertex in $B \setminus D"(S)$.

By definition of a bipartite graph, there is no edge from B to A, and hence no edge from $I \cap [B \cap D"(S)]$ to $J \cap (A \setminus S)$. Thus, the cut (I, J) can separate only the source $S_+ \in I$ from the vertices in $A \setminus S$, and the sink $S_- \in J$ from the vertices in $B \setminus J$:

$$I = \{S_+\} \cup (A \cap S) \cup [B \cap D"(S)],$$

$$(A \setminus S) \cup [B \setminus D"(S)] \cup \{S_-\} = J.$$

Therefore, the capacity $C(I, J)$ of the cut (I, J) equals the sum of the weights of the edges from the source S_+ to $A \setminus S$, and from $D"(S)$, to the sink S_-:

$$\begin{aligned} C(I, J) &= \sum_{X \in (A \setminus S)} W(S_+, X) + \sum_{Y \in D"(S)} W(Y, S_-) \\ &= \#(A \setminus S) * 1 + \#[D"(S)] * 1 \\ &= [\#(A) - \#(S)] + \#[D"(S)] \\ &= \#(A) - [\#(S) - \#[D"(S)] \\ &= \#(A) - E(S). \end{aligned}$$

However, the capacity of the cut (I,J) cannot exceed the capacity of any other cut (K,L) in T, with $S_+ \in K$ and $S_- \in L$. Indeed, if the cut separates any vertex $X \in A$ from any vertex $Y \in B$, then $C(K,L) \geq C(I,J)$ because

$$\begin{aligned} C(K,L) &\geq W(X,Y) \\ &= \#(A)+1 \\ &> \#(A)-E(S) \\ &= C(I,J). \end{aligned}$$

If the cut (K,L) does not separate any two vertices from A to B, then

$$\begin{aligned} D"(A \cap K) &\subseteq (B \cap K), \\ D"(A \cap L) &\subseteq (B \cap L). \end{aligned}$$

Hence the cut (K,L) separates only the source $S_+ \in K$ from a subset of vertices $P \subseteq A \cap L$, and from $P \subseteq L$ it follows that $D"(P) \subseteq B \cap L$, and it also separates a subset of vertices $Q \subseteq B \cap K$ from the sink $S_- \in L$. Thus,

$$\begin{array}{ccccccccc} K & = & \{S_+\} & \cup & (A \setminus P) & \cup & Q, & & \\ & & & & P & \cup & (B \setminus Q) & \cup & \{S_-\} \quad = \quad L. \end{array}$$

Again by definition of a bipartite graph, there is no edge from B to A, and, in particular, no edge from Q to P. Consequently,

$$\begin{aligned} C(K,L) &= \#(P)+\#(Q) \\ &= \#(A \setminus K)+\#(B \cap K) \\ &\geq \#(A \setminus K)+\#[D"(A \cap K)] \\ &= [\#(A)-\#(A \cap K)]+\#[D"(A \cap K)] \\ &= \#(A)-E(A \cap K) \\ &\geq \#(A)-E(S) \\ &= C(I,J). \end{aligned}$$

Thus, to each subset $S \subseteq A$ with a maximum excess supply $E(S)$, there corresponds a minimum cut (I,J) in T, to which there corresponds a maximum flow F, with total flow equal to the capacity of the cut:

$$\#(S) = \sum_{Y \in S} 1 = \sum_{Y \in S} F(S_+,Y) = C(I,J) = \#(A)-E(S) = \#(A)-[\#(S)-\#D(S)].$$

Consequently,

$$[\#(S)-\#D(S)] = [\#(A)-\#(S)],$$

so that the selection of any one edge from each vertex of S produces a matching with shortage $[\#(A)-\#(S)]$ equal to the excess supply $[\#(S)-\#D(S)]$ of the graph.

Finally, every matching $M: R \to B$ with shortage of supply $\#(A)-\#(R)$ corresponds to a flow in the graph T, with flow 1 in each edge (S_+,Y) from the source S_+ to a vertex $Y \in R$, and hence total flow $\#(R) \leq \#(S)$, with a larger shortage of supply $[\#(A) - \#(R)] \geq [\#(A)-\#(S)]$. □

7.6.3 Exercises

Exercise 1161 Determine the number of complete matchings for the complete bipartite graph $K_{N,N}$.

Exercise 1162 Determine the number of complete matchings for the complete bipartite graph $K_{K,N}$ with $K > N$.

Exercise 1163 Determine the number of complete matchings for the complete bipartite graph $K_{K,N}$ with $K \leq N$.

Exercise 1164 Determine the number of maximum matchings for the complete bipartite graph $K_{K,N}$ with $K > N$.

Exercise 1165 Prove that every complete bipartite graph $K_{K,N}$ is path-connected.

Exercise 1166 Provide an example of a bipartite graph that is *not* path-connected.

Exercise 1167 Prove that every complete bipartite graph $K_{K,N}$ is a tree.

Exercise 1168 Prove that each path-connected component of every bipartite graph is a tree.

Exercise 1169 Determine which bipartite graphs have a Hamiltonian path.

Exercise 1170 Determine which bipartite graphs have an Eulerian path.

7.7 GRAPH THEORY IN SCIENCE

7.7.1 The Shape of Molecules

A molecule consists of atoms held together by various types of electronic bonds between some pairs of atoms. Thus one method to represent a molecule consists of a graph, with a vertex for each atom, and with an edge for each bond [109, Ch. 8].

Example 1023 Each hydrogen atom (H) has only one electron, but two hydrogen atoms can form a molecule (H_2) held together by one bond consisting of the combination of their two electrons. Such a molecule corresponds to the following graph.

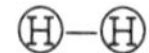

Hydrogen molecule H_2

A pair of atoms can form multiple bonds, representable by a multigraph.

Example 1024 Each oxygen atom (O) has eight electrons, with two electrons on an inner layer and six electrons on an outer layer. However, two oxygen atoms can form a molecule (O_2), held together by two bonds consisting of the combination of two pairs of electrons (one pair from the outer layer of each atom). Such a molecule corresponds to the following multigraph.

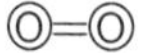

Oxygen molecule O_2

Example 1025 Each nitrogen atom (N) has seven electrons, with two electrons on an inner layer and five electrons on an outer layer. However, two nitrogen atoms can form a molecule (N_2), held together by three bonds consisting of the combination of three pairs of electrons (three electrons from the outer layer of each atom). Such a molecule corresponds to the following multigraph.

Nitrogen molecule N_2

Atoms of different elements can also combine into molecules.

Example 1026 In a water molecule (H_2O), one oxygen atom combines two of its outer electrons with each electron from two hydrogen atoms. Thus the water molecule holds together through two single bonds, corresponding to the following graph.

Water molecule H_2O

Graph theory thus provides methods to distinguish possible from impossible shapes of molecules. For instance, a molecule must correspond to a path-connected graph.

7.7.2 The Shape of Hydrocarbons

Hydrocarbon molecules consist entirely of atoms of carbon (C) and hydrogen (H). Each atom of carbon has six electrons: two electrons on an inner layer, and four electrons on an outer layer, which can form at most four bonds with other atoms [109, Ch. 22].

Example 1027 Each molecule of methane consists of one atom of carbon and four atoms of hydrogen (CH_4). Each atom of hydrogen shares its electron with one electron from the atom of carbon, so that the molecule holds together through four atomic bonds:

Methane CH_4

With at most four outer electrons available for at most four bonds, each atom of carbon lies at a vertex with a degree that cannot exceed 4. Similarly, with only one electron available, each hydrogen atom lies at a terminal vertex, with degree 1. Whence arises the problem of determining the largest number of hydrogen atoms in a hydrocarbon molecule with a specific number of carbon atoms. Such hydrocarbon molecules with the largest number of hydrogen atoms are called **alkanes** or **saturated hydrocarbons**.

Example 1028 Ethylene and ethane molecules are hydrocarbons with exactly two atoms of carbons. Each molecule of **ethylene** consists of two atoms of carbon and four atoms of hydrogen (C_2H_4). Each of the two carbon atoms shares two electrons with the other carbon atom, thus forming one double bond, corresponding to a double edge on the graph. Each of the two carbon atoms also shares two electrons with two hydrogen atoms, corresponding to two single edges on the graph.

Ethylene C_2H_4 Ethane C_2H_6

Ethylene is not saturated because ethane also has two carbon atoms but more hydrogen atoms. Each molecule of **ethane** consists of two atoms of carbon and six atoms of hydrogen (C_2H_6). Each of the two carbon atoms shares one electron with the other carbon atoms, thus forming one single bond. Each of the two carbon atoms also shares three electrons with three hydrogen atoms, corresponding to three single bonds.

The following theorem gives a formula for saturated hydrocarbons [20, Ch. 11].

Theorem 1029 *Each saturated hydrocarbon with exactly C carbon atoms has exactly $2C+2$ hydrogen atoms.*

Proof. A saturated hydrocarbon molecule with exactly C carbon atoms and H hydrogen atoms corresponds to a path-connected graph with C vertices of degree at most 4 and H vertices of degree exactly 1. One such graph consists of a sequence of $C-1$ edges connecting C vertices of degree exactly 4, with the first and last of these also connected to three terminal vertices, and all the $C-2$ other ones of these connected to two terminal vertices, for a total of $3+2(C-2)+3 = 2C+2$ terminal vertices.

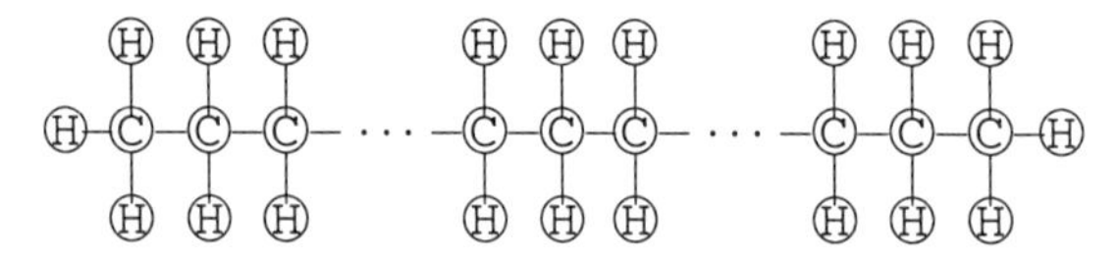

Hence the largest number of hydrogen atoms is at least $2C+2$. The proof that the largest number of hydrogen atoms cannot exceed $2C+2$ uses induction with C.

Initial step For $C := 1$ the graph has exactly one vertex with degree at most 4, and exactly H vertices with degree 1. Thus the sum s of the degrees in the graph cannot exceed $4+H$. Because the graph has exactly $C+H$ vertices, it has at least $C+H-1 = 1+H-1 = H$ edges. Thus the sum s of the degrees in the graph equals at least $2H$.

Consequently, $2H \leq s \leq 4+H$, whence $H \leq 4$. Yet $H = 4$ occurs with methane. Therefore $H = 4 = 2(1)+2$ is the largest number of terminal vertices for $C := 1$.

Induction hypothesis Assume that there exists $K \in \mathbb{N}^*$ such that the theorem holds for $C := K$, so that the largest number of terminal vertices equals $2K+2$.

Induction step For each path-connected graph $G = (V, D)$ with $K+1$ vertices of degree at most 4 and exactly $H \geq 2(K+1)+2 \geq 6$ terminal vertices, select any terminal vertex $Z \in V$. Then there exists at least one other vertex $Y \in D$ connected to Z by an edge $(Y, Z) \in D$, because G is path-connected. Moreover Y cannot be a terminal vertex, otherwise Y and Z would form a path-connected component of G with fewer than six terminal vertices. Consequently, $\deg_G(Y) > 1$, and there exists at least another vertex $X \in V$ connected to Y by an edge $(X, Y) \in D$. As for Y, X is not a terminal vertex. Thus G connects Y to X and at most three terminal vertices.

Removing Y, and all the terminal vertices other than Z connected to Y from G, of which there are at most two because $\deg_G(Y) \leq 4$, and replacing (X, Y) by (X, Z) yields a subgraph $F := (U, B)$ with exactly $K-1$ vertices of degree at most 4.

By induction hypothesis, F has at most $2K+2$ terminal vertices. Because the terminal vertices of F differ from the terminal vertices of G by at most two terminal vertices connected to Y, G has at most $(2K+2)+2 = 2(K+1)+2$ terminal vertices. □

Example 1030 Propane is another example of a saturated hydrocarbon: $2 * 3 + 2 = 8$.

Propane C_3H_8

Molecules of saturated hydrocarbons with the same number of carbon atoms can have various shapes called **isomers**.

Example 1031 Butane is another example of a saturated hydrocarbon: $2 * 4 + 2 = 10$.

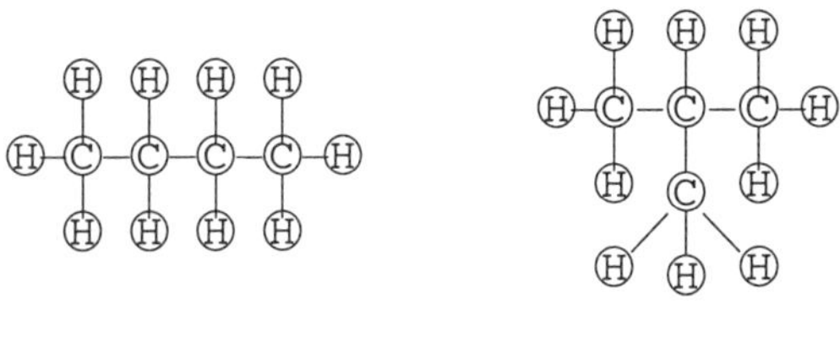

Butane C_4H_{10} Isobutane C_4H_{10}

Nevetheless, butane molecules exist in two isomers, called butane and isobutane.

7.7.3 Sequences of Radioactive Decays

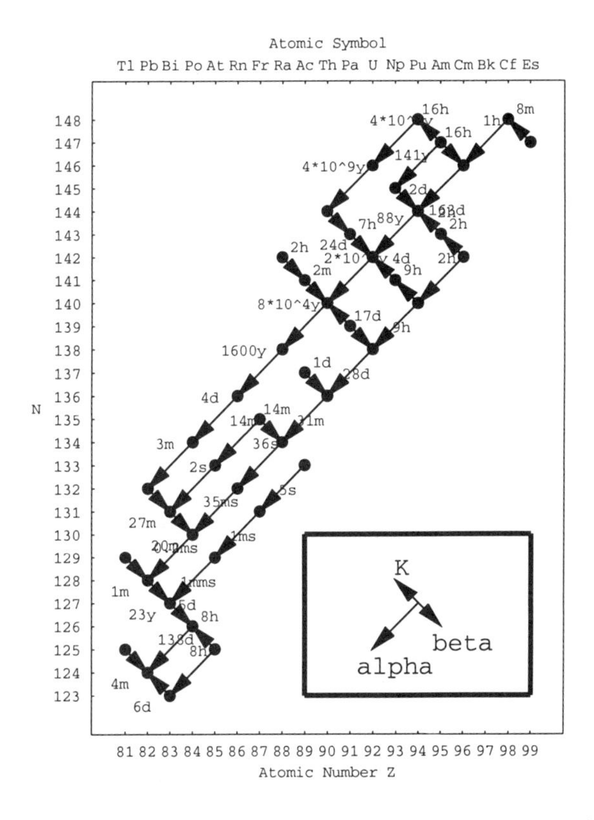

Figure 7.7: Half-lives and decays of heavy nuclei [69], [106, p. 300]. d: days, h: hours, m: minutes, ms: milli-seconds, mms: micro-seconds, s: seconds, y: years.

Heavy radioactive nuclei can go through several types of radioactive decay, as shown in figure 7.7, where N is the number of neutrons and Z the number of protons [106]. The sum $N + Z$ is the atomic mass number.

- In alpha-decay, a nucleus emits an α-particle, which consists of two protons and two neutrons (like the helium nucleus). Thus, alpha-decay decreases the mass of the nucleus by four atomic mass units: the atomic number Z (the number of protons) decreases by two, and the number of neutrons N also decreases by two.

- In the form β^- of beta-decay, within the nucleus a neutron is transformed into a proton, while an electron is emitted along with an anti-neutrino. Thus the atomic number Z increases by one, the number of neutrons N decreases by one, and the atomic mass number remains unchanged.

- In K-capture, a nucleus captures an electron and emits a neutrino, which transforms a proton into a neutron and thus decreases the atomic number Z by one.

Figure 7.7 displays a directed graph indicating that a nucleus X can decay into a nucleus Y by a directed edge (X, Y). The weight of each edge $W(X, Y)$ is the half-life of the isotope at the initial vertex X of the edge (X, Y); in other words, $W(X, Y)$ is the time it takes for one half of any amount of X to decay into Y.

7.7.4 Exercises

Exercise 1171 Prove that the graph in figure 7.7 has no loops.

Exercise 1172 Determine whether the graph in figure 7.7 is star-shaped.

Exercise 1173 For the graph in figure 7.7, determine a shortest decay time (a shortest path) from Americium (Am^{242}) at coordinates $(Z, N) := (95, 147)$ to inert lead (Pb^{206}) at coordinates $(Z, N) := (82, 124)$, in other words, a shortest decay sequence for Americium to become inert.

Exercise 1174 For the graph in figure 7.7, determine a shortest decay time (a shortest path) from Protactinium (Pa^{230}) at coordinates $(Z, N) := (91, 139)$ to inert lead (Pb^{206}) at coordinates $(Z, N) := (82, 124)$, in other words, a shortest decay sequence for Protactinium to become inert.

Exercise 1175 Prove that in the graph in figure 7.7 every maximal path terminates with the lead isotope Pb^{206} (with $Z := 82$ protons and $N := 124$ neutrons).

Exercise 1176 Draw three different shapes for a molecule of pentane, C_5H_{12}.

Exercise 1177 Draw several different shapes for a molecule of octane, C_8H_{18}.

Exercise 1178 Prove that the graph of each saturated hydrocarbon molecule is a tree.

Exercise 1179 Determine whether if the graph of an alkane is a tree, then it is saturated.

Exercise 1180 Draw a graph of a white phosphorus molecule P_4 with four atoms of phosphorus and a single bond between each pair of distinct atoms.

7.8 FURTHER ISSUES: PLANAR GRAPHS

Graphs without crossing within a plane arise in the design of printed circuit boards and other applications. Because the concepts of plane and crossing come from geometry, the theory of planar graphs relies to some extent on plane geometry [49].

Definition 1032 A subset $C \subset P$ of a plane P **separates** the plane P if and only if the complement $P \setminus C$ consists of two non-empty disjoint subsets.

Example 1033 A theorem of plane geometry states that every straight line L in a plane P separates the complement $P \setminus L$ into two non-empty disjoint subsets, called the region on "one side" and the region on the "other side" of the line [49, p. 8, thm. 8].

Example 1034 Another theorem of plane geometry states that every polygon G in a plane P separates the complement $P \setminus G$ into two non-empty disjoint subsets, called the region "inside" and the region "outside" of the polygon [49, p. 9, thm. 9].

Example 1035 An axiom of plane geometry specifies a concept of distance, whence it follows that for a circle C with center X and radius R, the complement $P \setminus C$ consists of exactly two regions. The "inside" consists of all the points at a distance smaller than R from the center X, while the "outside" consists of all the points at a distance greater than R from the center X [5, p. 133], [49, p. 166].

A formula attributed to Euler states that if $G = (V, D)$ is a non-empty path-connected planar graph with v vertices, e edges, and r regions, then

$$r + v - e = 2.$$

Example 1036 The graph ⓪ has $v = 1$ vertex, $e = 0$ edge, and its complement $P \setminus \{⓪\}$ has $r = 1$ region. For this graph, $r + v - e = 1 + 1 - 0 = 2$.

Example 1037 The graph ⓪ has $v = 1$ vertex, $e = 1$ edge, and its complement $P \setminus \{⓪\}$ has $r = 2$ regions. For this graph, $r + v - e = 2 + 1 - 1 = 2$.

Example 1038 The complete graph K_3 forms a triangle with $v = 3$ vertices and $e = 3$ edges; its complement consists of $r = 2$ regions: the "inside" and the "outside" of the triangle. For this graph, $r + v - e = 2 + 3 - 3 = 2$.

Example 1039 The complete graph K_4 is a planar graph with $v = 4$ vertices, $e = C_2^4 = 6$ edges, and its complement consists of $r = 4$ regions, as shown in figure 7.8.

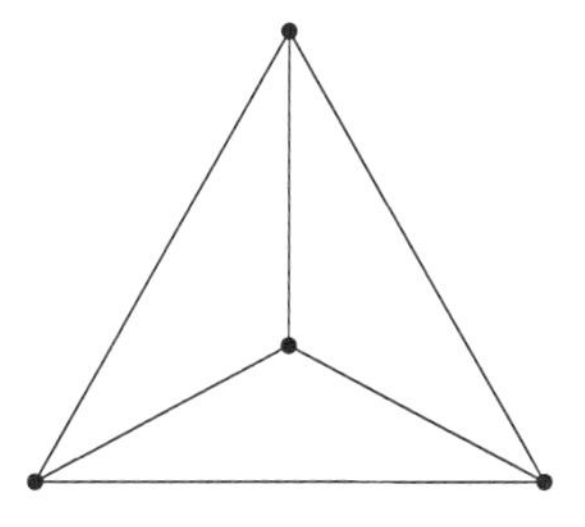

Figure 7.8: For the complete graph K_4 in the plane, $r + v - e = 4 + 4 - 6 = 2$.

Proofs of Euler's formula involve a precise definition of the concept of "region" in Euclidean plane geometry. For circles — in particular, for circular edges in graphs — the regions correspond to distances smaller than or greater than the radius. For all simple closed curves — in particular, for simple closed paths in graphs — the proof that they separate the plane into exactly two regions is known as the Jordan Curve Theorem and involves a substantial part of the mathematical field called topology. Some proofs of Euler's formula also uses material from an extension called algebraic topology [70, p. 208–210]. Alternatively, based on variants of a theorem attributed to István Fáry [21], other proofs of Euler's formula rely on a version of the Jordan Curve Theorem restricted to such polygonal curves as walks along straight edges in a graph [98]. The following statements summarize the principal results. (See the references for proofs.)

Theorem 1040 (Fáry Theorem) *Every loop-free planar graph admits a planar representation where all the edges are straight line segments [21].*

Theorem 1041 (Jordan Curve Theorem) *Every simple circuit separates the plane into exactly two regions [19, p. 361–363].*

Theorem 1042 (Euler Formula) *If $G = (V, D)$ is a non-empty path-connected planar graph with v vertices, e edges, and r regions, then [70, p. 208–210]*

$$r + v - e = 2.$$

Theorem 1043 (Kuratowski Theorem) *A graph is planar if and only if, after insertion or removal of vertices of degree 2, it does not contain any subgraph that is either the complete graph K_5 with five vertices or the complete bipartite graph $K_{3,3}$ [70, p. 212–220].*

7.9 PROJECTS

Project 15 Investigate algorithms to schedule classes in classrooms at certain times.

Project 16 Investigate algorithms to maximize traffic flows through traffic lights.

Bibliography

[1] Lars V. Ahlfors. *Complex Analysis*. McGraw-Hill, New York, NY, third edition, 1979. ISBN 0-07-000657-1; QA331.A45 1979; LCCC No. 78-17078; 515'.93.

[2] Friedrich L. Bauer. *Decrypted Secrets: methods and maxims of cryptology*. Springer-Verlag, Berlin Heidelberg, 1997. ISBN 3-540-60418-9; QA76.9.A25B38513 1997.

[3] Paul Bernays. *Axiomatic Set Theory*. Dover, Mineola, NY, 1991. ISBN 0-486-66637-9; QA248.B47 1991; LCCC No. 90-25812; 511.3'22–dc20.

[4] Garrett Birkhoff and Saunders Mac Lane. *A Survey of Modern Algebra*. Macmillan, New York, NY, fourth edition, 1977. ISBN 0-02-310070-2; QA162.B57 1977; LCCC No. 75-42402; 512.

[5] George David Birkhoff and Ralph Beatley. *Basic Geometry*. Chelsea Publishing, New York, NY, third edition, 1959. QA455.B552 1959.

[6] Bolzano. *Paradoxien des Unendlichen*. Leipzig, 1851. Cited in [17, p. 64].

[7] William L. Briggs and Van Emden Henson. *The DFT: An Owner's Manual for the discrete Fourier Transform*. Society for Industrial and Applied Mathematics, Philadelphia, PA, 1995. ISBN 0-89871-342-0; QA403.5.B75 1995; LCCC No. 95-3232; 515'.723–dc20.

[8] Lloyd A. Brown. *The Story of Maps*. Dover, New York, NY, 1979.

[9] Alonzo Church. Correction to a note on the Entscheidungsproblem. *Journal of Symbolic Logic*, 1(3):101–102, September 1936.

[10] Alonzo Church. A note on the Entscheidungsproblem. *Journal of Symbolic Logic*, 1(1):40–41, March 1936.

[11] Alonzo Church. *Introduction to Mathematical Logic*. Princeton Landmarks in Mathematics and Physics. Princeton University Press, Princeton, NJ, 1996. Tenth printing; ISBN 0-691-02906-7.

[12] Ruel V. Churchill and James Ward Brown. *Complex Variables and Applications*. MacGraw-Hill, New York, NY, 1984. ISBN 0-07-010873-0; QA331.C524 1984; LCCC No. 83-9384; 515.9.

[13] Vaşek Chvátal. *Linear Programming*. W. H. Freeman, New York, NY, 1983. ISBN 0-7167-1587-2; T57.74.C54 1983; LCCC No. 82-21132; 519.7'2.

[14] Paul J. Cohen. The independence of the continuum hypothesis. *Proceedings of the National Academy of Sciences, U.S.A.*, 50:1143–1148, 1963.

[15] Paul J. Cohen. The independence of the continuum hypothesis. II. *Proceedings of the National Academy of Sciences, U.S.A.*, 51:105–110, 1964.

[16] George B. Dantzig. *Linear Programming and Extensions*. Princeton University Press, Princeton, NJ, 1963. ISBN 0-691-08000-3; L. C. Card 59-50559.

[17] Richard Dedekind. *Essays on the Theory of Numbers*. Dover, New York, NY, 1963. ISBN 0-486-21010-3; 63-3681. (Also Open Court Publishing, Chicago, IL, 1924; QA248.D3.).

[18] J. L. E. Dreyer. *A History of Astronomy from Thales to Kepler*. Dover, New York, NY, second edition, 1953.

[19] James Dugundji. *Topology*. Allyn and Bacon, Boston, MA, 1966. ISBN 0-205-00271-4; LCCC No. 66-10940.

[20] Susanna S. Epp. *Discrete Mathematics with Applications.* Brooks/Cole, Pacific Grove, CA, second edition, 1995. ISBN 0-534-94446-9; QA39.2.E65 1995; LCCC No. 94-31184; 511–dc20.

[21] István Fáry. On straight line representation of planar graphs. *Acta Universitatis Szegediensis. Acta Scientiarum Mathematicarum*, 11(4):229–233, 1948.

[22] Solomon Feferman. *In the Light of Logic.* Logic and computation in philosophy. Oxford University Press, New York, NY, 1998. ISBN 0-19-508030-0; QA9.2.F44 1998; LCCC No. 97-051336; 511.3–dc21.

[23] Solomon Feferman. Does mathematics need new axioms? *American Mathematical Monthly*, 106(2):99–111, February 1999.

[24] Jo Ann Fellin. Adders and their design. *UMAP Journal*, 6(3):71–111, Fall 1985. Also reprinted as UMAP Module 668, COMAP, Lexington, MA, 1985.

[25] Richard Phillips Feynman, Robert B. Leighton, and Matthew L. Sands. *The Feynman Lectures on Physics*, volume III. Addison-Wesley, Redwood City, CA, commemorative edition, 1989. ISBN 0-201-51005-7; QC21.2.F49 1989; LCCC No. 89-433; 530–dc19.

[26] Richard Phillips Feynman, Robert B. Leighton, and Matthew L. Sands. *The Feynman Lectures on Physics*, volume II. Addison-Wesley, Redwood City, CA, commemorative edition, 1989. ISBN 0-201-51004-9; QC21.2.F49 1989; LCCC No. 89-433; 530–dc19.

[27] Gerald B. Folland. *Real Analysis: Modern Techniques and their Applications.* John Wiley, New York, NY, 1984. ISBN 0-471-80958-6; QA300.F67 1984; LCCC No. 84-10435; 515.

[28] A. Fraenkel. *Abstract Set Theory.* North-Holland, Amsterdam, 1953.

[29] G. Frege. *Begriffsschrift, eine der arithmetischen nachgebildete Formelsprache des reinen Denkens.* Halle, 1879.

[30] Joseph Gallian and Steve Winters. Modular arithmetic in the marketplace. *American Mathematical Monthly*, 95(6):548–551, June-July 1988.

[31] Kurt Gödel. Über formal unentscheidbare Sätze der Principia Mathematica und verwandter Systeme. *Monatshefte für Mathematik und Physik*, 38:173–198, 1931. Leipzig.

[32] Kurt Gödel. The consistency of the axiom of choice and of the generalized continuum hypothesis. *Proceedings of the National Academy of Sciences, U.S.A.*, 24:556–557, 1938.

[33] Kurt Gödel. *The Consistency of the Axiom of Choice and of the Generalized Continuum Hypothesis with the Axioms of Set Theory*, volume 3 of *Annals of Mathematics Studies.* Princeton University Press, Princeton, NJ, 1940. QA9.G54.

[34] Kurt Gödel. *On formally Undecidable Propositions of Principia Mathematica and Related Systems.* Dover, Mineola, NY, 1992. ISBN 0-486-66980-7; QA248.G573 1992; LCCC No. 91-45947; 511.3–dc20.

[35] Winfried Grassmann and Jean-Paul Tremblay. *Logic and Discrete Mathematics: A Computer Science Perspective.* Prentice Hall, Upper Saddle River, NJ, 1996. ISBN 0-13-501296-6; QA76.9.M35G725 1996; LCCC No. 95-38351; 005.1'01'5113–dc20.

[36] Nathaniel Grossman. *The Sheer Joy of Celestial Mechanics.* Birkhäuser Boston, Cambridge, MA, 1996.

[37] Richard K. Guy. Nothing's new in number theory? *American Mathematical Monthly*, 105(10):951–954, December 1998.

[38] Paul Guyer and Allen W. Wood. *Immanuel Kant Critique of pure reason.* Cambridge University Press, Cambridge, UK, 1998. ISBN 0-521-35402-1; B2778.E5G89 1998; LCCC No. 97-2959; 121–dc21.

[39] Peter Hamburger and Raymond E. Pippert. Simple, reducible venn diagrams on five curves and hamiltonian cycles. *Geometriae Dedicata*, 67:1–20, 1997.

[40] Peter Hamburger and Raymond E. Pippert. Venn said it couldn't be done. *Mathematics Magazine*, 71(2):105–110, April 2000.

[41] Peter Henrici. *Essentials of Numerical Analysis with Pocket Calculator Demonstrations*. Wiley, New York, NY, 1982. ISBN 0-471-05904-8; QA297.H42; LCCC No. 81-10468; 519.4.

[42] Edwin Hewitt and Karl Stromberg. *Real and Abstract Analysis*. Springer-Verlag, Berlin-Heidelberg, 1965. LCCC No. 65-26609.

[43] Hewlett-Packard Co., Corvallis Division, 1000 NE Circle Blvd., Corvallis, OR 97330, USA. *HP-15C Owner's Handbook*, March 1982. Part Number 00015-90001.

[44] Hewlett-Packard Co., Corvallis Division, 1000 NE Circle Blvd., Corvallis, OR 97330, USA. *HP-28C Reference Manual*, first edition, September 1986. Part Number 00028-90021.

[45] Hewlett-Packard Co., Corvallis Division, 1000 NE Circle Blvd., Corvallis, OR 97330, USA. *HP-28S Reference Manual*, first edition, October 1987. Part Number 00028-90068.

[46] Hewlett-Packard Co., Corvallis Division, 1000 NE Circle Blvd., Corvallis, OR 97330, USA. *HP 48SX Scientific Expandable Calculator Owner's Manual*, first edition, January 1990. Part Number 00048-90001.

[47] Hewlett-Packard Co., Corvallis Division, 1000 NE Circle Blvd., Corvallis, OR 97330, USA. *HP 48G Series User's Guide*, 7th edition, March 1994. Part Number 00048-90126.

[48] Nicholas J. Higham. *Accuracy and Stability of Numerical Algorithms*. Society for Industrial and Applied Mathematics, Philadelphia, PA, 1996.

[49] David Hilbert. *Foundations of Geometry*. Open Court, La Salle, IL, tenth edition, 1971. Eighth printing, 1996. Translated by Leo Unger and revised by Paul Bernays. ISBN 0-87548-164-7; LCCC No. 73-110344.

[50] David Hilbert and Wilhelm Ackerman. *Principles of Mathematical Logic*. Chelsea Publishing Company, New York, NY, 1950. BC135.H514. (Translation of David Hilbert & Wilhelm Ackerman, *Grundzüge der theoretischen Logik*, Julius Springer, Berlin, Germany, 1928 & 1938.).

[51] Francis Harry Hinsley and Alan Stripp, editors. *Codebreakers: The inside story of Bletchley Park*, Oxford, UK, 1993. Oxford University Press. ISBN 0-19-285304-X; D800.C88C63 1994; LCCC No. 94-882; 940.54'8641.

[52] Robert V. Hogg and Allen T. Craig. *Introduction to Mathematical Statistics*. Macmillan, New York, NY, fourth edition, 1978. ISBN 0-02-355710-9; QA276.H59 1978; LCCC No. 77-2884; 519.

[53] Thomas W. Hungerford. *Algebra*. Holt, Rinehart and Winston, New York, NY, 1974. ISBN 0-03-086078-4; QA155.H83; LCCC No. 73-15693; 512.

[54] P. T. Johnstone. *Notes on Logic and Set Theory*. Cambridge University Press, Cambridge, UK, 1987. ISBN 0-521-33502-7; QA9.J64 1987.

[55] David Kahn. *The Codebreakers: The Story of Secret Writing*. Macmillan, New York, NY, 1967. Z103.K28 1967.

[56] M. Karnaugh. The map method for synthesis of combinatorial logic circuits. *Transactions of the AIEE*, 72:593–598, November 1953.

[57] Hannu Karttunen, Pekka Kröger, Heikki Oja, Marku Poutanen, and Karl Johan Donner, editors. *Fundamental Astronomy*. Springer-Verlag, New York, Berlin, Heidelberg, second enlarged edition, 1994.

[58] Linda Keen. Julia sets. In *Chaos and Fractals: The Mathematics Behind the Computer Graphics*, pages 57–74, Providence, RI, 1989. American Mathematical Society.

[59] John L. Kelley. *General Topology*. Springer-Verlag, New York, NY, 1975. ISBN 0-387-90125-6; QA611.K4 1975; LCCC No. 75-14364; 514'.3. (Reprint of 1955 edition, Van Nostrand, New York.).

[60] Stephen G. Kellison. *The Theory of Interest*. Richard D. Irwin, Homewood, IL, 1970. HB539.K28; LCCC No. 79-98251; second edition: McClelland & Stewart, Toronto, CA, 1991, ISBN 0256091501.

[61] Murray S. Klamkin, editor. *Mathematical Modeling: Classroom Notes in Applied Mathematics*, Philadelphia, PA, 1987. Society for Industrial and Applied Mathematics. ISBN 0-89871-204-1; LCCC No. 86-60090.

[62] Murray S. Klamkin, editor. *Problems in Applied Mathematics: Selections from SIAM Review*, Philadelphia, PA, 1990. Society for Industrial and Applied Mathematics. ISBN 0-89871-259-9; QA43.P76 1990; LCCC No. 90-49670; 510'.76–dc20.

[63] Władysław Kozaczuk, editor. *Enigma: How the German Machine Cipher Was Broken, and How It Was Read by the Allies in World War Two*, Frederick, MD, 1984. University Publications of America. Translated by Christopher Kasparek. ISBN 0-89093-547-5; D810.C88K6813 1984; LCCC No. 83-16691; 940.54'85.

[64] Kenneth Kunen. *Set theory: An Introduction to Independence Proofs*. North Holland, Amsterdam, 1980. ISBN 0-444-85401-0; QA248.K75 1980; LCCC No. 80-20375; 510.3'22.

[65] Edmund Landau. *Foundations of Analysis*. The Tan Chiang Book Co., 1951.

[66] Serge Lang. *Algebra*. Addison-Wesley, Reading, MA, 1965. LCCC No. 65-23677.

[67] Cooper H. Langford and Ralph A. Beebe. *The Development of Chemical Principles*. Dover, New York, NY, 1995. ISBN 0-486-68359-1; QD31.2.L34 1995; LCCC No. 94-40923; 541.2–dc20.

[68] Pierre Simon de Laplace. *Traité de Mécanique Céleste*. Duprat, Paris, 1798–1825. (Tranlated with commentary by Nathaniel Bowditch, Boston, MA, 1829; Chelsea, Bronx, NY, 1966).

[69] David R. Lide, editor. *CRC Handbook of Chemistry and Physics*. CRC Press, Boca Raton, FL, 80th edition, 1999. ISBN 0-8493-0480-6; QD65.H3 1999/2000; LCCC No. 13-11056.

[70] C. L. Liu. *Introduction To Combinatorial Mathematics*. McGraw-Hill, New York, NY, 1968. ISBN 07-038124-0; LCCC No. 68-22763.

[71] Joseph Malkevitch and Gary Froelich. *Loads of Codes*, volume 22 of *HistoMAP Module*. COMAP, Lexington, MA, 1993. ISSN 0889-2652.

[72] Joseph Malkevitch, Gary Froelich, and Daniel Froelich. *Codes Galore*, volume 18 of *HistoMAP Module*. COMAP, Lexington, MA, 1991. ISSN 0889-2652.

[73] Yu. I. Manin. *A Course in Mathematical Logic*. Springer-Verlag, New York, NY, 1977. ISBN 0-387-90243-0; QA9.M296; LCCC No. 77-1838; 511'.3.

[74] V. Wiktor Marek and Jan Mycielski. Foundations of mathematics in the twentieth century. *The American Mathematical Monthly*, 108(5):449–468, May 2001. ISSN 0002-9890.

[75] Benson Mates. *Elementary Logic*. Oxford University Press, New York, NY, second edition, 1972. BC135.M37 1972; LCCC No. 74-166004.

[76] E. J. McCluskey. Minimization of Boolean functions. *The Bell System Technical Journal*, 35(5):1417–1444, November 1956.

[77] Alko Meijer. Digital signatures. *UMAP Journal*, 15(1):69–80, Spring 1994.

[78] Alko Meijer. Groups, factoring, and cryptography. *Mathematics Magazine*, 69(2):103–109, April 1996.

[79] J. Donald Monk. *Mathematical Logic*. Springer-Verlag, New York, NY, 1976. ISBN 0-387-90170-1; QA9.M68.

[80] Bruce Pourciau. Reading the master: Newton and the birth of celestial mechanics. *The American Mathematical Monthly*, 104(1):1–19, January 1997.

[81] John H. Reif and Robert E. Tarjan. Symbolic program analysis in almost-linear time. *SIAM Journal of Computing (Society for Industrial and Applied Mathematics)*, 11(1):81–93, February 1981.

[82] Marian Rejewski. Appendix C: Summary of our methods for reconstructing ENIGMA and reconstructing daily keys, and of german efforts to frustrate those methods. In Kozaczuk [63], pages 241–245. Translated by Christopher Kasparek. ISBN 0-89093-547-5; D810.C88K6813 1984; LCCC No. 83-16691; 940.54'85.

[83] Marian Rejewski. How the Polish mathematicians broke Enigma. In Kozaczuk [63], pages 246–271. Translated by Christopher Kasparek. ISBN 0-89093-547-5; D810.C88K6813 1984; LCCC No. 83-16691; 940.54'85.

[84] Marian Rejewski. The mathematical solution of the Enigma cipher. In Kozaczuk [63], pages 2472–291. Translated by Christopher Kasparek. ISBN 0-89093-547-5; D810.C88K6813 1984; LCCC No. 83-16691; 940.54'85.

[85] Abraham Robinson. *Non-standard Analysis*. Princeton University Press, Princeton, NJ, revised edition, 1996. ISBN 0-691-04490-2; QA299.82.R6 1995; LCCC No. 95-43750; 515'.1–DC20.

[86] Claude Elwood Shannon. A symbolic analysis of relay and switching circuits. *Transactions of the American Institute of Electrical Engineers*, 57:713–723, 1938.

[87] Carl Ludwig Siegel and Jürgen K. Moser. *Lectures on Celestial Mechanics*. Classics in Mathematics. Springer-Verlag, Berlin Heidelberg New York, NY, 1995. ISBN 0-387-05419-7; LCCC No. 71-155595.

[88] Raymond M. Smullyan. *First-Order Logic*. Dover, Mineola, NY, 1995. ISBN 0-486-68370-2; QA9.S57 1994; LCCC No. 94-39736; 511.3–dc20.

[89] John Parr Snyder. *Map Projections — A Working Manual (U.S. Geological Survey Professional Paper 1395)*. United States Government Printing Office, Washington, D.C. 20402, 1987.

[90] Michael Spivak. *Calculus*. Publish or Perish, Wilmington, DE, second edition, 1980. ISBN 0-914098-77-2; LCCC No. 80-82517.

[91] Josef Stoer and Roland Bulirsch. *Introduction to Numerical Analysis*. Springer-Verlag, New York, NY, second edition, 1993.

[92] Abram Aronovich Stolyar. *Introduction to Elementary Mathematical Logic*. Dover, Mineola, NY, 1983. ISBN 0-486-64561-4; BC135.S7613 1983; LCCC No. 83-5223; 511.3.

[93] Alan Stripp. The Enigma machine: Its mechanism and use. In Hinsley and Stripp [51], pages 83–88. ISBN 0-19-285304-X; D800.C88C63 1994; LCCC No. 94-882; 940.54'8641.

[94] K. F. Sundman. Recherches sur le problème des trois corps. *Acta Soc. Sci. Fennicae*, 34(6), 1907.

[95] Patrick Suppes. *Axiomatic Set Theory*. Dover, Mineola, NY, 1972. ISBN 0-486-61630-4; LCCC No. 72-86226.

[96] Alfred Tarski. *Introduction to Logic and to the Methodology of Deductive Sciences*. Oxford University Press, Oxford, UK, 1941.

[97] Angus E. Taylor. *Introduction to Functional Analysis*. Wiley, New York, 1958. LCCC No. 58-12704.

[98] Carsten Thomassen. The Jordan-Schönflies theorem and the classification of surfaces. *American Mathematical Monthly*, 99(2):116–130, February 1992.

[99] Peter R. Turner. Will the "real" real arithmetic please stand up? *Notices of the American Mathematical Society*, 38(4):298–304, April 1991.

[100] Bartel Leenert van der Waerden. *Geometry and Algebra in Ancient Civilizations*. Springer-Verlag, Berlin & Heidelberg, 1983. ISBN 3-540-12159-5; QA151.W34 1983; LCCC No. 83-501; 512'.009.

[101] John Venn. *Symbolic Logic*. Chelsea Publishing Company, Bronx, NY, undated. ISBN 0-8284-0251-5; LCC No. BC135.V4 1971b; LCCC No. 79-119161; Second edition London, UK, 1894.

[102] John von Neumann. Zur einführung der transfiniten Zahlen. *Acta Szeged*, 1:199–208, 1923. Cited in [95, p. 129].

[103] Robert John Walker. *Algebraic Curves*. Springer-Verlag, New York, NY, 1978. ISBN 0-387-90361-5; QA564.W35 1978; LCCC No. 78-11956; 516'.352. (First published by Princeton University Press, Princeton, NJ, 1950.).

[104] André Weil. *Number Theory: An approach through history from Hammurapi to Legendre*. Birkhäuser, Boston Basel Stuttgart, 1984. ISBN 0-8176-3141-0; QA241.W3418 1983; LCCC No. 83-11857; 512'.7'09.

[105] Gordon Welchman. *The Hut Six Story: Breaking the Enigma Codes*. McGraw-Hill, New York, NY, 1982. ISBN 0-07-069180-0; D810.C88W44; LCCC No. 81-13657; 940.54'86'41.

[106] Eyvind H. Wichmann. *Quantum Physics*, volume 4 of *Berkeley Physics Course*. McGraw-Hill, New York, NY, 1967. LCCC No. 64-66016.

[107] Oscar Zariski and Pierre Samuel. *Commutative Algebra*. Van Nostrand, Princeton, NJ, 1958. LCCC No. 58-7911.

[108] E. Zermelo. Untersuchungen über die Grundlagen der Mengenlehre, i. *Mathematische Annalen*, 65:261–281, 1908. Cited in [3, p. 21].

[109] Steven S. Zumdahl. *Chemistry*. D.C. Heath, Lexington, MA, 1986. ISBN 0-669-04529-2; LCCC No. 85-60981.

Index